Fundamentals of
ELECTRONICS

PRENTICE-HALL SERIES IN ELECTRONIC TECHNOLOGY

Dr. Irving L. Kosow, editor

Charles M. Thomson, Joseph J. Gershon, and Joseph A. Labok
consulting editors

Fundamentals of
ELECTRONICS

THIRD EDITION

MATTHEW MANDL

PRENTICE-HALL, INC.

Englewood Cliffs, New Jersey

Library of Congress Cataloging in Publication Data

MANDL, MATTHEW.
  Fundamentals of electronics.

  1. Electronics. I. Title.
TK7815.M26 1973   621.381   72-7353

10 9 8 7 6 5 4 3 2 1

PRENTICE-HALL INTERNATIONAL, INC., *London*
PRENTICE-HALL OF AUSTRALIA, PTY., LTD., *Sydney*
PRENTICE-HALL OF CANADA, LTD., *Toronto*
PRENTICE-HALL OF INDIA (PRIVATE) LTD., *New Delhi*
PRENTICE-HALL OF JAPAN, INC., *Tokyo*

Printed in the United States of America

To my sister, Fay

Contents

Preface

Fundamentals of Electronics is a single-volume text covering not only the basic theory of electronics but also the practical aspects. It has been specifically written for those planning a career in any of the various branches of electronics, including industrial control, automation, radar, microwave electronics, computer systems, communications, or allied fields. In this third edition greater emphasis has been given to solid-state circuitry and systems, as well as inclusion throughout of the latest standards in terminology, abbreviations, and symbols. As with the last edition, the topics have been divided into three primary sections:

1. Review of Electronic Fundamentals
2. Principles of Electronics
3. Applications and Components

The first section reviews electron theory; it includes a study of subshells and atomic bond factors that provide a firm foundation toward an understanding of solid-state devices and transistors covered later. Included in the first section are the basic topics of current flow, fundamental circuit analysis, magnetism, and other material needed for acquiring the necessary groundwork for advanced studies. The topic features of the original chapters have been retained from the first edition, though the material has been reorganized, expanded, and rearranged to improve topic sequence and upgrade the text structure.

The second section, which starts with vacuum-tube principles, also contains discussions of solid-state fundamentals, transistors, power supplies, amplification systems, oscillators, pulse factors, and miscellaneous electronic circuitry. Portions of the original material have been revised and brought up to date. Many chapters have been enlarged since the first edition to cover new topics. The transistor section, for instance, has been updated to include MOSFET devices and Y-parameter discussions have been added to the others for aid in analysis of FET characteristics. Integrated circuitry has also been included. Additional solid-state components are discussed in the power-supply section, and other subsequent chapters contain representative examples of practical usage of solid-state devices.

Topics in the third section include receiver circuitry, transducers, test instruments, switching and gating systems, magnetic amplifiers, and signal-frequency multiplication and division. Receiver and transmitting principles are covered since their basic circuitry is also encountered in other branches of electronics and because this information is essential for those who will ultimately be engaged in some aspects of research and design in the communications field. In addition, these principles round out the foundation acquired earlier and unify some of the application factors. New material in this section included stereo transmitting and receiver principles, multiplex practices, SCA systems, color-television transmission and reception, and special circuitry.

Topics such as transducers, switching and gating, logic circuitry, and control amplification have been retained, with new material added wherever appropriate. In *all* sections, the emphasis on circuit and component analysis has been continued, and the aim has been to present explanations as clearly and as thoroughly as possible.

The review questions given to the student at the end of each chapter for home and classroom usage have been expanded to meet the increased text coverage. The questions are so worded that a rereading of the chapters involved will funish the necessary answers. In most chapters, the number of practical problems has been increased to better illustrate practical applications and provide for added experience in solving typical circuit analysis and design equations. Most of the practical problems require numerical answers as opposed to the essay answers required for the review questions. Hence, self-checking facilities are provided by the answers to practical problems in the appendix.

The appendix lists the various reference data so essential to both the study of electronics and electricity. Included are a summary of the principles of logarithms along with appropriate tables; right-angle factors; and a table of trigonometric ratios, color coding, and other necessary information.

The author wishes to thank the practicing engineers, technicians, and electronic field men who have outlined the general scope of electronic knowledge that industry usually looks for in new employees. Grateful acknowledgment is also expressed herewith to the instructors in many colleges and institutes who have contributed valuable suggestions regarding the scope and topic sequence found most successful in electronic education.

Yardley, Pennsylvania   MATTHEW MANDL

Review of
Electronic   I
Fundamentals

Electrons, Charges, and Fields   1

1-1. Introduction

Electronics (which stems from the word *electron*) deals with the electric utilization of solid-state devices or vacuum and gas tubes to perform tasks of signal sensing, routing, amplification, visual display, measuring, and other related functions. This application of *electronic devices* distinguishes *electronics* from electrical practices confined to power generation and distribution.

All phases of electricity and electronics, however, are related to the movement of electrons through some conducting medium to constitute what is known as *current flow*. Hence the study of electronics must embrace the same fundamentals as that of electricity. In addition, electron movement also generates *fields* that must be understood for their relation to the construction of such items as electric generators, transformers, capacitors, antennas, and numerous other electronic units. We must also learn how to alter the chemical composition of basic crystal elements to modify electron movement characteristics for designing items such as transistors, photo-diodes, and other widely used devices.

The essential factors relating to the electron theory are presented in this chapter to serve as a foundation for the subsequent studies involving more advanced aspects of current flow, circuit behaviors, and component applications.

1-2. Atomic Charges

The atom is a closely knit structure consisting of a central core called a *nucleus* and one or more electrons revolving around the core. The electrons are referred to as *planetary* electrons because the atomic structure resembles the solar planetary system. In our solar system each planet orbits at a different distance from the sun; likewise an atom with a number of electrons has them revolving around the nucleus in orbits that are spaced at different distances from the central nucleus.

In a fashion similar to the gravitational pull between the sun and its planets, there is also an attraction between the nucleus of an atom and its planetary electrons. The potential energy represented by the attraction between the nucleus and any particular electron of the atom is known as a *charge*. To distinguish between the charge of an electron and that of the nucleus, the charge of the electron is designated as *negative* (sometimes referred to as *minus*). The nucleus, on the other hand, is said to have a *positive* charge (also referred to on occasion as a *plus* charge). The positive charge of the nucleus, however, simply indicates the predominant or primary charge of the nucleus. It may be composed of a number of positively charged particles (known as protons) as well as some particles with no charge, which are known as *neutrons*. Other particles also exist, but for purposes of simplification only the positively charged particles of the nucleus, the protons, will be considered here.

In a normal atom, the total value of the positive charge of the nucleus is equal to the total negative charge established by the planetary electrons surrounding the nucleus. Since the nucleus has a positive value and the electrons have a negative value, they are equal though opposite in their charge relationship, and the atomic structure as a *whole* may be considered as having a *neutral* charge.

Because the nucleus has a positive charge and the electrons have a negative charge, an attraction is created between the nucleus and the planetary electrons. This attraction is in conformity with one of the most basic laws of electricity; that is, *unlike charges attract and like charges repel*. Hence, the individual electrons, having like charges (negative), repel each other; yet all are attracted to the positive nucleus.

1-3. Elements and Compounds

There are many types of atoms, some having electrons in a single orbit and others having electrons in several orbits. The number of electrons in an atom (as well as the composition of the nucleus) determines its type. The various types of atoms form the fundamental structure of *all* substances or matter that exist on earth and in the universe. When identical atoms are grouped together, they combine to form an *element*, and all matter is made up of either a single element or a combination of various elements.

The proton has considerably greater mass and weight than an electron. If the proton mass is taken to represent one, the mass of the electron is virtually negligible (actually less than $\frac{1}{1800}$ of the proton weight). Thus, even though the electrons revolve around the nucleus at high speeds, they constitute only a fractional portion of the total mass of an atom.

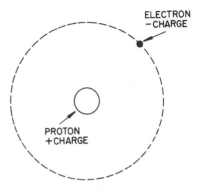

FIGURE 1-1. Atom of hydrogen.

The simplest atom is that of the gas *hydrogen*. An atom of this gas consists of one proton and one electron, as shown in Fig. 1-1. This is a simplified illustration since the electron can rotate at any angle and its orbit may not be perfectly circular, as shown here. Because there is only one electron and one proton in the hydrogen atom, there is very little mass; hence the element formed by combining such atoms has very little weight. For this reason, the light-weight gas elements are composed of atoms with few protons and electrons.

The next most simple atom is that of the gas *helium*, shown in Fig. 1-2. Here there are two electrons in the single orbit around the nucleus. Also, there are two protons present, as well as two neutrons. When protons and neutrons coexist in the nucleus, they are

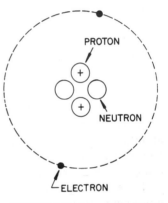

FIGURE 1-2. Atom of helium.

tightly bound together and the neutrons contribute to the total mass and weight of the nucleus. With the helium atom, as with the hydrogen atom, if each proton and neutron is considered to have a mass of one, the two neutrons with a mass of two and the two protons with a mass of two would represent a total weight of four.

Atoms with a larger number of electrons revolving around the central nucleus form other elements. The first orbit beyond the nucleus can accommodate only two electrons so that additional orbits are present in most elements. The element *lithium*, for instance, has two electrons in the first orbit and another electron in the second orbit, as illustrated in Fig. 1-3. In many cases, more than two orbits are present, as shown later. The orbits are sometimes referred to as planetary *rings* or *shells*. An increase in the number of electrons around the nucleus forms other element atoms, such as *beryllium*

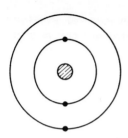

FIGURE 1-3.   Atom of lithium.

with four electrons, *boron* with five electrons, and *carbon* with six electrons. A greater number of electrons forms such elements as gold, silver, copper, lead, and oxygen, among many others. There are over 100 known elements to date. The number of protons in a normal atom determines the atom's position in the so-called atomic series or list of standard atomic numbers. (see appendix).

Because there are over 100 basic elements, they can be combined in virtually an infinite number of ways to form what are known as *compounds*. Examples of compounds are such items as water (two parts hydrogen and one part oxygen) and table salt (sodium and chlorine). Mixtures of compounds form glass, stone, cloth, wood, and countless other substances. Since there are millions of the various atoms in even a small segment of a compound, one can recognize a compound even when the particles may be so small that a microscope is required to observe them. If a compound is divided into smaller and smaller sections, however, a point is finally reached where any additional division results in the dividing of the fundamental elements themselves. Since these make up the element structure, their division results in the loss of the identifiable compound. Thus, the smallest particle of a substance that is still identifiable as such is known as a *molecule*, and it contains the minimum number of the various atoms forming the identifiable compound.

An element, even though made up of similar atoms, can also be considered as having a molecular structure. The molecule of an element such as atmospheric oxygen or hydrogen gas consists of two atoms only. (As shown later in this chapter, certain characteristics are established when atoms are brought close to each other to form elements. As also discussed, the atom itself may undergo some change with respect to its electrons when it is combined with other atoms to form an element such as iron.) In comparison to the element, the compound is made up of one or more similar atoms plus one or more dissimilar atoms, and the smallest combination of such various atoms is known as the molecule. Combinations of such molecules in large numbers form materials such as paper, plastic, and bronze.

1-4.   Shells and Subshells

As a foundation for the study of current flow, magnetics, solid-state devices, and other electronic factors, we must next consider the aspects of electron orbit characteristics. As mentioned earlier, the various orbital paths of the electrons around the nucleus are sometimes referred to as *shells* or *rings*. Each primary shell or ring can accommodate only a certain amount of electrons. All shells, except the first shell near the nucleus, are composed of two or more *subshells*.

As shown in Fig. 1-4, the shell is a solitary one and can accommodate only 2 electrons. The second primary shell is composed of two subshells, the one nearest the nucleus being capable of accommodating 2 electrons and the other subshell being able to contain up to 6 electrons. The third primary shell can consist of three subshells, the first of these being able to hold a maximum of 2 electrons; the second, 6 electrons; and the third, 10 electrons. The fourth primary shell may have no more than four subshells. As with the others mentioned, the first subshell can hold 2 electrons; the second, 6, and the third, 10. The fourth subshell of the fourth primary ring can hold up to 14 electrons. Note that, in any primary shell group, the second subshell can contain 4 *more* electrons than the first subshell. Each succeeding subshell can contain 4 more electrons than the preceding subshell. In the next primary shell group, however, the first subshell again starts with 2 as the maximum number of electrons that it can contain.

All the outer shells need not, of course, be completely filled. With the lithium atom, for example, the first primary shell contains 2 electrons and the second primary shell contains only 1 electron. This single electron revolves in an orbit representative of the first subshell of the second primary shell group (see Fig. 1-3). *In a solitary atom* each shell and subshell, because of the strong nuclear attraction from the core, must be filled successively before subsequent shells can hold electrons. It must be emphasized here, however, that this rule only applies to a solitary atom that *is not part of an element*. This factor is illustrated later in Fig. 1-12 and covered in greater detail in the related discussions.

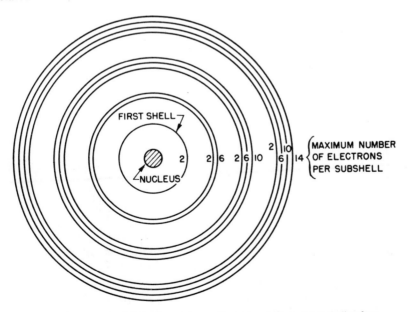

FIGURE 1-4. Sequential arrangement of primary shells and subshells of an atom.

The electrons in orbit nearest to the nucleus are those representative of the *lowest-energy levels* because they are firmly bound to the nucleus, particularly if the shells are filled. Electrons within the outer rings represent the highest-energy levels and are those most readily influenced and capable of being removed from the atom.

From the foregoing, it is evident that the first *primary* ring can contain 2 electrons; the second primary ring, a total of 8 electrons, the third primary ring, a total of 18 electrons, and the fourth primary ring, 32 electrons. Beyond the fourth primary ring are two additional rings but these are never entirely filled, and it is not known how many they could hold since there are only about 100 known basic elements.

The number of electrons in the outer shell of an atom determines the stability of the atom. If the quota of electrons in an outer shell is filled, the atom is stable and exhibits virtually no chemical reactions. Such a stable element is known as an *inert* element. Helium is inert since it has only two electrons and these completely fill the first primary shell. Neon is another inert element since it has two electrons in the first primary shell and eight electrons in the second primary shell, as shown at A of Fig. 1-5. Argon, shown at B, is still another inert gas. Note, however, that while in the argon atom the first two primary shells are filled completely, the third primary shell is filled only with respect to its first two *subshells*. The third subshell is empty. Here, however, we still have an inert element because the first two subshells are completely filled. The inert elements just mentioned will not combine with any other elements.

When the outer shell of an atom has less than its full quota of electrons, it can readily gain or lose electrons. In particular, atoms lacking only one or two electrons in their *outer* shell can easily acquire the additional electrons

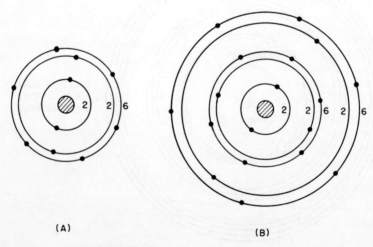

(A)   (B)

FIGURE 1-5. Atoms of neon and argon.

to fill the quota of the outer ring. Atoms having only one or two electrons in the outer shell can readily lose such electrons.

1-5. Electron Movement

Elements that have the ability to lose electrons are usually metals, while elements with a capacity for acquiring electrons are nonmetallic. Calcium, barium, and strontium each have 2 electrons in the outer shell, and hence these elements all have the ability to lose such electrons. These elements are often used as the sources of electrons in vacuum tubes because they will emit electrons freely when heated. Metals such as aluminum, copper, and silver have incomplete outer shells with only a few electrons of the full quota present. Hence, such electrons can be removed easily from tne outer orbit. Copper, for instance, has a total of 29 electrons in each atom of the element. As shown in Fig. 1-6, the first three primary shells of the copper atom are completely filled; that is, the first three shells contain 2, 8, and 18 electrons, respectively, for a total of 28 electrons. The fourth shell has only 1 electron (actually in the first subshell of the fourth orbital ring). This solitary electron of the copper atom is a *free* electron and hence can be removed rather easily. For this reason, copper is often utilized in electricity and electronics because its electrons can be influenced. If copper is formed into a length of wire, electron flow can be produced from one end of the wire to the other by applying pressure to the electrons through use of a battery or other source of

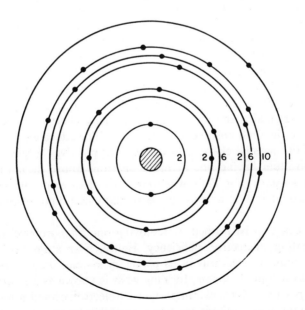

FIGURE 1-6. Atom of copper.

electric force. In copper, as in other elements, millions of atoms make up the structure. The single electron in the outer shell, as it travels in its orbit, may become equidistant between its own atom and an adjacent one. When this happens, it is not under the exclusive attraction of either nucleus and can leave its own orbit to travel and enter another orbit of an adjacent atom. Hence, the electric pressure that is applied to such a free electron causes it to leave one atom and move on to the next. The second atom thus has a force applied to it, in the form of an arriving electron, and in turn transfers a free electron to the next atom. This progression of electrons through the copper wire is known as *current flow*.

When the outer shell of an atom is completely filled (or almost completely filled), electron movement *through* the element is difficult to produce and consequently the substance becomes an *insulator*, which means that it is a nonconductor of electric current. When only a few electrons are part of the outer shell, that is, when the outer shell lacks a number of electrons to reach its full *quota*, the substance has the ability to permit current to flow through it on application of electric pressure and thus is known as a *conductor*.

1-6. Ions and Plasma

As mentioned earlier, because of the equality existing between the positive charge of the nucleus and the negative charge of the electrons of an atom, the net charge is neutral. In various branches of electronics, however, there are occasions where conditions are created that either remove or add electrons to the normal quota surrounding the nucleus of an atom. To define this altered atomic characteristic, the word *ion* is applied to an atom that is no longer neutral but has either gained or lost one or more electrons from its original state. When an atom has more than the normal amount of electrons, an unbalanced condition is created between the planetary electrons and the nucleus because the excessive electrons now cause the atom to be predominantly negative. Originally, the total negative electron charge was equal, though opposite, to the total positive nucleus charge. The added electrons, however, increased the negative charge above that of the positive charge of the nucleus. Hence, an atom having one or more electrons above its normal quota is known as a *negative ion*. Thus, when a number of atoms of this type are utilized, they are referred to as ions and serve many useful purposes in electronic devices.

Ions can also be formed by removing one or more electrons from an atom, creating an electron deficiency. Hence, if an atom has less than the normal amount of electrons in its planetary system, the positive charge of the nucleus will predominate, forming what is known as a *positive ion*. Ions of both types are of particular importance in electronic applications using gas-filled vacuum tubes, cathode-ray tubes, as well as transistors, as shown later.

The term *plasma* is also applied to ionization, particularly to ionized gases forming a cloud of ions in a highly agitated state. As such, the plasma can be considered the equivalent to an electrical conducting fluid that can be acted on by magnetic fields. Because it is actually neither a true solid nor a liquid, it has been referred to as the fourth state of matter. The word plasma stems from the Greek word *plassein*, meaning "to form" or "to mold." While plasma is usually formed by gas ionization, it also applies to ionization in solid-state devices. Practical applications apply to all gas-filled electron tubes, thermonuclear fusion, missile reentry problems, industrial processes for application of refractory coatings to base metals, as well as some aspects of metal shaping and welding processes.

Plasma temperatures start at approximately 10,000°F, and gas ionization causes molecular structures to break up into individual atoms with a loss of normal charge neutrality. In industrial utilization and research, magnetic fields are utilized as containers of the plasma since solid materials disintegrate because of the extremely high temperatures. Plasma engineering and research are continuing to explore the potentials of this ionization cloud formed by basic atomic elements.

1-7. Types of Fields

There are two types of fields: the electrostatic and the electromagnetic. Electrostatic fields are created when a condition of either insufficient or excessive electrons exists. Magnetic fields are created by the alignment of the orbital spins of electrons. Electrostatic fields can be produced by applying friction to objects, which also produces electron movement. If, for instance, a silk cloth is rubbed against a glass rod, it will tear electrons away from the rod and absorb them into its own atomic structure. As a result, the affected atoms of the glass rod will have an electron deficiency and hence the rod is said to be *positively charged.* As mentioned earlier, an atom with one or more of its normal quota of electrons removed becomes a positive ion and the net charge of the atom is positive. A group of such atoms, constituting a particular element, will cause that element to have a positive charge. Thus, rubbing the glass rod with silk caused an electron movement from the glass to the silk and created a positive charge in the rod and a negative charge in the silk.

A charged particle produces *fields* that emanate from it as shown at A of Fig. 1-7. Here a charged particle has been suspended in air and the dotted lines indicate the fields that exist around such a particle. These fields can be likened to tentacles of electricity that, though invisible, extend out from the charged particle in all directions and are capable of influencing other charged materials. Thus, if two charged glass rods are suspended as shown at B of Fig. 1-7, the two will repel each other since the polarities of the fields are alike.

A common term used to describe electrostatic fields (as well as the

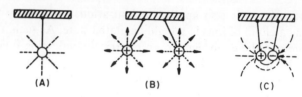

FIGURE 1-7. Electrostatic lines of force in charged particles.

magnetic fields discussed later) is *lines of force.* This is a concept originating with Michael Faraday (1791–1867) the British physicist and chemist who visualized the fields as having tension or stress. Though based on imagery, the lines of force idea has proved of considerable usefulness as a convenient reference phrase and as a base for appropriate mathematical expressions.

If a piece of flannel is used to rub a hard rubber rod, electrons are torn away from the flannel by virtue of the friction produced. The electrons leave the flannel and are transferred to the rubber, setting up a negative charge in the latter. Thus, if two rubber rods are charged in this manner and suspended near each other, they will repel each other just as the glass rods did because in either case the rule mentioned earlier, regarding the repulsion between like poles, applies here.

If the charged glass rod (positive) is suspended near the charged rubber rod (negative), the unlike poles will create an attraction, as shown at C of Fig. 1–7 causing the rods to swing together. Note that the lines of force of a positively charged particle are shown with arrows pointing away from the particle, while the lines of force of a negatively charged particle are shown with arrows pointing toward the particle. The lines of force are referenced as originating from the north pole (or positive charge) and terminating at the south pole (negative charge).

If the particles are permitted to touch each other, the excessive amount of electrons in the negatively charged particle will be transferred to the particle having an electron deficiency, thus canceling the respective charges and making each item neutral again. A charged particle will hold its charge in a dormant or static state until it is discharged. The terms *electrostatic charge* and *electrostatic lines of force* are based on the word *static.*

Charles Coulomb (1736–1806) was the first to state the law of charged bodies:

> The force between two electrically charged bodies is inversely proportional to the square of the distance between the two, and directly proportional to the product of the two charges.

The electrostatic charge unit is called the *coulomb* in honor of this French scientist, and the unit is the quantity of charge carried by electrons in the amount of 6.28 times the product of 10 multiplied by itself 18 times. Thus, to represent such a large number of electrons, a long string of zeros would be necessary; hence an abbreviated method of writing such large numbers is used, as more fully detailed later.

While friction is not used for the generation of electric current in practical electronics, charged bodies and electrostatic lines of force are encountered frequently and thus are an important phase of electronics.

Magnetic fields can also be demonstrated easily. If, for instance, a thin cardboard is placed over a bar magnet, and iron filings are sprinkled over the cardboard, the magnetizing force of the magnet will, in turn, magnetize the particles of iron. If the cardboard is now tapped gently, the iron filings will arrange themselves in a pattern representative of the magnetic field. The

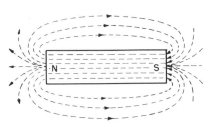

FIGURE 1-8. Fields of bar magnet.

magnetic field is shown in Fig. 1-8, with the lines indicating a direction from the north pole of the magnet to the south pole. The lines form loops around the magnet (the open lines indicate that insufficient magnetism is present to influence the iron filings).

If two bar magnets are brought near each other, as shown at A of Fig. 1-9, the fields of one interacts that of the other and, thus, the two influence each other. If the north and south poles of the two magnets are brought into close proximity, the unlike poles will *attract* each other and *lines of force* are set up between them. If both north poles are brought close to each other, as at B of Fig. 1-9, the lines of force are such that a *repelling* action occurs. Lines of force are not closed loops, like the fields around a single magnet, but instead extend from one pole to the other as shown between the two magnets at A of Fig. 1-9. Thus, the law previously stated, regarding the attraction between unlike charges (and the repulsion between like charges), also holds for the poles of magnets.

The force of either attraction or repulsion between two magnetic poles follows Coulomb's law with respect to the force between two charged bodies and hence is inversely proportional to the square of the distance between the

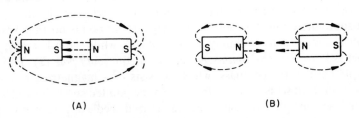

(A)   (B)

FIGURE 1-9. Aiding and opposing fields of bar magnets.

two and directly proportional to the product of the strength of the two charges. This law, expressed mathematically, then becomes

$$\text{Force} = \frac{M_1 \times M_2}{D^2} \tag{1-1}$$

where M_1 and M_2 represent the strength of each pole and D represents the distance in centimeters between the poles.

The strength of M_1 or M_2 is expressed in terms of the *unit pole*. A unit pole has a strength such that it will exert a force of 1 dyne upon an equal pole in air (or vacuum) when the distance between the poles is 1 centimeter. The *dyne* is the unit of force in the centimeter-grams-second (cgs) system and is equal to the force required to produce, in a 1-gram weight mass, an acceleration of 1 centimeter per second for every second that the force is present. The dyne is an extremely small unit, 980 dynes equaling only 1 of force.

As an example of the use of this formula, assume the north pole of a magnet with a unit pole strength of 100 is brought within 5 cm of the south pole of another magnet having a unit pole strength of 50. The calculation would then be

$$\frac{100 \times 50}{5 \times 5} = \frac{5000}{25} = \text{a force of 200 dynes}$$

1-8.   Theory of Magnetism

Before undertaking an explanation of the electron-spin principle that produces magnetic lines of force, some factors relating to magnets in general should be discussed.

When a magnet is dipped into a pile of iron filings, a considerable quantity of the filings will cling to the ends of the magnet but only a few iron particles will be attracted and held by the center portion of the magnet, as shown at B of Fig. 1-10. This simple experiment proves that the magnetism is concentrated in the two ends of the magnet. These two areas or regions of concentrated magnetism are known as the *poles* of the magnet, with each pole possessing the same magnetic strength as the other.

If a bar magnet is suspended from the center by a string, it will turn until it is aligned with the natural magnetism of the earth, that is, in a general north-south direction. The pole of the magnet that points north is known as the *north-seeking pole* or simply *north pole* (N), while the other pole of the magnet is known as the *south-seeking pole* or the *south pole* (S). This is the principle used in the compass, which consists of a magnetized needle-type indicator pivoted so as to rotate freely with the pointed end polarized to form the north-seeking pole. Actually, the magnetized needle of the compass points to the magnetic poles of the earth and not to the geographic north and south poles. The north *magnetic* pole at 71°N latitude, 96°W longitude, is approxi-

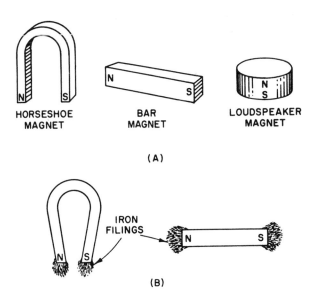

FIGURE 1-10. (A) Magnet shapes (B) Filings show poles.

mately 1300 m distant from the geographic north pole. The south *magnetic* pole, on the other hand, is 72–73°S latitude, 156°E longitude. Because of its magnetic poles, the earth can be considered a huge natural magnet.

As with other basic principles of electronics and electricity, early scientists focused their attention on magnetism and attempted to evaluate and explain it, as well as to set up fundamental laws governing its behavior. One of these earlier scientists was the German physicist Wilhelm Weber (1804–1891), who made the proposition that each molecule of a magnetic substance was a permanent magnet in itself. Later, Sir James Ewing (1855–1935) the Scottish engineer and physicist offered an explanation of magnetism based on Weber's earlier theory. In consequence, the Weber-Ewing concepts are known as the *molecular theory of magnetism.* According to this theory, the magnet molecules of magnetic material such as iron and steel are normally in a random arrangement, as shown at A of Fig. 1–11, and their effect on each other is a neutralizing one; that is, the magnetic material has no overall polarity and no attraction for other magnetic materials. When the material

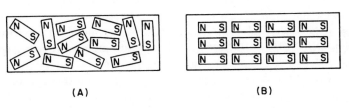

FIGURE 1-11. Molecular theory of magnet formation.

is magnetized, the molecules are in orderly arrangement, as shown at B, and one magnetic particle aids another to form a north polarity at one end and a south polarity at the other.

If the magnetic material shown at A is iron, the magnet molecules can be aligned as shown at B by bringing a permanent magnet close to the iron bar. The symmetrical arrangement of the molecules will then cause the iron bar to become a magnet. Upon removal of the permanent magnet, however, the molecules in the iron bar shift and again become disorganized. In consequence, the iron bar loses its magnetic properties. If a steel bar is used, however, the molecules will remain positioned in their orderly and organized manner and a permanent magnet results.

The modern concept of magnetism goes deeper than the older Weber-Ewing theory and is based not on molecular structures but on the *electron-spin* principle. This relates to the theory that the individual electrons of an atom not only revolve around the nucleus in orbital paths but also spin around their own axes just as the earth, in revolving around the sun, also turns on its own axis. Some of the atom's electrons spin in one direction and others in the opposite direction. If as many electrons are spinning in one direction as the other, they tend to neutralize each other insofar as magnetic properties are concerned. Thus, iron, with 26 electrons to the atom, will be nonmagnetized if 13 electrons spin in one direction and 13 in the other.

As discussed earlier, the electrons around the nucleus of an atom have primary rings for their orbital travel as well as subshells. Thus, iron, with 26 electrons, would have an atom wherein the first ring contains 2 electrons; the second ring consists of two subshells of 2 and 6 electrons, and the third ring consists of three subshells of 2, 6, and 8 electrons, for a total of 26 electrons. According to the modern concept, when the atoms are brought close together to form a solid metal, the outer subshell is disturbed, and free electrons are liberated (which are the ones that constitute current). Hence, *an atom that is part of an element is somewhat different from an atom isolated from the element*, just as in the case of the oxygen atom when combined with others to form atmospheric oxygen, as discussed later.

An iron atom (as part of the element iron) is shown in Fig. 1-12, wherein a fourth ring is indicated containing 2 free electrons. The number of electrons in each ring and subshell is shown with the positive sign indicating electron spin in one direction and the negative sign indicating electron spin in the other. Here it is the third subshell of the third primary ring that is responsible for creating magnetic property. As shown, the iron atom is in a magnetized state because there are 15 electrons spinning in one direction (+) and only 11 spinning in the other direction (−). If the iron atom were nonmagnetized, there would be 3 electrons spinning in one direction (+) in the third subshell of the third ring and also 3 electrons spinning in the other direction (−), forming, as already mentioned, 13 in one direction and 13 in the other.

The iron element atom shown in Fig. 1-12 (in its magnetized state)

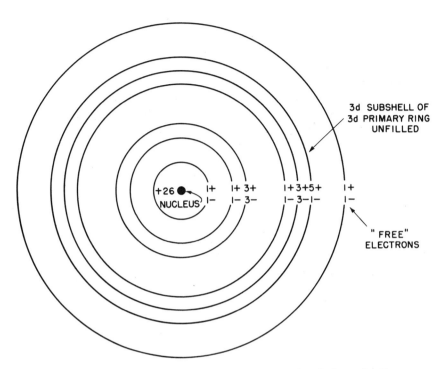

FIGURE 1-12. Electron spins in iron atom. Polarity signs indicate direction of spin; electrons are still negative with respect to nucleus.

acts as an individual (though infinitely small) permanent magnet. When a number of such atoms are grouped together to form the element iron, there is an interaction between the magnetic forces of the various atoms. Such interaction only occurs when atoms are in close proximity to each other. The interaction among the atoms making up the element permits the electron spins of one atom to influence the electron spins of adjacent atoms and thus to produce alignment. The alignment of the electron spins is confined to a small area or section of the material known as a *domain*. The atoms making up one domain all have parallel electron spins and hence the domain is magnetized in a particular direction. The fields of an external magnetizing force will tend to align all the domains in an element in a direction to produce a uniform magnetic field.

1-9. Atom Bonds

The electrons in the *outer* orbit of the atom's planetary system are capable of binding the atom to that of an adjacent atom. This bonding property of atoms is responsible for the cohesive characteristics of matter that permit the formation of the materials and substances found in nature. The outer ring electrons that have such bonding properties with respect to the outer electron shell of adjacent atoms are known as *valence electrons.*

The number of electrons in the outer ring or subshells has an important bearing on the ability of the individual atoms of an element or compound to cling together and form solid substances such as copper, silver, and gold. Knowledge of how the number of electrons influences the binding characteristics of substances is of material aid in understanding current flow, transistor behavior, and other important electronic phenomena.

The binding characteristics of the atom fall into three general classes, the *ionic*, the *metallic*, and the *covalent*. The ionic bonds are those present in compounds. The metallic bonds are those established in metals, while the covalent bonds are the type encountered in the formation of transistors and other crystal structures.

Ionic bonds are created when metallic atoms (with only a few electrons in the outer shell) combine with atoms that have almost a full quota of electrons in their outer shells. With atoms having almost a full quota of outer ring electrons, there is a tendency to absorb electrons from metallic atoms because a few electrons in the outer shell are easily lost. A typical example of this is the formation of ordinary table salt (sodium chloride) by combining the element *sodium* with the element *chlorine* to form a compound, as shown in Fig. 1-13. Note that the sodium atom has only 1 electron in its third ring (which could normally accommodate 18 electrons). The chlorine atom, on the other hand, has 7 electrons in its third ring. Thus, the first two subshells of the third ring have almost their full quota of 8. Hence, when the sodium atom and the chlorine atom are brought closely together, an ionic bond is

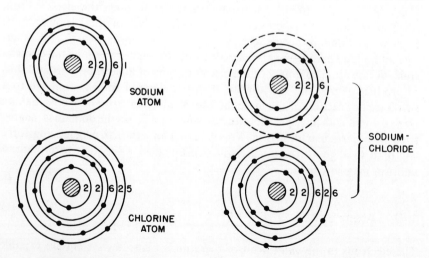

FIGURE 1-13. Ionic bond formed by chlorine atom capturing a sodium valence electron.

formed because the chlorine atom captures a sodium valence electron. This results in the compound sodium chloride, wherein the sodium atom is now a positive ion and the chlorine atom is a negative ion. The two ions are firmly bound together because of the attraction set up between the opposite charges of the two ions. This conforms to one of the laws of electricity, mentioned earlier, that like charges repel each other and unlike charges attract each other.

In the case of the metallic sodium atom and the chlorine atom, the bond formed when sodium chloride is produced results in the outer shells of each atom having a full electron quota. The consequence is that this compound (if it is absolutely pure) is an insulator (a nonconductor of electricity). Each chlorine ion is attracted to a neighboring sodium ion and the various bonds that are formed create a cubic arrangement of atoms as shown in Fig. 1-14. Such a symmetrical arrangement forms a crystal and the structure shown in the illustration is known as a *crystal lattice network*. Crystals formed by other elements have atoms arranged in other geometrical patterns depending on the molecular structure formed.

As already mentioned, atoms having only a few electrons present in their outer rings tend to give up such electrons readily, and such atoms usually form metals. When atoms having only a few electrons in the outer ring are combined, *metallic* bonds are formed to create solid materials. Since the outer rings or subshells are not completely filled, free electrons are present that can move from one atom to another. An outer ring electron spinning in its orbit may easily move into an orbit of an adjacent atom. This is particularly true at the instant when the free electron is at a point between the two atoms and hence virtually equidistant from each nucleus. Therefore, if a free electron leaves one atom and enters an orbit of the next, it will leave behind a positive ion. This condition is only temporary because the positive

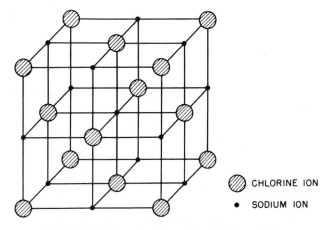

                                                      CHLORINE ION

                                               •   SODIUM ION

FIGURE 1-14. Ion atoms in sodium chloride crystals.

ion will capture an electron from another adjacent atom. Thus the free electrons can move about within the element at random. Upon application of electric power, however, all the free electrons can be made to move in orderly fashion from one atom to another, in a direction established by the manner in which the electric pressure is applied.

Covalent bonds are those in which outer ring electrons of similar atoms combine. A typical example of this is the hydrogen atom. Normally, the single atom does not have its outer ring completely fixed (the ring can accommodate two electrons and only one is present). When such atoms are brought closely together, however, a bond is formed because each nucleus not only has its own electron revolving around it but also shares the electron of its neighbor, as shown at A of Fig. 1-15. Neither nucleus can attract the adjacent electron more than the other nucleus; in this way the inert gas hydrogen is formed.

Another example of the covalent bond is that of normal atmospheric oxygen. The oxygen atom, by itself, has two electrons in the first ring and six in the second ring, as shown at B. When the two atoms are brought together, there is a sharing of two of the electrons by each atom nucleus. The first two subshells of the outer ring are, for all practical purposes, filled to their quota of eight, and a bond is formed.

Another example of covalent bonding of electrons occurs in carbon, germanium, or silicon. Since covalent bonding is of prime concern in solid-state diodes, transistors, and other such items, it is discussed more fully in Chapter 12, "Solid-State Fundamentals."

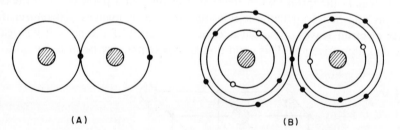

(A)   (B)

FIGURE 1-15. Bonding of hydrogen atoms and of oxygen atoms.

Review Questions

1-1. Briefly define the terms *electron, neutron, proton,* and *atom.*

1-2. What are *planetary* electrons?

1-3. Explain the relationships of positive and negative charges with respect to the nucleus of an atom and its planetary electrons.

1-4. List the number of protons in each of the following atoms: atomic No. 3; No. 6; No. 8; No. 30.

1-5. What are the essential differences among elements, compounds, and molecules?

1-6. List five elements and five compounds.

1-7. If an atom has an atomic No. 28, how many electrons does it contain? How many electrons are in each subshell of the second and third primary shells?

1-8. If an atom has 33 electrons, how many electrons are there in the outer ring subshells?

1-9. (a) In an atom, which electrons have the lowest-energy levels and which have the highest?
(b) Why do they have such characteristics?

1-10. What factor in relation to electrons determines the stability of an atom? Explain briefly.

1-11. (a) What is meant by a *free electron*?
(b) In what substances are free electrons usually found?

1-12. Explain briefly what constitutes a positive ion and what constitutes a negative ion.

1-13. How is electronic plasma created? Explain briefly.

1-14. What are the basic characteristics of electrostatic and electromagnetic fields?

1-15. Define the term *line of force*.

1-16. By simple drawings of two bar magnets, show which magnetic fields repel each other and which attract each other, and indicate which are the north and which are the south poles.

1-17. Briefly explain the older molecular theory of magnetism, and compare it to the newer theory of electron spins.

1-18. Briefly define the term *unit pole*, and show the related formula for finding the *magnetic force* in dynes.

1-19. What are *valence electrons*, and what are their basic characteristics?

1-20. Explain what factors cause some substances to be conductors of electricity and others to behave as insulators.

1-21. Explain the essential differences between *ionic* and *covalent* bonds.

1-22. Briefly define a *crystal-lattice network*.

Practical Problems

1-1. The north pole of a magnet with a unit pole strength of 300 is brought within 2.5 cm of the north pole of another magnet with a unit pole strength of 150. What is the force of repulsion in dynes between the two?

1-2. If the same magnets in Problem 1-1 were separated by 5 cms, what would be the force of repulsion in dynes between the two?

1-3. The south pole of a magnet is positioned within 2 cm of the north pole of another magnet. If the distance of separation is decreased to 1 cm, would the force of attraction double, triple, or quadruple? Explain your answer.

1-4. The north pole of a magnet with a unit pole strength of 90 is brought within a certain distance of the south pole of another magnet having a unit pole strength of 40. The force of attraction is 400 dynes. What is the separation between the two magnets?

Electrical Sources and Units   2

2-1. Introduction

The electric pressure that must be applied to cause electron movement is called *voltage*. When such a voltage is applied to a conducting medium, free electrons move progressively from atom to atom and constitute what is known as *current flow*. In electronic and electric practices, a continuous path for current flow is formed by interconnecting transistors, coils, resistors, and other components to form what are known as *circuits*. Such circuits serve many purposes, among them the generation of low- or high-frequency signals, the amplification of signals being processed, and the modification of certain signals as required for proper system operation.

The required voltage necessary to force current through an electric circuit may be obtained from a number of sources. One is the power supply (discussed fully in Chapter 14) that converts the a-c (alternating-current) line potential of the power mains to d-c (direct-current) for application to transistors, tubes, and other units of receivers, transmitters, or other electronic devices. Additional sources of electric power include various batteries that are used in portable electronic equipment. A battery consists of a number of cells so wired to provide the voltage value required. The cells create electron movement by a chemical reaction that occurs when the cell is wired into a circuit having a closed-loop path for the current flow.

This chapter discusses these factors in detail and introduces the unit values of voltage, current, power, and other related characteristics. In addition, the Standard System of Scientific Notation is covered in conjunction with other abbreviated forms as a necessary foundation for the study of more advanced circuit behaviors and functions.

2-2. Primary Cells

In each cell of a battery, two dissimilar conducting elements are brought into contact with a chemical composition known as an *electrolyte*, and the chemical reaction is such that electrons are forced out of one cell terminal and into the other, if an external path is provided for them. Without an interconnection between battery terminals, no current flows, and there is little chemical action. As shown in Fig. 2-1, a flashlight bulb, or other device that completes the circuit, allows the electrons to flow from the negative terminal of the cell, through the bulb, and back into the cell via the positive terminal. The chemical reaction within the cell that causes electron movement can be likened to pressure forcing water through a pipe. Hence, the term *electromotive force* (emf) is applied to indicate the force that could cause electron movement and current flow if the circuit were completed. Voltage is another name for electromotive force.

A battery, in its strictest sense, is a combination of basic energy-producing units known as cells. Each cell has available at its terminals a certain voltage and when two or more such cells are placed in series, they form a battery of increased voltage. A primary cell is the type commonly used in flashlights and for some portable transistor radios. The cell principle in the chemical types is based on the fact that when two dissimilar metals (such as zinc and copper or zinc and carbon) are placed in a solution of an acid, they act as a base. Electrons are then torn away from one metal element (the positive

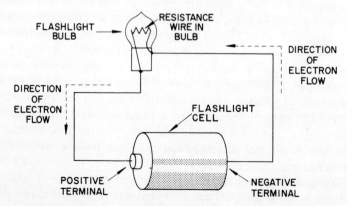

FIGURE 2-1. Electrons flow from the negative to the positive terminal.

terminal) and flow internally and accumulate on the zinc element, making the latter negative. If zinc and carbon elements are employed, the zinc element is called the *negative electrode* and the carbon section is called the *positive electrode*.

There are a number of battery types: some use mercury as the primary chemical ingredient, others use an alkaline solution, and many have special compositions designed for specific applications. Some are readily recharged or rejuvenated for continuous usage, while others are not easily renewed and must be discarded after prolonged usage. To illustrate the basic chemical principles involved, however, our discussions will be confined to the common primary cell utilized for flashlights, wherein a chemical paste is used, made up of sal ammoniac and zinc chloride as the principal active ingredients. The chemical paste (or the solution in a wet-type battery) is called an *electrolyte*.

The chemical action of tearing electrons from the carbon plate and depositing them on the zinc electrode creates a potential difference between the two terminals of the cell. If a circuit is formed by connecting a resistor to the negative and positive terminals, the closed circuit will permit electrons to flow. The electrons flow out of the negative terminal of the cell, through the resistor, and into the positive electrode. Thus, the cell converts chemical energy into electric energy.

The *size* of the electrodes in such a cell has little influence in determining the potential developed. The emf of a single primary cell is 1.5 volts when new, and this will gradually decrease as the cell is utilized, or as it ages. Some special cell types have an initial voltage of 1.25.

Fig. 2-2 shows a cross section of the basic construction of a flashlight-type cell. The outside shell consists of a zinc can that, as previously mentioned, serves as the negative electrode. When the cell circuit is completed by adding a resistor across the terminals, the current flow through the cell consumes some of this zinc because of the action of the electrolyte chemicals.

The separator consists of a pulpboard paper lining, coated with a layer of electrolyte paste made up of sal ammoniac, zinc chloride, and other material to form a thick consistency. This layer of electrolyte also separates the zinc from the depolarizing mixture but permits the electrochemical action to continue between them. The depolarizing mixture contains manganese dioxide to combine with hydrogen as it accumulates and carbon to provide conductivity. Other chemicals such as graphite are added and the mixture also contains some of the sal ammoniac and zinc chloride previously mentioned.

The depolarized mixture aids in giving the cell much longer life than would normally be the case. Depolarization is a chemical reaction that the cell undergoes; this reaction occurs while the cell is not in use and thus tends to "rejuvenate" the cell to some extent. For this reason, it is advantageous to use the cell intermittently.

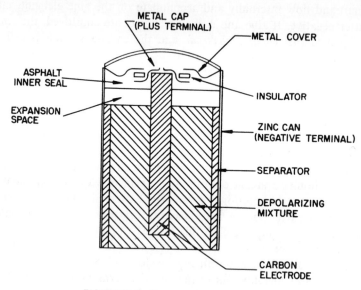

FIGURE 2-2. Cross section of battery cell.

The carbon electrode rod through the center is the *positive* terminal of the cell and provides a conducting path for the current flow. This carbon electrode is composed of powdered carbon particles bonded together and made rigid by baking at a very high temperature.

An inner expansion space is provided to permit the cell contents to expand without causing the cell to bulge during use. The cell top is closed at the center by an asphalt inner seal. A metal cover is usually employed, which also closes the cell at the top and sides, minimizing breakage and bulging. This metal cover section is insulated from the metal cap that makes contact with the carbon positive terminal at the center top of the cell.

These primary cells can be combined in parallel to make a large battery that furnishes more power than a single cell or they can be used in series to provide higher voltage batteries. Cell combinations are described more fully in Chapter 3.

Battery cells have an internal resistance determined by the type of chemicals used. The internal resistance of a primary cell is very low when the cell is new but increases as the cell ages and also as it is used. As the internal resistance increases (because of chemical changes), there is also a decrease of voltage across the terminals and a decline in the power available from the cell. Internal cell or battery resistance is not easily measured with accuracy and, in practical applications, internal resistance values need not be known. Compensation for internal resistance is made by *bypassing*, a process described more fully in Chapter 13.

2-3. The Secondary Cell

The secondary cell is a so-called *storage* cell to which electric energy is applied and converted into chemical energy. This chemical electric power is stored in the cell and reconverted into electric energy when the cell circuit is completed by placing a resistance material across it to permit current flow.

The storage battery used in automobiles is made up of a lead-acid type of secondary cell. In such batteries, the negative electrode of each cell is usually composed of a spongy lead, while the positive electrode is composed of lead peroxide. The two elements are immersed in an electrolyte of sulfuric acid. When current flows through the cell due to a closed circuit, some of the diluted sulfuric acid is transferred from the electrolyte chemical to the lead plates. Since sulfuric acid is heavier than water, there is a change in weight when the battery is fully charged, in relation to its discharged condition. A device known as a *hydrometer* is utilized to determine the condition of the cell by measuring the density of the electrolyte.

Actually, the storage battery does not "store" electric energy but stores *chemical* energy. The latter is converted to electric energy when the circuit is completed by applying a resistor or resistive load to the battery. The amount of chemical energy that the battery is capable of storing depends on the area of the plates. Three secondary cell types are commonly employed to obtain a 6-V battery, such as shown in Fig. 2-3. Six secondary cells are necessary for the 12-V car battery. The entire battery is encased in a hard rubber or plastic case, with provision available, in the form of removable caps, for checking the water density versus electrolyte density in each cell.

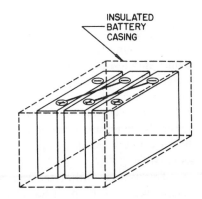

INSULATED
BATTERY
CASING

FIGURE 2-3. Cells in series for storage battery.

The hydrometer has a small float, which is marked with a scale usually starting at 1000 and ending at 1300. Some scales also have color sections that are labeled *fully charged, half charged,* and *discharged.* At full charge, the electrolyte is heavier since it contains a greater amount of acid, and the specific gravity of the electrolyte is approximately 1280, which corresponds to the markings on the hydrometer. Thus, when the battery is fully charged, the hydrometer will read near the 1300 mark, while a completely discharged battery will be indicated by a hydrometer reading of approximately 1100. A typical hydrometer is shown in Fig. 2-4.

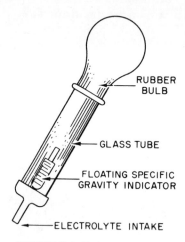

RUBBER BULB

GLASS TUBE

FLOATING SPECIFIC GRAVITY INDICATOR

ELECTROLYTE INTAKE

FIGURE 2-4. Typical hydrometer.

The lead storage cell, such as is used in automobiles, can be recharged after use. The charging device consists of a unit that furnishes at its terminals a voltage somewhat greater than the battery voltage. This higher voltage then forces current to flow through the battery in the opposite direction to the current flow that occurs during the time the battery is normally discharging through a load.

The storage battery, as previously mentioned, has a chemical storage capacity proportionate to the activity and size of the plates utilized. This is generally expressed in *ampere hours*, which indicate how many hours the battery can deliver a given amount of current measured in amperes (the ampere is discussed in greater detail later in this chapter). Most batteries are rated at 8 hr for a 100-amp-hr battery. This would indicate that such a battery can supply 12.5 amp of current continuously for 8 hr (12.5 × 8 = 100 amp-hr).

The charging rate depends on the amount of current forced through the battery in the opposite direction to what would flow during normal operation of the battery. The higher the charging potential, the greater the charging current that will flow.

Too high a charging rate will not only increase battery temperature but will also generate an excessive amount of gas. Some gas is generated even at lower charging rates and, for this reason, no lighted matches or other open flames should be brought near a battery that is being charged. The 100-amp-hr battery can be charged at a rate ranging between about 10 to a fraction of an ampere. Some battery chargers start at a higher rate and then gradually reduce the charging rate in amperes as the battery nears full charge. A lower charging rate is recommended wherever possible. Higher rates should be used only when it is urgent to put the battery into service again as soon as possible.

The average life of the lead-type storage battery can be extended for several years by keeping the electrolyte at a good level above the plates. This means the periodic addition of pure water (or distilled water), as well as occasional checking to see that the charge level has not decreased to too low a value. The top of the battery should be kept clean and the terminals may be coated with petroleum jelly to minimize the corrosion that is set up because of acid leakage.

When the battery is fully charged, there is a minimum danger of its freezing when exposed to extremely cold weather. When the battery is near the discharge level and the electrolyte reads about 1100 on the hydrometer, the freezing point is approximately 20° above zero. When the hydrometer reads 1280 to indicate a fully charged battery, the freezing point is approximately 60° below zero.

2-4. Other Electric Sources

Magnetism is also used for producing current flow through conductors and circuits. Both the permanent magnets and the so-called electromagnets (formed by passing current through a coil) produce the magnetic lines of force discussed in the preceding chapter. When a conductor, such as a wire (or a coil composed of wire), is moved through magnetic lines of force, a current flow is produced. This process is more fully described in Chapter 7. The principles of producing current flow by utilizing magnetic lines of force are employed in transformers, generators, and a variety of other electronic devices as subsequently detailed in Chapters 7 and 8.

Thermal (heat) processes are also used for creating electron movement. One thermal method consists in joining the ends of two dissimilar metals and heating the joined ends, as shown at A of Fig. 2-5. A voltage appears at the open ends and if these open ends are connected to a suitable current path (a conductor), electric current will flow. This device, known as a *thermocouple*, finds applications in industry through its use for reading high temperatures (in commercial ovens or kilns) and as a temperature-sensitive device for high-frequency radio signal measurements, and also for use as a protective safety device.

Another use of the thermal process for production of electron movement is in an evacuated tube, as shown at B of Fig. 2-5. By bringing a filament within the tube to a high temperature, a cloud of free electrons is produced, as more fully explained in Chapter 11.

Photoelectric emission is a term applied to still another process for creating electron movement and hence current flow. This method consists of

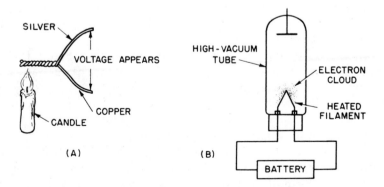

FIGURE 2-5. Thermal process for electron movement.

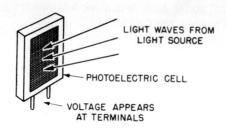

LIGHT WAVES FROM
LIGHT SOURCE

PHOTOELECTRIC CELL

VOLTAGE APPEARS
AT TERMINALS

FIGURE 2-6. Photoelectric emission.

permitting light to strike certain sensitive materials such as selenium or cesium, which in turn will then emit electrons (see Fig. 2-6). Again, if a current path is provided, the amount of current that flows can be used as an indication of the amount of light striking the photosensitive surface. The uses of photoelectric emission are numerous and find application in burglar alarms, devices for turning lights on at night, television cameras, motion picture "sound on film" techniques, and other devices. Heinrich Hertz (1857–1894) the noted German scientist is credited with the discovery of photoelectric emission.

2-5.   Units of Voltage and Current

As mentioned earlier, electric pressure is needed to move electrons through a conductor to cause current flow. Such electric pressure is know as *electromotive force* and is abbreviated *emf.* The unit of emf is the volt (V) named after Alessandro Volta (1745–1827) the Italian researcher who first built a cell which provided electromotive force and which was the forerunner of our modern battery. Hence, either V or emf (E) designates voltage (electric pressure), and sometimes the word *potential* is also used. All these terms have the same meaning. *A volt is the quantity of electromotive force that will cause one ampere of current to flow through one ohm of resistance.*

The source of electrons, from a battery or other electric generating device, is referred to as the negative terminal or negative section of the unit. The terminal toward which electrons flow is designated as the positive terminal. Instead of referring to the two different types of potentials as charges, the term *polarity* is more often used. Thus, a flashlight cell may have a potential (emf) of 1.5 V, with one terminal having a negative polarity and the other terminal a positive polarity.

In practical electric or electronic applications, voltages of a fractional value will frequently be encountered, as well as voltages having values up in the thousands, depending on the amount of electric pressure necessary to force current flow through the resistances encountered in the various circuits and devices. Hence, fractional unit values of voltage are often in terms of a *millivolt* (one-thousandth of a volt) or *microvolt* (one-millionth of a volt). Thus, 0.001-V can be expressed as 1 millivolt (1 mV), while 0.00003-V can be expressed as 30 microvolt (30 μV). High voltages are often designated as *kilovolts* (kV), to indicate thousands of volts. Thus, 10 kV represents 10,000 V. Such terms simplify the designation of fractional voltages or high-value voltages by substituting a prefix for the number of zeros that would have to be employed, as shown more fully later in this chapter.

As mentioned in Chapter 1, a current-flow path is provided by wires or other metals and thus form *conductors* of electricity. Hence, the ability of a substance to conduct electric current is termed *conductivity*. Current can also flow in substances such as liquids, gases, or materials whose composition offers opposition to the flow and limits the amount of current to definite quantities. The unit of electric current is known as the *ampere*, named after André Ampère (1775–1836) the famous French experimenter and scientist. One ampere of current represents the exact quantity of electrons that flows past a given point in one second and is equal to one coulomb. (See the earlier discussion on charges in Chapter 1.) The symbol for current is the capital letter *I*, with the symbol *A* for ampere.

While many circuits have a current flow of one or more amperes, there are also numerous electronic devices wherein only a fractional portion of an ampere of current flows. The amount of current that flows in any electronic unit or device can be measured by means of a current-reading meter, as more fully detailed in Chapter 5.

Years ago, when electric phenomena were first observed, the battery polarities were thought to indicate that current flowed from the positive terminal to the negative terminal. This is still known as the *conventional current-flow theory* and is still referred to in some of the literature. Many texts on electronics, however, recognize the direction of electron flow as constituting also the direction of the current flow (from negative to positive). The electron-flow direction is unquestionably from negative to positive since we have observable and measurable proof of this in vacuum tubes (electrons emitted from a hot cathode and flowing toward a positive metal plate) as well as in television tubes (electrons leaving the negative cathode structure and forming a scanning beam as it also is attracted by a positive potential), etc.

To avoid conflict and confusion, however, electron flow will be specified herein when referenced with respect to direction of flow. Otherwise, current flow will be mentioned for discussing factors relating to quantity, power, type, etc.

2-6. Units of Resistance and Conductance

All substances do not provide the same degree of conductivity since the number of free electrons present depends on the atomic structure. Thus, even various metals offer differing opposition to current flow, with some providing good conductivity and others opposing current flow to a considerable degree. The opposition of a substance to the flow of current is known as *resistance*, and the unit of measurement for such resistance is termed an *ohm*, in honor of George Ohm (1787–1854) the German professor who formulated the basic law relating to current flow, resistance, etc., described later and known as *Ohm's law*. The symbol for resistance is the capital letter *R*. The capital Greek letter Ω (omega) is used to indicate ohm or ohms.

The standard that has been established for 1 ohm is the resistance provided at 0°C, by a column of mercury having a cross-sectional area of 1 sq mm and a length of 106.3 cm. When a substance is specially prepared to offer a given amount of resistance, it is known as a *resistor*. Resistors of all types are widely used in virtually all electric and electronic devices. Examples of types, plus practical factors, are discussed in Chapter 3.

The measure of how well a substance will permit current flow is known as *conductance*. Because conductance is functionally opposite to resistance, it is the reciprocal of resistance and is therefore equal to the numeral *one* divided by the value of resistance, as expressed by the formula $1/R$. Thus, if a particular resistance is 1000 Ω, the conductance is $\frac{1}{1000} = 0.001$. Because conductance is the opposite of resistance, the unit for conductance is expressed as the word *ohm* spelled backwards, which becomes *mho*. Hence, the conductance for the last example is 0.001 mho. Generally, a fractional measurement of mho is used, known as the *micromho*. This is one-millionth of a mho. The symbol for conductance is the capital letter *G*.

2-7.   Ohm's Law

The values of current, voltage, and resistance are all related because the amount of current that flows is dependent on the amount of electric pressure (emf) applied and the opposition (resistance) encountered by the electron movement. These relationships were established by George Ohm and were set down as specific formulas whereby an unknown quantity of *R*, *I*, or *E* can be found when the values of any two of these are known. One of these formulas shows how the value of current (*I*) can be ascertained when voltage (*E*) and resistance (*R*) are known.

$$I = \frac{E}{R} \quad \text{or} \quad \frac{V}{R} \qquad (2\text{-}1)$$

This formula indicates that the amount of current flowing in a circuit is equal to the quotient obtained when the voltage value is divided by the resistance value. Thus, if there are 10 V impressed across a 5-Ω resistor, the current flow through the resistor is 2 amps.

An unknown value of either voltage or resistance can also be found by rearrangement of the formula. Voltage, for example, can be obtained if the values of current and resistance are known.

$$V \text{ or } E = IR \qquad (2\text{-}2)$$

As this formula shows, an unknown voltage can be found by multiplying the current value by the resistance value. Thus, if the current flow through a 3-Ω resistance is 20 amps, the voltage across the resistance is 60.

If the value of the resistance is desired, the known values of current and voltage are used for solving the problem by use of the formula

$$R = \frac{E}{I} \quad \text{or} \quad \frac{V}{I} \tag{2-3}$$

As an example, assume that the voltage across a resistance is 50 and the value of the current flow is 10 amps. Dividing the E value by the I value indicates a resistance of 5 Ω.

2-8.   Units of Electric Power

When voltage is applied to a conductor, current will flow and the amounts of current flow and voltage represent a quantity of power. Such electric power can be used for heating purposes, for operating a motor, or in other applications of electric energy. Since we cannot get power for nothing, a battery or other power source must be used to generate the energy needed. The electric power (symbol W) is measured by the amount of voltage multiplied by the quantity of current flow.

$$W = EI \tag{2-4}$$

The unit of power is the *watt*, named after the Scottish inventor James Watt (1736–1819). One watt of power is equal to one ampere of current flow produced by one volt of electric pressure. Because the watt unit relates to electric power, the symbol P has been used as well as W. For the new SI system (see Sec. 2-12) the symbol is W.

When power is calculated in terms of time, the unit of energy is the *joule*. This is also known as the *watt-second* and represents one watt of power for one second. In the measurement of ordinary electric power consumed in homes, the *kilowatthour* (kWhr) is utilized, and this refers to 1000 W for 1 hr. In many electronic applications, however, only fractional power units are used, and the term *milliwatt* (mW) is then utilized for convenience, to express one-thousandth of a watt. Thus, 0.0005-W = 0.5 mW.

Formula 2-4 solves for the amount of energy consumed in terms of unit watts. Hence, if 20 V are present across a resistance and 2 amp of current flow, the amount of energy consumed equals 40 W. If E is unknown but I and R are known, the following formula can be used:

$$W = I^2 R \tag{2-5}$$

Thus, if 2 amp of current are flowing and the R value is 10 Ω, the amount of power is

$$W = 2 \times 2 \times 10 = 40$$

Power can also be found by

$$W = \frac{E^2}{R} \quad \text{or} \quad \frac{V^2}{R} \tag{2-6}$$

Example: A voltage of 20 is measured across a resistance of 4 Ω. What is the power in watts?

Solution: Twenty squared equals $20 \times 20 = 400$. When this product is divided by the value of R (4 ohms), the answer is 100 watts.

Because power is also related to Ohm's law, the amount of power used can be employed in formulas for finding unknown values of current of voltage, as shown by Formulas 2-7.

$$I = \frac{W}{E} \qquad R = \frac{W}{I^2} \qquad E = \sqrt{WR} \qquad I = \sqrt{\frac{W}{R}} \tag{2-7}$$

Example: If the power consumed in a circuit is 50 W and the circuit resistance is 800 Ω, what is the applied voltage?

Solution: Using the third equation of Formula 2-7,

$$E = \sqrt{50 \times 800} = \sqrt{40,000} = 200 \text{ volts}$$

Circular or rectangular groupings are often used as a convenient method for displaying the various Ohm's law equations. A typical example is shown in Fig. 2-7, where an unknown is shown with its related component. Thus, $EI = W$, as does I^2R; or $I = W/E$, etc.

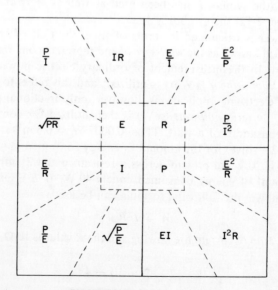

FIGURE 2-7. Ohm's law square

2-9. Units of Magnetism

In electricity and electronics, certain terms and unit values are used to define characteristics of magnetic materials. The types of magnetic materials also have names applied to them for easy recognition and reference. The following list, with definitions, covers the most commonly encountered. Explanations of the differences between the *cgs* and *mks* systems, as well as data on the new International System of Units (SI) are given in Sections 2-11 and 2-12.

Magnetic flux. This term relates to the number of magnetic lines in a given area of magnetism. In the *cgs* system, the unit of magnetic flux is the *maxwell* and represents one line of the magnetic flux, symbolized by the lower case Greek letter phi (ϕ). The maxwell is named after James Maxwell (1835–1879), the famed Scotch theoretical physicist. The unit in the *mks* system is the *weber* (Wb) and one maxwell $= 10^{-8}$ webers.

Flux density. This is a term applied to the number of lines of force that pass perpendicularly through a square centimeter and is symbolized by the capital letter *B*. The unit of flux density is the gauss, named after the German mathematician Karl Gauss (1777–1855).

Magnetic field intensity. This relates to the force (magnetic field strength) exerted by the field, and the unit is the *oersted*, named after Hans Oersted (1777–1851) the Danish physicist. An oersted represents the intensity of the magnetic field at a 1-cm distance from the unit magnetic pole (in air or vacuum). The symbol for the oersted is the capital letter *H*.

Magnetic induction. The term applies to the magnetizing of a magnetic material by inducing into it the lines of force from a magnet. Thus, a bar of iron when rubbed with a permanent magnet can become magnetized through magnetic induction.

Permeability. This is a measure of the conductivity of magnetic flux through a material. Permeability is symbolized by the lower-case Greek letter mu (μ). It is the ratio of the flux which exists when using a certain material to the flux which would be present if air were used instead. The permeability of air is thus considered as unity (one), and all other materials have varying degrees of permeability above one. Soft iron, for instance, has better conductivity for lines of force than steel because it has greater permeability. Other materials with high permeability include cobalt, ferrite, and certain alloys. High permeability materials find extensive use as cores of transformers.

Reluctance. The opposition offered by material to the magnetic flux is referred to as its *reluctivity* or *reluctance*, and it corresponds to the resistance to current flow in electric circuits. The symbol is the script capital letter $\mathcal{R}$.

Permeance. This term, seldom used, refers to the property that determines the magnitude of the flux in a material and is equal to the reciprocal of reluctance; the symbol is the script capital letter $\mathcal{P}$.

Retentivity. The ability of a material to retain magnetism after withdrawal of the magnetizing force is known as its *retentivity.* Steel, for example, having a higher retentivity, will retain more magnetism than iron.

Instead of classifying materials simply as magnetic and nonmagnetic, modern practices group materials into three primary classifications, as follows:

Diamagnetic materials. This is the term used for materials that become only slightly magnetized, even though under the influence of a strong magnetic field. When a diamagnetic material is magnetized, it is in a direction opposite to that of the external magnetizing force, as shown at A of Fig. 2-8. Diamagnetic materials have a permeability of less than unity and include such metals as copper, silver, gold, and mercury.

Paramagnetic materials. These materials become only slightly magnetized, even though under the influence of a strong magnetic field, as with the diamagnetic materials. The difference between the two, however, is that the paramagnetic materials become magnetized in the same direction as the external magnetizing field, as shown at B of Fig. 2-8. Permeability of the materials is greater than unity, though low compared to the ferromagnetic materials. Paramagnetic materials include aluminum, chromium, manganese, platinum, and also, air.

Ferromagnetic materials. The most important materials used in magnetic applications in electricity and electronics are the ferromagnetic materials. These are characterized by a very high permeability and will become very strongly magnetized by a field relatively much weaker than that required for the other two types discussed. The direction of the induced magnetism is identical with that of the magnetizing field, as shown at B of Fig. 2-8, as with the paramagnetic materials. The permeability of ferromagnetic materials is not constant but varies with the strength of the magnetizing field. Among the ferromagnetic materials are iron, steel, cobalt, magnetite (lodestone), and

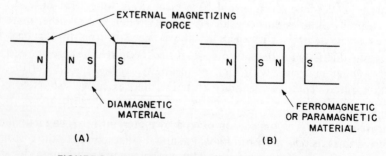

FIGURE 2-8. Direction of magnetism for various materials.

alloys such as alnico and Permalloy. Alnico derives its name from the metals used in the alloy: *al*uminum, *ni*ckel, and *co*balt (with some iron added). Permalloy (*perm*anent *alloy*) contains iron and about four times as much nickel. Such newer alloys produce powerful magnets and find extensive application in industry. Compared to soft iron, with a permeability of about 2000, these alloys have permeability ratings ranging, under certain conditions, as high as 50,000 or more.

2-10. Abbreviated Forms

To simplify explanations hereafter, it is expedient at this time to discuss the mathematical abbreviations utilized by engineers and technicians as a means for eliminating the necessity of writing long strings of zeros in very small fractional unit values or extremely large unit values. An understanding of these abbreviations will overcome the need for employing cumbersome numbers in referring to specific measured values.

As an illustration, the unit of electric current, the ampere, was previously indicated to consist of 6.28 electrons times the product of the figure 10 multiplied by itself 18 times. This enormous figure can be represented in much more simple form by setting it down as

$$6.28 \times 10^{18}$$

This method of expression is known as the *Standard System of Scientific Notation* and is also called *engineers' shorthand* because it is a method for writing large numbers in simple (abbreviated) form. Thus, if 10 is to be multiplied by itself, it is represented by 10^2 (*10 to the second power*). This expression, because it indicates that the 10 is to be multiplied by itself, represents 100. The expression 10^3 indicates the following multiplication process:

$$10 \times 10 \times 10 = 1000$$

The basic number, here the "10" in 10^3, is called the *base*, and the raised (or superior) number is called the *exponent*.

Example: $6.28 \times 10^3 = 6280$

Note that the decimal point in 6.28 has been moved three places to the right in the answer. In using engineers' shorthand in the foregoing example, the initial figures were not changed but only the decimal point. Similarly, 4.8×10^4 equals 48,000. Here the decimal has been moved four places to the right. The "shorthand" feature is more apparent when the exponent is a larger number: 6×10^8 equals 600,000,000. Thus, the engineers' shorthand expression is much simpler than writing the higher number out.

Scientific notation or *power of 10*, as it is also known, is also used with a minus sign. A minus power such as 10^{-2} moves the decimal two places to the left because the exponent is -2.

Example: $8 \times 10^{-2} = 0.08$

Again, the advantages of this kind of abbreviated notation are more apparent when the exponent is a larger number, such as 6.3×10^{-6}, which, when written out, equals

$$0.0000063$$

The expression 5×10^{-10} becomes

$$0.0000000005$$

Since both fractional values in millionths and whole numbers in millions are encountered frequently in electronics, engineers' shorthand is quite useful in eliminating a long string of zeros:

$$10 = 10^1$$
$$100 = 10^2$$
$$1,000 = 10^3$$
$$10,000 = 10^4$$
$$100,000 = 10^5$$
$$1,000,000 = 10^6$$

$$0.1 = 10^{-1}$$
$$0.01 = 10^{-2}$$
$$0.001 = 10^{-3}$$
$$\text{one-millionth} = 10^{-6}$$
$$\text{one-millionth of a millionth} = 10^{-12}$$

$$100^0 = \text{one (1)}$$
$$100^1 = \text{one hundred (100)}$$
$$100^2 = \text{ten thousand (10,000)}$$
$$100^3 = \text{one million (1,000,000)}$$

In electronic work, it is not always convenient to use the unit expression of ampere because only fractions of such current may be present. In such cases, the term milliampere and microampere are employed. *Milliampere* means one-thousandth of an ampere, and *microampere* means one-millionth of an ampere. The following list gives additional terms. These apply not only to current but also to other units, as detailed in later chapters.

Another abbreviated form that has been internationally adopted is the word *hertz* to replace the phrase *cycles per second*. Hertz is named after Heinrich R. Hertz the German physicist and researcher, mentioned earlier in Sec. 2-4. The abbreviation for hertz is Hz and one thousand cycles per second would be expressed as 1 kHz; two million cycles per second as 2 MHz, and 60 cycles per second as 60 Hz. Thus the "Hz" not only shortens the expression but is a universal word that eliminates the many differences for the expression in the various languages.

Prefix	Symbol	Value	Submultiples and multiples
atto	a		1×10^{-18}
femto	f		1×10^{-15}
pico	p	one-millionth millionth	1×10^{-12}
nano	n	$\frac{1}{1000}$ of a millionth	1×10^{-9}
micro	μ	one-millionth	1×10^{-6}
milli	m	one-thousandth	1×10^{-3}
centi	c	one-hundredth	1×10^{-2}
deci	d	one-tenth	1×10^{-1}
deca	da	ten	1×10^{1}
hecto	h	one hundred	1×10^{2}
kilo	k	one thousand	1×10^{3}
mega	M	one million	1×10^{6}
giga	G	one thousand million	1×10^{9}
tera	T	one million million	1×10^{12}

2-11.   The *mks* and *cgs* Units

As mentioned in Sec. 2-9, several unit systems have been used in electric and electronic references. The designation *mks* refers to the *meter-kilogram-second* system of units, while the *cgs* is the *centimeter-gram-second* system. A newer system is the International System of Units (SI) described in Sec. 2-12. The basic differences between the *mks* and *cgs* systems are shown in the following listing of some of the more common units:

Quantity	Symbol	*mks* unit	*cgs* unit
Capacitance	C	farad	farad
Conductance	G	mho	mho
Current	I	ampere	ampere
Electric charge	Q	coulomb	coulomb
Electric potential	V	volt	volt
Force	F	newton	dyne
Inductance	L	henry	henry
Length	l	meter (m)	centimeter (cm)
Magnetic field intensity	H	ampere-turn/*m*	oersted
Magnetic flux	ϕ	weber	maxwell
Magnetization	M	weber/m^2	
Magnetomotive force	mmf	ampere-turn	gillert
Mass	*m*	kilogram	gram
Permeability	μ	henry/*m*	gauss/oersted
Permeance	$\mathcal{P}$	weber/amp-turn	maxwell/gilbert
Power	P	watt	watt
Resistance	R	ohm	ohm
Reluctance	$\mathcal{R}$	amp-turn/weber	gilbert/maxwell
Time	*t*	second	second
Energy-work	W	joule	joule

2-12.　International System of Units (SI)

The International System of Units are newer standards and the basis is the meter, thus representing a modernized version of the metric system. The official abbreviation is SI in all languages, even though derived from the French phrase *systéme internationale*. Because this system has been adopted by international agreement, it is now the basis of all national measurements throughout the world, and integrates such measurements for science, industry, and commerce. Thus, there is only one unit for a particular quantity, whether thermal, electrical, or mechanical.

The SI can be considered an *absolute system*, using absolute units for simplification in engineering practices. Thus, the unit of force, for instance, is defined by acceleration of mass (kg·m/s²) and is unrelated to gravity. The SI is formed on a foundation of 6 base units of *length, mass, time, temperature, electric current*, and *luminous intensity*. Four of these are independent: *length, mass, time*, and *temperature*. The remainder require use of other units for definition, with multiples and submultiples expressed in decimal system. Two supplementary units are also used: *radian* (for measurement of plane angles) and *steradian* (for measurement of solid angles). These two are termed *supplemental* because they are not based on physical standards, but rather on mathematical concepts.

The standards for the 6 base units are those defined by international agreement. The prototype for mass is the only basic unit still defined by a rigid physical device. Thus, the kilogram is a cylinder of platinum-iridium alloy housed at the International Bureau of Weights and Measures in France, with a duplicate at the National Bureau of Standards in the United States.

The meter is defined as a specific wavelength in vacuum of the orange-red line of the spectrum of krypton 86. Time has for its unit the *second* defined as the duration of specific periods of radiation corresponding to the transition between two levels of cesium 133. Kelvin for temperature is defined as 1/273.16 of the thermodynamic temperature of the triple point of water (the latter approximately 32.02°F). Ampere for current is defined as the current flowing in two infinitely long parallel wires in vacuum, separated by 1 m, and producing a force of 2×10^{-7} newtons p/m of a perfect radiator at the temperature of freezing platinum, 2024 °K.

The following listing includes the base units, the supplementary units, and the derived units. The latter are produced without having to use conversion factors. Thus, a force of 1 N acting for a length of 1 m produces 1 J of energy. If this force is maintained for 1 s, the power is 1 W.

Base SI Units and Symbols

Quantity	Symbol	SI Unit	Derivation
Length	m	meter	
Mass	kg	kilogram	
Time	s	second	
Temperature	°K	degree Kelvin	
Electric current	A	ampere	
Luminous intensity	cd	candela	candela

Supplementary Units

Plane angle	rad	radian	
Solid angle	sr	steradian	

Derived Units

Area	m^2	square meter	
Acceleration	m/s^2	meter per sec. squared	
Angular acceleration	rad/s^2	meter per sec. squared	
Angular velocity	rad/s	radian per sec.	
Density	kg/m^3	kilograms per cubic meter	
Electric capacitance	F	farad	$(A \cdot s/V)$
Electric charge	C	coulomb	$(A \cdot s)$
Electric field strength	V/m	volt per meter	
Electric resistance	Ω	(V/A)	
Energy, work, quantity of heat	J	joule	$(N \cdot m)$
Flux of light	lm	lumen	$(cd \cdot sr)$
Force	N	newton	$(kg \cdot m/s^2)$
Frequency	Hz	hertz	(s^{-1})
Illumination	lx	lux	(lm/m^2)
Inductance	H	henry	$(V \cdot s/A)$
Luminance	cd/m^2	candela per sq. meter	
Magnetic field strength	A/m	ampere per meter	
Magnetic flux	Wb	weber	$(V \cdot s)$
Magnetic flux density	T	tesla	(Wb/m^2)
Magnetomotive force	A	ampere	
Power	W	watt	(J/s)
Pressure	N/m^2	newton per sq. meter	
Velocity	m/s	meter per sec.	
Voltage, potential difference, electromotive force	V	volt	(W/A)
Volume	m^3	cubic meter	

Review Questions

2-1. Briefly define in your own words the term *primary cell.*

2-2. Briefly explain in what manner the secondary cell differs from the primary cell.

2-3. What is the purpose of using a depolarizing mixture in electric cell construction?

2-4. When is the lead-type storage battery more subject to freezing than otherwise? Explain your answer.

2-5. What two other processes besides using a battery can create electron movement?

2-6. (a) What is meant by *conventional current flow*?
(b) In what direction do electrons flow in a circuit in relation to terminal polarities?

2-7. Define the terms *resistance* and *conductance*, and specify what the unit values are.

2-8. With respect to voltage and current, which relates to quantity of electron flow and which is analogous to pressure? Explain your answers.

2-9. Define Ohm's law, and give two equation examples of its application.

2-10. What determines the amount of electric power consumed by a resistance?

2-11. Briefly describe what is meant by *magnetic flux* and *flux density*.

2-12. Define the terms *magnetic field intensity* and *magnetic induction*.

2-13. What word describes the opposition offered by a material to magnetic flux (corresponding to the resistance to current flow), and what symbol identifies it?

2-14. Briefly define the terms *reluctance* and *retentivity*.

2-15. Briefly describe what is meant by permeability, and name three materials having high permeability.

2-16. Define the term *ferromagnetic materials*, and compare their characteristics with those of diamagnetic materials.

2-17. What is meant by the *Standard System of Scientific Notation*?

2-18. (a) What whole number does 3.14×10^3 equal?
(b) What whole number does 0.870×10^4 represent?

2-19. (a) What whole number does 15×10^6 equal?
(b) What fractional number does 24×10^{-3} equal?
(c) What fractional number does 340×10^{-4} equal?
(d) What fractional number does $20,000 \times 10^{-6}$ represent?

2-20. Briefly explain the essential differences between the cgs and mks systems.

2-21. Explain the basic factors relating to the SI.

Practical Problems

2-1. In an electronic fabrication control circuit, 845 V appeared across a 1300-Ω resistor. What amount of current flows through this resistor?

2-2. In an industrial electronic circuit a 9-Ω resistor was used to decrease the voltage applied to the filament of a tube. If the current through the resistor is 0.5 amp, what voltage drop appears across the resistor?

2-3. In a test instrument for automation processes 2000 V dropped across a resistor that had 2.5 A of current flow through it. What is the value of the resistance?

2-4. In a plastic fabrication plant a heating element resistance of 500 Ω had 0.5 amp of current flow through it. What is the power in watts dissipated by the resistor? What is the voltage drop across the resistor?

2-5. In a portable transmitter a 400-Ω resistor had a 20-V drop across it. What wattage is dissipated by the resistor? What amount of current flows through the resistor?

2-6. If a resistor dissipated 1000 W of power and the current through the resistor is 5 A, what is the value of the resistor? What voltage appears across the unit?

2-7. In a solid-state electronic device the voltage measured across a resistor was 0.02. What would this value be when expressed in millivolts?

2-8. A test instrument utilized for obtaining visual characteristics of electronic equipment uses 5000 V. How would this voltage be expressed in kilovolts?

2-9. A test instrument indicates that 0.03 A of current is flowing in an electronic circuit. In recording this information, how would the fractional current be expressed in milliamperes?

2-10. A delicate instrument is applied to a low-power device and reads 0.0004 amp. Express this in milliamperes.

2-11. A broadcast station is rated as having a power of 50 kW. How many watts of power does this represent?

2-12. A certain transistor radio has an audio output rating of 200 mW. What is this in fractional wattage?

2-13. In measuring an antenna system, it is indicated that 250 μV is delivered to a receiver. Which of the following values also correctly expresses this voltage?

$$0.250 \quad V$$
$$0.0025 \quad V$$
$$0.00025 \ V$$

2-14. (a) In a computer, one pulse signal has a duration of 26 nanoseconds. As a fractional value, how many zeros actually precede the 26?
  (b) Express 0.00000918 in nanoseconds.

2-15. A component in a signal-filter network had a designation of 0.000015 micro units. What would this value be in pico units?

2-16. An FM radio station transmits on 90 MHz. How many cycles per second does this represent?

2-17. An AM radio station broadcasts on 1.2 MHz. Express this in kilohertz and hertz.

2-18. A stereo hi-fi amplifier has an audio-response rating (in terms of cycles per second) of 30 to 35,000. Express this range in *hertz*.

Basic Resistor Circuits   3

3-1. Introduction

The word *circuit* has already been introduced in Chapter 2, with a basic type illustrated in Fig. 2-1. A circuit is a combination of two or more electric or electronic components that are wired together in some specific manner to perform a required task. Circuits may be *closed* or *open* types, depending on their usage as needed at a given time. A flashlight with a switch, for instance, represents an open circuit when the switch has not been thrown to light the bulb. When the switch is closed, the circuit becomes a closed (or completed) circuit, and the amount of current flow is determined by the battery voltage and the resistance rating of the bulb.

An undesired characteristic is the *short circuit*. This comes about when the circuit resistance is abnormally low in comparison to the power source used. If, for instance, a copper wire is placed directly across a battery, the virtually zero resistance of the wire would permit such high current conductivity that all the current that the battery can furnish would flow. The copper wire, in such an instance, "shorts out" the battery and hence we have the term *short circuit*. Short circuits blow fuses, open circuit breakers, and impose a severe and possibly damaging load on the power source.

In this chapter the basic aspects of series- and parallel-resistor circuits are reviewed. Before analyzing these circuit characteristics, however, the practical factors regarding resistors should be understood and hence are covered initially.

Because the standard symbols were covered in Sec. 2-12, their usage will begin with this chapter to familiarize the reader with standard notational practices.

3-2. R-E-I Relationships

The voltage (electric pressure) required to cause a given amount of current to flow through a conductor or resistance is dependent on the ohmic value of the resistance. This is exemplified by the application of Ohm's law in a circuit where a resistance value of 10,000 Ω permits only 0.005 A (5 mA) of current to flow at a potential of 50 V. Raising the voltage to 100 doubles the amount of current flow to 0.01 A (10 mA). At 50 V, however, the amount of current flow can also be increased by reducing the resistance. Hence, in practical electric and electronic circuits, the amount of current is not only established or adjusted by voltage values but also by using electric opposition in the form of resistors.

The amount of resistance within a conductor or a resistor depends on the composition of the unit, and its cross section, length, and temperature. Wire usually has a high resistance at higher temperatures, while many non-metallic substances offer lower resistance as the temperature rises. Pure soft-metal elements, such as silver and copper, have a very low resistance—so low in fact that when silver or copper wire of any sizable diameter is used, it can be considered as a conductor whose resistance to current flow is practically negligible and usually not worth considering. (The current-carrying capacity of wire is covered more fully later in this chapter.) Harder elements or compounds, such as iron, metal alloys of iron (steel, for instance), or nichrome wire, have a measurable resistance, which can be made quite high if the wire diameter is made small.

Plastics, ceramic tile products, wood, glass, and many other materials offer such a high opposition to current flow that their resistance can be considered to be virtually infinite. Such materials can be used for supports or separators of current-carrying conductors and, because of their very high resistivity, will prevent the currents from straying from their assigned paths. Hence, such items act as an electric *insulation* between two sections that have current flow and thus are known as *insulators*. (See Fig. 3-1.)

Commercial resistors differ widely in size and composition, in order to match the various requirements of their countless applications. Resistors are rated according to both their ohmic value and their wattage. The ohmic value is, of course, an indication of the actual resistance expressed in ohms, such as 20 Ω, 1000 Ω, or several megohms. (A *megohm* represents 1 million ohms; in some applications, values as high as 5 or 10 MΩ are used.) The wattage rating of a resistor is directly related to the amount of heat generated and indicates how much heat is permitted in the resistor before the point is reached where there is danger of resistor damage. The amount of heat that

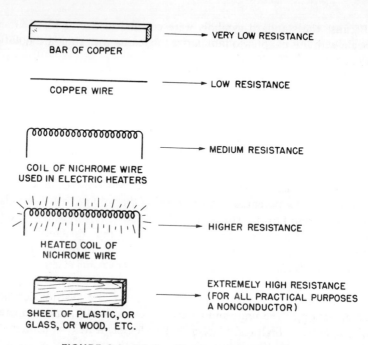

FIGURE 3-1. Relative resistances of common items.

a particular resistor will tolerate depends on the composition of the resistor, its heat dissipating factors, and the general cross-sectional area. Thus, if a resistor is designated as a 50-W resistor, this resistor can withstand that wattage (when suspended in free air) without overheating excessively and burning out. In actual usage, however, a larger wattage-rating resistor than the one calculated is usually employed, even though this is somewhat more costly. The larger-rating resistor will operate at lower temperatures and will thus have a longer life.

Various combinations of voltage and current values can provide a given wattage rating. For instance, a resistor consumes 50 W when the applied voltage is 2 and the current flow is 25 A. On the other hand, if 10 A of current is flowing and 5 V are applied, the amount of energy consumed will again be 50 W; 50 W will also be consumed if 50 V are applied and 1 A of current flows.

Resistors are not only obtainable in a variety of shapes and sizes but can also be formed from special wire, carbon alloys, or other materials. Wire-wound resistors use a special wire such as *nichrome*, which has inherent resistive characteristics, as mentioned earlier. Such wire is capable of handling a considerable amount of current, and for this reason resistors fashioned from nichrome wire are employed where considerable wattage is encountered. Nichrome, or other special wire, is also used for the resistive elements of electric heaters and electric irons. In these devices, sufficient current is made to flow through the resistive element to generate the required heat. Such heating devices consume much more electric power than do ordinary light bulbs and consequently have a much higher wattage rating.

In electronic devices, the most popular type of resistor is the carbon-composition resistor. Such resistors use a mixture of carbon and other substances to establish a fixed value of resistance. Carbon resistors are small units and are available in wattage ratings starting at $\frac{1}{4}$ W and continuing through intermediate wattage sizes up to 2 W. The carbon resistors are less expensive than the nichrome-wire types and therefore are extensively employed in receivers, high-fidelity systems, and low-power industrial devices.

Two basic types of fixed resistors are illustrated in Fig. 3-2. The one

FIGURE 3-2. Fixed resistors. (Courtesy International Resistance Co.)

shown at the top is a wire-wound resistor, available in various wattage ratings from a few to well over 50 W. In radio and television receivers, high-fidelity systems, and other comparable electronic devices, only a few wire-wound high wattage resistors are employed and rarely are such resistors used with wattage ratings as high as 50 W. Generally when it is necessary to use wire-wound resistors, ratings ranging from 5 to 20 W are utilized.

The wire-wound resistor shown at the top of Fig. 3-2 has an adjustable center terminal, which connects to the internal resistance wire. This center terminal can be adjusted to procure an intermediate value of resistance below the value established between the two outer terminals.

It must be emphasized that the ohmic value of the resistor is independent of its wattage rating. A 20-Ω resistor can be purchased with a 1-, 10-, or a 50-W rating, as desired. Conversely, a 30-W resistor could be purchased with an ohmic value of 10, 100, or 1000 Ω.

The lower resistor of Fig. 3-2 is the standard carbon unit previously mentioned and is used extensively in electronic devices. In order to facilitate identifying the resistance value in ohms, bands of color are painted on the

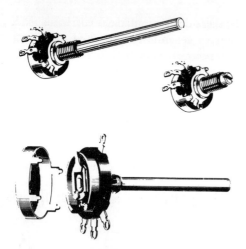

FIGURE 3-3. Variable resistors. (Courtesy International Resistance Co.)

body of the resistor. By *coding* a resistor in this manner, the necessity for measuring its value with a meter is eliminated. The fixed resistance value is indicated by several bands of color shown on the left of the resistor. Such identification is in accordance with the standard color code given in the appendix. The carbon resistors are generally available in resistance value tolerances of 5 and 10%, as shown by the chart in the appendix. The wattage ratings of such resistors are between $\frac{1}{4}$ and 2 W, as required by the circuits in which the resistors are used.

Frequently, the necessity arises for using a resistor that has provision for varying the resistance manually. Such variable resistors are required for volume controls in audio amplifiers and receivers, for level controls in test equipment, and for other applications described later. Variable resistors of this type are usually manufactured by depositing the carbon composition on a circular strip of paper, though there are occasions where resistance wire is wrapped around a circular form, when higher-wattage variable resistors are needed. In both the carbon and wire types, a sliding contact permits a change of the resistance value. Several variable resistor types are illustrated in Fig. 3-3. The one shown at the upper left has three terminals, with the center terminal representing the sliding arm. This type of variable resistor is usually furnished with a long shaft that may be cut off to the length desired.

Another type of variable resistor is shown in the upper right of Fig. 3-3. This variable resistor has a short slotted arm. Such resistors are often found in circuits where the resistance value need only be adjusted occasionally for balancing the circuit. Since continuous manual variation is not necessary, a screwdriver slot is provided for adjustments when required. These resistors are often used in high-fidelity amplifiers for hum balancing or for voltage adjustments in critical circuitry as subsequently covered.

The variable resistor at the bottom center of Fig. 3-3 shows the rear cover removed, thus exposing the sliding contact arm and the resistance strip. This is a wire-wound resistor. The internal appearance of the carbon type would be similar, except that a flat circular strip of carbon-coated paper or fiber would be used. The additional terminal shown at the top is a fixed tap sometimes utilized for "bass compensation" as discussed in Chapter 15.

Variable resistors (such as those illustrated in Fig. 3-3) are often provided with a switch arrangement so that the unit can be combined with an "on-off" switch. The switch is fastened to the back of the control, and a hole in the back cover of the variable resistor allows the variable arm of the resistor to trip the switch mechanism. The switch mechanism is insulated from the resistor and is actually a completely separate circuit.

The manner in which resistors are indicated in electronic diagrams is shown in Fig. 3-4. The number of points in the zigzag formation is unimportant and is left to the draftsman's decision. A fixed resistor is shown at the top of the drawing. The two lower drawings show how a variable resistor may be designated.

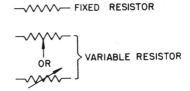

Variable resistors are sometimes referred to as *rheostats* or *potentiometers*. A rheostat is usually the term used when

FIGURE 3-4. Symbols for resistors.

a wire-wound variable resistor is employed for controlling the amount of voltage applied to a circuit or electronic device. In most instances, only the variable arm terminal and one outside terminal are employed. When such a resistor is placed in series with the current flow, as shown at A of Fig. 3-5, the amount of voltage drop that occurs across the variable resistor can be regulated by moving the variable arm. With a higher-resistance value, a larger voltage drop develops across the rheostat and less voltage is applied to the terminating circuit or electronic device. Since a higher resistance also decreases the series current, a reduction in output voltage occurs. When the variable arm is adjusted so that the series rheostat has less resistance, more current flows to the output circuit and a larger-voltage drop occurs across it. With less resistance in the rheostat, output voltage will increase and a smaller voltage drop develops across the variable resistor.

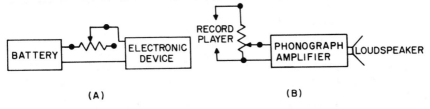

FIGURE 3-5. (a) Use of variable resistor as rheostat (B) Use of variable resistor as potentiometer.

A potentiometer is usually employed for controlling signal levels and in most instances a low-wattage unit is chosen. The two outside terminals, as well as the variable arm terminal, are used. When the two outside terminals are connected to a voltage source, the amount of output voltage is a function of the setting of the variable arm furnishing such voltage as shown at B of Fig. 3-5.

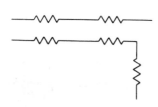

FIGURE 3-6. Symbols for series resistors.

When two or more resistors are joined together, they are drawn as indicated in Fig. 3-6. Resistors may be combined in three ways: in series, in parallel, or in series parallel. When a number of resistors or a combination of resistors and other components such as coils are interconnected in one drawing, the latter is referred to as a *schematic*. The resistor combinations shown in Fig. 3-6 are wired in series.

3-3. Resistors in Series

When resistors are wired up in series, the total circuit resistance increases because each resistor contributes opposition to the circuit's current flow. Thus, if a 10-Ω resistor is placed in series with another 10-Ω resistor, the total resistance contributed by the two is 20 Ω, double what one resistor offers in ohmic opposition. Similarly, if three 10-Ω resistors are in series, the total resistance is 30 Ω. Hence, the formula for resistors in series indicates a simple additive process:

$$\text{Total resistance} = R_1 + R_2 + R_3 + R_4 + \cdots + R_n \qquad (3\text{-}1)$$

Figure 3-7 illustrates the principles of series resistors as applied to resistors of 4, 8, and 3 Ω. The total resistance of the circuit is, therefore 15 Ω (4 + 8 + 3 = 15).

As shown, a current-measuring meter inserted in the series circuit indicates that 2 A of current are flowing. Also, since the three resistors are connected to a 30-V battery, it is evident that the 30 volts must be present across

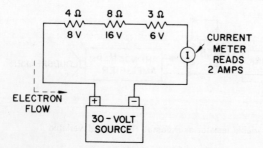

FIGURE 3-7. Series circuit with current meter.

the *combination* of resistors. Ohm's law proves this because if the total resistance is $15\,\Omega$ and the current is 2 A, then $E = IR = 2 \times 15 = 30\text{ V}$. Because the individual resistors have different values, *the amount of voltage across each* resistor differs. Across the 4-Ω resistor, there is a drop of 8 V (2 A multiplied by 4 Ω). Across the 8-Ω resistor, there is a drop of 16 V, and across the 3-Ω resistor a drop of 6 V exists. When these individual voltages are added together, they will equal the total voltage previously calculated.

Current	×	resistance	=	voltage
2	×	4	=	8
2	×	8	=	16
2	×	3	=	6
2	×	15	=	30

Total current in this example is 2 A. Because the resistors are in series, the amount of current at any point in the circuit is the same. It must be remembered that current is a measure of the amount of electron flow past a given point for a certain time interval. This amount would not change unless the applied pressure (voltage) or the resistance value (ohms) changes. Thus, current is always the same through any portion of a given series circuit, though the voltage across an individual resistor is proportional to the resistance value. In a series circuit, it is necessary only to solve for the current through *any one* of the several resistors to obtain the *current* value for the *entire* circuit.

Once the total resistance value has been found, we could prove the validity of the meter reading of Fig. 3-7 by Ohm's law.

$$I = \frac{E}{R} = \frac{30}{15} = 2\text{ A}$$

Once two values are known, other unit values can be found by using appropriate Ohm's law equations. For Fig. 3-7, the power dissipated in each resistor, as well as total power, can be solved by use of the formula I^2R.

Current²	×	resistance	=	wattage
4	×	4	=	16
4	×	8	=	32
4	×	3	=	12
4	×	15	=	60

Additional proof can be had by utilizing any of the formulas previously given. Thus, if the voltage is to be calculated from the power and resistance in the circuit, we would employ the formula $E = \sqrt{PR}$. Since the power previously calculated is 60 W and the total resistance is 15 Ω, the following calculation would be performed:

$$PR = 15 \times 60 = 900$$

$$E = \sqrt{PR} = \sqrt{900} = 30 \text{ V}$$

While current in any part of a series circuit is the same as in any other part of the circuit, it must be remembered that the current depends on the amount of resistance as well as voltage. If *either* the resistance or the voltage in a circuit is changed, the former current value will also change. Thus, if the same values of resistors are employed as in Fig. 3-7, but twice the voltage is applied across them, as shown in Fig. 3-8, a new value of current results. This new current value can be calculated by using the current formula ($I = E/R$). Thus, the new voltage value of 60, divided by the total resistance of 15, equals 4 A, showing that when the voltage is doubled, the current also doubles. In the circuit shown in Fig. 3-8, the new value of current (4 A) is the same in any part of the circuit. To calculate the new voltage drops across each resistor, we shall apply the voltage formula ($E = IR$). Hence, across the 4-Ω resistor there is now 16 V because we are multiplying 4 A times 4 Ω of resistance. The same calculation is applied to the other resistors:

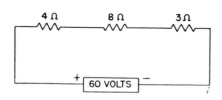

FIGURE 3-8. Series circuit with 60-V battery.

Current	×	resistance	=	voltage
4	×	4	=	16
4	×	8	=	32
4	×	3	=	12
4	×	15	=	60

If the wattage across each resistor is recalculated, it is found that the wattage has now increased by four. Since the wattage is equal to the $I^2 R$, the new value of current (4 Ω) is multiplied by itself to give a product of 16. This number is now used to multiply the ohmic value of each resistance, which gives a new set of wattage values:

Current^2	×	resistance	=	wattage
16	×	4	=	64
16	×	8	=	128
16	×	3	=	48
16	×	15	=	240

Note that the wattage for each resistor is now *four times* as great as it was. In consequence, the total wattage is also four times the 60 W previously shown.

Proof can again be obtained by utilizing any one of the formulas previously given. Using $E = \sqrt{PR}$, multiply the 240 W by the total resistance, 15 Ω. This gives a value of 3600. The square root of 3600 is 60, indicating that the voltage, which produces 240 W for a total resistance of 15 Ω, equals 60 V.

Ohm's law can also be employed to find the total battery voltage if it is not known or to calculate the value of a resistance if the voltage drop across the resistance and the current through it are known. Figure 3-9 shows a typical problem of this type. Here the value of the resistor R_1 is given as 100 Ω but the value of resistor R_2 is unknown. The voltage drop across R_2 is 8 V, and a current-indicating device shows that 0.02 A is flowing in the circuit. The problem consists in solving for the resistance value of R_2, as well as for the total voltage.

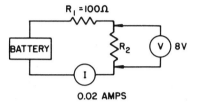

FIGURE 3-9. Measurement of voltage and current in a circuit.

The value of R_2 is solved as follows:

$$R_2 = \frac{E}{I} = \frac{8}{0.02} = 400 \ \Omega$$

After the resistance of R_2 has been ascertained, the total resistance is known since it is the sum of the two resistors and equals 500 Ω. The total voltage is now

$$0.02 \times 500 = 10 \text{ V}$$

Another method for finding the total voltage is to solve for the voltage across resistor R_1, as follows:

$$E_{R_1} = IR = 0.02 \times 100 = 2 \text{ V}$$

The total voltage in a series circuit is the sum of the voltage drops; hence the total voltage for Fig. 3-9 equals $8 + 2$ and is therefore 10 V, as established earlier.

3-4. Cells in Series

When more voltage is required than is available from a single cell, two or more cells are wired in series. This procedure is followed in the manufacture of batteries, where a number of cells are connected together to procure the necessary voltage. Thus, a 4.5-V battery is constructed by employing three 1.5-V cells. When cells or batteries are connected in series, the positive terminal of one unit connects to the negative of the other.

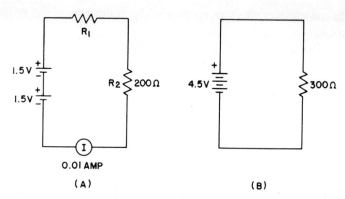

FIGURE 3-10. Cells in series.

The symbol for a single cell is shown at A of Fig. 3-10. Two cells are shown wired in series; hence the total voltage is three. The symbol for a battery is usually an expansion of the single-cell drawing, as shown at B. Often, however, the single-cell symbol is used interchangeably for either the cell or battery.

The total voltage obtained by combining cells must be taken into consideration when calculating for unit values in circuits. At A, for instance, the voltage drop across R_2 is

$$E = IR = 0.01 \times 200 = 2 \text{ V}$$

Since total battery voltage is three, the obvious voltage drop across R_1 is 1-V. Resistance value is, therefore,

$$R = \frac{E}{I} = \frac{1}{0.01} = 100 \ \Omega$$

Thus, total resistance is 300 Ω. Note that the circuit at B also has 300 Ω of resistance. Because of increased voltage, however, its current is higher.

$$I = \frac{4.5}{300} = 0.015 \text{ A}$$

3-5.   Resistors in Parallel

Resistors are also used in a parallel-circuit arrangement as shown in Fig. 3-11. Here two resistors of 8 Ω each are in parallel with a 32-V source. When resistors are in parallel, the voltage source connects directly to each resistor; hence the total source voltage is present across each resistor.

The total value of resistance can be solved by the following:

$$R_{\text{Total}} = \frac{R_1 R_2}{R_1 + R_2} \tag{3-2}$$

Applying this formula to Fig. 3-11, we obtain

$$R_{\text{Total}} = \frac{8 \times 8}{8 + 8} = \frac{64}{16} = 4\,\Omega$$

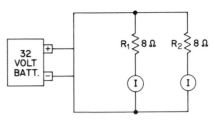

FIGURE 3-11. Parallel resistor circuit.

Thus, when two parallel resistors are equal in value, the total resistance is half that of a single resistor. (For more than two resistors, Formula 3-3 is used, as discussed later.) Since the two resistors are equal in value, each one will have the same amount of current flowing through it. Because the two resistors are placed across the power source, the current divides, and the current through the first resistor is E/R $= \frac{32}{8} = 4$ A. Since the second resistor also draws 4 A, the two parallel resistors draw a total of 8 A.

To solve for the total power consumed by the two resistors, we use Equation 2-5:

$$P = I^2R = (8 \times 8)4 = 64 \times 4 = 256 \text{ W}$$

The wattage dissipated by each single resistor can also be calculated by the same equation. When the two wattages are added together, their sum equals the total wattage.

$$P \text{ in } R_1 = 16 \times 8 = \underline{128 \text{ W for } R_1}$$
$$P \text{ in } R_2 = 16 \times 8 = \underline{128 \text{ W for } R_2}$$
$$256 \text{ W}$$

In many branches of electronics, the values of current are often in fractional amperes and fractional values are also encountered in wattages. In consequence, the terms *milliamperes* and *milliwatts* are frequently utilized. A typical parallel-resistor circuit, where a fractional value of current is involved, is shown in Fig. 3-12. Let us assume that it is necessary to calculate

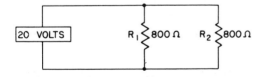

FIGURE 3-12. Parallel resistors of equal value.

the total current in *milliamperes* and the total power in *milliwatts*. The total current is found by dividing the voltage value (20 V) by the total value of resistance. Using the formula for parallel resistors previously given, the total resistance value is

$$\frac{800 \times 800}{800 + 800} = \frac{640,000}{1600} = 400\,\Omega$$

Knowing the potential (20 V) and the total resistance (400 Ω), we can solve for the amount of current flow by Equation 2-1:

$$I = \frac{E}{R} = \frac{20}{400} = 0.05 \text{ A (or 50 mA)}$$

For finding the *total* power consumed by this circuit, we can use any one of the several formulas given earlier in Chapter 2. The most convenient is Equation 2-4:

$$P = EI = 20 \times 0.05 = 1 \text{ W}$$

In many instances the individual resistors of a parallel group have dissimilar values. A typical circuit of this type is shown in Fig. 3-13. For finding

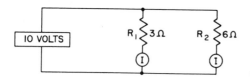

FIGURE 3-13. Parallel resistors of unequal values.

the current through resistor R_1, we divide the potential (10 V) by the resistance (3 Ω) and obtain a value of approximately 3.33 A. Similarly, for R_2, we divide 10 by the 4-Ω resistance value and find that approximately 1.666 A of current flow through this resistor.

Total current flow for this circuit is found by dividing the potential (10 V) by the total value of the resistance. In this instance the total resistance (found by using Equation 3-2) is 2 Ω, giving us a total current flow of 5 A. A point to remember is that when two or more resistors are placed in parallel, the *total* resistance is *always less than the* value of the *lowest-valued* resistor. Since additional resistors are shunting the lowest-value resistor, additional current paths are provided and thus the total resistance is reduced below the ohmic value of the smallest-value one. (This rule does not apply to *series-*resistor circuits since the addition of resistors there *increases* total resistance value.)

When more than two resistors are placed in parallel, the following can be employed to find the total value of resistance.

$$R_{\text{Total}} = \frac{1}{1/R_1 + 1/R_2 + 1/R_3 + \cdots + 1/R_n} \tag{3-3}$$

A typical circuit of this kind is shown in Fig. 3-14. Utilizing the previous formula, the calculation for the total resistance is as follows:

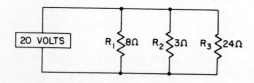

FIGURE 3-14. Parallel circuit with three resistors.

$$R_{\text{Total}} = \frac{1}{\frac{1}{8} + \frac{1}{3} + \frac{1}{24}} = \frac{1}{\frac{3}{24} + \frac{8}{24} + \frac{1}{24}}$$

$$= \frac{1}{\frac{12}{24}} = \frac{24}{12} = 2\,\Omega$$

Here the resistance values are placed into the formula, as shown, giving a reciprocal function of $\frac{1}{8} + \frac{1}{3} + \frac{1}{24}$. The least common denominator is 24, and the other fractional values of $\frac{1}{8}$ and $\frac{1}{3}$ are consequently converted to the common denominator. This gives us $\frac{3}{24}$, $\frac{8}{24}$, and $\frac{1}{24}$. When these fractions are added together, their sum is $\frac{12}{24}$. Since a reciprocal function is involved, the $\frac{12}{24}$ can be inverted and divided. When this is done, the total ohmic value is shown to be $2\,\Omega$. In accordance with the statements made earlier, this result is verified because the total resistance value indicated is less than that of the smallest resistor in the parallel circuit. If the calculation had indicated a resistance value equal to or in excess of the lowest resistor in the circuit, this would have shown that the calculation was in error.

Once the total resistance has been found, the total current can again be calculated by dividing the total voltage by the total resistance. Since 20 V are indicated and a total resistance of $2\,\Omega$ has been found, the total current for Fig. 3-14 equals 10 A.

The current through each individual resistive branch of the parallel-resistor circuit shown in Fig. 3-14 can also be calculated on the basis of the voltage across the resistor (20 V) divided by the resistance value.

Since the values of the individual resistors are fixed, the total resistance established by the parallel-resistor combination will also be fixed, regardless of the voltage applied. Thus, if the voltage for the circuit shown in Fig. 3-14 is increased to 40 V, the current through each resistor doubles and the total current becomes 20 A. Total resistance is then equal to

$$\frac{40\ \text{V}}{20\ \text{A}} = 2\,\Omega$$

This is the same total resistance value obtained when the problem was solved on the basis of 20 V applied.

Equation 3-3 need not be used if both the total current and the voltage are given. For Fig. 3-14, let us assume that the voltage is 50 V and the total current is 25 A. Since $R = E/I$, the calculation is simple and indicates that the total resistance is $2\,\Omega$. If, however, the individual resistance values were not given, each resistor would have to be calculated on the basis of the current flowing through it and the voltage across it.

If the resistance values are given as shown in Fig. 3-14 but the voltage is not given, the voltage can be measured before calculation. If total resistance is to be solved for, a voltage can be *assumed* and the current through each resistor calculated. Total resistance would then be obtained by dividing the assumed voltage by the total calculated current. This will give the correct total resistance of the circuit since the resistance values remain fixed, regardless of voltage. The assumed voltage may be far removed from the correct voltage but the total resistance value procured will be accurate. Since, however, the voltage is an assumed value, the *individual* currents through the resistors will be *incorrect* because they are based on a false voltage.

As an illustration of this process, let us suppose that it is necessary to know the total resistance of the circuit shown in Fig. 3-14 but that there is no ready means for measuring the voltage. In such an instance, assume the total voltage is 48 V. Current through the resistors will then be

$$I_{R_1} = \tfrac{48}{8} = 6$$

$$I_{R_2} = \tfrac{48}{3} = 16$$

$$I_{R_3} = \tfrac{48}{24} = 2$$

$$\text{Total current} = 24 \text{ A}$$

Since $R_{\text{Total}} = E/I$, the problem becomes

$$R_{\text{Total}} = \frac{\text{assumed voltage}}{\text{total current}} = \frac{48}{24} = 2 \ \Omega$$

3-6.   Cells in Parallel

When cells are connected in parallel as shown at A of Fig. 3-15, the current availability increases, though the total voltage output remains the same as for a single cell. The amount of current that will flow from a battery depends

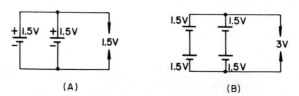

(A)   (B)

FIGURE 3-15. Cells in series and parallel combinations.

on the ohmic value of the resistor placed across it. As the resistance value is lowered, more current is drawn from the battery. As more and more current flows, the load on the battery increases and eventually a point is reached where an excessive amount of current flow overheats the battery and damages it. At the point where overheating occurs, the battery is considered to be *partially shorted*. A complete short circuit is the condition where a length of wire is placed directly across the battery terminals. Unless the wire is extremely thin, a section of copper or aluminum wire has such a low resistance that it would permit the maximum amount of current to flow that the battery is capable of delivering. Since the only resistance would be the low internal resistance of the battery, the excessive current flow through this internal resistance would cause the battery to heat up. The result would be a damaged battery.

When constructing batteries that must be able to provide higher currents than a simple cell, additional cells are placed in parallel with the original cell, as shown at A of Fig. 3-15. On the other hand, a parallel arrangement can also be utilized to extend the life of a battery. Thus, if the load resistance is such that a single cell would not be overloaded in terms of current drawn, an additional cell can be placed in parallel and the two cells virtually double the life of the battery system and reduce replacement intervals. The amount of current drawn from the combination for a given load resistance would still be the same since the *voltage* of the parallel arrangement shown at A has not been increased.

If it is necessary to increase the voltage available from a battery, as well as provide for higher available current, the arrangement shown at B of Fig. 3-15 can be employed. Here, four cells are used in a series parallel combination. This grouping would furnish twice the voltage obtainable from a single cell, with an increase in the available current because of the parallel arrangement. If the load resistor placed across the battery is 500 Ω, the amount of current flowing in the circuit would be 0.006 A, or 6 mA. It is this 500-Ω resistance value, in conjunction with the 3 V, that determines the current flow and not the particular battery arrangement.

3-7.   Current Capacity of Wire

It was mentioned earlier that fixed and variable resistors are formed by using materials such as carbon composition or special alloy wire. Such resistors are utilized when it becomes necessary to increase opposition to current flow in order to reduce voltages or signals. There are occasions, however, when it is desirable to have *high conductivity*, such as in the interconnections of the various components of electronic devices, in forming circuits. Thus, the hookup wire used between electronic parts must have a very low resistance (high conductivity). To keep resistance so low that it does not affect circuit function, copper wire is usually employed because of its inherent low resistance as compared to other metallic conductors. Thin wire and long lengths, however, will increase resistance even in copper wire; hence such factors must be considered in selecting wire of suitable size to carry current with no appreciable loss.

Wire with a larger cross-sectional area can carry more current than wire with a smaller cross-sectional area, just as a larger water pipe will permit a greater amount of water to flow past a given point during a certain time interval. In electronics, it is usual to find interconnecting circuit wires having cross sections of only a fraction of an inch in diameter. Hence, it is standard practice to designate wire diameters in units equal to *one thousandth* of an inch. The name of this measure is the *mil*. Because the most common type of wire is round, the cross-sectional area of round wire has been assigned a unit of area known as the *circular mil;* and this is shown at A of Fig. 3-16, where

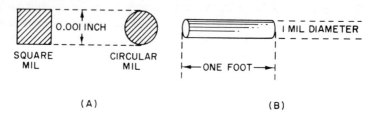

FIGURE 3-16. The circular mil of wire.

it is compared to the square mil. With the diameter in mils, the circular mil is found by simply squaring the diameter. Thus, if a diameter is 2 mils (0.002 in.), the circular mil area is $2 \times 2 = 4$ circular mils.

In the United States, a standard of reference with respect to copper wire sizes has been established, known as the American Wire Gauge, where gauge numbers have been assigned to wires of various diameters. The higher numbers, such as 30, 35, or 40, refer to very thin wire, while the lower numbers, 18, 12, 10, refer to progressively thicker wire, ranging down to below zero, where several zeros are used to indicate very thick wire, such as 0000. The larger sizes (starting from approximately #18 and increasing in wire size to #0000) are used for carrying large amounts of current, such as in home and factory wiring, industrial electronic applications, and the feed wires from electric plants. Hookup wire used for electronic circuits having only milliamperes of current flowing through them usually consists of wire sizes from #22 to #28 or smaller-diameter wire, depending on how many milliamperes of current will flow. On the other hand, wire to various tube filaments may have to carry several amperes of current, in which case appropriate wire sizes range from #18 to #20. A table giving the ratings of copper wire for sizes usually encountered in electronic circuitry is given in the appendix.

In this wire table, the ohmic value of the wire of any specific gauge is given in terms of a 1000-ft length. This is based on the standard unit of wire size known as the *circular mil foot* illustrated at B of Fig. 3-16. As shown, a wire having a cross-sectional area of 1 circular mil and a length of 1-ft is referred to as a *mil foot*, or *circular mil foot*, of wire.

Knowing the gauge number, reference can be made to the table in the appendix for the resistivity of copper wire. For instance, size 20 wire has an ohmic resistance of 10.35 per 1000 ft. Hence, 2000 ft would have twice the resistance (20.7 Ω), while 500 ft would have only half the resistance (5.175 Ω).

Review Questions

3-1. Briefly explain what factors determine the wattage rating of a resistor.

3-2. Define the terms *rheostat* and *potentiometer*.

3-3. Give two uses for potentiometers.

3-4. A circuit consists of three resistors in series (each with a different value) connected to a voltage source. Is the current flow identical through each resistor? Explain your answer.

3-5. When two cells are placed in series, what occurs with respect to the available voltage and current?

3-6. In a parallel-resistor circuit, is the current through each resistor the same when each resistor has a different ohmic value? Explain your answer.

3-7. How is the ohmic value of the total resistance calculated in a parallel-resistor circuit?

3-8. In a parallel-resistive network, is the total resistance always lower or always higher than the lowest-value resistor in the network? Explain.

3-9. Show how two resistors, each of different ohmic value, can consume a similar amount of energy in watts.

3-10. What is the principal advantage of placing cells in parallel?

3-11. Explain how a combination of cells can be connected to provide both higher voltage and greater current.

3-12. Define the terms *partial short circuit* and *short circuit.*

3-13. What factors determine the current-carrying capacity of wire?

3-14. Explain what is meant by *circular mil.*

3-15. Briefly explain what is meant by a *mil foot.*

3-16. In what applications is wire size $\#0000$ generally used?

Practical Problems

Note: Draw a diagram of the circuit of each problem to gain practice in schematic work and to aid in visualizing each problem.

3-1. A 60-W lamp bulb is operated at 120 V. What current flows through the lamp? On the basis of the calculated current, what is the resistance of the bulb?

3-2. In a television receiver, two resistors are in series. The first resistor (R_1) has a voltage drop of 200 across it. The second resistor (R_2) has a drop of 50 V across it. Current in the circuit is 20 mA (0.02 A). What is the value of each resistor? What is the total resistance?

3-3. In Problem 3-2, what wattage is dissipated by each resistor?

3-4. Across a 300-V power supply, in an electronic device, there are four resistors in series. Resistor values are 20,000, 5000, 4500, and 500. What is the current, in milliamperes, that flows through this network?

3-5. In a radio there are three resistors in series: R_1 has a 1.5-V drop across it; R_2 has a 3-V drop across it; and R_3 has an 0.6-V drop across it. Resistor

R_3 is 20 Ω. What is the current in the circuit, and what are the resistance values of R_1 and of R_2?

3-6. In a transmitter, a resistor has burned out and the exact value was not available immediately for replacement purposes. The technician replaced the resistor with a 30,000-Ω resistor in parallel with a 60,000-Ω resistor. What was the value of the original resistor? If there is a voltage of 1000 across the parallel-resistor circuit, what is the total current flowing through the two resistors in combination?

3-7. Across a power source, there are three parallel resistors as follows: R_1 or 2000 Ω, R_2 or 500 Ω, and R_3 or 1000 Ω. The power source is 150 V. What is the current through each resistor? What is the total current? What is the total resistance?

3-8. Calculate Problem 3-7 on the basis of an assumed voltage of 300 V. Under this condition, what is the calculated total resistance?

3-9. What is the total power dissipated in the two resistors of Problem 3-6?

3-10. What is the total power dissipated for the circuit described in Problem 3-7?

3-11. What would be the total power dissipated if the assumed voltage of 300 were the correct voltage in Problem 3-8?

3-12. Two resistors, R_1 and R_2, are in parallel. Resistor R_2 has a 20-volt drop across it and a current through it of 2 A. The total current drawn by both resistors equals 6 A. What are the values of R_1, R_2, and total resistance?

3-13. In what manner must a group of 1.5-V cells be combined to obtain 22.5 V?

3-14. Six cells of 1.25 V each were wired as in Fig. 3-17. What is the total voltage of the combination?

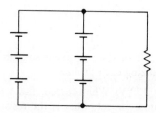

FIGURE 3-17. Illustration for Problem 3-14.

3-15. In a television receiver, the transmission line contains size 20 copper wire. If the wire length is 250 ft, what is its ohmic value?

3-16. For Problem 3-15, what is the resistance value if a wire size of 22 were used?

Circuit Analysis (DC)   4

4-1. Introduction

To understand how electronic circuits function, it is necessary to acquire a good foundation in basic procedures used to analyze circuits. Series- and parallel-resistor combinations, as well as other components discussed later, can usually be reduced to simple equivalent circuits to aid the analysis. When a circuit that appears complex has been reduced to a simple equivalent, mathematical procedures are applied to find unit values of current, voltage, resistance, and power. Certain theorems are also applied on occasion to help understand circuit makeup. This chapter applies basic principles of circuit analysis to d-c resistor circuits composed of series, parallel, and series-parallel combinations. Initially, Ohm's law calculations are shown and then some of the fundamental theorems used in more advanced circuit analysis are introduced.

4-2. Series-Parallel Combinations

In many electronic circuits, series and parallel resistors are combined to perform specific tasks. One of the basic combination circuits of this type is shown at A of Fig. 4-1. Here, resistor R_1 is in parallel with the series branch R_2 and R_3. Since the total value of resistors in series is found by adding their

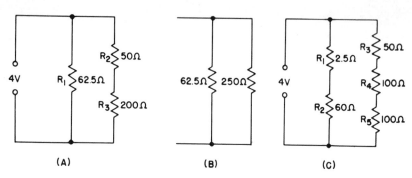

FIGURE 4-1. Series-parallel circuits.

ohmic values, the series section (R_2 and R_3) at A is equal to 250-Ω. Thus, the equivalent circuit would be as shown at B where a 62.5-Ω resistor is in parallel with a 250-Ω resistor. The 4-V power source (either a battery or a generator of electric power) would be across both resistors.

The circuit shown at C of Fig. 4-1 is a variation of that shown at A and draws the same amount of current. The equivalent circuit for C is also the one shown at B. Note the first series string of R_1 and R_2 again has an ohmic value of 62.5, while the second string is equal to 250 Ωs. Thus, these two series strings in a parallel arrangement provide the same total resistance as the circuit at A.

Once the individual ohmic values of the series branch resistors have been added together, the circuit can be assumed to be a simplified equivalent, such as shown at B of Fig. 4-1. Now Equation 3-3 for calculating the total resistance (R_{Total}) of a parallel circuit is again utilized. Solving for the two resistances shown at B produces the following:

$$R_{\text{Total}} = \frac{1}{\frac{4}{250} + \frac{1}{250}} = \frac{1}{\frac{5}{250}} = \frac{250}{5} = 50 \ \Omega$$

If it were now necessary to solve for total current (I_{Total}), this would be done by the Ohm's law

$$I_{\text{Total}} = \frac{E}{R} = \frac{4}{50} = 0.08 \ \text{A}$$

If it is desired to know the current in each branch, it then becomes necessary to solve for the current through the 62.5-Ω resistance, based on the fact that 4 V appears across it. Since the voltage in a parallel circuit is the same across each parallel branch, the current for the second resistor would also be solved on the basis of the 4 V.

$$
\begin{aligned}
I \text{ through first string} &= \tfrac{4}{62.5} = 0.064 \ \text{A} \\
I \text{ through second string} &= \tfrac{4}{250} = 0.016 \ \text{A} \\
I_{\text{Total}} &= \overline{0.080 \ \text{A}}
\end{aligned}
$$

Another typical series-parallel combination is shown at A of Fig. 4-2.

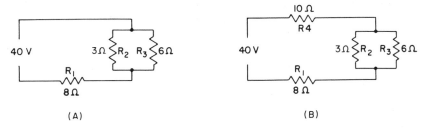

FIGURE 4-2. Series-parallel combination.

Here again, the resistances are reduced to the most simple resistance values for ease in obtaining total values. In this instance, the method for solving begins with the calculation of the total value of parallel resistors. Because the resistors in the parallel branch are dissimilar in value, the standard formula for solving parallel resistors would be utilized, as shown below:

$$R_{\text{Parallel}} = \frac{1}{\frac{1}{3} + \frac{1}{6}} = \frac{1}{\frac{3}{6}} = \frac{6}{3} = 2\,\Omega$$

(Again note that the total resistance in a parallel circuit is always less than the lowest ohmic value of resistance in the parallel circuit.)

Total resistance in the circuit shown at A of Fig. 4-2 would now be obtained by adding the 8-Ω series resistor R_1 to the sum of the parallel branch, which is 2 Ω.

$$R_{\text{Total}} = 8 + 2 = 10\,\Omega$$

The total current for this circuit is equal to the total voltage (40) divided by the total resistance.

$$I_{\text{Total}} = \tfrac{40}{10} = 4\,\text{A}$$

The voltage across the single 8-Ω resistor, R_1, and the voltage across the parallel branch are solved for next.

$$E_{R_1} = 4 \times 8 = 32\,\text{V}$$
$$E_{\text{Parallel}} = 4 \times 2 = 8\,\text{V}$$
$$(E_{\text{Total}} = 32 + 8 = 40\,\text{V})$$

Current through the 8-Ω resistor R_1 is equal to 4 A, since the total current of the circuit must flow through this series resistor. This total current will, however, divide at the parallel branch since more current will flow through the 3-Ω resistor than through the 6-Ω resistor. To solve for the currents through each resistor (R_2 and R_3), divide the voltage across the parallel branch (8 V, as found in the last example) by the individual resistor values.

$$I_{R_2} = \tfrac{8}{3} = 2.666\,\text{A}$$
$$I_{R_2} = \tfrac{8}{6} = 1.333\,\text{A}$$
$$I_{\text{Total}} = \overline{3.999}\,\text{(or 4) A}$$

Another series resistor could be placed in the circuit, as shown at B of Fig. 4-2. Here, a 10-Ω resistor (R_4) has been added to the circuit. Since the parallel combinations equal 2 Ω and the R_1 resistor equals 8 Ω, to give a total of 10 Ω, the additional R_4 resistor gives a grand total resistance of 20 Ω. The resistance has been doubled so that the current will now be only one-half of the former value. The individual currents through R_2 and R_3 will be decreased proportionately.

A still more complex type of series-parallel combination circuit is shown at A of Fig. 4-3. Here again, in solving, the parallel branches are reduced to

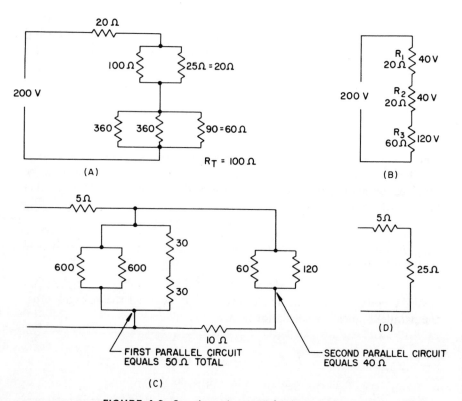

FIGURE 4-3. Complex series-parallel resistor circuits.

equivalent single resistive values and then are considered to be in series with each other. Thus, the total resistance of the lower parallel branch is calculated on the basis of the formula and found to be 60 Ω. The parallel branch above it is equal to 20 Ω and the two parallel branches are also in series with a single 20-Ω resistor. This circuit is equivalent to that shown at B, which is a simple one having two resistors of 20 Ω each and one resistor of 60 Ω, each in series with the other, for a total of 100 Ω.

Since voltages across the resistors of a series circuit are proportionate to the resistance values, the individual voltage drops can be solved for on a *proportionate* basis.

$$20\% \text{ voltage across } R_1' = 40 \text{ V}$$
$$20\% \text{ voltage across top parallel} = 40 \text{ V}$$
$$60\% \text{ voltage across other parallel} = 120 \text{ V}$$
$$\text{Total voltage, } E_{\text{Total}} = \overline{200} \text{ V}$$

or

$$I = \frac{E}{R} = \frac{200}{100} = 2 \text{ A}$$

$$E_{R_1} = IR = 2 \times 20 = 40 \text{ V}$$

As shown in B, this circuit provides 40 V across the first 20-Ω resistor, 40 V across the second, and 120 V across the lower 60-Ω resistor. Total current, as given by these calculations, is equal to 2 A.

Instead of finding the voltage drops by using the proportionate drops based on resistance values, the individual voltage drop across each resistor can be calculated by multiplying the resistance value by the current through the resistor (see last example). Since the total current is 2 A, this same current is flowing through each resistor. Thus, to find the voltage across R_1, we would multiply the 20 Ω by 2 A, which would result in a voltage drop of 40 V. The same is done for R_2, again resulting in 40 V. For R_3, multiply the 60 Ω by the 2 A to obtain a voltage drop of 120.

Another complex series-parallel circuit combination is shown at C of Fig. 4-3. Again, the individual parallel circuits are solved to give a single resistance value. In this case, the first parallel circuit, composed of two 600-Ω resistors in parallel, indicates that these two resistors would give a single resistance of 300 Ω. This opposition is in parallel with two series resistors of 30 Ω each, or 60 Ω. Using 300 as the common denominator, the total resistance value of the first parallel circuit at C is indicated as being 50 Ω.

$$\frac{1}{\frac{1}{300} + \frac{5}{300}} = \frac{300}{6} = 50 \ \Omega \text{ (for first parallel combination)}$$

The second parallel circuit has a total resistance of 40 Ω. This circuit, however, has another resistor of 10 Ω in series with it, which means that a total of 50 Ω is shunting the first parallel circuit. Thus, since both the first and the second parallel circuits each represent a total of 50 Ω, again solve for the single resistance value of these two parallel combinations. As two 50-Ω resistors in parallel equal 25 Ω, the two parallel branches (plus the 10-Ω resistor in series with the second parallel branch) can be represented by a single resistor of 25 Ω. Since this single resistor of 25 Ω is also in series with a 5-Ω resistor (as shown at D of Fig. 4-3), the total resistance is 30 Ω for this circuit.

4-3.   Practical Applications

Equation 3-3 for finding the total resistance of several resistors in parallel can be employed in reverse order when it is necessary to find what two resistors would have the same value as a single resistor. Let us suppose that a 60-Ω resistor needs to be replaced but that no other 60-Ω unit is available. In such an instance, two 120-Ω resistors can be placed in a parallel arrangement to give the desired 60 Ω. Assume, however, that only dissimilar resistors were available—what combination would give 60 Ω?

The 60 can arbitrarily be multiplied by any number to find what resistors in parallel will still give 60 Ω. Suppose the 60 were multiplied by 3 to give 180. Since 3 was used to multiply the 60, we would have $\frac{3}{180}$, or $\frac{1}{180} + \frac{2}{180}$. In reciprocal expressions, these would become $\frac{180}{1} + \frac{180}{2}$. The former expression of 180 divided by 1 becomes 180, which is one value of resistor to be used. The second expression becomes 180 divided by 2, which equals 90. This is the second resistor to be used. Thus, a 180-Ω unit in parallel with 90 Ω would give 60 Ω. This can be proved by using the standard formula for parallel circuits on these 180- and 60-Ω resistors (the reverse of the procedure just undertaken).

Since we had $\frac{3}{180}$ as the basis for our analysis, we could have stated this as $\frac{1}{180} + \frac{1}{180} + \frac{1}{180}$, which upon inversion would have indicated that three parallel resistors, each of 180 Ω, could also have been employed to obtain the necessary 60 Ω.

If these resistors had not been available, we could have multiplied 60 by 5 to obtain another set of values. This could have been set down as $\frac{1}{300} + \frac{4}{300}$, a total of $\frac{5}{300}$. Inverting the first expression gives $\frac{300}{1}$, which indicates 300 Ω for one of the required resistors. The second expression, when inverted, is $\frac{300}{4}$; therefore, 75 Ω are required for the second parallel resistor for both to furnish a total of 60 Ω.

Two additional examples of the foregoing method follow.

Example: What two values of resistors in parallel can be used to give a total of 12 Ω?

Solution: Multiply 12 Ω by some other number, for instance 5: $12 \times 5 = 60$, or $\frac{5}{60}$, which can be split up into $\frac{1}{60}$ and $\frac{4}{60}$. When these terms are inverted to $\frac{60}{1}$ and $\frac{60}{4}$, we find that one resistor should have a value of 60 Ω and the other a value of 15 Ω.

Example: What three values of resistors can be employed to give 6 Ω when the resistors are in parallel?

Solution: $6 \times 3 = 18$, or $\frac{3}{18}$; inverting to $\frac{18}{3}$, we find that three 18-Ω resistors can be used. (If two resistors were asked for, values of 9 and 18 Ω would suffice.)

4-4. Kirchhoff's Laws

Gustav Kirchhoff (1824–1887), the German scientist, formulated two important laws concerning electric circuits. These are known as *Kirchhoff's laws*
and may be stated as follows:

1. The current (or sum of currents) flowing into any junction of
   an electric circuit is equal to the current (or sum of currents)
   flowing out of that junction.
2. The power source voltage (or sum of such voltages) around
   any closed circuit is equal to the sum of the voltage drops
   across the resistances around the same circuit.

The first law, also known as the *current law*, is often stated as "the
algebraic sum of all currents at any point in a circuit is *zero*." Thus, at
a certain point, the current flowing toward the point has an opposite direction to that of the current flowing away from that point (one positive, the
other negative). Hence, the algebraic sum indicates zero from the standpoint
of analysis, even though a definite amount of current flows.

The second law, also known as the *voltage* law, is stated as "the algebraic
sum of all the voltages around the circuit is zero." Consider, for instance,
the circuit shown at A of Fig. 4-4. Starting at point *x* and going in the direc-

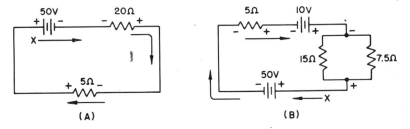

FIGURE 4-4. Analysis using Kirchhoff's law.

tion of electron flow, mark down the voltages encountered, giving them
a polarity as indicated by the first polarity found in either a battery or a resistor. For Fig. 4-4, the first unit is the battery and this is set down as $+50$.
The next unit is the 20-Ω resistor and because the voltage drop across this is
R times I, it is set down as $-20I$. The next voltage drop is across a 5-Ω
resistor and this is set down as $-5I$, bringing us back to the starting point
x. We now have

$$50 - 20I - 5I = 0 \quad \text{or} \quad 50 - 25I = 0$$

By algebraic processes, we solve for I and find that $I = 2$ A.

For most of the simple circuits, Ohm's law provides a more direct and
easy method for solving unknowns; hence Kirchhoff's laws are not usually

employed unless the circuit is much more complex, At A of Fig. 4-4, for instance, the current can be found by $I = E/R = 50/(20 + 5) = 2$ A.

When two power sources are in the series circuit, as shown at B, one will oppose the other if the source of electrons from the negative terminal of one battery faces the negative terminal of the other battery. Thus, if one battery is 50 V and is opposed by another of 10 V, the resultant total battery potential is only $50 - 10 = 40$ V. This is proved by applying Kirchhoff's voltage law, again starting at any point and continuing once around the circuit. Starting at x, for instance (with the parallel branch reduced to its single resistance value), gives

$$50 - 5I - 10 - 5I = 0$$

$$40 - 10I = 0$$

$$I = 4 \text{ A}$$

By Ohm's law also, $I = E/R = \frac{40}{10} = 4$ A because the 50-V battery is opposed by the 10-V, giving a total of 40 V, and the total resistance is 5 Ω + 5 Ω = 10 Ω. Both Ohm's law and Kirchhoff's law can be used, one to prove out the other, in circuit analysis. Either law can also be used to verify the results obtained from meter readings.

To illustrate the use of Kirchhoff's laws for solving of branch currents in a more complex circuit, consider the circuit shown in Fig. 4-5, where two

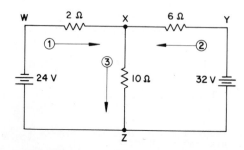

FIGURE 4-5. Dual-voltage circuit.

voltage sources are present in a circuit with three resistors. The *closed circuit* mentioned in the second law at the beginning of this section applies to both sections of Fig. 4-5 and the term *closed loop* is often used. Thus, from points w to x to z and back to the battery forms a closed loop and hence a continuous electron-current path from one battery terminal to the other.

For the circuit shown, the arrows indicate the direction of electron flow, though a reverse direction can be used and the same values will be obtained. By Kirchhoff's first law, the electron flow from points x to z equals the combined values of electron flow through the 2- and 6-Ω resistors. Thus, current flow through the 10-Ω resistor is $I_1 + I_2$. The closed-loop voltage equations for the second law set down expresses I_3 as $I_1 + I_2$ as follows:

$$2I_1 + 10I_1 + 10I_2 = 24$$

or

$$12I_1 + 10I_2 = 24$$

For the right-hand loop we obtain

$$6I_2 + 10I_1 + 10I_2 = 32$$

or

$$16I_2 + 10I_1 = 32$$

Now we have obtained two simultaneous equations:

$$12I_1 + 10I_2 = 24$$
$$10I_1 + 16I_2 = 32$$

In order to eliminate the I_2 expression, we now multiply the first equation by 8 and the second by 5 so both I_2 values coincide. With both I_2 values preceded by a plus sign, subtraction eliminates the quantities. If one sign had been opposite, addition would have eliminated the I_2 values:

$$96I_1 + 80I_2 = 192$$
$$50I_1 + 80I_2 = 160$$
$$\overline{46I_1 \qquad\quad = \quad 32}$$

By dividing 32 by 46, we obtain a value of 0.695 A for I_1. We can substitute this value in one of the initial equations for finding I_2. Thus, for the equation obtained for the left-hand loop earlier, we use the $0.695I_1$ value but multiply it by 12 since this is indicated in the equation ($12I_1 + 10I_2 = 24$). Thus we obtain

$$(12 \times 0.695) + 10I_2 = 24$$

or

$$8.34 + 10I_2 = 24$$

Hence

$$10I_2 = (24 - 8.34)$$

and

$$10I_2 = 15.66 \text{ A}$$

so

$$I_2 = 1.566 \text{ A}$$

Knowing both the I_1 and I_2 values, we find that the total current through the 10-Ω resistor is

$$I_1 + I_2 = 0.695 + 1.556 = 2.261 \text{ A}$$

4-5.   Thévenin's Theorem

The characteristics of a circuit can be analyzed by measurement of voltage and current values even though the components are not readily accessible. One method is by use of *Thévenin's theorem*, which states essentially that *a given network, with constant voltages and resistance, produces a current flow in the load resistor equal to that which flows if the load resistor were applied across an equivalent circuit that has*

1. An internal resistance measured at the terminals of the circuit with the voltage source replaced by its equivalent internal resistance.
2. A voltage at the terminals equal to that existing in the original circuit after removal of the load resistor.

If circuit components are not accessible for measurement, it is as though the circuit were completely enclosed in a container (a box, for instance) and two or more terminals are the only available points for circuit analysis. The box concept is often encountered in circuit design analysis and is usually referred to as a *black box* when discussing this method of evaluating circuit characteristics.

The load resistor mentioned in the theorem can be an actual resistor or some other network combination of components representative of a resistive load. As an example of the application of Thévenin's theorem, a network of resistors with a battery is shown at the left in Fig. 4-6 with the load resis-

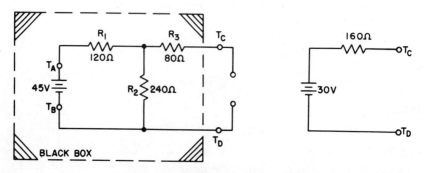

FIGURE 4-6. Applying Thévenin's theorem.

tance removed from terminal T_C and T_D. Applying a meter to these terminals to read voltage, a value of 30 V is obtained. This occurs because the meter reads the voltage drop across R_2 only. Because of the open circuit, no current flows through R_3 and hence no voltage drop occurs across it. (The assumption here is the voltmeter has such a high resistance that the terminals T_C and T_D are still a virtual open circuit.)

The values in the black box are given in Fig. 4-6, and we can prove that the voltmeter would read 30 V by Ohm's law. In actual circuit analysis of this type, however, we would not, of course, know the internal unit values of voltages and resistances.

$$I = \frac{E}{R} = \frac{45}{120 + 240} = 0.125 \text{ A}$$

Knowing the current flow through R_1 and R_2, the voltage drop across R_2 is found.

$$E = IR = 0.125 \times 240 = 30 \text{ V}$$

The voltmeter is now removed and an ammeter placed across terminals T_C and T_D. The ammeter, having almost zero resistance (see next chapter), acts as a short across the output terminals and would read 0.1875 A. This can also be proved by Ohm's law. With the terminals closed, R_2 and R_3 are in shunt (parallel circuit) with a total resistance value of 60 Ω.

$$\frac{240 \times 80}{240 + 80} = 60 \text{ Ω}$$

This 60-Ω value is added to that of R_1, 120 Ω, for a total of 180 Ω. Now, total current is

$$I = \frac{E}{R} = \frac{45}{180} = 0.25 \text{ A}$$

Of this 0.25-A current, only one-third as much current flows through R_2 as through R_3 because of the higher R_2 value.

$$I \text{ through } R_2 = 0.0625$$
$$I \text{ through } R_3 = \underline{0.1875}$$
$$I_{\text{Total}} = \overline{0.2500}$$

Once current through the shorted terminals is known, the *equivalent* circuit resistance can be found.

$$R = \frac{E}{I} = \frac{30}{0.1875} = 160 \text{ Ω}$$

Hence, the *equivalent* voltage and resistance forms an equivalent circuit as shown at the right in Fig. 4-6. With these values we can ascertain the amount of current flowing through any load resistor applied across the two terminals, as well as the voltage drop that would occur and the power consumed by the load resistance. If, for instance, a load resistance of 140 Ω were placed across T_C and T_D, we would solve for load current in the equivalent circuit by the following formula:

$$I_L = \frac{E}{R_e + R_L} \qquad (4\text{-}1)$$

where R_e is the equivalent circuit resistance and R_L is the value of the applied resistor.

Using Equation 4-1 for the values in Fig. 4-6, we have

$$\frac{30}{160 + 140} = \frac{30}{300} = 0.1 \text{ A}$$

Proving Thévenin's theorem by Ohm's law does, of course, lengthen the process. In actual practice a voltmeter is applied initially for the voltage reading and next an ammeter for current. Dividing the voltage by the current immediately gives the equivalent circuit resistance. The latter, when applied to Equation 4-1, indicates the load current for any value of load resistance.

If, in Fig. 4-6, the battery were outside the black box, we would have a four-terminal network, with input terminals T_A and T_B available in addition to the output terminals T_C and T_D. Now if the battery were removed and the terminals T_A and T_B shorted, the process would be simplified. Next, a resistance-reading meter (ohmmeter) is placed across the output terminals and the resistance read directly. The value for Fig. 4-6 would again be 160 Ω. (With the input terminals shorted, R_1 shunts R_2 for a total resistance value of 80 Ω. With this value in series with R_3, the total resistance is again 160 Ω.) If the battery has an appreciable internal resistance, this should be indicated in the calculations and the T_A and T_B terminals not shorted but replaced with the equivalent internal battery resistance. Similarly, with the first procedure, the battery resistance would have to be included as part of the R_1 resistance.

4-6.   Norton's Theorem

Norton's theorem is based on a source voltage producing a *constant current* as opposed to the *constant voltage* theorem of Thévenin. The impedance of the equivalent circuit is considered to be in parallel with the load resistance. Other than this difference, Norton's theorem also states that any resistive-voltage network can be replaced by a single voltage and resistance as an equivalent.

To show that the same solution is obtained with Norton's theorem, Fig. 4-6 is again used. Initially, T_C and T_D are shorted, which places R_3 in shunt with R_2 for a combined resistance of 60 Ω. When this is added to the 120-Ω resistor, a total of 180 Ω is obtained. Circuit current is, therefore,

$$I_c = \frac{45}{180} = 0.25 \text{ A}$$

With an ammeter used as the shorting component, the current would read 0.1875 A. This can be proved by Ohm's law or by considering proportionate values. Since 0.25 A is the total current, this value divides across the parallel network of R_2 and R_3. Since R_3 is one-third the value of R_2, it will have three times the current flow through it. Now the resistance is determined at T_C and T_D by the same method employed for Thévenin's theorem. The battery is removed and battery terminals T_A and T_B are shorted.

Again the resistance value will be found to be 160 Ω. Current through the load resistor is found by the following equation involving circuit resistance (R_c) and the load resistance (R_L):

$$I_L = I_c \times \frac{R_c}{R_c + R_L} \tag{4-2}$$

If the same value load resistance is used as for the example with Thévenin's theorem, we have the same current value obtained with Thévenin.

$$I_L = 0.1875 \times \frac{160}{160 + 140} = 0.1 \text{ A}$$

Review Questions

4-1. Draw a circuit with a single resistor in series with a parallel branch having two resistors each. Assign an ohmic value to each resistor, and show the calculation for total resistance.

4-2. For the circuit drawing in the preceding question, assign a battery potential across the network, and calculate the voltage drop across each resistor, on the basis of the applied voltage.

4-3. Briefly explain how the formula for finding the total resistance in parallel can be employed in reverse order for finding what individual values two resistors must have so that their combination will provide the same value as a single resistor.

4-4. Give an example of using two resistors in series to provide the same ohmic value as a single resistor.

4-5. Show how three resistors can be used in series to provide the same ohmic value as a single resistor.

4-6. Give a typical example of using two resistors in parallel to provide the same ohmic value as a single resistor.

4-7. Give a typical example of employing three resistors in parallel to provide the same ohmic value as a single resistor.

4-8. Explain in your own words what is meant by Kirchhoff's *current* law.

4-9. Explain briefly how the *voltage law* differs from the current law.

4-10. Explain what is meant by a *closed loop*.

4-11. What occurs in a series circuit if two batteries of dissimilar values oppose each other?

4-12. What theorem applies to a constant-voltage network, and what is the value of using such a theorem?

4-13. What theorem applies to a constant-current network, and what is its advantage, if any, over constant E?

4-14. Briefly explain what is meant by the *black box* concept in electronics.

4-15. What measuring instrument can be used as a shorting element in utilizing the principles of Norton's theorem? Briefly explain how the instrument would be used.

Practical Problems

Note: Draw a diagram of the circuit of each problem. This will give you practice in schematic work and aid you in visualizing each problem.

4-1. In an electronic circuit, a resistor R_1 of $100\ \Omega$ is in series with two parallel resistors. The latter, designated R_2 and R_3, are $600\ \Omega$ each. The voltage applied to the network is 800 V. What is the total resistance, and what is the voltage drop across each resistor?

4-2. A 700-Ω resistor R_1 had in series with it a parallel-resistor circuit composed of a 600-Ω resistor R_2 shunted by a 300-Ω resistor R_3. Also in series with this parallel combination was another resistor R_4 having a value of $100\ \Omega$. The voltage measured across the R_4 resistor was 50. What are the total current, total resistance, and total voltage applied to the complete circuit?

4-3. In the cathode circuit of a vacuum tube, a 100-Ω resistor burned out. This value of resistance is not immediately available. What resistor can be placed in parallel with a 500-Ω resistor on hand to obtain a resistance value equal to the original 100-Ω resistor?

4-4. Give three values of resistance that can be used in parallel to produce a total of $26\ \Omega$.

4-5. In a voltage-divider network, there is an applied voltage of 80 and the circuit consists of two parallel resistors, one having a value of $12.5\ \Omega$, and the other having a value of $50\ \Omega$. In series with this parallel circuit is another parallel circuit having two resistors of $36\ \Omega$ and $18\ \Omega$. From the second parallel circuit, a single 18-Ω resistor completes the circuit to the voltage source. What is the total resistance value, and what is the voltage drop across the 18-Ω resistor?

4-6. For the circuit shown in Fig. 4-7, solve for current I_1, I_2, and I_3, using Kirchhoff's laws.

4-7. For Problem 4-6, what power in watts is consumed by the 20-Ω and the 8-Ω resistors? (Refer to Chapter 2 for appropriate equations.)

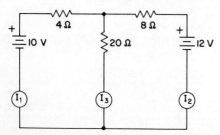

FIGURE 4-7. Circuit for Problem 4-6.

4-8. For Fig. 4-7, what voltages appear across the the 4-Ω and 20-Ω resistors?

4-9. For the circuit shown at Fig. 4-8, use Thévenin's theorem to solve for the equivalent circuit resistance R_e and I_L, using a load resistance value of 3422 Ω.

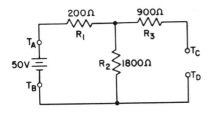

FIGURE 4-8. Circuit for Problem 4-9.

4-10. For Fig. 4-8, solve for I_L using Norton's theorem and an R_L of 3422 Ω.

D-C Measurements 5

5-1. Introduction

Values of current, voltage, resistance, and wattage in various electronic circuits are measured by specially designed instruments that have dials calibrated to indicate the unit values of the quantities being measured. The meter that measures current is known as an *ammeter*. When fractional values of current are to be measured, a *miliammeter* is employed. For measuring voltage, a *voltmeter* is used, which has several ranges for measuring low, intermediate, or high values. Meters are also available for measuring resistance values and wattages. There are also combination meters that are capable of reading voltages, currents, and resistances, as more fully described in Chapter 22.

In order to provide a particular meter with the ability to make measurements over a useful range of values, it is necessary to employ series- and parallel-resistor combinations based on principles given in .this and the previous chapters. The basic meter movement involves a permanent magnet and a moving coil, as well as electromagnetic fields (as more fully discussed in subsequent chapters). This chapter is concerned with the basic principles of extending meter ranges by use of resistors.

5-2. The Ammeter

For an ammeter to measure current, it is necessary for the current *to flow through* the meter coil that moves the indicator needle. Hence, an ammeter must be connected *in series* with the circuit components in which the current is to be measured. For this reason, the ammeter must have a low d-c resistance so that the meter will not offer opposition to the current flow in the circuit. If the ammeter has any appreciable resistance, it will decrease the current flow of the circuit and an accurate reading of the amount of current that flows in the circuit without the meter will not be obtainable. Hence, the ammeter must be constructed using wire having such a low resistance that its effect on the circuit is negligible. The same factor of low internal resistance also applies to milliammeters. For electronic design or testing of radio and television receivers, recording and playback equipment, and other solid-state units where low currents are present, the milliammeter finds greater usefulness than the ammeter. In transmitters and industrial-electronic equipment or in other such devices where high power is involved, ammeters must be used for current readings.

A basic ammeter or milliammeter employs a needle indicator that has a deflection proportional to the current flowing through the instrument, up to a maximum needle deflection. Hence, if the maximum needle deflection is 1 mA, lesser currents will appear as a fraction of that value and the meter will read that amount as a maximum. For increasing the range of such an instrument, parallel resistors are shunted across the meter movement. For each such parallel resistor, an additional scale must be provided on the meter to read the current flow within the range provided by the shunt resistor. In commercial types, a rotary switch with a number of terminals is used for convenience in selecting the proper shunt resistor and, hence, the range desired.

Figure 5-1 shows the circuit of a typical milliammeter. If the switch is in an open position, the milliammeter will read current values up to 1 mA when the test probes are placed in series with the circuit to be measured. Since an ammeter must be placed *in series* with the circuit, the latter must be opened for insertion of the meter.

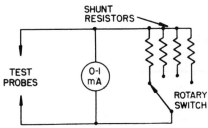

FIGURE 5-1. Basic milliammeter circuit.

The ranges established by the resistors are determined by the internal resistance of the meter. The following formula is employed for calculating the value of the shunt resistor required for full-scale deflection.

$$R_s = \frac{R_m}{(N-1)} \qquad (5\text{-}1)$$

where

R_s = the value of the shunt

R_m = the meter resistance

N = the value of the number by which the scale is to be increased.

Thus, if a 0–1 milliammeter with an internal R of 100 Ω is to have a maximum current reading of 5 mA, the formula would be used as follows:

$$R_s = \frac{100}{5-1} = 25 \ \Omega$$

Hence, a 25-Ω resistor, in shunt with the meter, would permit current readings up to 5 mA.

Another type of current-reading meter is that known as the *galvanometer*. This instrument is primarily a laboratory device with the zero reading at the center of the meter dial permitting observation of proportional changes of current flow in either the negative or positive direction. Usually the galvanometer is not intended to read the actual amount of current flow but only relative proportions of positive or negative polarity. The instrument finds greatest usage in the bridge circuits described later.

5-3. The Voltmeter

A voltmeter is employed for measuring the voltage drop *across* a resistor or some other component of a circuit. For this reason, the voltmeter is *not* placed in series with the resistor or other component, but *across* it. Since the voltmeter is shunted across the place where voltage is to be read, the meter should have a high internal resistance so that it will not behave as a low-shunt resistor. If the meter resistance is low, some of the current flowing into the resistor (or other component across which voltage is to be measured) will branch into the meter and upset true voltage readings.

Voltmeters are constructed by using a basic milliammeter, plus a series resistance, to reduce the current flowing through the meter to a value within its range. The ohmic value of the external resistor will establish the range of the voltmeter.

If the 0–1 milliammeter is placed across a circuit to be measured instead of in series with it, the milliammeter acts as a voltmeter, though it will be a very low resistance type that is capable of measuring only fractional voltage values. If the milliammeter has 100 Ω of internal resistance, it is evident that 0.1 V will cause full needle deflection since voltage is a function of the maximum current (0.001 A) times the resistance (100 Ω).

If a voltage range of 5 V is desired, then 4.9 V must develop across the external series resistor and 0.1 V across the meter resistance to give a total voltage drop of 5 V. This 5-V potential would cause the meter to deflect fully because the meter movement would have 0.1 V applied to it, which would cause 1 mA of current to flow through the 100-Ω internal resistance of the meter coil. Since 1 mA of current is the maximum that may flow through the meter, higher-voltage scales would require the series resistors always to have all but 0.1 V of the total voltage across them. Thus, if the maximum voltage that is to be read is 500 V, 499.9 V must drop across the series external resistor so that again only 0.1 V is applied to the internal meter movement. Figure 5-2 illustrates the basic voltmeter circuit. The various resistors are again selected by a rotary switch. The resistors are also referred to as *multipliers*. The following formula can be used to calculate the value of the series resistors for a desired voltage range:

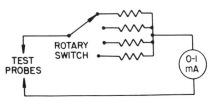

FIGURE 5-2. Basic voltmeter circuit.

$$R_s = R_m(N - 1) \qquad\qquad (5\text{-}2)$$

where

R_s = the value of the series resistor (multiplier)
R_m = the resistance of the meter
N = the value of the number by which the scale *is to be increased.*

As an example, assume the meter is to read 5 V full scale. To apply Equation 5-2, it is necessary to determine the value of N. To do this, divide the desired full-scale reading by the voltage necessary to deflect the meter fully. Thus, 5 V divided by 0.1 equals 50. The number 50 is then placed in the formula and, when solved, it indicates that the required resistance for a full 5-V deflection is 4900 Ω.

$$R_s = 100(50 - 1) = 4900 \ \Omega$$

The required value of the multiplier needed for a certain voltage scale can also be found by using Ohm's law to solve for total resistance. Since 5 V is the maximum scale deflection and 0.001 is the maximum current in amperes that will flow in the series circuit composed of the resistor and meter, the total resistance indicated is 5000 Ω.

$$R_{\text{Total}} = \frac{5}{0.001} = 5000 \ \Omega$$

Because the meter resistance is known to be 100, however, this value must be subtracted from the figure of 5000 above to obtain the required series resistance (4900 Ω) that must be used for a maximum scale reading of 5 V.

In the foregoing, the external resistor was 4900 Ω and the internal resistance was 100 Ω, which provided a 5000-Ω total resistance for a 5-volt scale. It will be found, with an 0–1 mA meter, that 1 V full scale would require 900 Ω plus 100 Ω for the meter, or a total of 1000 Ω. Hence, the meter has an internal resistance of *1000 Ω per volt* of full-scale deflection.

Such a meter will have some influence on the circuit when voltages are read because the internal resistance is rather low. Most of the inexpensive meters use 1-mA meter movements (with internal R values ranging from 25 to 100 Ω) as the basis for the voltmeter. Higher internal resistances can be obtained by using a more sensitive milliammeter. For instance, some 20-μA meters provide 50,000 Ω per volt and thus affect the circuit much less, in terms of current shunting, than the 1000-Ω-per-V meter. Other voltmeters are available that have internal resistances of 1 megohm or higher, such as the special type voltmeters described in Chapter 22.

5-4. The Ohmmeter

The ohmmeter measures the ohmic value of resistors by employing a voltage source within the meter cabinet and measuring the amount of current flow in the circuit formed by the resistance to be measured and the meter circuit. The ohmmeter is usually incorporated in the same housing with voltmeters, as more fully described in Chapter 22. A typical ohmmeter circuit is shown in Fig. 5-3. Again, as with meters discussed earlier, a 0–1 milliammeter can be used as the basic meter movement. Two resistors are placed in series with the meter, one being a fixed resistor to limit the current within the range desired and the other a variable resistor so that the meter needle can be adjusted for a zero-ohm reading. A voltage source from flashlight cells in series is usually employed, as shown. When the test probes are shorted together, the ohmmeter circuit is closed and current from the battery will flow through the meter and resistors. The variable resistor is then adjusted so that the needle deflects fully (indicating 1 mA of current flow within the meter). Since the test probes are to be placed across the resistor to be measured, it is obvious that when the test probes are shorted together, a condition of zero resistance is present. Because the needle deflects to full scale, however, the ohmmeter scale must indicate *full needle deflection as zero ohms*.

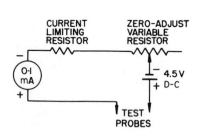

FIGURE 5-3. Basic ohmmeter.

Open test probes indicate an infinite resistance and because the ammeter circuit is also opened, the needle moves to the left. Thus, zero resistance is shown at the right side of the scale and increasing resistance toward the left, with maximum resistance at the extreme left position of the needle.

When the test probes are placed across a resistor, the amount of current flow will be inversely proportional to the ohmic value of the resistor. Hence, a high-value resistor permits less current flow than a low-value resistor. The meter scale is calibrated to indicate such ohmic values.

Since the meter scale is calibrated on the basis of the current values in the meter circuit, the accuracy of the device is also dependent on the voltage source. Besides, as the battery ages and its voltage declines, it will be necessary to readjust the variable resistor for zero-ohm reading when the test leads are shorted together.

5-5. Wheatstone Bridge

The Wheatstone bridge is a method of employing resistor combinations in a so-called "balanced bridge" arrangement, for finding the value of any resistor whose resistance is unknown. The Wheatstone bridge principle can also be applied to finding the values of coils and other components, as more fully described later. It is named after Sir Charles Wheatstone (1802–1875), the English physicist who first stressed the importance of this balanced circuit.

The Wheatstone bridge principle is based on the polarities that are established in a resistive network similar to that shown at A of Fig. 5-4.

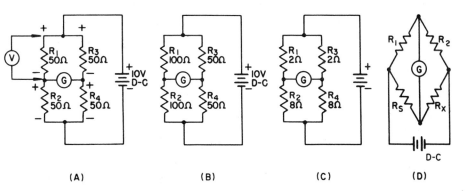

FIGURE 5-4. Characteristics of resistive bridge.

Here are four resistors, each having a value of 50 Ω, and the potential applied across the network is 10 V. Under these conditions, there is a 5-V drop across each resistor. If a voltmeter is placed across resistor R_1, as shown, the plus test probe of the voltmeter would be put to the top of R_1 and the minus probe of the voltmeter to the bottom of R_1. If the voltage is to be read across resistor R_2, the voltmeter is moved down, and the plus probe applied to the top of R_2 and the minus probe to the bottom of R_2.

It is obvious, from the foregoing, that the center junction of R_1 and R_2 is either plus or minus, depending on whether the measurement is taken from across R_1 or R_2. (It must be remembered that polarity is relative. A plus polarity, for instance, by itself has no meaning. Such a polarity is plus only with respect to some minus potential.)

An inspection of the circuit consisting of R_3 and R_4 indicates that the center junction of these two resistors is also *either* plus or minus, *depending on whether reference is made to the top or the bottom resistor.* If a sensitive current-indicating device, such as the galvanometer (*G*), is placed across the two junctions, as shown, there is no reading because the potential difference at the junction of R_1 and R_2 is identical with the potential difference that exists at the junction of R_3 and R_4. Since the polarities are identical, there is no *voltage difference* between these two points.

This principle of a balanced circuit is the basis for forming a Wheatstone bridge. It is logical, if R_2 is of a value different from the other resistors, that the zero potential difference between the two junctions will no longer exist, and a current flow is then established between these two points.

Another condition can be set up, as shown at B of Fig. 5-4. Here, even though resistors R_1 and R_2 are of a different value from resistors R_3 and R_4, a balanced circuit is still obtained. The voltage drop across each group of registors in a parallel circuit is the same, even though the individual resistors have different ohmic values. Therefore, across R_1 and R_2 in combination, the voltage drop is still 10 V (5 V across each resistor), even though the total resistance is 200 Ω. Similarly, there is also a voltage drop of 10 V across R_3 and R_4, even though this total resistance is only one-half that of the other branch. Therefore, halfway down each branch the voltage drop will be identical and across these two junctions no potential difference exists.

The circuit arrangement shown at C will also function as a balanced bridge network since the voltage drop across the upper resistors will be 2 V and the drop across the two lower resistors will be 8 V. Again, across the junctions there is no potential difference.

The bridge circuit is usually illustrated as shown in Fig. 5-5. For measurement of an unknown resistance, the resistors marked R_1 and R_2 are of a fixed value. Resistor R_s is the *standard* or known value that is employed to determine the unknown value of R_X. The unknown can be solved on the basis of the following formula:

$$R_X = R_s \frac{R_2}{R_1} \qquad (5\text{-}3)$$

The unknown resistor is placed in the circuit shown at R_X and the resistor R_s is varied until the bridge balances. The difference in ohmic value between R_2 and R_1 (if any) is then multiplied by the value of the known resistance so as to find the unknown. The variable resistor employed for R_s would have to be calibrated in terms of resistance variation. Instead of having

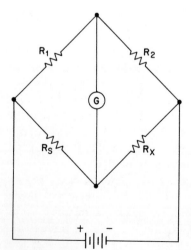

FIGURE 5-5. Wheatstone bridge.

a variable resistor for R_s, various resistors of known value can be placed into the circuit at R_s until the bridge is balanced, as indicated by a zero reading on the galvanometer. (A galvanometer, with its zero indication at the *center* of the scale, will read an unbalance in either the minus or plus polarity direction).

5-6. The Decibel

In addition to methods for measuring unit values of current, voltage, etc., in electronics, a means for *comparing* power, voltage, or current levels is also available. This system is based on a value known as the *decibel* (abbreviated dB). Because the decibel involves *comparisons* among powers, voltages, or currents, the decibel is not, therefore, a unit *measurement*. The decibel value is based on the manner in which the human ear hears sounds of different intensities.

The human ear responds to changes in sound intensity in logarithmic fashion. (A summary of the theory of logarithms, plus tables, will be found in the appendix. Reference should be made to this section if the subject matter is new to the reader or if a review of the principles is indicated.) Hence, the ear is much more responsive to changes in low-intensity sound levels than high-intensity sound levels. Thus, for practical purposes, a means for comparing power changes in audio and electronics should also be based on logarithms. For this reason, the word *decibel* is used to indicate the change in volume level that occurs when the average ear is barely able to perceive a difference in a gradually changing sound amplitude.

The decibel is one-tenth of a Bel, the unit expression being named after Alexander Graham Bell (1847–1922) the famed American scientist and inventor.

Mathematically, the decibel is a function of the following expression:

$$dB = 10 \log \frac{P_1}{P_2} \tag{5-4}$$

Thus, the ratio of two powers is taken and the logarithm (see appendix) of this ratio is multiplied by 10. Ordinarily, however, the mathematical expression need not be used if one simply remembers that a *doubling* of power represents *3 dB*. Thus, if the power output from an audio amplifier is 2 W and the volume control is turned up to produce 4 W, the change in decibels is three. Because this is a unit of comparison, however, the 3 dB would only indicate that the power had been doubled but would not express the *amounts* of power involved. Thus, if a public-address system were delivering 10 W and this were increased to 20 W, the difference would also be 3 dB. When there is an increase in power, this represents *plus* decibels. If there is a decrease in power, the expression becomes *minus* decibels. If, for instance, a stereo amplifier had an output of 60 W per channel and this model were redesigned for economy reasons to produce only 30 W per channel, it would represent a change of -3 dB per channel. Similarly, a -3 dB change occurs when the signal strength from a television station drops from 500 to 250 mW at a given distance from the transmitter.

If the power of a device is increased 10 times, the difference in decibels would also be equal to 10. Thus, if sound power were increased from 1 to 2 W, this would represent 3 dB. If the 2 W were now increased to 4, it would represent another 3 dB. Decibel increases are *additive*. Thus, the change in power from 1 to 4 W would be equal to 6 dB. If the power is increased from 4 to 8 W (double again), another 3 dB would be added, to indicate a total change of 9 dB, etc. Because the decibel represents a multiplication of the logarithm of power ratio by 10, any power ratio of 10 would be equal to 10 dB, while a power ratio of 100 would be equal to 20 dB, 1000 to 30 dB, 10,000 to 40 dB, etc.

The decibel expression is also used in reference to voltage and current changes, though in such instances a doubling of either the voltage or current would represent 6 dB instead of 3. Assume, for instance, that a resistor of 5 Ω has a voltage drop of 10 across it. According to Ohm's law, the voltage divided by the current would indicate that 2 A of current is flowing through this resistor. Hence, multiplying the voltage (10) by the current (2) shows that 20 W of power is being consumed by the resistor. If the voltage across the resistor is now doubled, it would represent a 6-dB difference since a voltage or current doubling is equivalent to a 6-dB change. The proof of this is given when the amount of power is calculated, on the basis of the voltage change. Since the voltage was doubled, the current through the resistor would now be 4 A (20 V divided by 5 Ω); hence, the power is now

$$P = EI = 20 \times 4 = 80 \text{ W}$$

Since the power dissipated in the resistor is now 80 W, as compared to the former 20-W value, a power change of 6 dB has occurred. This is the case because a change from 20 to 40 W would be 3 dB, and a change from 40 to 80 W would be *3 dB* more—a total of 6 dB.

Similarly, consider the case of

$$R = 10 \text{ Ω}$$
$$I = 5 \text{ A}$$
$$P = I^2R = 25 \times 10 = 250 \text{ W}$$

Now, if the current is doubled to become 10 A, again a 6-dB difference would occur. If the current is doubled to become 10 A, the power would be

$$P = I^2R = 100 \times 10 = 1000 \text{ W}$$

In terms of wattage, the following decibel change is evident when we compare the original 250 W to the new value of 1000 W:

$$
\begin{aligned}
250 \text{ to } 500 \text{ W} &= 3 \text{ dB} \\
500 \text{ to } 1000 \text{ W} &= 3 \text{ dB} \\
\hline
\text{Total} &= 6 \text{ dB}
\end{aligned}
$$

Thus, for dB calculations involving either voltage or current values, Equation 5-4 must be altered as follows:

$$dB = 20 \log \frac{E_1}{E_2} \qquad (5\text{-}5)$$

$$dB = 20 \log \frac{I_1}{I_2} \qquad (5\text{-}6)$$

As with Equation 5-4, ratios are involved and for convenience the larger value is selected to represent E_1 or I_1.

Reference levels have been established on occasion to simplify analysis of power or voltage changes. In telephone work, a zero level of 0.006 W had been commonly used as a reference. This was originally chosen because the 0.006 W (6 mW) was the output power of a vacuum tube often used in telephone repeaters. Thus, if a certain unit was designated as plus 10 dB, it meant that it had 10 times the output of the so-called telephone repeater tube of 6 mW. This level has also been used by several amplifier companies as a reference level.

Review Questions

5-1. Briefly explain how the ranges of ammeters and voltmeters can be altered to provide for higher readings.

5-2. Describe the basic characteristics of the galvanometer, and designate its primary usage.

5-3. Define the term *meter sensitivity.*

5-4. Describe the basic circuit principles of the ohmmeter.

5-5. Name two component values that can be found by using the Wheatstone bridge.

5-6. Describe the basic circuit principles of the Wheatstone bridge.

5-7. If a change in electric power is a multiple of 2, why can the decibel value be ascertained easily? Explain briefly.

5-8. How do current and voltage changes differ in decibel values compared to power changes?

5-9. Does a change from 2 to 4 A have the same decibel value as a change from 60 to 120 V? Explain your answer.

5-10. Show two examples of a -6-dB change in power in an electric circuit by giving the value of the original power and the final power causing the decibel difference.

Practical Problems

5-1. A 0–20 microammeter has a d-c resistance of 2000 Ω. What voltage is required to deflect the meter needle to full scale?

5-2. A 0–50 microammeter requires 0.1 V for full-scale deflection. What is the meter resistance?

5-3. A meter has a maximum current reading of 5 mA, but a maximum reading of 6 mA is required. What must be the value of the shunting resistor if the basic 0–1 milliammeter movement has an internal resistance of 100 Ω?

5-4. A voltmeter, reading 5 V full scale, has a series resistor of 4900 Ω. The basic meter movement is 0–1 mA at 100 ohms. What must be the value of a new series resistor to change the meter to 50 V full scale?

5-5. For the Wheatstone bridge circuit shown in Fig. 5-5, what is the value of the unknown resistor when R_1 is equal to 1000 Ω; R_2 is equal to 2000 Ω; and R_s, the standard or known resistor, has a value of 4000 Ω?

5-6. A stereo FM receiver had a maximum audio-power output of 5 W per channel. The manufacturer redesigned this unit to one having an output of 20 W per channel. By how many decibels has the converted unit been increased in power per channel compared to the original?

5-7. During repair of a television-station transmitter, the output power was reduced to one-tenth of its former value. What was the approximate decibel difference between the reduced power and the original?

5-8. The voltage across a resistor was increased from 2.5 to 20 V. What is the decibel difference?

5-9. In an industrial control amplifier the power was increased from 6 to 18 W. What is the decibel difference?

5-10. In a radar system the voltage across a resistor dropped from 23.4 to 7.8 V. What is the decibel change that occurred?

5-11. A cassette tape deck had an output of 500 mW. How does this compare in decibels to a tape amplifier with an output of 32 W?

5-12. In a communications satellite a failure of a component caused a change of −9 dB in the transmitted power. By what percentage was the power reduced?

Inductors and Capacitors   6

6-1. Introduction

Besides resistors, two other circuit components widely used in electronics are the inductor and the capacitor. The inductor (coil) finds applications in filter circuits, transformers, and other electronic circuits and devices. The capacitor (also referred to as a condenser) is employed for signal bypass purposes, filtering, signal transfer, and other uses described later. Both the capacitor and inductor are capable of storing energy and in combination form many important circuits. The characteristics of these two items in relation to direct current are reviewed in this chapter. The effects of alternating current and practical applications are discussed in Chapter 8.

6-2. Electromagnetism

Oersted, while demonstrating electric principles to a class of students, noticed that a magnetized needle was deflected each time it was brought near a current-carrying wire. The needle, instead of being attracted or repelled, assumed a position that was perpendicular to the wire. This accidental discovery led to the recognition of the link between magnetism and electricity and proved that current flow through a conductor sets up, about the conductor, fields

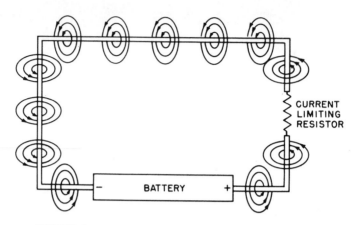

FIGURE 6-1. Magnetic field around current-carrying wire.

that have characteristics *identical with the fields surrounding a permanent bar magnet.*

The magnetic field around a current-carrying wire is illustrated in Fig. 6-1. Separate circular sections are shown; though actually there are no gaps along the wire and the fields exist along the entire length of the wire. The electron flow through the wire causes an alignment of the electron spin of the individual atoms in the conductor, thus magnetizing the metal, as discussed earlier with respect to the magnetization of iron. The field intensity is greatest at the wire.

The strength of the magnetic field (H), in oersteds, can be found at any point on a straight conductor by the following formula:

$$H = \frac{2I}{10d} \qquad (6\text{-}1)$$

where I is the current in amperes and d is the distance from the conductor in centimeters.

From the foregoing formula, it is obvious that the field strength around a straight conductor is proportional only to the current flow through the conductor (for a fixed distance from the conductor). Hence, if the battery voltage for the circuit shown in Fig. 6-1 were increased, more current would flow and the field intensity would also increase.

The direction of the magnetic lines of force are related to the direction of current flow. If the battery terminals are reversed, the direction of the magnetic lines of force shown in Fig. 6-1 will also change direction. With the direction of the magnetic field known, the direction of current (or electron flow) can be ascertained. Similarly, the direction of the magnetic field can be found if we know the current direction. One method, based on the direction of electron flow, is known as the left-hand rule and is illustrated in Fig. 6-2. With electron-flow direction determined by the battery-terminal

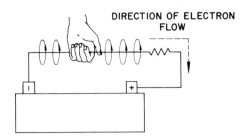

FIGURE 6-2. Application of left-hand rule.

polarity, the wire is grasped with the left hand so the thumb points in the same direction as electron flow. The fingers now point in the direction of the magnetic field.

6-3. Magnetic Fields Around a Coil

Coils are formed by wire loops, and a simple coil can be constructed by wrapping wire around a tube of cardboard, with each turn of wire placed next to (or near) the previous turn. (The term *solenoid* is also applied to a coil, particularly if the latter has a length greater than its radius.) In industry, a variety of coils (and transformers) are used, as discussed and illustrated more fully in Chapter 8. For this introductory explanation, however, assume a simple coil has been formed and volt-age is applied, as shown in Fig. 6-3. When current flows through the coil, the individual fields of the turns com-bine to form an *electromagnet* with characteristics similar to a bar mag-net. Hence, the coil will have a north and south pole, as shown in Fig. 6-3, and is capable of attracting magnetic materials or repelling or attracting a similar coil (or bar magnet).

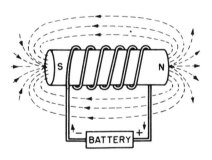

FIGURE 6-3. Magnetic fields of coil.

Figure 6-4 indicates how the individual turns of the coil act to combine the fields and form an electromagnet. At A is shown the cross-sectional areas of two of the wire turns forming the coil. With a separation between

(A)   (B)

FIGURE 6-4. Combined fields of turns of coil.

the wires, as shown, the individual fields have the same direction since the current flow through each turn of wire is also of like direction. When the fields are brought together, as would be the case with close-wound wire, the fields, though in the same direction, oppose each other between the turns of wire and tend to neutralize each other. Hence, with a number of closely spaced turns of wire, as shown at B of Fig. 6-3, the outer portions of the individual fields combine, increasing the total field *parallel with the coil length*. Consequently the magnetic fields are formed to produce an electromagnet.

The left-hand rule also applies to a coil. If the left hand grasps the coil so the fingers point in the direction of electron flow, the thumb will point to the coil's *north pole*. For the coil shown in Fig. 6-3, the left hand is laid palm over the coil with fingers pointing upward in the direction of electron flow. Now the thumb points to the north pole of the coil.

When a soft-iron core is inserted into the coil, as shown in Fig. 6-5, the magnetic lines of the field are increased greatly, forming a more powerful electromagnet. The field intensity does not increase (unless the current flow through the coil were made to increase) but the increase in magnetic lines is caused by magnetization of the iron core. The central core, becoming magnetized, produces fields of its own that are added to the lines of the coil, resulting in increased field strength. The iron, with its high permeability, provides for increased conductivity of the magnetic flux through the core with the result that a more definite pole area is created at the ends of the coil. In practical applications, iron and other metal cores are used extensively, as more fully detailed in Chapter 8.

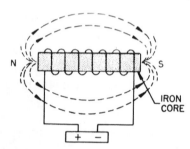

FIGURE 6-5. Increasing magnetic lines of force in a coil by use of metal core.

In coils such as shown in Fig. 6-3 and 6-5, the magnetic fields exist only while current is flowing through the coil. Current, on the other hand, flows because of the applied electric pressure (voltage). Hence, the creation of the magnetic fields is caused by what is termed *magnetomotive force* (mmf), just as current flow in a resistive circuit is caused by electromotive force (emf). Magnetomotive force, however, not only can be derived from current flow but may also be in the form of a magnetized external unit, such as a magnet or another electromagnet. The unit of magnetomotive force in the *cgs* system is the *gilbert*, named after William Gilbert (1540–1603), the English researcher in magnetism.

Increasing the number of turns in a coil adds to the strength of the magnetizing force, over and above what is obtained when the current is increased. Thus, the equation of magnetomotive force, in gilberts, is expressed as follows:

$$\text{Magnetomotive force } (mmf) = 1.256NI \qquad (6\text{-}2)$$

where

    $1.256 = 0.4\,\pi$
      $N = $ the number of turns of wire
      $I = $ the current in amperes.

A gilbert may also be defined as the magnetomotive force required to produce a flux of one maxwell in a magnetic circuit in which the reluctance is one unit. The product NI (number of turns times current) is also known as the *ampere turns* in the *mks* system. As with Ohm's law for electric circuits, the relationships among flux, reluctance (see Chapter 2), and magnetomotive force in magnetic circuits are expressed by the equations

$$\text{mmf} = \phi R. \qquad \phi = \frac{\text{mmf}}{R}, \qquad R = \frac{\text{mmf}}{\phi} \qquad (6\text{-}3)$$

6-4. The Hysteresis Loop

Cores used in practical coils consist of a variety of materials, including soft-iron sheets, powdered iron, silicon steel, and a crystal-metallic substance known as *ferrite* (the latter finding use as antenna cores and in special core applications in high-frequency techniques and computer systems). Practical considerations of core materials are covered in Chapter 8 and other subsequent chapters. The core characteristics are based, however, on the laws of magnetism and, hence, their fundamental aspects are discussed in the present chapter.

    When current is made to flow through a coil such as the one in Fig. 6-5, the soft-iron core becomes magnetized and the magnetic characteristics set up in the individual atoms are aligned as discussed previously. If the battery is removed, however, the current flow through the coil stops and the magnetic domains of the iron tend to assume a random arrangement again. Some alignment remains, however, because of the retentivity of the material. The slight magnetism that remains in the core material is known as *residual magnetism*. When the magnetizing force is plotted on a graph against the flux density, the residual magnetic characteristics of the core material are clearly indicated.

    A circuit for varying the flux density is shown in Fig. 6-6. A coil is wound around a soft-iron bar and is connected to a battery, as shown. A variable resistor R is used so that the voltage impressed on the coil can be varied from zero to the

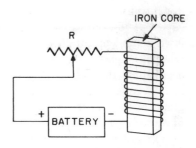

FIGURE 6-6. Circuit for varying flux density.

maximum produced by the battery. As the voltage is increased, current flow through the coil increases and the magnetizing force (*H*) thus produced is graphed along a horizontal axis, as shown at A of Fig. 6-7. The flux density (*B*) created is graphed along the vertical axis, as shown. With zero battery voltage (and hence zero current), the iron bar is in an unmagnetized state and the representation on the graph is at the zero point of intersection between the vertical and horizontal axes.

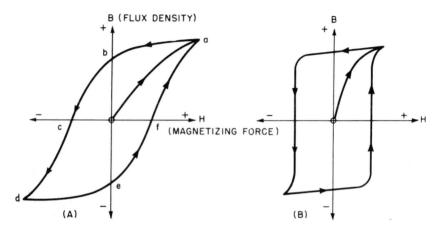

FIGURE 6-7. Hysteresis loops (ordinary magnetic material vs ferrite).

When the resistor *R* in Fig. 6-6 is varied to apply voltage to the coil, current flow occurs and the magnetizing force is applied to the core. One oersted of magnetizing force produces one gauss of flux density and as the current flow through the coil increases, flux density rises from zero toward the point marked *a* on the graph. As the current through the coil increases, the flux density eventually reaches point *a* (the leveling off region). An additional increase of the magnetizing force no longer increases flux density, and the state known as core *saturation* has been reached. If the magnetizing force (by virtue of the current through the coil) is now decreased, it will be found that the flux density will not decline along the initial upward curve but now retraces along the line *a* to *b*. Thus, when the magnetizing force is zero (no current through the coil), flux density still exists (indicating residual magnetism in the core). As seen from the graph, the flux density *B* *lags* behind the magnetizing force *H*, and this lag characteristic is known as *hysteresis* (from the Greek verb *hysterein*: "to be behind, to lag"). The graph of the hysteresis of the core material is known as a *hysteresis curve* or *hysteresis loop*, or also as a *B-H curve*.

In order to decrease the residual magnetism (flux density) to zero, it would be necessary to *reverse* the battery shown in Fig. 6-6 so as to change the polarity of the magnetizing force. Applying such reversed magnetizing force to the core (from zero to *c* in Fig. 6-7) will bring the flux density back to zero. The magnetizing force used to bring the flux density to zero again is known as the *coercive force*.

If the reverse-polarity magnetizing force is increased by an additional amount, the curve of flux density goes from *c* to *d* as shown. Once more, decreasing the magnetizing force to zero brings the flux density to the point *e*, which again represents a residual magnetism. To bring the residual section of the curve to zero, the polarity of the magnetizing force must be reversed, as before, to create the necessary coercive force. Increasing the magnetizing force again brings the curve to point *a* and the ferromagnetic core has gone through a complete magnetic cycle. In order to rid the core of the residual magnetism, the core would have to be demagnetized by subjecting it to a strong external a-c field of the type subsequently described. (Coils used to demagnetize cores are known as *degaussing coils* and are utilized on occasion to rid tape recording heads, or other devices, of residual magnetism, in such cases where its effects on electronic processes would be harmful.)

The shape of the hysteresis loop obtained depends on the type of material, with a higher permeability material producing a narrower curve than a lower permeability material, for a given magnetizing force. Ferrite material has a hysteresis loop that is almost rectangular, as shown at B of Fig. 6-7. Here removal of the magnetizing force results in only a slight change in the flux density of the core material. When the reverse-polarity magnetizing force reaches a critical value, there is a more abrupt decline of the curve than is the case with the loop shown at A. The rectangular loop is useful when it becomes necessary to switch the magnetic state rather abruptly, as is the case of switching and gating systems discussed later.

6-5. Electromagnetic Induction

Michael Faraday (1791–1867), the English scientist, discovered in 1831 that electric energy can be induced from one circuit to another by utilizing magnetic lines of force. This principle is illustrated in Fig. 6-8, showing that when a conductor is moved through a magnetic field, a difference of potential is set up between the ends of the conductor and an electromotive force is induced. This voltage exists only *during* the time when the conductor is in *motion* through the magnetic field. Thus, the current flow caused by the voltage is also present only

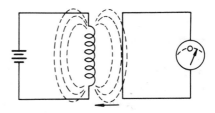

FIGURE 6-8. Conductor wire cutting magnetic field.

during the time when the conductor cuts the lines of force by movement. The current that is caused to flow is known as *induced current*.

The conductor can remain stationary and the lines of force can be shifted so that they will cut across the conductor. When this is done, an electromotive force is again induced. Thus, it is evident that the voltage may be induced in a conductor by *moving the conductor* through magnetic lines of force or by *moving the source* of the magnetic lines of force so that they cut across the conductor.

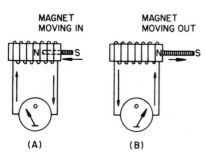

FIGURE 6-9. Effects of bar magnet on coil.

A permanent magnet can be used, as shown in Fig. 6-9, and the conductor may be a coil. When the magnet is moved into the coil, the meter reading will be in one direction, as shown at A. When the magnet is removed, however, the meter needle will deflect in the other direction, as shown at B. As the electric field cuts through the stationary conductor, the induced current establishes another magnetic field across the conductor. The relationship between the magnetic lines of force around the conductor and those of the magnetic field that induces the voltage were observed by Heinrich Lenz (1804–1865), the German physicist. In 1834, he established the law that now bears his name, *Lenz's law*.

An induced current set up by the relative motion of a conductor and a magnetic field always flows in such a direction as to form a magnetic field that opposes the motion.

Figure 6-9 also serves to illustrate Lenz's law. As the north pole of the magnet is inserted into the core area of the coil, the magnetic lines of the magnet cut across the conductor (the wires of the coil) and the induced voltage starts a current flow in the coil, causing the meter needle to deflect. The direction of the meter needle movement indicates the electron flow is in the direction shown by the arrows. The left-hand rule indicates that the electromagnet formed by the coil has its north pole at the end that faces the north pole of the bar magnet. Because two like poles *repel*, energy must be imparted to the magnet to overcome the repulsion between the north poles. After the magnet has been inserted into the coil, there is no longer any induced voltage and current flow stops. Since movement is no longer present, the fields will not be cut. Now the meter needle drops back to the zero position.

As the bar magnet is removed, its electromagnetic field lines again cut across the conductor wires of the coil and the induced voltage again causes an induced current to flow. Now the current-flow direction is opposite to that prevailing at the time the magnet was inserted into the coil. The meter needle now deflects in the other direction to indicate a change of polarity. Now the left end of the coil becomes the south pole and hence attracts the north pole of the magnet. This holding characteristic must be overcome by energy applied during the removal of the magnet from the coil center. (A magnet that produces a stronger magnetic field permits more lines of force to be cut within a certain time interval. The magnitude of the induced voltage is proportional to the number of magnetic lines of force that are cut by the conductor per second. To induce an electromotive force of 1 V, 100 million magnetic lines of force must be cut per second.)

The induced voltage can also be increased by increasing the *speed* with which the magnetic lines of force cut the conductor or *increasing the number of conductors* that are cut. The latter process involves increasing the number of turns of wire in the coil.

Because of the induction characteristics, a coil is often referred to as an *inductor* or *inductance*. With a steady dc flowing through a coil, however, the fields represented by the magnetic lines of force remain at a fixed distance from the wire. The magnetic lines of force represent *stored energy* and if the applied voltage is removed, the lines of force collapse back into the wire, returning the energy to the latter. An inductor with dc flowing through it behaves just as a resistor, with only the resistance of the wire offering opposition to the flow of dc. In practical applications, ac is used and its effect on coil characteristics differs radically from that of dc. These factors are discussed in Chapter 8, where unit values, series and parallel connections, and other related data are covered.

6-6. Characteristics of Capacitors

A capacitor is formed when two conductors are brought within close proximity to each other without touching. In such an instance, air would be the medium of separation between two such conductors and the air thus acts as an insulator. Other types of insulation can also be employed, such as mica, paper, ceramic, and plastic. The insulation between the two conductors that form a capacity is called the *dielectric*. Capacitors take many forms, some utilizing metal plates for the conducting surfaces, while others use metal foil. The various commercial types of capacitors are discussed in greater detail later. A capacitor effect is also formed between two wires that are close together or between a length of wire and another metal surface, such as the circuit chassis. Such capacity effects are often undesirable and provisions must be made for reducing their unwanted effects in circuitry. The capacitors that are *desired* are usually of the physical type specifically designed for applications in electronic circuits. The discussion that follows is primarily concerned with the capacitors deliberately formed by metal plates or foil. Subsequent discussions will treat the capacities formed inadvertently by circuit wiring.

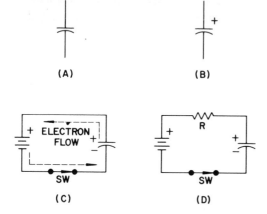

FIGURE 6-10. Capacitor symbols and charging factors.

Typical capacitor symbols are shown in Fig. 6-10. At A is the standard symbol for a fixed-value capacitor. At B is shown a polarized capacitor— one where the correct polarity must be observed.

The charging characteristics of a capacitor are shown at C of Fig. 6-10. Here a battery and a switch are connected in series with a single capacitor, as shown. When the switch is closed, electrons flow from the negative side of the battery and at the same time electrons flow into the positive terminal. The electron flow from the negative battery terminal places electrons on the lower plate of the capacitor and such an accumulation of electrons, beyond the normal amount that would exist on this plate, creates a negative charge here. The electrons drawn into the positive side of the battery from the upper plate of the capacitor create a deficiency of electrons and, thus, the upper capacitor plate becomes positively charged.

At the instant when the switch is closed, a large amount of electrons flow and very little voltage is required. As electrons crowd on the lower plate, however, electromotive force is required to force additional electrons onto it and, in consequence, there will be a sharp rise of voltage across the capacitor. When the voltage across the capacitor reaches the battery voltage, no more electrons can be forced on the lower plate nor drawn away from the upper plate, and current flow to the capacitor ceases. The capacitor is now said to be *charged* and a stable condition exists wherein the applied voltage is constant and current flow is zero.

If a resistor is placed in series with the capacitor shown at D, there will be a delay in the charging rate of the capacitor since the resistor offers opposition to the flow of current. The effect of the resistor in the circuit, with respect to the charging rate of the capacitor, is known as the *time constant*, as is more fully explained later.

The dielectric of the capacitor has considerable influence on the *amount* of electrons which can be stored and hence on the *charge* which can be placed on the capacitor. The reason for this is indicated in Fig. 6-11. Here, the capacitor has been charged so that its upper plate has an excessive amount of

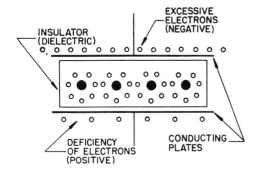

FIGURE 6-11. Electron placement in charged capacitor.

electrons (a negative charge), while the lower plate has a deficiency of electrons (a positive charge). The electrostatic fields of these two charges cut across the dielectric (insulator) and influence the planetary electrons of the atoms forming the dielectric material. Because an insulator has a minimum of free electrons that can be set in motion, the electrostatic fields of the charges on the plates will distort the orbit of the planetary electrons but will not remove them from such an orbit. The negative plate of the capacitor will repel the negative electrons near it in the dielectric, while the positive plate will attract the electrons that lie near it in the dielectric. Thus, the planetary atomic orbits are *distorted*, as shown in Fig. 6-11.

Each orbital distorted atom is referred to as an *electric dipole* or *induced dipole* and as such is comparable to the atoms or magnetic dipoles that exist in magnetized material, as previously explained. In the capacitor, the induced dipoles of the dielectric material are aligned symmetrically by the electrostatic lines of force, as shown in Fig. 6-11, just as the magnetic domains and electron spins in magnetic material are symmetrically aligned. When symmetrically aligned induced dipoles are created in the dielectric, the latter is also referred to as being *polarized*. When the charge on the capacitor is removed, the electrostatic fields collapse and the atoms are no longer distorted.

In a charged capacitor, the atomic distortion created in the dielectric material by the electrostatic lines of force will create an additional electrostatic field that opposes the original field around the charged plates and tends to neutralize them. Hence, additional electrons can be forced onto the negative plate and more can be drawn away from the positive side. In such a manner, the dielectric is influential in increasing the storage capacity of the unit. The degree to which the capacity can be increased by the dielectric depends on the nature of the dielectric. Air has the least influence in increasing capacity and, hence, the *dielectric constant* of air is assigned the numeral one. All other materials thus have a higher dielectric constant and a greater influence toward increasing the ability of the capacitor to assume a greater charge. Glass, for instance, has a dielectric constant of from six to nine and mica is from six to seven.

Since the amount of charge is also dependent on how many electrons can be forced on the negative plate and how many can be drawn from the positive plate, the amount of charge is also influenced by the *area* of the conducting plates of the capacitor. A larger area means that more electrons can be accommodated by the negative plate and that more free electrons are available for withdrawal from the positive plate. Another factor that affects the amount of charge is the *spacing* between the capacitor plates. The closer the plates, foil, or other conductive material of the capacitor, the larger the capacity. A closer spacing of the plates means that the electrostatic fields of the negative and positive charges have a greater influence on the dielectric. Thus, the opposing fields of the dielectric have a greater influence on increasing capacity.

The unit of capacity is the *farad*, named after the eminent British scientist Michael Faraday. A farad is the capacity to store one coulomb of charge at the emf of 1 V. The equivalent of a coulomb, as mentioned earlier, is

$$1 \text{ coulomb} = 6.28 \times 10^{18} \text{ electrons} \tag{6-4}$$

From this definition of a coulomb, the formula for capacity can be stated as follows:

$$C \text{ (in farads)} = Q \text{ (coulombs)} = E \text{ (volts)} \tag{6-5}$$

The farad, however, represents too large a capacity for use in electronics. The amount of capacity employed in ordinary electronic circuits is expressed in millionths of a farad and such capacitors are rated in microfarads or picofarads (see Sec. 2-10). As an example, a capacitor may have a rating of 2 μF, which means that it is two-millionths of a farad. Another capacitor may be represented as having 20 pF of capacity, or twenty-millionths of a microfarad.

6-7.   Capacitors in Series and Parallel

When capacitors are placed in series, the total capacity will be decreased. Hence, a formula for series capacitors is

$$C_{\text{Total}} = \frac{1}{1/C_1 + 1/C_2 + 1/C_3 + \cdots + 1/C_n} \tag{6-6}$$

When capacitors are connected in parallel, each additional capacitor placed in parallel adds more capacity since more plate area is also available for the withdrawal of electrons from the positive plates. The formula for parallel capacitors thus indicates simple addition of the individual capacities.

$$C_{\text{Total}} = C_1 + C_2 + C_3 + \cdots + C_n \tag{6-7}$$

At A of Fig. 6-12 three capacitors are shown in series across a battery. The electron flow from the negative side of the battery causes an accumulation of electrons on the lower plate of C_3 and thus produces a negative charge. Electron flow into the battery from capacitor C_1 places a positive charge on

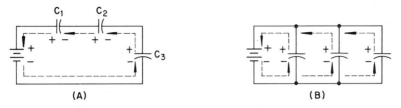

FIGURE 6-12. Charge polarities in capacitor circuits.

the plate nearest the positive terminal of the battery. Electrons moving away from the upper plate of C_3 place the latter at a positive potential, and these electrons flow to the right-hand plate of C_2 and produce a negative charge. The electrons on the left-hand plate of C_2 leave; hence the latter is charged at a positive polarity and these electrons accumulate on the right-hand plate of C_1 producing negative polarity. Thus, the voltage drops around this series circuit resemble the voltage drops that would occur across several resistors in series.

At B of Fig. 6-12 a parallel arrangement of capacitors is shown. Here the voltage drops are similar to resistors in parallel; that is, each capacitor has across it the source voltage, regardless of the individual capacitor value.

6-8. Time Constant

The time constant of a circuit is a calculation used to find the charging rate of a capacitor (or a coil) when a series resistor is in the circuit. The time constant of a resistance-capacitance circuit has for its symbol RC, which is the algebraic expression of resistance multiplied by capacitance, with the resistance value in ohms and the capacity value expressed in farads. For coils, the symbol L/R gives the time constant of the inductance value in henrys divided by the resistance value in ohms. Because coils are used primarily with ac, their characteristics with respect to time constants are discussed in Chapter 8. The following discussion covers the RC time constants.

The chart in Fig. 6-13 shows the percentages of maximum voltage or current for any particular time constant. Hence, the two curves shown on the time-constant chart represent the changes of values that occur upon application of voltage and current to coils and capacitors. The curves rise in an exponential manner; that is, they ascend with a sharply changing amplitude initially and then the rate of change gradually tapers off.

The curve that has its beginning at the lower left-hand corner at zero represents the voltage rise in a capacitor upon application of energy to a circuit composed of a capacitor and a resistor in series. This curve is also representative of the inductive current in a coil. Thus, the lower left-hand curve indicates the rise time of the voltage as the capacitor is being charged or the current rise time in an inductance. The curve that has its beginning

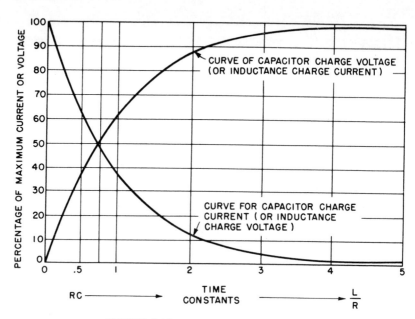

FIGURE 6-13. Universal time constant chart.

at the upper left starts at maximum amplitude and represents the capacitor charge current when voltage is first applied and also represents the inductance charge voltage. The upper curve thus shows that capacitor current flow starts at a maximum when voltage is first applied to the *RC* circuit. It also shows that the capacitor current flow gradually decreases as the voltage across the capacitor rises. In an inductance, the voltage is maximum initially, and current starts from zero and rises as shown.

The usefulness of the time constant and the chart (also known as the *Universal Time Constant Chart*) is that charging rates of capacitors (and current rises in coils) can be found readily. The time constant of an *RC* circuit is the time required to charge the capacitor to approximately 63% of its final voltage level. Hence, the time constant *in seconds* for a capacitor (in an *RC* circuit) to reach 63% of full value is found by multiplying the resistance value by the capacity value.

Thus, if a 3-MΩ resistor is in series with a capacitor of 5 μF and a 100-V source is used, the time constant (*RC*) equals

$$3,000,000 \times 0.000005 = 15 \text{ s}$$

Hence, it would take 15 s for the capacitor to charge to 63% of 100 V, or 63 V. Since the current to the capacitor decreases inversely as the voltage increases, the current after 15 s would be 37% of the initial amount of current at the time the circuit is closed and the capacitor begins to charge.

If a capacitor of 0.04 μF is in series with a 5000-Ω resistor, the time constant would be

$$0.04 \times 5{,}000 = 200 \ \mu s$$

This indicates that the capacitor will be charged to 63% of its full value in 200 μs. After five time constants have elapsed, the capacitor is considered to be fully charged with zero current flow. The time-constant calculation can also be employed to ascertain the time it takes to discharge a capacitor to 37% of its charged value.

6-9. Charge and Discharge Factors

If a capacitor has been charged with a battery, the latter can be removed and the capacitor will retain the charge for an indefinite period, unless some internal leakage exists. In the latter instance, the electric energy stored in the charged plates will leak across the insulator and, eventually, there will be as many electrons on one plate as on the other. When the latter state is reached, no potential difference exists and the charge is no longer present. With a good-quality capacitor having a minimum of leakage, the charge may last for hours or for days. If the leads from the capacitor are touched together, as shown at A of Fig. 6-14, the capacitor will discharge since a path is now provided for the accumulated electrons on the negative plate to flow to the positive side where a deficiency exists and where electrons are demanded. This shorting together of the leads discharges the capacitor and neutralizes the effect of an excessive amount of electrons on one plate with respect to the other.

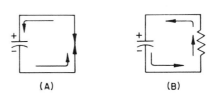

(A)   (B)

FIGURE 6-14. Capacitor discharge.

If a resistor is placed in series with the discharge path, as shown at B of Fig. 6-14, the rate of discharge will be slowed down and the time constant (*RC*) is again a factor. As discussed earlier, this calculation will give the amount of discharge (63%) that occurs in 1 s.

If an excessive amount of voltage is employed when placing a charge across a capacitor, the electrostatic fields will be so great that the insulation will be punctured and the capacitor will have an internal short. When the latter occurs, the accumulated electrons on the negative plate will rush through the puncture and discharge the capacitor. Since the insulation has been punctured, the resultant short will also appear across the battery because it will permit an excessive amount of electrons to flow *through* the capacitor.

Note that in the foregoing discussion of capacitor discharge, the discharge path is opposite to the charging path and that the electrons, in order to discharge, must flow in the direction opposite to that which they followed when the capacitor was being charged. If an ammeter were placed in series with a capacitor being charged, the needle would swing in one direction and after a full charge had been reached, it would drop back to zero. If the capacitor were now discharged, the ammeter needle would swing in the opposite direction, indicating the reversal of electrons during the discharge process. Thus, the charge and discharge of the capacitor exhibits the properties of ac insofar as the ammeter is concerned, since the ammeter needle swings in one direction and then in the other, indicating a plus and a minus polarity.

Review Questions

6-1. Give the unit for the strength of a magnetic field, and explain how this strength can be ascertained when the current in amperes and the distance from the conductor are known.

6-2. (a) Briefly explain what is meant by the left-hand rule for finding the direction of current flow in a conductor.
(b) Explain how the left-hand rule is also applied with respect to finding the north pole of an electromagnet.

6-3. Briefly explain what is meant by magnetomotive force, and show how this value in gilberts can be found by use of a formula.

6-4. Briefly explain what is meant by residual magnetism, and draw a typical hysteresis curve.

6-5. Given a conductor and a magnetic field, explain how an electromotive force may be induced into the conductor.

6-6. If the ends of a coil are attached to a current-reading meter, explain what occurs when a magnet is inserted into the coil and then removed.

6-7. In your own words, explain what is meant by Lenz's law.

6-8. What factors determine the amplitude of induced voltage?

6-9. Briefly explain how the dielectric of a capacitor contributes to total capacitance.

6-10. Define a *farad*.

6-11. What expression is now used for the older term *micro-microfarads*?

6-12. When capacitors are placed in series, is the equation for finding total capacitance similar to the one used for finding total resistance in a series-resistor circuit or the equation for total resistance for parallel resistors?

6-13. Briefly explain the voltage distribution that occurs for series capacitors and parallel capacitors when connected to a voltage source.

6-14. Explain briefly what is meant by the *time constant* of a circuit.

6-15. What may occur when an excessive amount of voltage is impressed across a capacitor?

6-16. Compared to the charge path of a capacitor, in what direction is the *discharge* path?

Practical Problems

6-1. A single conductor had 22.5 mA of current flowing through it. What is the strength of the magnetic field in oersteds at 1.5 cm from the wire?

6-2. What is the magnetomotive force in gilberts of a single-layer coil with an air core if the number of turns is 50 and the current flow is 50 mA?

6-3. If the number of turns in Problem 6-2 were increased to 100 and the current doubled, what would be the resultant magnetomotive force in gilberts?

6-4. In a single-layer coil the *ampere turns* equaled 40. What is the magnetomotive force in gilberts?

6-5. An inductor had 300 turns of wire and 2 mA of current flow through it. What is the magnetomotive force in giberts?

6-6. In a laboratory power supply, three capacitors were placed in series to increase their voltage-handling capability. One capacitor had a value of 90 μF; the second, 45 μF; and the third, 30 μF. What is the total capacitance?

6-7. In an electronic tracking system, two capacitors were in parallel, one having a capacitance of 0.001 μF and the other, 0.002 μF. An additional capacitor of 0.047 μF was added in parallel. What is the final capacitance?

6-8. In a computer circuitry a capacitor became defective and no single unit of proper value was available. The required capacitance was obtained by using three capacitors in parallel, with values of 20 pF, 0.005 μF, and 0.0008 μF. What was the total capacitance thus obtained?

6-9. Express the following values in picofarads: 0.00002 μF, 0.0000018 μF, and 0.000135 μF.

6-10. In a signal switching circuit a 0.003-μF capacitor was in series with a 5-MΩ resistor. What is the time constant in seconds?

6-11. In a high-fidelity system a 0.02-μF capacitor has in series with it a 3,000-Ω resistor. An emf of 200 V is impressed across this circuit. What is the time in microseconds for the capacitor to reach full charge?

6-12. A resistor, placed across a capacitor, discharged the latter to 94.5 V in one time constant. What was the full charge voltage on the capacitor?

Alternating Current 7

7-1. Introduction

In all the many areas of electric and electronic circuit applications, both the direct and the alternating type of currents are extensively used. Direct current is designated as *dc* and alternating current as *ac* and they have a broad reference. Thus, if we say that a circuit uses dc, we mean that a single-direction current is utilized and the voltage polarities are also dc in nature; that is, they do not undergo a transposition. Similarly, ac is also used as an adjective to describe *alternating voltage* and it has become common practice to refer to alternating voltage as a-c voltage. Actually, this is a redundant description since a-c voltage means "alternating-current voltage." Popular usage, however, has held to this form and for this reason its usage will be accepted as valid in this text.

The d-c type power is used in transistor or tube amplifiers for processing signals to bring their amplitude to that required for practical employment. The a-c type electric is that present in the power mains coming into our homes and supplying industry. For use in electronic equipment, the voltage must be altered in many instances and usually it is necessary to convert the ac to dc for specific applications. Speech, music, and other sounds have a-c characteristics, as do the transmitted carrier signals that arrive at receivers. The basic factors relating to the a-c type signals are covered in this chapter.

7-2.   Generation of ac

The insertion of a magnet into a coil, as described in the preceding chapter, actually produces ac. Prior to the movement of the magnet into the coil, zero voltage exists. As the magnet approaches the inductor, the induced energy builds up from zero to a maximum value. When the magnet is withdrawn, voltage again builds up but now it is of opposite polarity. After the magnet is withdrawn, the induced voltage (and current) again drops to zero.

The *waveform* of voltage induced across the inductor has a characteristic as shown in Fig. 7-1. The gradual rise of amplitude from zero to maximum, with its subsequent decline to zero represents one-half of an a-c wave. The rise to maximum again in the opposite direction and thence to zero again is the second half of the ac wave. The two combined form a *sine wave*. The induced voltage of sine waveform also creates an induced current flow that has a waveshape similar to that of the voltage. The respective amplitudes of the current and voltage may differ depending on the internal resistance of the coil, the number of magnetic lines of force of the magnet, and the coil turns. Both current and voltage waveshapes, however, are sinusoidal.

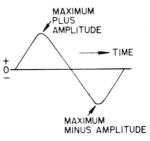

FIGURE 7-1. Sinewave of ac.

There are a number of practical methods for generating ac; the most common form used is the *generator*. This device utilizes the inductive principles discussed earlier, where a conductor rotates to cut magnetic lines of force. The basic form of a generator is shown in Fig. 7-2. Here a permanent magnet is used in conjunction with a wire loop. The wire loop is between the north and south poles of the permanent magnet and, thus, is within the highly concentrated magnetic lines of force existing between the pole pieces. The wire loop of the generator has its ends terminating at collector rings so that the loop can be rotated while the ends still make electric contact with the collector rings. Two so-called *brushes* are utilized, which are formed of a composition having a carbon base and are thus capable of conducting electricity. These brushes make physical contact with the collector rings and slide along them as the latter are rotated. The brushes are connected to the output terminals of the generator. In actual generators, instead of the single wire loop shown in Fig. 7-2, a number of turns is used to form a coil that rotates within the magnetic lines of force. The rotating coil section is known as an *armature*.

The manner in which this simple device illustrated in Fig. 7-2 generates ac is shown in Fig. 7-3. When the armature coil is in a vertical position, as

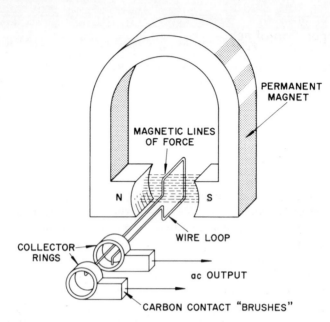

FIGURE 7-2. Basic generator.

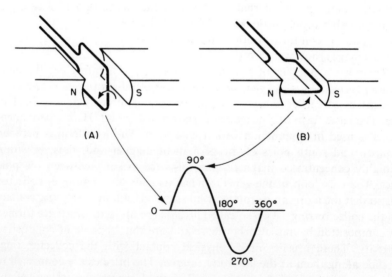

FIGURE 7-3. Function of a basic generator.

shown at A, the least number of lines of force are intercepted. If the coil is stationary, no voltage is induced into the coil and, hence, none appears at the collector rings and output terminals. As the coil is rotated, as shown at B, the rotation causes the coil to assume a horizontal plane and as it enters such a horizontal plane, it cuts more and more lines of force and, consequently, a greater voltage is induced into the coil and at the output terminals. As the coil is rotated, it again moves toward a vertical position and as it leaves the horizontal position, the voltage that is induced still has the same polarity and the current flow is in the same direction, but it now diminishes and decreases to zero when the coil again assumes a vertical position. This rotation of the coil has now produced *one-half* of the sine waveform previously mentioned because the coil loop has made *one-half turn*. Each conductor of the loop is now in a reversed position with respect to its starting position. As the coil is once more made to rotate toward a horizontal plane, the electromotive force is again induced and gradually increases in amplitude from zero to a maximum but with an opposite polarity than previously. During the last quarter turn of the coil, the voltage decreases from a maximum negative value back to zero, at which point the coil conductors will have assumed their original positions. If the shaft of the generator is connected to a motor so that the armature coil is rotated at a rapid rate, an alternating current will be generated where the positive and negative voltage peaks are produced in rapid succession.

7-3.   Characteristics of ac.

When the a-c waveform initially begins at zero, reaches maximum, and then drops to zero again, this portion of the sine wave is known as an *alternation*. The sine waveform shown in Fig. 7-1 has, therefore, two alternations. Since the third alternation would be identical to the first, and subsequent alternations would be repetitions of the initial two, *any two successive alternations are called a cycle*. Thus, one cycle of ac may be represented as at A of Fig. 7-4, where the first alternation is negative, or one cycle may be in the form shown at B, where the first alternation is positive. The next two alternations that follow a cycle would be another cycle since the waveform would be repeated.

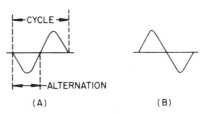

(A)   (B)

FIGURE 7-4. Terms applied to an ac sinewave.

   The *frequency* of an a-c waveform is determined by *the number of cycles that occur per second*. As mentioned in Sec. 2-10, the standard for cycles per second is *hertz* (Hz). Thus, 60 Hz refers to the frequency commonly used in the power mains. Because there are 60 cycles of ac occurring each second, there are 120 *alternations* per second.

   Since the armature coil of the generator must make a complete revolution to produce one cycle, the armature coil rotates through 360°. The relative degrees of rotation of the armature, versus the sine waveform produced, is shown in Fig. 7-3. Reference to this illustration makes it clear that maximum values of voltage (or current) are reached at the 90° point of the first alternation and at the 270° point of the second alternation. Zero points are reached at 0°, 180°, and 360°.

Battery voltages are of a steady-state value under normal conditions and if a resistor is placed across the battery, a fixed value of current flows. In ac, however, the values of voltage and current are always changing, starting at zero, rising, and then declining again. Hence, for one cycle of ac, the average value would be zero because the positive amplitude at its peak value is equal to the negative peak value amplitude and these equal and opposite polarity values would cancel in terms of an average value. The term *average value*, however, usually refers to approximately one-half of the peak amplitude of an alternation and is equivalent to 0.636 times the maximum or peak amplitude. The average value is encountered in rectification and pulse work but, in general, a more commonly encountered value is the *effective value*.

The effective value is also known as the *root-mean-square* value (rms) and is equal to 0.707 times the maximum or peak value of voltage or current. The term *root-mean-square* is derived from the fact that its calculation is based on adding the instantaneous values that have been squared and dividing this sum by the number of instantaneous values taken. The effective value is, therefore, equal to the square root of the average of the instantaneous values.

The effective value of ac is equivalent to the value of dc required to do the same amount of work. Thus, if a light bulb is designed for 110 V rms (effective value), it would mean that this 110-V a-c bulb would light as brightly and consume as much power when used at 110 V dc. Thus, the effective value is the commonly indicated value given to the power mains furnishing ac to the homes and it is also the value generally read by the average ac voltmeter.

If the effective value is known, the peak value of the a-c waveform may be found by multiplying the effective value by 1.41. This figure, 1.41, is approximately equal to the square root of 2. Thus, it is evident that the effective voltage is equal to the peak voltage divided by the square root of 2. The 110 V of ac supplied by the power mains consequently has a peak amplitude of 155 V. Since the peak amplitudes are present only for short intervals, they will not perform the same amount of work that could be furnished if this peak voltage were maintained constantly. The effective value, therefore, is equal to the same d-c value that would have to be utilized to perform the same amount of work.

7-4. Phase

In the a-c generators previously discussed, reference was made to induced voltages and induced currents. If a resistor is placed across the wires at the output of the armature coil, both the voltage and current would start simultaneously and both would reach their positive amplitudes at the same time, as well as their respective zero and negative amplitudes. When the voltage and current peaks and zero points coincide in time, the voltage and current are said to be *in phase*. Under some circuit conditions, however, the voltage and current may not reach peak values at the same time and when this is the case, the current and voltage are referred to as being *out of phase*. At A of Fig. 7-5, the voltage and current are shown in phase, while at B an out-of-

(A)   (B)

FIGURE 7-5. Voltage and current phase relationships.

phase condition is indicated. Relative amplitudes of voltage and current at A may differ since current could be higher than voltage or vice versa. Relative amplitudes have no bearing on phase, however.

The out-of-phase condition shown at B indicates that the voltage reaches its maximum 90° ahead of the current maximum. Under this condition, the voltage also reaches zero 90° ahead of the time when the current reaches zero. The voltage is said to *lead* the current by 90° or it can also be stated that, under this condition, the current is *lagging* the voltage by 90°. In other words, the phase angle (the difference in degrees between voltage and current) is equivalent to 90°. The symbol for the phase angle is the Greek lowercase letter theta (θ). Sometimes the lowercase Greek letter phi (ϕ) is also used.

7-5. Practical Generator

The basic a-c generator illustrated earlier in Fig. 7-2 indicates the fundamental principles of operation. In actual practice, however, the a-c generators are more complex and efficient and, instead of a single loop of wire, coils are wound around the armature frame. Also, additional coils are utilized to form an electromagnet of considerable strength to produce a high-power a-c output. The coils that produce the magnetic fields are referred to as *field coils* and are supplied with dc from an external source. A typical commercial generator of this type is shown in Fig. 7-6.

In the basic generator discussed earlier, the armature was the rotating element and the magnetic fields produced by the magnet surrounded the armature. As the armature rotated, the coil cut the stationary lines of force.

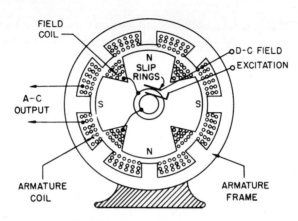

FIGURE 7-6. Practical ac generator.

As discussed in Chapter 6, however, the magnetic fields could be moved and induction would still be established. Thus, in an a-c generator, the armature coils can be kept stationary and the field coils rotated so that their magnetic lines of force move and cut across the armature coil to produce ac. Thus, in practical applications the a-c generators mostly use rotating fields because of the simplicity of design. Also, the slip rings no longer need to carry the high-amplitude currents that the generator furnishes the load.

As shown in Fig. 7-6, the center cylindrical section (a laminated metal drum) has appropriate slots for accommodating the field coils. This center section has four poles and the field coils are in series, with the two terminals connected to the slip rings. The d-c field voltage (*excitation*) is applied to these two collector rings by carbon brushes that slide over the collector rings as the center field coil section rotates. The armature, which is now stationary, has its individual coils also connected in series, with the ac that is generated produced at the output terminals as shown. Such a generator is a single-phase type and it generates a single a-c waveform. The 110–120 V ac found in the homes is the single-phase, 60-Hz variety furnished with a two-wire line. When 240-V ac is furnished to homes, a three-wire line is used, with one conductor representing the ground wire. Either of the other wires, with respect to the ground wire, would furnish 120 V. This type ac is still *single phase*, however, and must not be confused with the three-phase commercial type discussed next.

7-6.   Three-Phase ac

In industrial plants and in many commercial applications where large motors must be operated and high power furnished to other equipment, the common type of ac is the three phase as distinguished from the single phase furnished to residential areas by the power mains. (Two phase is only used on rare occasions for specific industrial applications.) The three-phase generator

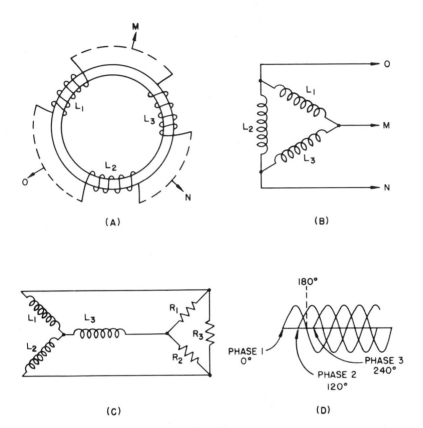

FIGURE 7-7. Three-phase ac.

has the armature coils divided into three sections, as shown at A of Fig. 7-7 (this generator also employs a stationary armature as does the commercial-type generator previously discussed). Thus, the armature coils, regardless of their number, are grouped into sets so arranged that the induced voltage for one set of coils differs from the others by a third of a cycle (120°). Such a generator is termed a *polyphase* type since it has more than one phase of ac.

With the armature coils divided into three sections, six terminals are available as shown at A. By wiring the coils as shown at B or C, however, the number of terminals is reduced to three, as shown. The connection shown at B is known as *delta* since it resembles the Greek letter positioned sideways. At C the so-called *star* or *Y* connections are shown. (This is sometimes called the *Wye* type.) The delta and *Y* windings can also be employed on transformers, with either serving as the primary or secondary. Similarly, a delta generator (or transformer secondary) can feed a *Y* load or vice versa. At C is shown a *Y* winding feeding a delta load composed of resistors R_1, R_2 and R_3. This load could consist of a motor specifically designed for three-phase operation. The load, represented by the three resistors, is a balanced type; that is, each resistance value is identical to the others.

The three-signal waveforms produced are shown at D. Since one complete cycle (two alternations) of ac takes in 360°, one alternation takes 180° as shown. For the three-phase ac, phase 2 is displaced with respect to phase 1 by 120° (one-third cycle), and phase 3 is displaced by 240° with respect to phase 1 (and by 120° with respect to phase 2). With three-phase ac, current at any instant may be flowing out of one wire and returning through the other two. At another instant, the current flows through two wires and returns through the third.

In the distribution of power to industry, a neutral (ground) wire is often used with the three-phase system as illustrated in Fig. 7-8. Such a wire

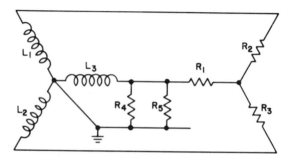

FIGURE 7-8. Three-phase four-wire system.

is connected to the junction of the Y winding as shown and constitutes what is known as a *three-phase, four-wire* system. The neutral wire permits single-phase ac to be obtained from the three-phase system. If, for instance, each of the inductors shown in Fig. 7-8 has a 120-V drop across it, this voltage is obtained with respect to the neutral wire and any one of the other wires. For the system shown in Fig. 7-8, the voltage drop across L_3 is utilized to furnish 120 V *single-phase* to the load circuits R_4 and R_5. These may be electric fans, heaters, or other devices requiring only 120-V single-phase ac. At the same time, a balanced load (R_1, R_2, and R_3) is furnished the three-phase ac as shown.

Review Questions

7-1. Draw a sine wave, and label the alternations, the peak amplitudes, and the time base.

7-2. Define words *generator* and *armature*.

7-3. Explain how a generator produces ac.

7-4. Explain what is meant by the *effective* value of ac, and indicate what percentage of the peak value this represents.

7-5. Explain what is meant by the *average* value of ac, and indicate what percentage of the peak value this represents.

7-6. Draw two waveforms, and identify one as *current* and the other as *voltage*. Show the voltage with approximately twice the amplitude of the current, with the voltage *leading* the current by 90°.

7-7. Repeat Problem 7-6, but show the current *leading* by 45°.

7-8. If the potential in an ac line reaches a peak value of 155 V, what is the root-mean-square value, and which value (peak or rms) is equal to the same d-c voltage that would have to be utilized for equivalent electric power?

7-9. If the effective value of a voltage is known, how may the peak voltage be found?

7-10. Why do a-c generators employ rotating coils?

7-11. Draw a schematic of a three-phase delta generator winding connected to a Y-type balanced load circuit.

7-12. Explain what is meant by a four-wire, three-phase system.

Practical Problems

7-1. In a transmitter, the carrier generator produced sine wave *alternations* at a rate of 2 million per sec. What is the carrier frequency in hertz?

7-2. How many cycles per second are there in 20 kHz and 2 MHz?

7-3. A 60-W light bulb consumes 0.5 A rms of current on a 120-V line. What are the peak values of current and voltage?

7-4. An electric light bulb is used in a 32-V d-c system. What effective value of a-c voltage is necessary to have this bulb light as brightly as it does on dc?

7-5. In a public-address amplifier circuit there is a 300 peak sine wave a-c voltage drop across a resistor. What value of voltage will an rms type of a-c voltmeter read?

7-6. In a certain community the a-c power mains gave a reading of 125 V rms. What was the peak value of this voltage?

7-7. What is the effective value of the sine wave voltage of power main system when the peak voltage is 340?

7-8. In an industrial control circuit, two resistors are in series with a sine wave a-c source of 200 V rms. Only one resistor is readily accessible and this has 50 V (effective value) across it. What is the *peak* voltage that appears across the other resistor?

7-9. In the schematic of a radar system a pure sine wave is shown .having a voltage value from the minus peak of one alternation to the plus peak of the next alternation (peak-to-peak) of 126 V. What is the rms value of this a-c voltage?

7-10. In a four-wire, three-phrase system, 240 V rms appears across each inductance of a delta winding. What is the peak value of the voltage impressed across the load resistor connected between the neutral wire and one of the other output terminals?

Reactance and Impedance 8

8-1. Introduction

When dc flows through a circuit, the unidirectional current is opposed by resistance or blocked by a capacitor (which permits current to flow to and from it without letting it pass through). The only power consumption is that caused by resistance, and basic Ohm's law principles apply. For ac, however, inductors and capacitors exhibit a variety of characteristics, including opposition without power consumption, phase changes between voltage and current, and the combined effects of these factors with the normal resistance. Thus, in a-c circuitry, coils and capacitors have a profound influence on signal amplitudes, particularly in reference to their frequency. These factors, plus the basic mathematical relationships among resistance, inductance, and capacitance, are covered in this chapter.

8-2. Inductance and Reactance

When ac flows through a coil, in contrast to dc, the magnetic lines of force are constantly building up and collapsing, and this action produces a changing field. Thus when current starts to flow in one turn of the coil, the buildup of the magnetic lines of force that represent a changing voltage amplitude will induce a voltage into the next turn. As described earlier, the induced voltage causes a current to flow, which opposes the initial current. Thus, an opposing current is built up in the adjacent turn, which tends to reduce the initial flow of current which was established through the coil by the voltage applied.

This same opposition factor occurs throughout the coil; that is, as the changing fields induce voltages and currents in adjacent turns, a counterelectromotive force is developed by virtue of the *new* magnetic lines of force generated in adjacent turns. The induced emf always opposes the applied emf of the coil and, for this reason, it is known as *back-electromotive force* (*back-emf*) and also as *counterelectromotive force*. The characteristic of a coil that relates to such a *counter-emf* and represents opposition to a change of current is known as the *inductance* of the coil. The counter-emf is established during the time when the applied ac is building up and also occurs after the a-c peak voltage has been reached, when the latter starts to decline. As the magnetic fields decline, the collapsing field again cuts across adjacent coil turns, thus inducing an electromotive force that is opposite in polarity to the applied voltage.

The unit for inductance is called the *henry*, named after the American scientist Joseph Henry (1797–1878). The unit of inductance also relates to the principle known as Lenz's law (mentioned earlier in Chapter 6), which may be stated as follows:

When a changing current is caused to flow through a coil by applying a voltage, the magnetic lines of force produced by the current establish an induced voltage which opposes the changing current which produced it.

Thus, an inductance is *that property of a circuit that opposes any change in the amount of applied current.*

The henry is the amount of inductance present in a coil when a current change of 1 A per second produces an induced voltage of 1 V. (The henry is often employed in fractional expressions such as millihenry, which is one-thousandth of a henry, and microhenry, which is one-millionth of a henry.) The symbol used for inductance is L, and the symbol for henry is H.

A resistor opposes the flow of ac just as it does dc. Thus, the resistance of the wire making up the coil opposes the flow of ac. In addition, however, the coil's inductive characteristics also provide opposition to ac and is called *inductive reactance*. The symbol for reactance is X and for inductive reactance the symbol becomes X_L.

The opposition to a-c flow by inductive reactance has for its unit value the ohm, as for resistance. Resistance, however, not only opposes the flow of dc and ac but dissipates some of the electric energy in the form of heat. Inductive reactance, however, *does not* consume electric energy, even though it opposes current flow. Actually, electric energy is *stored* in the magnetic field of the coil, and when opposition is offered to the current flow, the current is reduced but the unused energy is returned to the voltage source. Thus, while inductive reactance is an indication of the amount of opposition to the current flow and hence indicates the current decrease that results by virtue of such reactance, it is not used to calculate the power consumed in the coil.

The inductive reactance can be calculated by Ohm's law, and the formula is similar to that used in solving dc problems. These formulas can be set down as follows:

$$X_L = \frac{E}{I}, \qquad I = \frac{E}{X_L}, \qquad E = IX_L \qquad (8\text{-}1)$$

When no resistance is present in the coil, there is a 90° phase angle between the applied a-c voltage and the current flow through an inductance. This phase difference (with voltage leading) also indicates that no power is consumed because when the voltage is at its peak, the current flow at the instance is zero.

The inductive *reactance* that is established in a coil depends on the amount of inductance in henrys (or fraction of a henry), as well as the rate of change of the ac. The rate of change, of course, depends on frequency and, for this reason, determines the rate at which the current (and hence the magnetic field) is changed. Thus, the amount of inductive reactance can also be calculated by the following formula, which takes the foregoing factors into consideration:

$$X_L = 2\pi f L \qquad (8\text{-}2)$$

Here f relates to the frequency in hertz of the ac applied to the coil, while L refers to the value of the inductance in henrys. The symbol π (Greek lower-case letter pi) that follows the 2 has an approximate value of 3.1416.

In Sec. 2-6, it was pointed out that conductance is the reciprocal of resistance. Similarly, with reactance, the reciprocal form is used and is termed *susceptance* (symbol, B) and indicates the ability of an inductance to pass alternating current. As with conductance, the unit is the *mho*. Thus, $B = 1/X$ and on occasion the inductance relationship is indicated as $B_L = 1/X_L$ to distinguish this particular susceptance from that obtained for capacitors as more fully discussed in Sec. 8-14.

8-3. Angular Velocity

In a-c calculations, the expression 2π (two pi) is used extensively. π multiplied by 2 is approximately 6.28, and this value is used in conjunction with the frequency and inductance value for solving inductive reactance, without knowledge of the current or voltage values. Reference to Fig. 8-1 will help clarify the usage of 2π.

If the radius were bent around the circumference, as shown at A of Fig. 8-1, and if another line were drawn as shown by the dotted line, which is a second radius, it would be found that the angle between these two radius

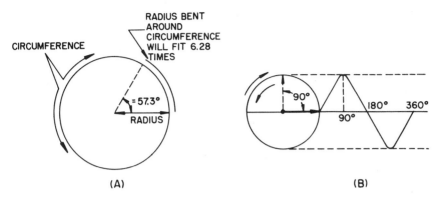

FIGURE 8-1. Relative degrees in ac waveform.

lines would equal approximately 57.3°. This angle, which is established when the radius is bent around the circumference once, is known as the *radian*. Since 1 radian is created each time the radius is bent around the circumference, in a 360° circle, there will be 6.28 radians, just as 6.28 radius lines can be laid around the circumference without overlapping.

If a circle is drawn such as shown at B, and if it is assumed that this is a wheel that rotates in a *clockwise* direction, as indicated by the arrow, an additional representation can be made of the manner in which ac functions and of the characteristics which it presents. If a radius line is drawn horizontally in the circle, as shown, and if this radius line revolves *counterclockwise*, a clear understanding of the relation of frequency and velocity to reactance can be gained. Assume that if the wheel (circle) rotates *clockwise* once through 360° the horizontal radius line also makes a complete revolution but in a *counterclockwise* direction. Thus, if the circle is rotated once, assume that a gear mechamism automatically turns the radius line completely around once in a direction opposite to that of the circle. If a marker were now fastened to the arrow point of the radius line, it would be found that, for a complete rotation of the circle, the radius line would draw a sine waveform, such as shown at B of Fig. 8-1. This is a *vector* representation of the a-c characteristics. The radius is now called a *vector arm* since it will draw out a complete sine wave during a counterclockwise rotation of the arm while the circle rotates for one complete revolution in a clockwise direction. Thus, if the vector arm moves counterclockwise until it is vertical, it will represent an angular change of 90° in the circle and at the same time will draw one-half of an alternation, or 90° of the sine wave.

The speed with which the circle rotates is directly related to the frequency of the sine wave, in cycles per second. Since the vector arm, in its vertical position, represents the maximum amplitude of the sine wave produced, this arm also represents *instantaneous* values of a sine wave for positions other than the vertical. Thus, if the vector arm is at a 45° angle with respect to the horizontal zero reference line, it will indicate the instantaneous voltage at the 45° point (or instantaneous current of the sine wave) since the vector arm at 45° will be above the zero reference line by the amount of voltage or current representing the instantaneous value.

The angular velocity of a vector arm as it rotates indicates the change of degrees as well as the amplitude; hence it is also an indication of the velocity of the sine wave. Thus, the expression $2\pi f$ is known as the *angular velocity*. Since 2π is equal to 6.28, the angular velocity may also be written 6.28f. When the angular velocity is multiplied by the inductance value in henrys, the value of the inductive reactance is found.

As shown in Fig. 8-1, 6.28 rad equals 360°. Because 1 rad equals 57.3° and because this represents the angle between the extremes of the radius laid along the circumference, 6.28 times the radius would indicate the complete circumference, and thus 6.28 radi equals 360°. Here basic principles of trigonometry are obviously utilized and this branch of mathematics is useful in analyzing and understanding electronc circuits. Ratios and right-angle factors are given in the appendix for reference.

8-4.   Instantaneous Values

Since the vector line, or radius, has a length equal to the peak or maximum value of either voltage or current, the *instantaneous value* can be ascertained if the *angle* of the vector line with respect to the zero voltage line is known. The instantaneous value of either voltage or current indicates the value of such voltage or current that prevails at the instant when the vector arm is considered to be at one place and thus has a definite angle with respect to the zero line. If the maximum value of a voltage or current is then multiplied by the *sine* of the angle, the instantaneous value of the voltage or current will be ascertained. The formula for finding the instantaneous values of either the voltage or the current is

$$e = E \sin \theta \quad \text{or} \quad i = I \sin \theta \qquad (8\text{-}3)$$

where

   e or i = the instantaneous value
   E or I = the maximum (peak) value
   θ = the angle formed by the vector (radius) and the zero (horizontal) line.

The application of this simple formula is shown in Fig. 8-2. At A the vector line is shown at an incline of 45° with respect to the zero voltage line. If the vector arm represents 50 V, this potential is multiplied by the sine of the angle. The sine of a 45° angle is 0.707. Hence, if the 50 V of peak value is multiplied by 0.707, the result will be 35 V.

At B is another instance, where the peak voltage is 40 V and the angle of incline is 30°. Here the sine of the 30° angle is 0.5 and when the 40 V is

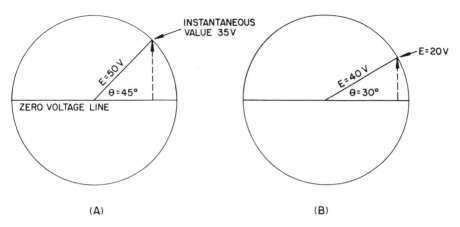

FIGURE 8-2. Instantaneous value designations.

multiplied by the sine of the angle, we find that the instantaneous value of the voltage is 20 V.

The voltage representations in Fig. 8-2 could also be current values, and the methods for solving the instantaneous value would still apply. Note that at the 30° angle at B the instantaneous value is exactly one-half of the peak value.

Example: The maximum value of an a-c voltage is 200. What is its value at the 135° vector angle?

Solution:
$$\sin 135° = 0.707 \text{ (same as 45°)}$$
$$0.707 \times 200 = 141.4 \text{ V instantaneous value}$$

Often voltage and current in an a-c circuit are not in phase. This occurs when an inductance or a capacitance is present. As previously mentioned, an inductance will cause the voltage to lead the current by 90°. A capacity will cause the current to lead the voltage by 90°. Combinations of these two, or when both factors are used in conjunction with a resistor, may cause other differences than 90°, as more fully explained later in this chapter. When, however, either the voltage or the current lags, the instantaneous values can still be ascertained by finding the degrees of difference in the angle between voltage and current. The formulas for finding the instantaneous values under such a condition are as follows:

Current lagging: $i = I \sin (\theta - \phi)$ (8-4)

where θ is the phase of voltage and ϕ is the phase difference between E and I, or angle of current lag.

Voltage lagging: $i = I \sin (\theta + \phi)$ (8-5)

Example: In a circuit having an inductance and resistance, the current lags the voltage by 60°. The maximum value of the current is 25 mA. What is the instantaneous value of the current when the voltage is at the 90° vector angle?

Solution:

$$i = I \sin (\theta - \phi)$$
$$= I \sin (90° - 60°)$$
$$= 25 \times \sin 30°$$
$$= 25 \times 0.5 = 12.5 \text{ mA}$$

The solving for instantaneous values does not have as much application in practical work as does the solving of peak values, effective values, etc. There are occasions, however, when a knowledge of the basic principles of instantaneous values will be helpful, and the foregoing explanations illustrate the mathematical factors that relate to a-c theory.

8-5. Power Factor

The formula for calculating the power for instantaneous values is similar to the formula for dc where $P = EI$ (or $W = EI$):

$$P_{\text{inst}} = E_{\text{inst}} \times I_{\text{inst}} \tag{8-6}$$

When solving for power, where effective values of voltage and current are employed, the formula can still be used, provided that the voltage and current are in phase. When current or voltage are out of phase, however, there are occasions during the cycle when the current is negative while the voltage is positive or vice versa. During such times when voltage and current have opposing polarities, the power consumed by the load circuit would be zero since the power would be returned to the generator. When current and voltage are only slightly out of phase, some power is used, depending on the degree of phase difference. Thus, less power would be consumed by the load than would seem to be indicated by use of the formula. Actually, the formula would only indicate the *apparent* power. The *true* power in such instances would be less. The ratio of the *true* power to the *apprent* power for out-of-phase effective values of voltage and current is known as the *power factor*. Thus, power factor may be solved by the following formula:

$$\text{Power factor} = \frac{\text{true power}}{\text{apparent power}} \tag{8-7}$$

This ratio can be expressed as a percentage or as a fractional value. Thus, the power factor can be 25% or one-fourth of the apparent power. When effective values of voltage and current are employed where there is no phase difference between the two, the apparent power is then equal to the true power. The actual power consumed by the load is equal to the product of the voltage multiplied by the current times the cosine of the angle between the current and the voltage. In the form of a formula, this becomes

$$P = EI \cos \theta \qquad (8\text{-}8)$$

The value of *the cosine of the angle is equivalent to the power factor* because when current is in phase with the voltage the angle is zero and hence the power factor would be the cosine of zero degrees, or one. (See the trigonometric table of ratios in the appendix.) If a 90° phase difference exists between current and voltage, the power factor is the cosine of 90°, which is zero, indicating no power is consumed because EI times zero equals zero. Hence, it is readily apparent that an in-phase E and I delivers maximum power to the load resistance but if the power factor is less than one (such as 0.8 or 0.6), only a portion of the voltage and current values produce actual power. Since only the true power shows all energy that performs work, it is necessary to ascertain such true power by considering the difference in the phase angle between voltage and current, as discussed more fully later in this chapter.

The power factor of a circuit can be ascertained practically by taking meter readings or calculated mathematically when the voltage, current, and the wattage of a circuit are known.

8-6. Impedance

For simplicity, it was assumed that the coils discussed earlier had no resistance, only inductance. Actually, however, all coils have some resistance. Since lengths of wire have resistance, it follows that coils formed by wire will also have some resistance, the amount depending on the wire size, the number of turns, and the composition of the wire. If copper wire is used, and the wire diameter is large, a coil of only a few turns would have such a low resistance that the resistance effect on circuit function could be ignored. If the coil has more than just a negligible resistance, however, the latter will alter the angle of phase difference that would exist without such coil resistance.

When a resistance is present in the inductance, it will decrease the angle of current lag to less than 90°. The amount by which such a phase difference is reduced depends on the *time constant* of the circuit (see Fig. 6-13). Depending on the amount of resistance present, there will be delay with respect to the amount of time required for the current to attain 63% of maximum. The time constant (τ) of an LC circuit can be calculated on the basis of the following formula:

$$\tau = \frac{L}{R} \qquad (8\text{-}9)$$

where L is in henrys and R is in ohms.

The difference in lead and lag will determine the actual amount of power consumed by the coil. When a coil has more internal resistance than another, more power is consumed. If the coil is a pure inductance and has no resistance, there would be a 90° current lag. The 90° phase difference would provide a cosine of zero, as shown on the table of trigonometric values in the appendix. With a cosine of zero, the product of the voltage times the current would always be zero when multiplied by a zero representing the cosine of the angle. (Assume that the voltage is 5 and the current is 2 A. The apparent power would be 10 W but the actual and true power would be zero.)

If, however, we have a pure resistance without inductance and the same formula is applied, the *cosine* of the zero *angle* would be one. Thus, the product of 5 V, 2 A, and one for the cosine is 10 W. Thus, when the cosine of the angle is one, the *apparent* power is also the *true* power.

If an inductance and a resistance are present, where the resistance causes the 90° angle to change to a 45° angle, the cosine of this angle would be 0.707. Now, when the formula is applied, the true power is derived as follows:

$$EI \cos \theta = 5 \times 2 \times 0.707 = 7.07 \text{ W}$$

When a resistance is present, the opposition to the current flow is no longer simple reactance or resistance. Rather, a new term must be employed to represent the condition where a combination of resistance and reactance is present. This term is *impedance*. The symbol for impedance is the capital letter Z. Since impedance is a function of the angle between 0 and 90°, it can be set down as a right triangle function, as represented in Fig. 8-3. Thus, if the inductive reactance is 4 Ω and the resistance is 3 Ω, the angle shown would be less than 90° and, therefore, the impedance would be proportional to the hypotenuse marked Z in Fig. 8-3. The electric formula for this hypotenuse calculation is as follows:

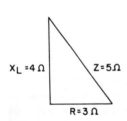

FIGURE 8-3. Impedance triangle.

$$Z = \sqrt{R^2 + X_L^2} \qquad (8\text{-}10)$$

Thus, the calculation for Fig. 8-3 would be

$$Z = \sqrt{4^2 + 3^2} = \sqrt{16 + 9} = \sqrt{25} = 5 \text{ Ω}$$

Thus, when reactance and resistance are involved, the reactance and resistance values cannot be added together for total opposition. If this had been done in the previous example, it would have indicated a value of 7 Ω for the opposition instead of the *true* value of 5 Ω.

The resistance of the coil can be represented schematically, as shown in Fig. 8-4. Here the coil resistance is indicated as a series resistor, for the purposes of calculation. The same schematic would also represent a coil with negligible resistance but in series with a fixed resistance value. The total current for the circuit in Fig. 8-4 can still be calculated by using Ohm's law.

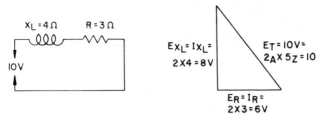

FIGURE 8-4. Impedance triangle using voltages.

$$I = \frac{E}{Z} = \frac{10}{5} = 2 \text{ A}$$

The same calculation holds true for finding the voltage drop across the resistor.

$$E_{X_L} = IX_L = 2 \times 4 = 8 \text{ V across } X_L$$

$$E_R = IR = 2 \times 3 = 6 \text{ V across } R$$

Total voltage can be found by multiplying the value of total current by the value of the impedance.

$$E_T = IZ = 2 \times 5 = 10 \text{ V}$$

The voltage drops across the individual units of inductance and resistance would be proportional to their respective values. Thus, if 2 A of current flow in the circuit, multiply the 4 Ω of inductive reactance by two to obtain the voltage drop across the inductance. The same multiplication factor of 2 A will also apply to the resistance. Thus, the voltage drop across inductive reactance differs with respect to the voltage drop across the resistance by the factor established by the multiplier two. This is shown in Fig. 8-4, where 8 V is indicated as dropping across X_L and 6 V across R. This can be solved by the next formula:

$$E_T = \sqrt{E_R^2 + E_{X_L}^2} = \sqrt{6^2 + 8^2} = \sqrt{36 + 64}$$

$$= \sqrt{100} = 10 \text{ V}$$

Thus, the total voltage in a *series* circuit composed of reactance and resistance can be solved by the vector formula, in similar fashion to the solving for total impedance.

The previous circuit can be made more complex if the inductance *reactance* is not known beforehand. If, for instance, only the *inductance* in henrys were given, as shown in Fig. 8-5, the initial calculation would consist in first solving for the amount of inductive reactance and then applying the answer to the vector formula for

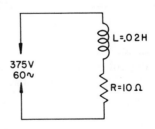

FIGURE 8-5. Series circuit, inductance and resistance.

finding total impedance. The inductive reactance amounts to 7.5 Ω and this brings the impedance to 12.5 Ω.

$$X_L = 2\pi fL = 6.28 \times 60 \times 0.02 = 7.5 \ \Omega$$

$$Z = \sqrt{10^2 + 7.5^2} = \sqrt{100 + 56.25}$$

$$= \sqrt{156.25} = 12.5 \ \Omega$$

Once the ohmic value of the total impedance is known, the current flow through the circuit can be found by applying Ohm's law.

$$I = \frac{E}{Z} = \frac{375}{12.5} = 30 \ \text{A}$$

When several series resistors are in the circuit, the total resistance is the sum of the individual resistor values.

As with the reciprocal values of resistance and reactance (conductance and susceptance), a reciprocal of impedance is also in use and is termed *admittance*. The symbol is Y ($Y = 1/Z$) and it is the ability of a circuit to pass ac. As with conductanc and susceptance, the unit value is the *mho*. Reciprocal forms are often useful for certain calculations involving circuit characteristics as shown in subsequent chapters.

8-7.   Parallel Reactance and Resistance Combinations

When a resistor shunts a coil in parallel, as illustrated in Fig. 8-6, Equation 8-10 does not apply. In a parallel circuit, the voltage drop across each unit is the same but the currents differ. For this reason, the calculation of impe-

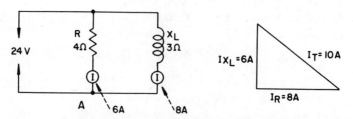

FIGURE 8-6. Parallel circuit with current triangle.

dance necessitates the calculation of total current. In order to find the total current, the current through the resistor and the current through the inductance must first be found, again using Ohm's law.

$$I_R = \frac{E}{R} = \frac{24}{4} = 6 \ \text{A}$$

$$I_{X_L} = \frac{E}{X_L} = \frac{24}{3} = 8 \ \text{A}$$

Once the currents through the individual units have been found, the total current must be calculated by vector addition.

$$I_{\text{Total}} = \sqrt{I_R^2 + I_{X_L}^2} = \sqrt{6^2 + 8^2}$$
$$= \sqrt{36 + 64} = \sqrt{100} = 10 \text{ A}$$
$$Z = \frac{E}{I} = \frac{24}{10} = 2.4 \, \Omega$$

From the foregoing, the total current was found to be 10 A, as proved by vector addition, and not 14 A, which would be the case if simple addition of currents were involved as for resistors. Once the total current has been found, the impedance can be ascertained by utilizing Ohm's law to give 2.4 Ω, as shown in the last example, for the circuit in Fig. 8-6. The vector diagram of the individual currents and the resultant total current, as represented by the hypotenuse, is also shown in Fig. 8-6.

When the values of resistance and reactance are known in a parallel circuit, the impedance can be ascertained, *even though the source voltage value is unknown.* The respective resistance and reactance values remain unchangd for various voltages impressed across them, though the current flow through each component will change for a change in applied voltage. Thus, impedance can be calculated by *assuming* a voltage and then solving for the current through the resistor and the current through the reactance. The vector sum of the currents will give the total current that would flow for the particular voltage impressed, and the impedance can then be found by dividing the assumed voltage by the total current.

As an illustration, suppose the voltage across the circuit shown in Fig. 8-6 is unknown. Assuming a 72-V source, the individual currents would solve as follows:

$$I_R = \tfrac{72}{4} = 18 \text{ A}$$
$$I_{X_L} = \tfrac{72}{3} = 24 \text{ A}$$

The vector calculation for total current would then be

$$I_{\text{Total}} = \sqrt{18^2 + 24^2} = \sqrt{324 + 576}$$
$$= \sqrt{900}$$
$$= 30 \text{ A}$$

Now when the assumed voltage is divided by the calculated current, the result indicates that the impedance is 2.4 Ω, the same value obtained when the calculation was performed with a 24-V source:

$$Z = \frac{E}{I} = \frac{72}{30} = 2.4 \, \Omega$$

While an assumed voltage can be used to calculate the total impedance, in actual practice such an impedance will be present only if the applied voltage is of the proper frequency. *Because a particular frequency had to be used initially to obtain a specific reactance value, any a-c voltage that is actually applied would thus have to be of the proper frequency.* For purposes of calculations, however, the frequency factor can be ignored because it is automatically taken care of, as proved by the foregoing illustration.

If several parallel resistors are employed, as shown at the left of Fig. 8-7, the current through each resistor can be found initially and then added

FIGURE 8-7. Parallel circuits composed of inductance and resistance.

together. The total resistive current (18 A) must then be used with the inductive current (1.2 A) in the vector addition to find total current. Instead of this procedure, however, the total resistance can be solved for and total resistance current calculated.

The inductive reactance of several inductors in parallel is calculated by the same formula used for parallel resistors. Thus, as shown at the right of Fig. 8-7, if the first inductance has 10 Ω of reactance and the second inductance also has a reactance of 10 Ω, the parallel combination gives a total inductive reactance of 5 Ω. Therefore, in a circuit composed of a number of parallel resistors plus a number of shunt parallel inductors, the total inductive current as well as the total resistive current must first be found and then applied to the vector addition of such currents. Once the total current has been established, the impedance can be found by simply dividing the total voltage by the total current.

8-8.   Practical Factors (Coils)

So far in this chapter, the coils illustrated were of the air-core type; that is, they had no metallic cores. Coils are considered to have an air core even though wound on some supporting coil form since the latter has virtually no effect on the characteristics of the coil. In practical schematic drawings, an air-core coil is indicated by a series of loops, omitting from the symbol for the coil whatever coil form, if any, is employed.

The air-core inductances find application as radio-frequency *choke coils* in FM, TV, and other communications-type shortwave receivers, as well as in the corresponding transmitters. A choke coil is usually placed in series with the power-supply voltage feed line for signal isolation purposes. Since the radio-frequency (RF) choke is designed to provide a high-inductive reactance for the frequencies of the signals involved in the circuit, the choke offers opposition to such energy and prevents it from leaking away from the circuits and into the power source. Such energy leakage represents a loss and, hence, the signals must be confined to their circuits as much as possible.

Metallic cores, though increasing the inductance of a coil and improving efficiency, give rise to hysteresis and eddy-current losses, which must be minimized. Hysteresis losses are due to the rapid magnetization and demagnetization of the core when ac is applied to the coil. The cyclic changes in ac involve a constant reversal of the field intensity polarity by a reversal of the magnetizing force, as would be the case if the hysteresis loop (Fig. 6-7) were made to change from one polarity to the other in quick succession. The rapid shift in the electron spin of the atoms making up the element generates heat, just as would be the case with a resistive loss that expends power. Hysteresis losses are reduced by using magnetic core material of better quality.

Eddy-current losses are set up by circulating currents within the core material, which are caused by lines of force cutting across the conductor in such a direction as to induce a voltage in the conductor. The currents that are created do not follow any well-defined path and, hence, are termed "eddy" currents. Eddy-current losses are reduced by using a core made up of thin sheets of metal, known as *laminations*. Laminations break up the solid core structure and reduce the effects of eddy currents and heat loss factors. During the manufacturing process, the laminations are heated so that they oxidize. The oxidized laminations thus have some insulating properties for additional isolation of the individual laminations. (Varnish is also used on occasion for insulating purposes with respect to core laminations.) The laminated cores do not help reduce the hysteresis losses mentioned earlier.

Soft-iron or silicon-steel laminated cores are used for coils that operate with low-frequency signals (such as the 60-Hz line voltage or audio signals from about 30 Hz to 20 kHz). At higher frequencies, such as those involved in the radio-broadcast spectrum (550 to 1600 kHz), as well as the shortwave frequencies above the radio-broadcast band, powdered metal or ferrite is used for coil cores, as detailed for the transformers discussed later in this chapter.

FIGURE 8-8. Symbols for inductance, showing variety of cores.

Symbols for various coils are shown in Fig. 8-8. At B an iron-core coil is shown, with the core indicated by two or three parallel lines (or dashed lines as at C). A coil having a movable iron core so that the inductance can be varied over a given range is shown at D and E. Either a single or a double arrow may be employed. A coil in which the inductance is variable, by virtue of a portion of the coil wound on a movable form, is shown at F.

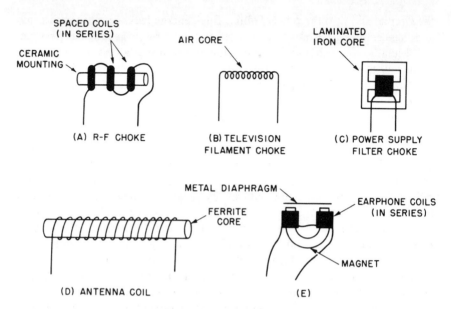

FIGURE 8-9. Typical commerical inductances.

Some typical commercial inductances are illustrated in Fig. 8-9. An RF choke consisting of three individual coils linked together to provide the necessary inductive reactance is shown at A. A nonmagnetic material is usually utilized as the supporting form. The coil shown at B has no core or coil form and is usually wound with a sufficiently heavy wire (size 18 or 20) to be self-supporting. At C is shown a typical filter choke. Note that the core has three legs with the coil positioned around the center leg. With this type of core, the laminations extend all around the core, for greater effectiveness in increasing inductance. The coil shown at D is of the type employed as the antenna in portable radios. It is wound over a ferrite core, for high permeability and efficiency. At E two coils are employed in conjunction with a magnet and a metal diaphragm, which make up the internal structure of an earphone.

8-9. The Transformer

When two coils are brought into close proximity so that the magnetic lines of force of one coil link with those of the other, as shown at A of Fig. 8-10, the device is known as a *transformer*. A transformer usually consists of a primary winding, to which the a-c energy is applied, and a secondary winding. The a-c energy can be the 60 Hz, 110 V derived from the power mains, or it can be in the form of audio- or radio-frequency signal energy. With ac applied to the primary, a voltage is induced into the secondary, and a current flow

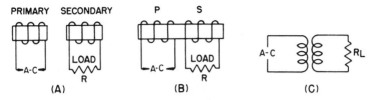

FIGURE 8-10. Basic transformers.

will occur if the secondary is a closed circuit as shown in Fig. 8-10. The induced voltage is generated by the magnetic lines of force cutting the coil turns of the secondary winding. Thus, energy is also drawn from the primary winding and this is the energy that is dissipated in wattage across the load resistor. The only other resistance that would consume energy in a circuit of this type is the internal resistance of the coil windings. As mentioned earlier, coil reactance does not consume power.

Transformers are extensively used in all electric and electronic applications. Transformers are employed in instances where it is necessary to have a higher voltage than is available from a given a-c source or in such cases where it is desirable to convert an existing a-c voltage to one of lower value. A transformer is also useful to couple the signal energy in one amplifier stage to that of another amplifier stage. In such an application, the transformer will isolate any dc between the circuits since a transformer will not pass dc. As dc does not produce a changing magnetic field that can cut across the secondary windings, it cannot induce a voltage into the secondary winding. Transformers are also useful for impedance matching, as more fully discussed later in this chapter.

While voltage is proportional to the number of turns, the available current is inversely proportional to the number of turns. As current availability is increased, voltage is decreased and vice versa. The power delivered by the secondary is equal to the power drawn from the a-c source only when the secondary winding is wound around or over the primary winding. This is known as *tight coupling* or *overcoupling* and all the lines of force cut both the primary and secondary coils. In such a transformer of good design, efficiency as high as 95% can be obtained.

When the primary and secondary windings are wound closely together so that the magnetic lines of each are linked (coupled), the total inductance depends on the individual inductances of the primary and secondary as well as the *mutual inductance*. The last represents the inductance established when the lines of force of one coil are permitted to cut across the turns of the coupled coil. Mutual inductance is defined as follows: *When ac of 1 A in the primary induces 1 V of ac in the secondary, the two inductances have a mutual inductance of 1 henry*. The formula, when all magnetic lines of the primary cut the secondary (such is the case with an iron-core transformer, where one coil layer is wound over the other), is as follows:

$$M = \sqrt{L_1 L_2} \qquad (8\text{-}11)$$

where M is the mutual inductance (when magnetic lines of force link both coils).

In coils such as shown at A and B of Fig. 8-10, the secondary can be spaced some distance from the primary. In this case, the coupling is known as a "loose coupling." Degrees of coupling have a pronounced effect on the amount of energy supplied to the secondary; this factor is discussed more fully in Chapter 10 when resonant circuits employing transformers are covered.

If loose coupling is employed (where all the lines of force of the primary *do not* cut through the secondary winding), the degree of coupling known as the *"coefficient of coupling"* (k) must be taken into consideration. The formula for mutual inductance now becomes

$$M = k\sqrt{L_1 L_2} \qquad (8\text{-}12)$$

where k is the coefficient of coupling.

The coefficient of coupling actually represents the *percentage* of coupling. Thus, if only one-fourth the lines of force of the primary intercept the secondary windings, the coefficient of coupling is 25%. The formula for the coefficient of coupling is

$$k = \frac{M}{\sqrt{L_1 L_2}} \qquad (8\text{-}13)$$

8-10.   Practical Factors (Transformers)

The transformer shown at A and B of Fig. 8-10 can be represented as shown at C. The schematic does not indicate on what form the coils are wound; it simply shows that some nonmagnetic material is used, such as cardboard, plastic, or rubber, as a support for the coil forms. Neither does the schematic representation indicate the *degree* of coupling. The number of turns, the degree of coupling, and other related data would have to be supplied in the text material that explains the schematic.

If an iron or other metal core is provided, the schematic representation of the transformer is as shown in Fig. 8-11. The usual practice is for the left winding to represent the transformer primary and the right winding, the secondary. When the secondary has more turns than the primary, the schematic representation is as shown at the left in Fig. 8-11. When the number of turns is equal for primary and secondary, the drawing is as shown at the right. When the secondary has less turns than the primary, the drawing at the left would be reversed, as shown later. When several secondary windings are present, they are shown schematically as in Fig. 8-12, again drawing them to the right of the primary winding.

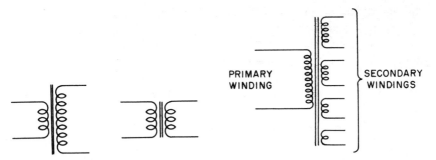

FIGURE 8-11. Metal-core transformer symbols.

FIGURE 8-12. Transformer with four secondary windings.

A transformer, with or without a core, can also be formed as shown in Fig. 8-13. Here a continuous winding is employed, which has a tap somewhere along the windings. This tap forms a junction with one a-c input lead and one side of a load resistor. The a-c voltage is thus applied across one portion of this transformer, at A and B. Since the lines of force, as generated in the winding from A to B, will cut across the winding from B to C, the section from B to C represents the secondary, and it is across this section that the load resistor, or any other device that receives the energy, is connected. This type of transformer is known as an *autotransformer.*

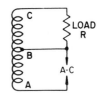

FIGURE 8-13. Autotransformer.

Various metal-core transformers are shown in Fig. 8-14. At A is shown

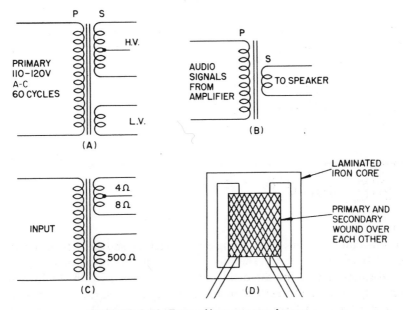

FIGURE 8-14. Types of iron-core transformers.

the type of transformer used in power supplies of electronic devices. This type of transformer is utilized to increase the power main ac to a higher voltage, for conversion to dc, as more fully explained in Chapter 14. An additional secondary winding is also employed to reduce the line voltage to a lower value, for application to transistors or to the filaments of vacuum tubes. In such a transformer, the primary and secondaries are wound over each other.

At B of Fig. 8-14 is shown a typical audio output transformer using a laminated metal core. The audio signals are impressed on the primary, and a low-voltage secondary is used for application to the loudspeaker. Such a transformer provides a match between the impedance of the vacuum tube or transistor circuit and the loudspeaker impedance. Some audio output transformers employ several secondary windings, as shown at C, so that the audio amplifier may be matched to any one of several speaker impedances. All the foregoing transformers employ a laminated iron core, as shown at D. These laminations are usually E-shaped and when the arms of the E sections are interleaved, they form the transformer core. Both the audio and power transformers are often encased in a heavy metal shield to minimize stray fields.

A transformer not only has the ability to step up or step down voltages but is also frequently used to provide an impedance match between two circuits. The turns ratio necessary to secure an impedance match can be calculated by use of the following:

$$\text{Turns ratio} = \sqrt{\frac{Z_1}{Z_2}} \qquad (8\text{-}14)$$

Typical air-core and variable-core transformers are shown in Fig. 8-15. The air-core types find primary usage in RF amplifier stages of communications equipment. The proper symbol designation is shown at A of Fig. 8-15,

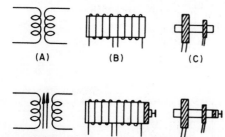

(A)   (B)   (C)

(D)   (E)   (F)

FIGURE 8-15. Air-core and iron-core transformers.

and this coil may consist of a single layer such as shown at B or a bunch-wound-type coil as at C. The efficiency and permeability of the air-core transformers can be increased by using a core, just as with the lower-frequency types discussed earlier. In RF amplifier circuits, the core consists of a circular "slug" of ferrite or powdered iron, which has been so shaped by using a binder material. Such a core is usually variable and can be adjusted to change its position with respect to the coil. By thus varying the transformer permeability, it can be "tuned," as more fully described in Chapter 10.

The symbol for a variable-core transformer is shown at D. Again, as with the air-core types, this may consist of a single-layer transformer, as shown at E or bunch wound as at F.

8-11. Coils in Series and Parallel

Coils can also be linked together by wire, without linking the magnetic lines of force that surround each coil. Two or more coils can thus be joined in series, as shown at A of Fig. 8-16. When such coils are strung in series, with

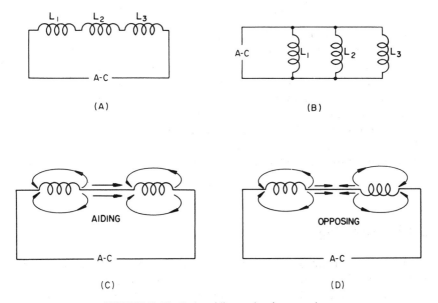

FIGURE 8-16. Series aiding and series opposing.

sufficient separation so that the magnetic lines of force do not intercept adjacent coils, the total inductance is the sum of the individual inductances, which is expressed mathematically in the following equation:

$$L_T = L_1 + L_2 + L_3 + \cdots + L_n \qquad (8\text{-}15)$$

Coils can also be placed in parallel, as shown at B. Again, they must be sufficiently separated so that the magnetic lines of force of one coil do not link with the others. The total inductance, in such an instance, is given by the formula:

$$L_{\text{Total}} = \frac{1}{1/L_1 + 1/L_2 + 1/L_3 + \cdots + 1/L_N} \qquad (8\text{-}16)$$

It will be noted that the calculations for total inductance of coils in series or in parallel are similar to the calculations for resistors in series or in parallel. The sum of the individual inductances gives the total inductance at A, in similar fashion as the sum of individual resistances is used to calculate the total resistance in a resistive circuit. Similarly, parallel inductances, as at B, are solved on the basis of the reciprocal function as with parallel resistors.

If the coils shown at A are close together so that linkage exists between the lines of force and adjacent coils, the mutual inductance that is established must be taken into consideration in order to solve for total inductance. Thus, the individual inductances of the coils are added together, and to this sum is added a factor consisting of twice the mutual inductance. The formula is as follows:

$$L_{\text{Total}} \text{ (series aiding)} = L_1 + L_2 + 2M \qquad (8\text{-}17)$$

where M is the mutual inductance. (See Equation 8-11.)

Formula 8-17 and the other equations use the henry as unit of inductance and, for two or more coils, the sum of the individual inductances must be added to the product of twice the mutual inductance, as shown. This formula indicates the condition known as *series aiding*. This occurs when all the coils are wound in the same direction so that the lines of force can aid each other, as shown at C of Fig. 8-16. If one coil is wound in the opposite direction to the other, as shown at D, the circuit is known as *series opposing*. The formula for calculating total inductance in the series-opposing circuit is as follows:

$$L_{\text{Total}} \text{ (series opposing)} = L_1 + L_2 - 2M \qquad (8\text{-}18)$$

8-12. Shielding

As mentioned previously, coils and transformers produce magnetic fields when current flows through them. The surrounding magnetic lines of force can cause undesired effects when intercepted by nearby circuit components. This comes about because of the unwanted coupling which is produced between circuits which should be electrically isolated for best performance. To minimize this condition to the point where such coupling is negligible, a metal container known as a *shield* is employed.

For transformers or individual coils handling low-frequency signals (power-line ac or audio), steel or iron shielding is usually employed. Shields of this type are more effective in intercepting the magnetic lines of force that surround the coil or transformer, and the magnetic lines of force of the latter induce into the shield an emf since the latter acts as a single-turn coil or closed circuit. Hence current flows through the shield and sets up a magnetic field that tends to oppose the original magnetic field of a coil. Thus, a shield is effective in minimizing the effects of the fields of the coil or transformer on other devices. A shield is placed as close to the coil or transformer as possible in order to conserve space. When this is done, however, other undesired effects may be created since the losses introduced by the shield may be severe. *Capacity* effects are also present, as more fully discussed later in this chapter.

Shields that are utilized for coils and transformers handling signals having high frequencies are usually made of aluminum, copper, brass, or some other nonmagnetic material of this type. These shields are more effective for the rapid rate of change encountered for signals in the high-frequency (VHF-UHF, etc.) ranges.

8-13. Effects of ac on Capacitors

In Chapter 6, reference was made to the use of a battery for charging a capacitor. If, instead of a battery, an a-c source were employed, the first alternation of the cycle would cause current to flow in one direction and charge the capacitor. This is indicated at A of Fig. 8-17, which represents the first alter-

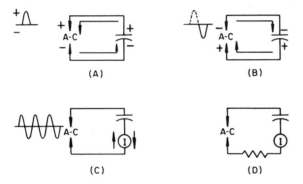

(A)　　　　　　　(B)

(C)　　　　　　　(D)

FIGURE 8-17. Effect of ac on capacitor.

nation as a positive one. The capacitor is charged, with the bottom plate negative and the top plate positive. At the second alternation, as shown, the top plate becomes negatively charged and the bottom plate positively charged. Thus the charge on the capacitor is completely reversed for each alternation of an a-c cycle. Also, the second alternation discharges the capacitor before charging it with the reverse polarity.

At C, the same circuit is shown with an a-c ammeter placed in series with the capacitor. An a-c *ammeter* will indicate the *amount* of ac flowing in a circuit by means of a needle moving up to the proper level of current indication *and remaining there*. In other words, an a-c ammeter indicator does not swing back and forth for the negative and positive polarity changes but moves in the same direction for either a negative of a positive current. Thus, the rapid rate of change in the ac at C would cause current to flow in both directions through the ammeter, but the ammeter needle would remain in a fixed position to indicate the amount of such current. If the ammeter alone were visible, and not the remainder of the circuit, the current reading of the ammeter would only indicate that a certain amount of current is flowing in the circuit, but there would be no indication of whether or not a capacitor or a resistor offered opposition to the current. Thus, inspection of the ammeter alone would *seem* to indicate that a closed circuit exists for the current and that the current flows through the ammeter and around to the voltage source and then reverses itself. This condition does not really exist but is only made to appear so by the capacitor. Electrons flow to one plate and away from the other. When the current is reversed, electrons again flow but in the opposite direction. *Actually, current does not flow through a capacitor* unless the capacitor is defective and thus has a resistive leakage permitting electron flow through it.

If a resistor were placed in series with a capacitor as shown at D of Fig. 8-17, the amount of current that flows would be reduced, and such current would flow *through the resistor* but only *to and from the capacitor.*

8-14.   Capacitive Reactance

Since the physical makeup of a capacitor (consisting of such factors as plate area, spacing, and dielectric constant) affects the capacity value of the unit, the amount of current flow to and from the capacitor is thus influenced by such factors. A smaller capacitor will permit less current flow to and from it and hence the ammeter reading of current will be lower. A larger capacitor allows more current flow to and from its plates and thus the ammeter indicates a higher current reading. Obviously, then, the capacitor is a limiting factor with respect to the current, and such a limiting factor or virtual opposition to a-c flow is known as *capacitive reactance.* As with inductive reactance, no power is consumed by the opposition offered by capacitive reactance and the latter only serves to limit the amount of a-c flow. Reference to the time-constant chart shown in Fig. 6-13 indicates that, as the capacitor is being charged, current flow starts from a high level and declines but voltage starts from a low level and increases. With ac applied to the capacitor, there is a consistent reversal of polarity across the capacitor and, in consequence, there is always a 90° phase difference between voltage and current. As with the inductances, no power is consumed in a pure capacity because $P = EI$ cos θ, and with a 90° angle the cosine is zero; hence power is zero. In a capacitor, the current *leads* the voltage by 90°.

With capacitive reactance, as with inductive reactance, consideration must be given to the angular velocity of the ac, as well as to the capacity in farads of the capacitor. An inductance tends to oppose a change of current, while a capacitance tends to oppose a change of voltage. For this reason,

capacitive reactance functions inversely to inductive reactance and the formula is

$$X_c = \frac{1}{6.28fC} \qquad (8\text{-}19)$$

Thus, the frequency (f) in Hz and the value of capacity (C) in farads will
determine the amount of capacitive reactance and also the current flow.
If the voltage drop across the capacitor is measured, as well as the current
flow, Ohm's law can be used to determine the value of the capacitive reactance. Similarly, if the voltage drop across the capacitor is known, as well
as the capacitive reactance of the capacitor, Ohm's law can be used to find
the amount of current flow, without the need for measurement with an
ammeter. Ohm's law can also be utilized to find the voltage drop, as well as
the impedance, as shown in Equation 8-20:

$$X_c = \frac{E}{I}, \qquad I = \frac{E}{X_c}, \qquad E = IX_c, \qquad Z = \frac{E}{I} \qquad (8\text{-}20)$$

A typical example of a series circuit is one composed of a resistor and
a capacitor, as shown at A of Fig. 8-18. Here, a 4-Ω resistor is in series with

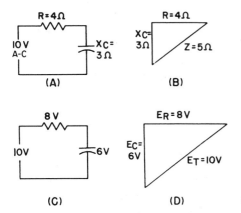

FIGURE 8-18. Series resistor-capacitor circuit and impedance triangle.

a capacitor having a capacitive reactance of 3 Ω. The vector representation
of this circuit is shown at B. As with inductive reactance in series with a
resistor, the opposition to the flow of ac in a circuit composed of a resistor
and a capacitor is no longer either resistance or reactance alone but *impedance*. Since the resistive component is at right angles to the reactive component, we solve for impedance, using a formula similar to the one employed
for inductive reactance.

$$Z = \sqrt{R^2 + X_c^2} \qquad (8\text{-}21)$$

As shown at B, the impedance for this circuit is 5 Ω, and by Ohm's law
we find the current flow equals 2 A. As with the series circuits employing an
inductance and a resistor, the series circuit composed of a capacitor and a

resistor will have, across the individual components, voltage drops of values determined by the current flowing in the circuit. Because current is the same throughout a series circuit, each resistance value and reactance value is multiplied by the current, and their product indicates the voltage across each component as shown at C. If these voltage values are known, but the applied voltage to the circuit is unknown, the total voltage can be found by use of the formula

$$E_{\text{Total}} = \sqrt{E_R{}^2 + E_C{}^2} \qquad (8\text{-}22)$$

and, as shown at D, this equals 10 V.

The power factor of the circuit can also be ascertained and would indicate the true power consumed by the circuit. The angle of the vector representation shown at D can be ascertained by solving for the tangent of the angle and then using the trigonometric table in the appendix. The tangent can be solved by dividing the side opposite by the side adjacent, as shown in the top triangle of Fig. 8-19. Thus, the tangent of the angle shown at D of Fig. 8-18 can be found, and the cosine as well as the angle (θ) can be ascertained from the table, as follows:

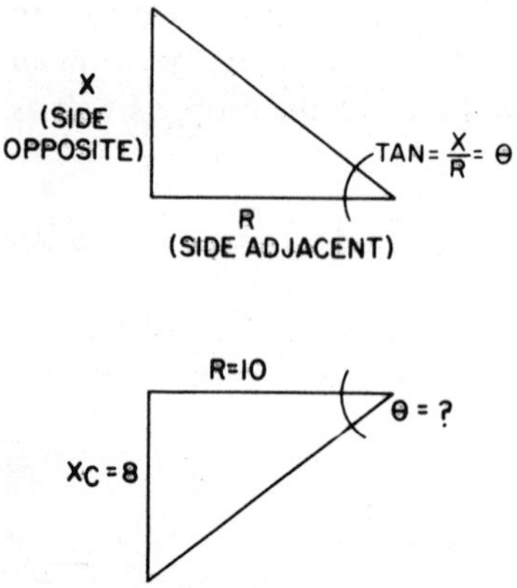

FIGURE 8-19. Solving for tangent.

$$\tan = \tfrac{6}{8} = 0.75$$

$$\cos = 0.7986$$

$$\theta = 37°$$

Using trigonometry, the impedance can also be found without the necessity for calculating by the formulas previously given. Since the cosine of an angle is the ratio of the side adjacent to the hypotenuse, the value of the side adjacent can be divided by the cosine to obtain the hypotenuse, as follows:

$$\text{Side adjacent} = 8$$

$$\text{Cosine} = 0.7986$$

and

$$Z\,(\text{hy}) = \frac{8}{0.7986} = 10\ \Omega$$

The tangent and the cosine functions of a right triangle are applicable whether an inductive reactance or a capacitive reactance is present. Thus, at the lower triangle or Fig. 8-19, the following calculation would be used, regardless of whether the 8 Ω represented capacitive reactance or inductive reactance:

$$\tan \theta = \frac{X}{R} = \frac{8}{10} = 0.8$$

and, from the appendix table, θ = approximately 39°.

When capacitors and resistors are in parallel, the method of solving for impedance is similar to that used for inductors in parallel. Initially, the current in each branch is calculated on the basis of the amount of voltage across the reactance divided by the value of the reactance. When the individual currents are known, the value of total current is found by

$$I_{\text{Total}} = \sqrt{I_R^2 + I_{X_C}^2} \tag{8-23}$$

The impedance is then calculated by dividing the value of the voltage source by the value of total current obtained from Equation 8-23.

A typical problem involving a capacitor and parallel resistance is shown in Fig. 8-20. Here two series resistors shunt a capacitor and the problem is to calculate the total impedance of the circuit.

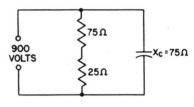

FIGURE 8-20. Capacity shunted by series resistor circuit.

Initially, the current through the resistive branch should be found. This is done by dividing the voltage by the resistance and the latter equals 100 Ω. Thus, the current flow through the resistive branch is 9 A.

The same voltage exists across the capacitor and when this is divided by the capacitive reactance, the current is found to be 12 A. Total current is the vector sum of the individual currents and is found as follows:

$$I_{\text{Total}} = \sqrt{9^2 + 12^2} = \sqrt{81 + 144} = \sqrt{225} = 15 \text{ A}$$

This indicates that 15 A of current flows to the parallel resistive-capacitive network. Impedance can then be calculated by Ohm's law, as follows:

$$Z = \frac{E}{I} = \frac{900}{15} = 60 \text{ } \Omega$$

As with parallel circuits using inductance, as discussed earlier in this chapter, the impedance can be obtained even if the voltage across the network is unknown. In Fig. 8-20, for instance, suppose the voltage were unknown. Any voltage can be assumed and the current through the resistive and capacitive branches solved. (These will be false currents but will be proportional to the opposition offered; hence they will yield accurate calculation of impedance.) Thus, if 1800 V is assumed, the current through the resistance would be equal to the assumed voltage divided by 100 Ω and would therefore be 18 A. The current value for the capacitive reactance equals 24 A. Vector calculation for total current gives 30 A. Dividing 1800 V by 30 A results in an impedance value of 60 Ω (the correct value).

It must be remembered, however, that this assumption of a source voltage only applies if *reactance* is known. If the reactance is unknown, it must be calculated, based on the value of capacity and the frequency of the a-c voltage. Once the particular reactance is found for the specific frequency used, the assumption of any voltage value will provide the right answer since the known reactance was derived for a particular frequency.

As with the inductive reactance described in Sec. 8-2, the reciprocal of capacitive reactance is also susceptance ($B_C = 1/X_C$) and the unit value is again *mho* as was the case for conductance and admittance. Thus, the susceptance of a capacitor indicates its ability to permit alternating current to flow in the circuit (remembering that current flows to and from the capacitor but not through it unless the capacitor is defective and has resistive leakage).

8-15.   Commercial Capacitors

There are various methods for producing capacitors for use in electronics. Some capacitors have dielectrics formed from mica, while others use ceramic, paper, or a chemical electrolyte. Others are specially processed to become an integral part of a printed circuit, chip, or module of the type discussed later. Such capacitors are referred to as *fixed capacitors*, which means that their capacitance value cannot be varied and each capacitor has a value that was established in the design by the manufacturer. The value of the capacitor is stamped on the larger capacitors but the smaller mica and ceramic capacitors have their value marked on them by a color code (see appendix).

Paper and electrolytic capacitors have the voltage as well as the capacitive value marked on them. The voltage rating is the maximum voltage that can safely be applied across the capacitor without danger of internal arcing and damage resulting from a short circuit. Electrolytic capacitors sometimes have two voltage ratings printed on the housing. One of these ratings is known as the *peak voltage* and refers to the maximum *short application* voltage that the capacitor can withstand. The other voltage rating is the *working voltage*. Two ratings are preferable because of the voltage variations encountered by these capacitor when used in power supplies, where the voltage rises to a high value (peak) when first turned on but drops to a "working" value after the tubes of the receiver, transmitter, or other electronic device warm up and draw current. Thus, the peak-voltage rating of a capacitor is the voltage that can be impressed on the capacitor for a short interval only, while the working voltage is the maximum under which the capacitor can operate continuously without danger of breakdown.

So-called mica and ceramic capacitors are low-value units ranging from a few pF upward to about 500 pF. Their primary function in circuitry is to bypass or couple high-frequency signals in the RF regions. For lower-frequency signal applications, including the audio type, paper-dielectric capacitors are sometimes used with values ranging as high as 0.5 μF. In solid-state circuitry where inherently low impedances require larger capacitances, low-voltage electrolytics are used, ranging from a few μF to 50 μF or more. The paper, ceramic, and mica capacitors are illustrated in that order in Fig. 8-21.

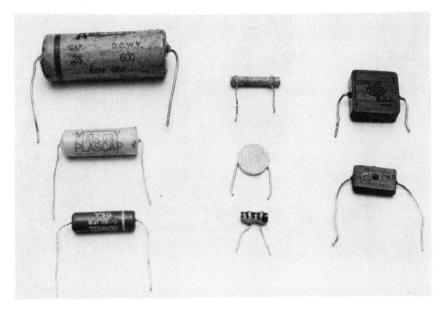

FIGURE 8-21. Paper, ceramic, and mica capacitors.

Electrolytic capacitors are also used as ripple filters in power supplies. Such capacitors have polarity markings and must not be connected into the circuit in opposition to the manner in which they are marked or serious damage will result. Electrolytic capacitors have an electrolyte of oxide or chemical composition and will pass current in one direction but very little current in the other (that is, they have a high resistance in one direction and low resistance in the other). Thus, particular attention must be paid to the polarities of electrolytic capacitors. Some paper capacitors are marked to indicate the lead that should go to the chassis since it is connected to the outer foil of the capacitor and, hence, will act as a shield. Reversal will not cause damage but may impair circuit function. Reversal of an electrolytic capacitor, however, will cause high current leakage and short circuit conditions. Typical electrolytic capacitors are shown in Fig. 8-22. Some are available in cardboard containers, while others have a metal can for shielding purposes. In some types of capacitor, the shield container can be used as the negative section. In other capacitors, the negative and positive terminals are brought from the capacitor and isolated from the shield. The black wire is usually the negative lead, while the red wire is the positive lead in electrolytics.

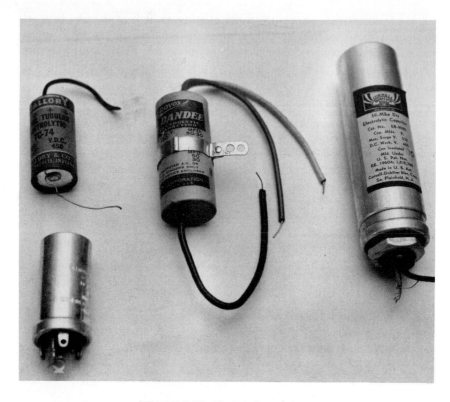

FIGURE 8-22. Electrolytic capacitors.

When a capacitor is needed in circuit applications where a higher voltage is present than that for which a particular capacitor is rated, two such capacitors can be placed in series to double their operating value. Thus if two 60-V capacitors are placed in series, the combination can withstand 120 V. When capacitors are placed in series, however, the total capacity is reduced. When two or more capacitors are placed in series for higher-voltage applications, each should have the same capacity value. Thus, two 0.04-μF capacitors can be placed in series and the resultant capacity will be 0.02 μF.

When increased capacitance is required over that available from a given capacitor, two or more can be placed in parallel. Thus, if two 10 μF units are paralleled, the total capacitance becomes 20 μF. The parallel combination, however, does not increase the voltage rating. Thus, two or three 35-V capacitors in parallel will still have a 35-V rating. Series parallel combinations can be used for increasing both the capacitance and the voltage, if required.

Variable capacitors are shown in Fig. 8-23 and are used for tuning purposes in transmitters and receivers. The miniature types are employed for high-frequency circuitry or for compensating for slight tuning differences in the larger capacitors. The small units can have the same capacitance as the

FIGURE 8-23. Variable Capacitors

larger by spacing plates closer together. For high-voltage operation, however, as in transmitters, wider plate spacings are required; hence larger units are necessary.

Variable capacitors have a set of stationary plates and a set of variable plates. The variable plates mesh with the stationary plates to a degree established by rotating the shaft. Air is the insulating dielectric utilized between the stationary and the rotating plates. The stationary plates are called the *stator* of the variable capacitor and the rotating plates are referred to as the *rotor*. Variable capacitors can be ganged, as shown in Fig. 8-23, so that control can be obtained over several circuits. The schematic representation of a single variable capacitor is shown at A of Fig. 8-24. The arrow identifies the variable arm, or rotor, of the capacitor (the rotor is usually placed at ground potential). When two or more capacitors are ganged together to permit simultaneous tuning, the ganging linkage is indicated schematically as shown at B by the dotted-line section. The older representation for the variable capacitor is as shown at C with the curved arrow indicating the variable portion.

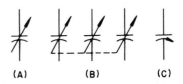

(A)   (B)   (C)

FIGURE 8-24. Variable capacitor symbols.

8-16.   Undesired Capacitances

As previously mentioned, a capacity can be established between any two conductors separated by an insulation. Thus, small capacity values exist between the hookup wire and adjacent wiring or the chassis of an electronic device. While such capacities are extremely small in value, they can have a low-shunt reactance at higher frequencies. Thus, the higher the frequency, the lower the capacitive reactance and, hence, the greater the amount of signal energy that may be shunted instead of transferred to the desired points. To minimize such *stray* or *shunt* capacitances, short leads are utilized at high frequencies and these leads are carefully spaced away from each other. The proper positioning of hookup wires in electronic devices is known as *lead dress*.

A coil or a transformer also has a considerable amount of internal capacity established between adjacent turns and between adjacent layers of wire. Since each turn of wire is insulated from the adjacent turn, the insulation material acts as a dielectric and the wires behave as miniature capacitor plates. Such capacity in a transformer is known as *distributed capacity* and represents another possible high-frequency signal loss. The higher the signal frequencies, the lower is the capacitive reactance they will encounter in the distributed capacity of a coil or transformer. Thus, a low value of capacitive reactance caused by distributed capacities has a considerable shunting effect on high-frequency signals.

The distributed capacitance in a coil also alters its inductive character-istics since the coil is no longer a pure inductance but now consists of induc-tance and capacity. Hence, the opposition to the flow of a-c energy must be calculated on the basis of the impedance established by the distributed capa-city, the inductance, and whatever resistance exists in the length of wire used in the coil section. (Combinations of *L, C,* and *R* are discussed more fully in the next chapter.)

When coils are shielded so that the magnetic fields of one coil do not interfere with those of an adjacent coil, the effects of capacity also influence circuit behavior. Because the metal shield is at ground potential, any capacity between the shield and the wire of the coil provides a shunt capacity; this shunting effect of some of the signal energy is particularly pronounced at the higher-signal frequencies because with increasing frequency the capacitive reactance decreases and offers less opposition to the transfer of signal energy. Even with adequate spacing between the coil and the shield, the capacity effects at higher frequencies must be considered during design.

8-17.   Measurement of *L* and *C*

The inductance of a coil, in henrys, or the value of capacitors, in microfarads, can be ascertained by utilizing the bridge circuit principle discussed in Chapter 5 and illustrated in Fig. 5-5 for resistance measurements. For measurements of *L* and *C*, however, an a-c source must be available, as shown in Fig. 8-25. The inductive circuit shown at A functions similarly to the resistive type discussed in Chapter 5, except that a-c voltage drops develop across the

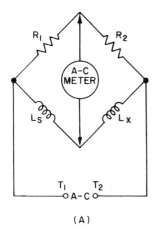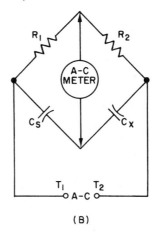

FIGURE 8-25. Wheatstone bridges for measuring *L* and *C*.

respective reactances of L_s and L_X (the standard and the unknown inductance). Instead of a galvanometer, an a-c reading device must be used. For the circuit at B the a-c voltage drops develop across the capacitive reactances of the standard capacitor C_s and the unknown capacitor C_X. For the inductive circuit, the unknown is found by multiplying the value of the standard inductance L_s by the ratio R_2/R_1 when the bridge is balanced. For the capacitive circuit, C_X is found by multiplying the value of the standard capacitor C_s by the ratio R_1/R_3 when bridge balance is achieved.

The formulas for the two bridge circuits differ because capacitive reactance acts inversely to inductive reactance. Even though the bridge balance is achieved by virtue of the voltage drops that occur across the resistances and reactances, the reading procured indicates the value of inductance and capacity. As an example, assume that for the circuit shown at A of Fig. 8-25, resistor R_1 has a value of 5 Ω and R_2 has a value of 10 Ω. The standard inductance L_s has a value of 2.5 henrys. The value of the unknown inductance L_X is then found by use of Equation 8-24, which indicates a value of 5 henrys:

$$\frac{R_2}{R_1} L_s = L_X$$

$$\frac{10}{5} 2.5 = 5$$

(8-24)

8-18. Comparison of *R, C,* and *L* Values

For comparison purposes, the calculations of total values of resistance, inductance, capacity, and reactance are listed by categories. These will aid in understanding the characteristics involved in series and parallel systems and will also be invaluable for review purposes.

TOTAL VALUE OBTAINED BY SIMPLE ADDITION

Resistance of resistors in series
Voltage of batteries in series
Inductance of coils in series (not coupled)
Inductance reactance of coils in series (not coupled)
Capacitive reactance of capacitors in series
Capacity of capacitors in parallel
Current of resistors in parallel

TOTAL VALUE OBTAINED BY USING RECIPROCAL FORMULA

Resistance of resistors in parallel
Capacity of capacitors in series
Capacitive reactance of capacitors in parallel
Inductive reactance of coils in parallel (not coupled)
Inductance of coils in parallel (not coupled)

TOTAL VALUE OBTAINED BY VECTOR CALCULATION

Impedance of inductance and resistance in series
Impedance of capacitance and resistance in series
Voltage across circuit composed of inductance and resistance in series
Voltage across circuit composed of capacitance and resistance in series
Current of a circuit composed of capacity and resistance in parallel
Current of a circuit composed of inductance and resistance in parallel

Review Questions

8-1. If the frequency of the ac across an inductance is increased, does the inductive reactance increase or decrease? Explain your answer.

8-2. Briefly explain what is meant by the *angular velocity*.

8-3. Briefly explain why the cosine of the angle of lead or lag between voltage and current must be employed in calculating true power.

8-4. What is the formula for finding the impedance of a circuit composed of a series resistor and inductance?

8-5. Explain how the impedance can also be found by using trigonometric relations of cosine, hypotenuse, etc.

8-6. In a circuit composed of a resistor and an inductor in parallel, what method can be employed for calculating the impedance of the circuit?

8-7. Where are RF chokes commonly employed?

8-8. A parallel circuit contains a known value of resistance and a known value of inductive reactives. Values of voltage and current are unknown. What process can be employed for finding the impedance?

8-9. Briefly explain what causes hysteresis and eddy-current losses. How may such losses be minimized?

8-10. Briefly explain what is meant by *mutual inductance.*

8-11. If ac of 10 V is applied to the primary of a transformer and 220 V is read across the secondary, what is the turns ratio of the transformer?

8-12. If three coils, each having a reactance of 30 Ω, are placed in parallel, what is the total reactance?

8-13. Briefly explain the necessity for shielding coils and transformers in some applications.

8-14. How does the addition of a capacitor to a resistive circuit alter the opposition to current flow?

8-15. Briefly explain what physical factors contribute to the specific capacitance of a capacitor.

8-16. If the *frequency* of an a-c signal across a capacitor decreases, why does the capacitive reactance increase? How does this differ from the results of a frequency decrease of an a-c signal across an inductance?

8-17. If two resistors and a capacitor are in parallel, explain what method can be employed for calculating the total impedance of the circuit.

8-18. (a) What are the two most common applications for capacitors having only a few picofarads of capacitance?

(b) Which capacitors have polarity markings, and what is the reason for this?

8-19. If three capacitors are placed in series, how do the voltage rating and capacitance of the group compare with a single capacitor?

8-20. What are two uses for minature variable capacitors?

8-21. Why are losses in circuit wiring and in a coil increased as the signal *frequency* is increased?

8-22. Of inductance or capacitance, which tends to oppose a voltage change, and which tends to oppose a current change?

Practical Problems

Note: Draw a diagram of the circuit of each problem to gain practice in schematic work and to aid in visualizing each problem.

8-1. What is the inductive reactance of a power-supply filter coil having an inductance of 20 henrys and operated on a frequency of 60 Hz?

8-2. In an electronic circuit, the current lags the voltage by 40°. The current at maximum value is 50 mA. What is the instantaneous value of the current at the 60° phase of the voltage?

8-3. In a circuit composed of an inductance and a resistance, there is a 45° phase difference. If the voltage source is 200 V and the current flow is 50 mA, what is the true power used?

8-4. A resistor of 10 Ω is in series with an inductance having a reactance of 7.5 Ω. What is the impedance?

8-5. What is the total inductive reactance if two coils are in a series circuit (not coupled) and one coil has a reactance of 720 Ω, while the other has an inductance of 50 mH? The frequency at which the circuit is operated is 20 kHz.

8-6. In a series circuit composed of a coil and a resistor, there is a 24-V drop across the resistor and an 18-V drop across the coil. What is the total voltage applied across the two?

8-7. In a parallel circuit consisting of a 24-Ω resistor and an inductance having a reactance of 32 Ω, what is the impedance of the circuit?

8-8. In a parallel circuit composed of a resistor shunted by an inductance, the current through the resistor is 36 mA and the current through the inductor is 48 mA. If the impressed voltage on the circuit is 210 V, what is the impedance?

8-9. A transformer primary has 32 mH of inductance and the secondary has 2 mH of inductance. All the magnetic lines of force of the primary cut the secondary. Under these conditions, what is the mutual inductance?

8-10. A transformer must match the impedance of a 4-Ω speaker to a circuit impedance of 14,400 Ω. What must be the turns ratio of the transformer?

8-11. An electronic circuit contained three inductors that were sufficiently separated so their lines of force did not interact. The inductance values were 39, 19.5, and 13 mH, respectively. What is the total inductance?

8-12. In an industrial control system, two coils were in series aiding. One coil had an inductance of 2 mH and the other had 13 mH, with a coefficient of coupling (*k*) of 0.2. What is the total inductance?

8-13. In an electronic testing device, a circuit contains a 250-pF capacitor in series with a 100 kΩ resistor, connected to a source voltage of 100. After the switch is closed, how long will it take for the capacitor to charge to 63 V?

8-14. In a cassette tape recorder circuit, two capacitors are in series, each rated at 100 μF, 30 working volts. What is the total capacitance and the working voltage of the combination?

8-15. In a power supply of an electronic device, there are two capacitors in parallel, one rated at 8 μF and the other at 16 μF. Each capacitor is designed for a maximum of 450 working volts. What is the total capacity and the working voltage of the combination?

8-16. In a transistor amplifier circuit, a certain capacitor should have no more than 5 Ω of reactance when a signal having a frequency of 5000 Hz is impressed across it. If the capacitor is rated at 30 μF, does it have a low enough capacitive reactance? What is the reactance value?

8-17. A particular tone control circuit consists of a resistor of 360 Ω in series with a capacitor that has a reactance of 480 Ω at 300 Hz. The total impedance is not to exceed 500 Ω. Is this condition satisfied? What is the impedance?

8-18. In a transmitter modulator, a bypass capacitor of 0.06 μF should be used. There are available three capacitors, two of 0.02 μF and one of 0.05 μF. In what combination will these provide the necessary 0.06 μF?

8-19. In an electronic circuit composed of a resistor in series with a capacitor, an a-c meter was placed across the resistor and a reading of 84 volts was obtained. The total voltage across the series circuit read 105. What voltage appears across the capacitor? What power is consumed for a current flow of 20 mA (0.02 A)?

8-20. In an electronic control circuit a resistor is in series with a capacitor. The resistor has a value of 1260 Ω and has a 25.2-V drop across it. The capacitor has an 18.9-V drop across it. What are the total voltage, total current, and impedance?

8-21. In a transistor circuit of a computer, a resistor is in parallel with a capacitor. An a-c milliammeter placed in series with the resistor reads 408 mA but when placed in series with the capacitor, the meter reads 544 mA. What is the total current? If the impressed voltage is 136, what is the impedance?

8-22. In a computer switching system a 15-pF capacitor was in series with a parallel branch of two capacitors, each rated at 60 pF. Only 100-V (working) capacitors were used. What are the total capacitance of the network and the total voltage rating?

8-23. A bridge circuit such as shown at A of Fig. 8-25 had an R_1 of 2000 Ω, an R_2 of 1500 Ω, and an inductance L_s of 16 henrys. The bridge was balanced. Under these conditions, what is the value of L_x, the inductor under test?

8-24. A bridge circuit such as shown at B of Fig. 8-25 was balanced and had the same resistor values as in Problem 8-23. C_s has a value of 16 μF. What is the value of C_x, the capacitor under test?

Circuit Analysis (AC)   9

9-1. Introduction

As discussed in the previous chapter, an inductance affects circuit characteristics by introducing a 90° phase difference between current and voltage, with voltage leading. A capacitor, on the other hand, also causes a 90° phase difference, except that current leads the voltage. Since the reactances established by inductance and capacity are opposite in their effect, a circuit which contains both an inductance and a capacitor will have a total reactance which is proportional to the difference between the two reactances. Such a difference, when combined vectorally with circuit resistances, indicates the impedance. In this chapter, circuits are analyzed that contain L, C, and R in various combinations generally encountered in electronics.

9-2. Series L-C Circuits

When an inductor and capacitor are placed in series with a resistor as shown at A of Fig. 9-1, the vector representation is as shown at B. The opposite effects of X_L and X_C cancel and the total reactance is, therefore,

$$100 - 55 = 45 \ \Omega$$

Because the capacitive reactance is greater than the inductive reactance,

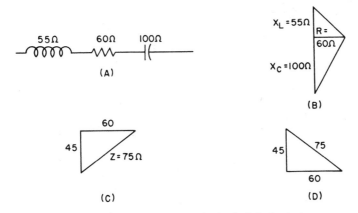

FIGURE 9-1. Vector triangles for **L-C-R** circuits.

the circuit will be predominantly capacitive, with the remaining 45 Ω representing capacitive reactance. Thus, the vector diagram would now be as shown at C. If the vector triangle is drawn to proportion, the hypotenuse length would indicate an impedance of 75 Ω.

Had the inductive reactance been 55 Ω and the capacitive reactance 100 Ω (opposite to the foregoing example), the lower value of capacitive reactance would have been subtracted from the higher value of inductive reactance. The vector representation would then be as shown at D, where the same impedance prevailed, except that the circuit is now predominantly inductive since some of the later reactance remained after subtracting from it the capacitive reactance. Formula 8-10 previously given for impedance can now be expressed as follows:

$$Z = \sqrt{R^2 + (X_L - X_C)^2} \tag{9-1}$$

This is the usual form in which the formula is indicated in textbooks, though a variation of this formula could be expressed as

$$Z = \sqrt{R^2 + (X_C - X_L)^2} \tag{9-2}$$

Either formula can be used because the parenthetical expression simply indicates that the sum representing the *difference* between the two reactances is squared.

Individual voltage drops across the reactances will still be a function of the amount of such reactances multiplied by the current flow through each of them. The voltages across the individual reactances may reach high values but since there is a difference in the phase angle between voltage and current, the total voltage is still the vector sum of the individual voltages. This is illustrated at A of Fig. 9-2. The total impedance is found by Equation 9-1 as follows:

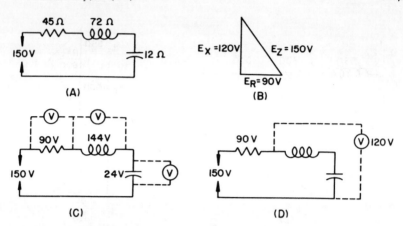

FIGURE 9-2. Series impedance circuits and voltage triangle.

$$Z = \sqrt{45^2 + (72 - 12)^2}$$
$$= \sqrt{2025 + 3600}$$
$$= \sqrt{5625}$$
$$= 75 \ \Omega$$

Since current is equal to voltage divided by impedance, the value of the current can be solved and the voltage drops indicated across each circuit component.

$$I = \frac{E}{Z} = \frac{150}{75} = 2 \ \text{A}$$

$$E_R = 2 \times 45 = 90$$

$$E_{X_L} = 2 \times 72 = 144$$

$$E_{X_C} = 2 \times 12 = 24$$

Here, the *additive* sum of the voltages would *exceed the source voltage* by a considerable amount but actually the sum of the voltages is a function of the vector formula.

$$E_{\text{Total}} = \sqrt{E_R{}^2 + (E_{X_L} - E_{X_C})^2} \qquad (9\text{-}3)$$

This formula takes into consideration the vector sum of the voltages and is similar to the impedance formula since the voltage across the lower reactance is subtracted from the voltage across the higher reactance. Thus, the total voltage in this instance will again equal the voltage indicated at A of Fig. 9-2. The vector representation of the voltage drops across the various components is shown at B. The voltage across the reactive component is 120 V since the 24-V drop across the capacitive reactance must be subtracted from the 144 V across the inductive component. Since there is a 90-V drop across the resistance, the vector calculation indicates that the voltage across the impedance is 150 V.

If an a-c voltmeter were placed across the individual components of the circuit shown at A, the voltages indicated at C would be read. If a voltmeter were now placed across the inductive and capacitive reactances, as shown at D, the total voltage read by the meter would only be 120 V because the 24 V across the capacitive reactance opposes and cancels out its equivalent portion of the 144 V across the inductive reactance.

Because $E = IX$, there are instances where the voltage drop across a reactance exceeds the source voltage. This would seem to indicate that more power is generated in the circuit than at the source but the actual voltage across the total reactance will be the difference between the various voltages. This is indicated in the circuit shown at A of Fig. 9-3. The total

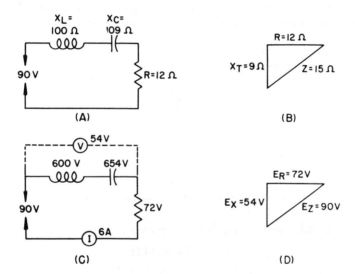

FIGURE 9-3. Series impedance circuits and impedance triangles.

reactance is only 9 ohms, as shown by the vector diagram at B. Solving for current and voltage drops across the individual components, however, indicates the following:

$$I = \tfrac{90}{15} = 6 \text{ A}$$

$$E_{X_L} = 6 \times 100 = 600 \text{ V}$$

$$E_{X_C} = 6 \times 109 = 654 \text{ V}$$

$$E_R = 6 \times 12 = 72 \text{ V}$$

Thus, the voltage drops around the circuit are as shown at C where the individual voltages across the inductive reactance and the capacitive reactance each exceeds the source voltage of 90 V. As shown at C, however, an a-c voltmeter across the two reactances would only indicate 54 V since the opposing phase differences cancel out 600 V. Thus, a vector representation of the voltage drops across the total reactance and the resistance would again indicate 90 V, the source voltage. This high-voltage drop across an individual reactive component is useful in electronics, however, since it provides a substantially high signal voltage for the particular frequency involved.

9-3. Multiunit Series Circuits

When several inductances, capacitors, and resistors are in series, the individual inductive reactances are added together to give the total reactance since reactances in series are additive. The additive principle also applies to the series capacitive reactances, as well as the series resistors. Thus, for a circuit such as shown at A of Fig. 9-4, the individual reactances and resis-

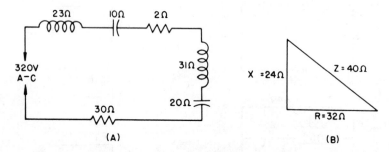

FIGURE 9-4. Complex series impedance circuit.

tances are added and then inserted in Equation 9-1.

$$X_L = 23 + 31 = 54 \; \Omega$$
$$X_C = 10 + 20 = 30 \; \Omega$$
$$R = 2 + 30 = 32 \; \Omega$$
$$Z = \sqrt{32^2 + (54 - 30)^2}$$
$$= \sqrt{32^2 + 24^2} = 40 \; \Omega$$

The vector representation of this circuit is shown at B.

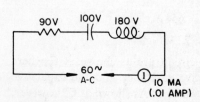

FIGURE 9-5. Series impedance circuit showing voltage drops.

By employing appropriate formulas, it is possible to solve for various unknown values if some known values are given. A typical example is the circuit shown in Fig. 9-5. The frequency of the source voltage is given but the amount of voltage is not indicated. Since the current flow is shown, however, the individual resistance and reactance values can be calculated. As the frequency of the ac is also given, the actual value of the capacity in microfarads and the inductance in henrys can be calculated. Solving for R *and* X_c,

$$R = \frac{E}{I} = \frac{90}{0.01} = 9000 \ \Omega$$

$$X_c = \frac{100}{0.01} = 10,000 \ \Omega$$

Since we know the frequency of the ac, we can transpose the formula for capacitive reactance and solve for the total capacity, as follows:

$$C = \frac{1}{6.28fX_c} = \frac{1}{376.8 \times 10,000} = 0.26 \ \mu F$$

The inductive reactance and the inductance can then be calculated in similar fashion:

$$X_L = \frac{180}{0.01} = 18,000 \ \Omega$$

$$L = \frac{X_L}{6.28f} = \frac{18,000}{6.28f} = \frac{18,000}{376.8} = 47.7 \ \text{henrys}$$

Thus, from the values given in Fig. 9-5, the individual reactances as well as capacities and inductances can be calculated. The total voltage can also be ascertained by solving for the vector sum of the individual voltages. The phase angle can be found by calculating for the tangent (dividing the total reactance by the total resistance) and when the tangent is known, the table of trigonometric values will show the degree of the phase angle involved.

$$\tan \theta = \frac{X}{R} = \frac{180 - 100}{90} = \frac{80}{90} = 0.8888$$

$$\theta = 41° \ \text{(approx.)}$$

When the frequency applied to an inductance is increased, the inductive reactance also increases, as shown by the inductive reactance Equation 8-2. Since capacitive reactance has a reciprocal function, an increase in the frequency of the a-c voltage to a capacitor will result in a decrease of capacitive reactance. When these factors are known, it is a relatively easy matter to ascertain the values of inductive reactance and capacitive reactance for frequencies double or half the given frequency. A chart can be made as follows:

Freq. (Hz)	X_L	X_C	X
60	18,000	10,000	8,000
120	36,000	5,000	31,000
240	72,000	2,500	69,500
30	9,000	20,000	11,000
15	4,500	40,000	35,500

In the foregoing, the frequency for the circuit of Fig. 9-5 is set down (60 Hz) and the inductive reactance for this frequency (18,000 Ω) is listed, as well as the capacitive reactance (10,000 Ω). The difference between these two reactances will give the total reactance, 8000 Ω. If this same circuit now had a 120 Hz ac applied to it instead of a 60-Hz ac, the inductive reactance would double and become 36000 Ω, while the capacitive reactance would be halved and become 5000 Ω. In the latter instance, the total reactance would be 31,000 Ω. A similar procedure can be undertaken for 240 Hz, which is double the 120-Hz frequency previously mentioned. Here again the inductive reactance would double and the capacitive reactance would decrease by one-half.

When the reactance values are not given, the solution of the problem is more lengthy since the individual values have to be ascertained from the inductance and capacity values. A typical example of a problem of this type is shown in Fig. 9-6. Here there are two inductances, two resistances, and two capacitors. Since both resistors have the same value, the total resistance will be 10 ohms.

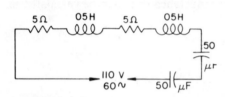

FIGURE 9-6. Complex series impedance problem.

Inductances in series provide a total inductance that is the additive sum of the individual inductances but for capacitors in series the total capacity will be less than any individual series capacitor. Thus, the total values of the individual components are as follows:

$$L = 0.05 + 0.05 = 0.1 \text{ henry}$$

$$R = 5 + 5 = 10 \ \Omega$$

$$C = \frac{50 \times 50}{50 + 50} = 25 \ \mu\text{F}$$

Once these values have been ascertained, the individual reactive values are calculated, using the formulas previously given:

$$X_L = 6.28fL = 6.28 \times 60 \times 0.1$$

$$= 37.68 \ \Omega$$

$$X_C = \frac{1}{6.28fC}$$

$$= \frac{1}{6.28 \times 60 \times 25 \times 10^{-6}}$$

$$= \frac{1}{942 \times 10^{-5}} = \frac{10^5}{942} = 106 \ \Omega$$

After the individual reactive components have been ascertained, they are combined with the resistance in the formula (Equation 9-1) giving the impedance in a series circuit composed of inductance, capacity, and resistance. This shows that approximately 68.6 Ω of impedance is present for the circuit shown in Fig. 9-6:

$$Z = \sqrt{100^2 + (37.68 - 106)^2}$$
$$= \sqrt{10,000 + 4668} = 121 \ \Omega$$

Once the value of total impedance is known, total current is found:

$$I = \frac{E}{Z} = \frac{110}{121} = 0.9 \text{ A}$$

Knowing this value, the voltage drops across the individual components can then be calculated.

The value of unknown reactance can also be found, provided the voltage source and the voltage drop across the other components are known. A typical example is shown in the simple series circuit of Fig. 9-7, where a 4-Ω resistor is in series with an inductance whose reactance is unknown. The source voltage and the voltage drop across the resistance are known, however, and hence the current is

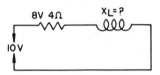

FIGURE 9-7. Simple series circuit with inductive reactance unknown.

$$I = \tfrac{8}{4} = 2 \text{ A}$$

Once it is known that 2 A of current flows, the impedance can be found by dividing the total voltage value (10) by the current (2). This indicates that the impedance is 5 Ω. The vector formula for solving for the unknown reactance can now be used, as follows:

$$Z = \frac{5}{\sqrt{R^2 + X_L{}^2}} = \frac{5}{\sqrt{16 + X_L{}^2}}$$

or

$$5^2 = 25$$
$$25 - 16 = 9 \ \Omega$$
$$\sqrt{16} = 4 \ \Omega \text{ for } R$$
$$\sqrt{9} = 3 \ \Omega \text{ for } X_L$$

The value of the resistance is 4 Ω and $4^2 = 16$. Since this 16 *plus another number* equals the square of 5, it is obvious that 16 plus the unknown number equals 25. Thus, the unknown number is 9 and the square root of 9 gives 3 Ω for the unknown inductive reactance.

9-4. Parallel *L, C,* and *R* Combinations

The method for solving a parallel circuit composed of resistance, inductance, and capacity is similar to the previous parallel-circuits method discussed, except that the opposing factors of inductive and capacitive currents must be considered. When the total reactive current is known, a vector sum of the resistive and reactive currents is taken to find the total current. The imped-ance is then a function of the total voltage divided by the total current.

A typical example of an a-c circuit employing resistance, inductance, and capacity in parallel is shown in Fig. 9-8. Since the voltage across each com-

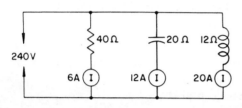

FIGURE 9-8. Parallel impedance cir-cuit giving current values with 240-volt ac source.

ponent is 240 V, the individual currents are calculated and the total current found as follows:

$$I_R = \frac{240}{40} = 6 \text{ A}$$

$$I_{X_c} = \frac{240}{20} = 12 \text{ A}$$

$$I_{X_L} = \frac{240}{12} = 20 \text{ A}$$

$$I_{\text{Total}} = \sqrt{I_R^2 + (I_{X_L} - I_{X_C})^2}$$
$$= \sqrt{36 + (20 - 12)^2}$$
$$= \sqrt{36 + 64} = \sqrt{100} = 10$$

The impedance as well as the phase angle can then be solved as follows:

$$Z = \frac{E}{I} = \frac{240}{10} = 24 \text{ } \Omega$$

$$\tan \theta = \frac{I_X}{I_R} = \frac{6}{8} = 0.7500$$

$$\theta = 37° \text{ (approx.)}$$

It is significant that the circuit of Fig. 9-8 will indicate the same imped-ance if the source voltage is changed. Since the current flow in each branch is proportional to the voltage divided by the opposition, the same relative proportions will hold, regardless of the source volt-age. In Fig. 9-9, the same circuit is shown except that the source volt-age has now been decreased to 120 V (one-half the original value). This

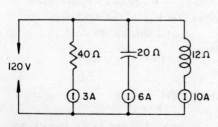

FIGURE 9-9. Circuit of Fig. 9-8 with 120-volt source.

will cause a proportionate decrease (by one-half) of the current through each branch. Thus, the current through the resistor is now 3 A, the current in the capacitive branch is 6 A, and the current through the inductive branch is 10 A. Solving for *I* and *Z*,

$$I = \sqrt{9 + 16} = \sqrt{25} = 5 \text{ A}$$

$$Z = \frac{120}{5} = 24 \text{ }\Omega$$

Thus, in a parallel circuit such as shown in Fig. 9-9, where the reactive and resistance values are given, the total impedance can be found by assuming a source voltage and calculating on the basis of the total current, using the vector form, and solving for impedance by dividing the assumed voltage by the calculated current. (It must be emphasized that the current values thus derived will hold only for the voltage used in the calculation, as in previous discussions of parallel circuits in Chapter 8. The current values will change for different source voltage values but the calculated impedance will be the true impedance and will be the same regardless of the value of the source voltage assumed for calculation purposes.)

When several components are in series and such series combinations are placed in parallel with other components, the individual reactances must be added together to solve for current. A typical example of such a circuit is shown in Fig. 9-10. Here resistors R_1 and R_2 form a series circuit and the total

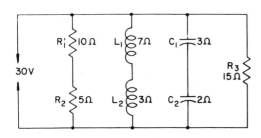

FIGURE 9-10. Parallel circuit composed of series branches.

resistance in this branch is 15 Ω. This is paralleled by another resistor (R_3), which is 15 Ω, so that the total resistance in the circuit is now 7.5 Ω. Therefore, in order to find the total resistive current, the voltage (30) can be divided by total resistance (7.5), which indicates 4 A of resistive current. Instead of this method, the current through the R_1 plus R_2 branch can be solved separately. Hence, for this combination, 30 V divided by 15 Ω gives 2 A. For R_3, 2 A is also found and adding the two resistive currents indicates the total resistive current is 4 A. The two inductive reactances of 7 Ω and 3 Ω total 10 Ω and the current thus would be 3 A. Since two capacitors in series are also in the circuit, the reactances of these two units are added together (3 + 2), which indicates that the current in the capacitive branch is 6 A (30 divided by 5). To solve for total current, the individual resistive and reactive currents are set down as follows and the total impedance is therefore 6 Ω:

$$I_{Total} = \sqrt{4^2 + (6-3)^2}$$
$$= \sqrt{16+9} = \sqrt{25} = 5 \text{ A}$$
$$Z = \tfrac{30}{5} = 6 \, \Omega$$

Review Questions

9-1. In a series circuit of resistance, inductance, and capacitance, the voltage drops across the reactances and resistances are known. How is the total voltage found?

9-2. In a series circuit of L, C, and R, how is the phase angle between voltage and current found?

9-3. (a) If the frequency and the capacitive reactives are known, what equation can be used for finding the capacitance?
(b) What formula is used for finding the capacity if the frequency and the capacitive reactance are known?

9-4. What formula can be employed for finding the inductance if the frequency and the inductive reactance are known?

9-5. What relationship of reactance to resistance will indicate the tangent of the phase angle?

9-6. How is the total a-c current calculated, on the basis of the individual currents, in the branches of a parallel circuit composed of L, C, and R?

9-7. How is the impedance calculated in a parallel a-c circuit containing L, C, and R?

9-8. Explain why the correct impedance can be found in some parallel a-c circuits containing L, C, and R, even though a source voltage value is assumed.

9-9. Explain what method is used for solving for total impedance of an a-c circuit composed of series-parallel combinations of L, C, and R.

9-10. Draw a typical circuit composed of two resistors in parallel, the latter in series with an inductance and capacitor in parallel. Assign reactance and resistor values, and show calculations for solving for impedance.

Practical Problems

9-1. In an electronic circuit identical to that shown in Fig. 9-4, it is necessary to find the total current and the phase angle. From the values given in Fig. 9-4, make these calculations.

9-2. An electronic device uses the same circuit as shown in Fig. 9-5, with identical values. Calculate the a-c voltage applied to the circuit.

9-3. A coil in an audio-filter network has an inductance of 20 henrys at 600 Hz. The distributed capacity has a reactance of 50,000 Ω at 600 Hz. What are

the *reactances* (capacitive and inductive) at the following frequencies?
(a) 150 Hz; (b) 300 Hz; (c) 600 Hz; (d) 1,200 Hz; (e) 2,400 Hz.

9-4. An electronic device employs the series L, C, R circuit illustrated in Fig.
9-11. What is the impedance of this circuit?

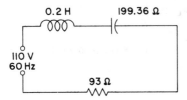

FIGURE 9-11. Illustration for Problem
9-4.

9-5. What is the apparant power and the true power for the circuit of Problem 9-4?

9-6. In an electronic device using the circuit shown in Fig. 9-12, the applied ac
had a frequency of 2 kHz. What is the total current flow in this circuit
(in milliamperes), and what is the total inductive reactance? Also, what is
the total impedance that this circuit presents to the voltage source?

9-7. What power is consumed by the 400-Ω resistor of Fig. 9-12?

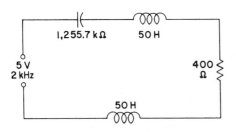

FIGURE 9-12. Illustration for Problem
9-7.

9-8. In a television receiver, the circuit shown in Fig. 9-13 was used. When a
signal of a certain frequency appears across this circuit, the opposition to
current flow is as shown. What is the total current flow in milliamperes?
What is the total impedance?

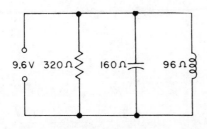

FIGURE 9-13. Illustration for Problem 9-8.

9-9. For Fig. 9-13, what power is consumed by this circuit?

9-10. The schematic for an industrial-electronic sensing device includes the network shown in Fig. 9-14. No input voltage is given. What is the total *impedance* of this circuit?

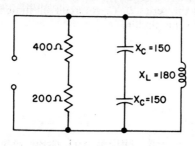

FIGURE 9-14. Illustration for Problem 9-10.

Resonance 10

10-1. Introduction

In many electronic circuits, the inductive reactance and the capacitive reactance are often equal in value. When this happens, the total reactance is zero because the inductance would cause the voltage to lead by 90° and the capacitor would cause the voltage to lag by 90°, and these two conditions oppose each other. Hence, the voltage would neither lead nor lag the current and the voltage and current would be in phase. This in-phase condition, when an inductance and capacitor are in the circuit, is known as *resonance*. This phenomenon is one of the most remarkable in electronics since it permits the selection of desired and specific signal frequencies while rejecting those not wanted. Thus, resonant circuits are extensively used in virtually all branches of electronics. Without resonance, electronic communications as we know it today would be impossible.

Resonance can be achieved in either a series or a parallel circuit composed of inductance, capacity, and resistance. When reactances are equal, the circuit not only exhibits marked selection characteristics for a specific signal or group of signals closely clustered around the resonant frequency, but it also tends to reject signals having frequencies removed from the resonant frequency, regardless of whether such rejected signals have frequencies which lie above or below the frequency which produces resonance.

This selection characteristic is known as *selectivity* and relates to the degree by which the circuit accepts desired signal frequencies and rejects undesired signal frequencies. The degree of selectivity is dependent on the amount of resistance present, as more fully detailed subsequently.

Signal reception, transmission, and other electronic applications that require the selection of a band of signal frequencies and the rejection of signal frequencies above and below this band all depend on the resonance principle. Since practically all transmitting and receiving circuits that handle RF signals employ either series- or parallel-resonant circuits, an understanding of the principles involved is of particular importance.

10-2.   Series-Resonant Circuits

At A of Fig. 10-1 is shown a simple series circuit consisting of a voltage source, a 1200-Ω resistor, an inductor having 1200 Ω of reactance, and a

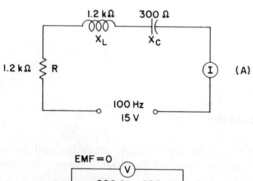

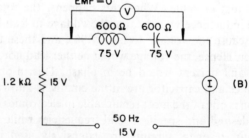

FIGURE 10-1. Series-resonant circuits.

capacitor of 300 Ω of reactance. Thus, this circuit has a total reactance of 900 Ω (1200 − 300) and when this reactive value is used in conjunction with the resistive value (in Equation 8-10), we find the *impedance* is 1500 Ω:

$$Z = \sqrt{1200^2 + 900^2} = 1500 \ \Omega$$

Now assume that this 15-V, 100-Hz signal is one of several signals applied to this circuit. For reference purposes this signal is designated as signal No. 1 and the other signals present are denoted as follows:

15-volt signal	Hz	X_L	X_C	X
No. 1	100	1200	300	900
No. 2	200	2400	150	2250
No. 3	50	600	600	0
No. 4	25	300	1200	900
No. 5	12.5	150	2400	2250

Since signal No. 2 is 200 Hz, the higher frequency will double the inductive reactance and will reduce the capacitive reactance by one-half. Total reactance is now 2250 Ω, as against 900 Ω for signal No. 1, and consequently the impedance will be still higher and less current will flow. Signal No. 3 has a frequency of 50 Hz and since this is one-half of the frequency designated for the circuit shown at A, the inductive reactance will decrease by half and become 600 Ω, while the capacitive reactance will double and also become 600 Ω. The total *reactance* is now zero and, in consequence, the impedance is equal to the resistance only, or 1200 Ω. Under this condition, the current flow is a function of the voltage (15 V) divided by the impedance (1200 Ω) and a higher current will now flow (12.5 mA) than for the other signal frequencies shown in the list. Signal No. 4 has a still lower frequency (25 Hz), which will cause the inductive reactance to decrease to a still lower value (300 Ω), and the capacitive reactance doubles again, becoming 1200 Ω. This gives a reactance of 900 Ω and again a higher impedance and lower current result.

For any frequency other than 50 Hz, the current will be lower than 12.5 mA, whether such frequency is higher or lower than 50 Hz. Thus, for the circuit shown at B of Fig. 10-1, resonance is achieved at a frequency of 50 Hz. Under this condition, maximum signal energy flows since neither the inductive reactance nor the capacitive reactance offers opposition to such flow. As the inductive reactance and the capacitive reactance are equal and opposite in their function, their effects are canceled to produce the resultant zero reactance. The voltage drop across the 1200-Ω resistor ($I \times R$) is 15 V. Since neither reactance is offering effective opposition, the full source voltage will develop across the series resistor. Voltage across each reactance is a function of IX and, therefore, amounts to 7.5 V for each. Such voltages could be read across each unit individually but a voltmeter across the two would read zero voltage since at any instant the voltage drop across the inductance is 180° out of phase with the voltage drop across the capacitance; in consequence, the voltage drop across the two combined is zero, as shown at B.

At resonance, the inductive reactance is equal and opposite to the capacitive reactance.

$$6.28fL = \frac{1}{6.28fC} \qquad (10\text{-}1)$$

This gives us the following expression:

$$6.28f^2 = \frac{1}{LC} \tag{10-2}$$

which is then converted to the following formula for finding the resonant frequency when the values of inductance and capacitance are known:

$$f = \frac{1}{6.28\sqrt{LC}} \tag{10-3}$$

Equation 10-3 indicates that the resonant frequency is established by definite values of inductance and capacitance since the product of *L* times *C* is utilized. Thus, for a *specific resonance frequency*, there can be only *one product of L times C*. Since this is a *product*, however, it is obvious that various values of *L* and *C* can be used to obtain the same product. For instance, assume that *L* is 2 μH and *C* is 4 μF, to give a product of eight. This product will produce a specific frequency even though the value of *L* is changed to 4 μH and the capacitance is changed to 2 μF since the product is still eight. Numerous other combinations of *L* and *C* to give a product of eight are possible and the chart shown below indicates a few of these.

L	C	LC
2	4	8
4	2	8
8	1	8
0.5	16	8

Even for the same product, however, a change in the values of *L* and *C* will alter circuit selectivity as detailed in the discussion that follows. The relationship between *L* and *C* that gives a constant product for a given frequency is known as the *L/C* ratio because the same product is procured for various ratios of *L* to *C*.

10-3.   Circuit Q

The *Q* of a circuit relates to the degree of selectivity realized from a combination of inductance, resistance, and capacity, and it is a figure of merit. The *Q* of a circuit is based on the effect that circuit resistance has on the ability of the circuit to reject frequencies on each side of the resonant frequency.

On occasion, it is desirable to have a graph of the resonant characteristics of a circuit so that the circuit current value with respect to a particular frequency is easily seen. If the resonant-frequency characteristics of a series-resonant circuit are plotted, the procedure consists in measuring the current for various frequency values, starting at a frequency considerably below resonance and gradually increasing the frequency to resonance, and then

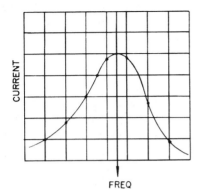 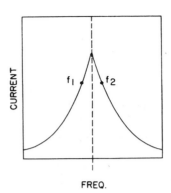

FIGURE 10-2. Selectivity curves.

above. For each frequency that is impressed on the circuit, the value of current would be read and a series of such points would be marked on the graph, as shown at the left of Fig. 10-2. When a line is drawn through all these points, a graph will be obtained that is known as the *response curve* of the series-resonant circuit. As seen from the response curve of the left graph of Fig. 10-2, the response characteristics may be such that several frequencies clustered around the resonant frequency may produce virtually the same high current flow. Thus, the circuit is said to be broadly resonant since it will have a high current flow for a group of frequencies and will reject those frequencies removed from this group. Another circuit, however, may produce a response curve such as shown in the right graph of Fig. 10-2, where virtually only one frequency is resonant, and frequencies on each side of the resonant frequency produce a considerably lower current flow. This type of resonant curve has a high degree of selectivity, while the curve at the left in Fig. 10-2 has low selectivity characteristics.

The response curve of the circuit is directly affected by the value of the resistance. Such resistance can be in the form of an external resistor or of any other resistance in the circuit, such as the resistance of the coil winding or the leakage resistance of the capacitor. Since the variable capacitors in radio or other electronic circuits are of the air-core variable type, they have virtually no leakage resistance. The inductance, however, always has some resistance in the wire and, for this reason, the figure of merit (circuit Q) is based on the ratio of the inductive reactance to the resistance:

$$Q = \frac{X_L}{R} \qquad (10\text{-}4)$$

Thus, the greater the inductive reactance with respect to the resistance, the higher the selectivity.

The resonant band pass of a circuit is defined as the width between two points on the graph designated as f_1 and f_2 that are the frequencies at which the amplitudes of the high- and low-frequency slopes are 0.707 of the peak amplitude. These f_1 and f_2 positions are sometimes referred to as *half-power* points. Thus, the Q of the circuit is also proportional to

$$\frac{f_r}{f_2 - f_1} \tag{10-5}$$

Since the 0.707 points of the response curve indicate the bandwidth and since the bandwidth is proportional to the Q of the circuit, the formula above is applicable. The aforementioned frequency points are indicated on the response curve shown in the right graph of Fig. 10-2.

A series-resonant circuit has a maximum amount of current flow at resonance (limited only by the resistive component of the circuit); hence the circuit will have a low impedance when the resonant-frequency signal is applied to it and higher impedance for signal frequencies either *above* or *below* resonance. The higher impedance on each side of resonance establishes a greater opposition to the current flow and it is this higher impedance, both above and below the resonant frequency, that tends to reject signal frequencies other than the resonant-frequency signal or the cluster of signals whose frequencies are near the resonant frequency.

Because the Q of the circuit is a function of the inductive reactance divided by the resistance, a higher Q can be obtained by increasing the L/C ratio (but maintaining the same product of L times C for the specific resonant frequency desired). Thus, if the inductive reactance were to be increased to obtain a higher Q, it would necessitate increasing the *inductance*. An increase in inductance would, however, lower the resonant frequency and, in order to keep the same resonant frequency as before (but still raise the Q), the series capacitor value has to be decreased. The decrease in capacity must be sufficient to compensate for the increase in inductance so that the same product of LC is maintained.

The Q of a circuit is important in establishing the degree by which a circuit will pass a group or band of signals having frequencies around the resonance point. Thus, the circuit Q establishes the amount of *band pass* required in a particular resonant circuit since in some applications perhaps only one signal frequency is involved, while in others several signal frequencies may be handled by the circuit. The higher Q not only narrows the band pass but offers a higher degree of selectivity for undesired signals having frequencies on each side of the resonant frequency because of the steepness of the band-pass characteristic. A low-Q circuit, on the other hand, while having a much wider band pass, also has more gradual inclines and slopes at the sides of the resonant curve and hence the degree of selectivity is reduced. To regain the high degree of selectivity that is lost when the Q is lowered, several resonant circuits can be employed in successive stages to improve the selectivity, as more fully detailed subsequently.

In a series-resonant circuit, a high Q produces a steep band-pass curve and the circuit has a lower impedance (and thus higher current) at resonance than would be the case with a circuit having a low Q. The circuit with considerable resistance in it produces a low Q.

Since the resistance in a series-resonant circuit is the only current-limiting factor and since low-Q circuits have a higher resistance than high-Q circuits, the series-resonant circuits with low Q limit the current flow because of the greater resistance that is present during resonance. Even though inductive reactance is equal and opposite to capacitive reactance and the opposite effects of the two cancel one another, the higher resistance that is left for a low-Q circuit is the current-limiting factor. With the low-Q circuit, however, the signal current at resonance is still higher than the current on each side of resonance. It is simply the *peak signal current* that has been reduced for the low-Q circuit.

There are several factors which tend to influence the amount of Q and which must be considered in circuit design. Since the Q is a factor of the inductive reactance divided by the resistance, it would seem that the Q could be raised by simply increasing the inductance and lowering the capacity, as previously mentioned. When the inductance is raised by increasing the number of turns of the coil, however, the d-c resistance of the wire that forms the coil will be increased since an additional length of wire is required, and thus added resistance is introduced into the circuit. The additional resistance would tend to decrease the Q. If the d-c resistance of the wire is appreciable, a larger conductor wire must be employed to reduce the d-c resistance established by the additional turns.

Another factor that must be considered when an attempt is made to raise circuit Q is the distributed capacity of an inductance. Since an inductance is composed of a number of turns of wire forming a coil, capacity will be established between adjacent turns of the wire since each turn acts as a capacitor plate, with the dielectric composed of the insulation around the wire. Capacity is also established between successive layers of coil when multilayer inductances are employed. Since capacity increases when the conductor areas are increased, coils employing larger wire will have more distributed capacity than those using relatively thin wire. For this reason, the increase in wire diameter mentioned in the previous paragraph to reduce the d-c resistance of a coil will still act adversely on the circuit Q since an increase in wire diameter will also mean an increase in circuit capacity. Because circuit Q is lowered with an increase in capacity (but raised with an increase in inductance), the increase in capacity of a coil employing larger wire must be compensated for by decreasing in proportion the physical capacity placed in series with the inductance.

Still another factor that affects circuit Q is the resistance encountered at very high frequencies. Such high-frequency resistance is known as *skin effect* and this subject will be treated more fully later.

The distributed capacity of an inductance must not only be kept at a minimum so as not to lower circuit Q but also to prevent attenuating or diminishing signal energy. A considerable amount of distributed capacity in an inductance has the same effect as an inductance being shunted by a capacitor. Thus, the higher-frequency signal energy would find a low capacitive reactance path shunting the inductance, which would cause some loss of such signal energy. Low coil capacity is of particular importance in RF choke coils, which are employed to isolate the signal energy present in one stage from another circuit. A choke coil is designed to have a high series inductive reactance, which will offer considerable opposition to the signal energy and prevent its leakage to circuits where it is not desired. The choke coil also preserves the signal energy contained in a circuit and minimizes its loss by preventing leakage of such signal energy away from the circuit where it is handled. The reactance of a choke coil can be made as high as required since inductive reactance for a certain frequency is proportional to the angular velocity multiplied by the frequency and the value of the inductance. If too high an inductive reactance is chosen, in an effort to provide maximum opposition to signal-energy leakage, the distributed capacity increases, and this will nullify the effect of the high inductive reactance since it will provide a low capacitive reactance path through the coil. Thus, choke coils are wound in a fashion that tends to decrease distributed capacity and, at the same time, a compromise must be established with respect to a minimum of capacitive reactance shunt and a maximum inductive reactance opposition.

In ultrahigh and microwave practices, resonant circuits are often established by employing physical inductances in conjunction with distributed capacities since the high frequency involved entails only a small amount of capacity for establishing resonance. Often the distributed capacity is already of a sufficient amount for resonance, without inclusion of an actual physical capacitor.

10-4.   RF Filters

Series-resonant circuits are useful in the design of electronic filters that pass signals of certain frequencies while rejecting others. One typical example of this is shown at A of Fig. 10-3, where two circuits are interconnected by a series-resonant circuit. Assume the first circuit contains signals of the following frequencies: 500, 800, and 1500 kHz. It is necessary to transfer the desired signal of 800 kHz to the second circuit but at the same time to prevent the entry into the latter circuit of the 500- and 1500-kHz signals.

A convenient design procedure that meets these requirements is to employ a series-resonant circuit to interconnect the two sections. This series-resonant circuit is tuned to the desired frequency, 800 kHz; hence it will have a low impedance and a maximum current flow for the desired signal.

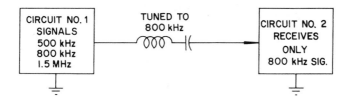

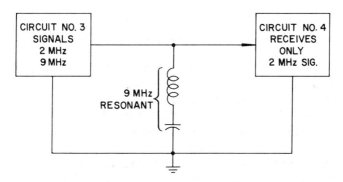

FIGURE 10-3. Series-resonant signal filters.

For signals having frequencies of 500 and 1500 kHz, however, the impedance will be considerably higher and thus offer a high opposition. Consequently, very little signal current flows for the undesired signals. Thus, the series-resonant circuit transfers the desired signal to circuit No. 2 while minimizing the coupling of undesired signals.

Such a series-resonant circuit is also convenient for shunting signals while permitting *desired* signals to pass between two sections. This is illustrated at B of Fig. 10-3, where a circuit contains two signals, one with a frequency of 2 MHz and the other of 9 MHz. If the 2-MHz signal is to be applied to the next circuit but the 9-MHz signal is to be kept out, the design at B meets these specifications. Here, a series-resonant circuit is tuned to the undesired signal of 9 MHz and placed in shunt across the circuit. Since the impedance is low at resonance, the undesired signal currents are shunted and have negligible amplitude at the output. The desired signal of 2 MHz, however, finds a high impedance in the shunt circuit and hence is not diminished to any appreciable degree.

A combination of the two methods can be used, as shown in Fig. 10-4. Here the first circuit again contains a signal of 2 MHz as well as one of 9 MHz, but only the 2-MHz signal is to be transferred to the second circuit. The series-resonant circuit consisting of L_1 and C_1 is tuned to the *desired* signal (2 MHz) and hence will transfer it since the resonant circuit has a low

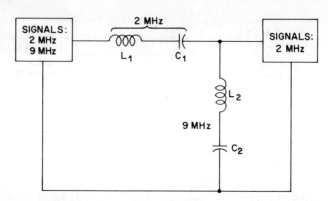

FIGURE 10-4. Dual series-resonant circuit signal filters.

impedance for the resonant-frequency signal. For the 9-MHz signal, however, the impedance is high and little signal energy reaches the output circuit.

The series-resonant circuit in shunt is composed of L_2 and C_2 and is tuned to the undesired signal of 9 MHz; hence it provides a low reactance shunt path for this signal. Thus it will, in effect, bypass the undesired signal and very little appears at the output section.

Series-resonant circuits are seldom used for receiver tuning purposes since the parallel-resonant circuits that are discussed next are more expedient. The series circuits are, however, used to a considerable extent in filter networks of the type previously described for trapping out undesired signals from circuits and otherwise getting rid of unwanted or spurious signals.

10-5. Parallel Resonance

When a capacitor parallels a coil as shown at A of Fig. 10-5, the L and C combination is resonant for a certain frequency that causes the inductive reactance to be equal to the capacitive reactance. When a signal of resonant

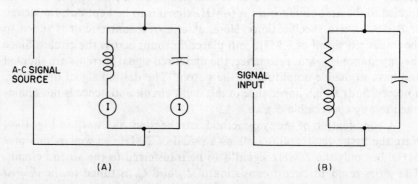

FIGURE 10-5. Parallel resonant circuits.

frequency is applied, the reactances are equal and the current measured in the inductive branch will have the same value as that in the capacitive branch since the same amount of voltage is applied to each component. Because these currents are equal and opposite, they would cancel out in the vector formula used for calculating total current in a parallel circuit composed of inductance, capacity, and resistance. Hence, in a parallel-resonant circuit, the impedance is a function of the total voltage divided by the total resistive current:

$$Z = \frac{E}{I_R} \qquad (10\text{-}6)$$

If no physical resistor shunts the parallel-resonant circuit, and assuming the coil has a negligible amount of resistance, the impedance will be very high since the only current consumed would be by what little resistance is present in the circuit. If the inductance has an appreciable resistance, it must be represented as a series resistance, as shown at B of Fig. 10-5, and the circuit will become a combination of a series circuit in parallel with a capacity.

With a negligible amount of circuit resistance, the voltage drop across the parallel-resonant circuit would be high since the impedance is high. For frequencies above and below resonance, the impedance would decrease because either the inductive reactance or the capacitive reactance would decrease below that established for resonance. This is illustrated in Fig. 10-6, where a parallel-resonant circuit is shown at A for a signal frequency of 10 MHz. With a specific L/C ratio, the inductive reactance is 50 kΩ (kΩ is abbreviation for kilo ohms) and since the circuit is resonant, the capacitive reactance is also 50 kΩ. Despite the fact that a coil (or a capacitor) with a reactance of 50 kΩ would normally permit current flow, the currents in the two branches are 180° out of phase and, being opposite, their effects cancel out.

If the input-signal frequency is now changed to 20 MHz (twice the resonant frequency), the inductive reactance becomes 100 kΩ since an increase in frequency will raise inductive reactance. For the 20-MHz signal, the capacitive reactance will decrease to 25 kΩ, as shown at B. As there is now a difference in the reactances between coil and capacitor, currents are no longer equal and opposite and will not cancel fully. Partial cancellation occurs since the lower current of the inductive reactance is subtracted from the higher current established by the lower capacitive reactance. The resultant current is combined with the resistive current in the vector formula for finding total current. Since a reactive current is now present, a higher circulating current results and the ammeter will now indicate a high current flow. Impedance being a function of the voltage divided by the current, the higher current would give a lower value of impedance than for the resonant circuit at A of Fig. 10-6. The lower value of capacitive reactance for the circuit at

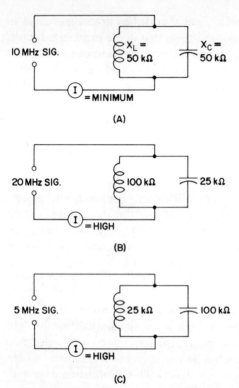

FIGURE 10-6. Effect of signal frequency on parallel-resonant circuits.

B establishes the high current flow since it now provides a low-shunt reactance across the inductance. Because the capacitor is instrumental in decreasing the impedance, the circuit at B is now primarily capacitive. If the frequency of the circuit shown at A is halved to 5 MHz, as at C, the inductive reactance shown at A will decrease by one-half since a decrease in frequency means a corresponding decrease in inductive reactance. Hence, the inductive reactance for the circuit at C is now 25 kΩ.

The *decrease* of signal frequency from that at A will cause the capacitive reactance of the circuit at C to increase to 100 kΩ since a decrease in signal frequency raises capacitive reactance. Again, as with the circuit shown at B, the off-resonance condition increases the current flow and lowers the impedance. As the inductive reactance has now decreased, it becomes the shunting factor that increases current; hence the circuit shown at C is predominantly an *inductive* circuit.

From the foregoing, it is evident that parallel-resonant circuits act in opposite fashion to series-resonant circuits. In a series circuit at resonance, the impedance for the signal frequency that establishes resonance is low and signal current is high. In a parallel-resonant circuit, the impedance for the signal frequency that establishes resonance is high, while the signal current is low.

The high impedance of a parallel-resonant circuit decreases the resonant-frequency signal current flow to and from the signal source but establishes a large voltage drop for the signal. For signals having frequencies either above or below the resonant frequency, circuit impedance is low. Thus, even though undesired signals are present in the circuit, they would develop only a low-value voltage drop across the parallel-resonant circuit.

At resonance, a parallel-resonant circuit acts as a storage device and, assuming negligible resistance, the signal energy is interchanged between the inductance and the capacity, at a rate corresponding to the frequency of the incoming signal. This interchange of energy would continue even though the signal source were removed since the capacitor would become charged by the signal energy and would discharge such energy into the coil. The collapsing field of the inductance would establish a back-emf that would recharge the capacitor. Since the resonant-frequency potential established the initial full charge, no more signal energy will be accepted by the circuit. This is similar to the charging of a capacitor by a certain potential. Once the capacitor is charged, current flow ceases since for a given emf and a fixed value of capacity only a specific charge can be established. A specific charge is also placed across the capacitor in the parallel-resonant circuit, except that the energy flows back and forth from capacitor to coil. Such an energy would continue the interchange indefinitely but for the fact that the signal energy would eventually be consumed by whatever resistance is in the circuit. The characteristic of energy exchange in a resonant circuit is known as the *flywheel* effect and is more fully discussed in Chapter 17.

The formula for ascertaining the frequency when the values of the inductance and capacity are known is similar to the formula employed for the series-resonant circuit.

$$f = \frac{1}{6.28\sqrt{LC}} \tag{10-7}$$

Because signals having frequencies in kilohertz are often encountered, however, the following equation can be used; it gives the answer directly in kilohertz rather than in hertz:

$$f \text{ (kHz)} = \frac{0.159}{\sqrt{LC}} \tag{10-8}$$

In circuits where wide ranges of signals are encountered, it is necessary to increase the band-pass characteristics of a resonant circuit by adding a physical shunting resistor across the circuit to broaden the range of frequencies handled, as shown in Fig. 10-7. As such a resistor shunts the high-impedance circuit, it will lower the impedance in pro-

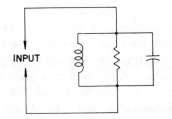

FIGURE 10-7. Resonant circuit with shunting resistance.

portion to its own value. A resistor having a very high resistance will have little effect on lowering the impedance, while a lower resistor will have an appreciable effect. In wideband circuits of this type, the circuit Q is proportional to the resistance divided by either the inductive reactance or the capacitive reactance. Since each of these reactances is the same at resonance, it is immaterial which is employed. Also, since the amount of such shunt resistance will determine the amount of impedance, the formula can be expressed as

$$Q = \frac{Z}{X} \tag{10-9}$$

10-6. Tuned Circuit Transformers

Parallel-resonant circuits are used extensively to couple the various circuits of RF signal generators, RF amplifiers, filter networks, and similar RF sections. This is usually done through a transformer arrangement, by coupling the inductance of one parallel-resonant circuit to the inductance of a second parallel-resonant circuit, as shown at A of Fig. 10-8. Since the induc-

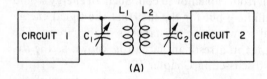

(A)

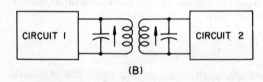

(B)

FIGURE 10-8. Parallel resonant circuits form coupling transformers.

tances generate magnetic lines of force, as previously explained, the placement of a second coil in close proximity to the first coil causes the former to intercept the signal energy present in the first resonant circuit. Because the coupling is by virtue of the magnetic fields of the inductances, it cannot be truly considered as the coupling of one parallel circuit to another since the individual capacitors of the parallel circuits are not coupled together; only the inductances are. For this reason, the transfer of energy from L_1, which forms the primary of the transformer, to L_2 which forms the secondary, must be considered equivalent to a series circuit. The induced voltage will cause the current to flow, which again is subject to the characteristics of the second parallel-resonant circuit composed of L_2 and C_3.

Variable capacitors can be utilized, as shown at A, so that the individual stages can be tuned to resonance. Another method for tuning is to employ metallic slugs that are moved in and out of the core of the coils. As previously mentioned, the core affects permeability and hence will change the inductance of a coil. When the core is inserted entirely into the coil, a high inductance results, which will decrease as the core is gradually removed from the coil.

When a variable core is used for tuning purposes, an arrow is often placed beside each coil, as shown at B of Fig. 10-8. Since both coils are wound on a single form, as shown in Fig. 10-9, a single metallic slug is sometimes utilized so that both the primary and secondary resonant circuits can be tuned simultaneously. The factors relating to the characteristics of such devices are covered more fully in subsequent discussions.

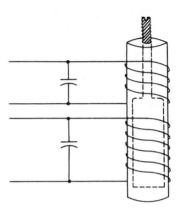

FIGURE 10-9. Iron-core slug for tuning purposes.

The degree of coupling, or how much the two coils are brought together, affects the band-pass characteristics of the circuits, as well as the amplitude of the signal trans- ferred. As mentioned earlier in the discussion on transformers, when two coils are coupled as closely together as possible, such coupling is known as *overcoupling*, or *tight* coupling. Under this condition, the loading effect of one coil on the other decreases the Q of the combined circuits to a considerable extent and has a pronounced influence on the impedance of the parallel-resonant circuits. At the resonant frequency, the impedance decreases but is still high on each side of the resonant frequency, as shown in Fig. 10-10. Thus, a double-hump resonant curve is produced when overcoupling is employed.

If the inductors are spaced some distance apart (loose coupling) but still sufficiently close so that some energy is transferred, not all the magnetic lines of force generated by the primary coil will be intercepted by the secon-

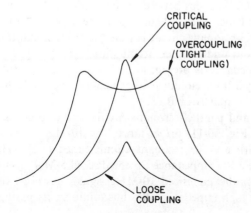

CRITICAL
COUPLING

OVERCOUPLING
(TIGHT
COUPLING)

LOOSE
COUPLING

FIGURE 10-10. Effects of the degree of coupling in transformers.

dary coil and considerably less signal energy is transferred than would be the case with overcoupling. Since there is less loading effect from one inductance to the other, the band-pass characteristics are narrow and selectivity is good, as shown in Fig. 10-10. At some point between loose coupling and tight coupling, the condition known as *critical* coupling exists. At this point, there will be a maximum transfer of signal energy from one parallel-resonant circuit to the other with much sharper selectivity, as shown in Fig. 10-10.

The factors relating to the coefficient of coupling also apply for the coupling of parallel-resonant circuits in transformer arrangement.

10-7. Parallel Filter Systems

As with the series-resonant circuits previously discussed in this chapter, the parallel-resonant circuits can also be employed as filter devices. If, for instance, a circuit contains a signal having a frequency of 5 MHz and an undesired signal of 10 MHz, as shown for circuit No. 1 at A of Fig. 10-11, the undesired 10-MHz signal can be shunted by a parallel-resonant combination utilized as shown so that only the 5-MHz signal reaches the second circuit. The parallel-resonant circuit is tuned to 5 MHz (the desired signal) and, hence, the high impedance prevents the desired signal from being shunted. Since the 10-MHz signal does not produce the same high impedance as does the resonant signal, the low impedance shunts the 10-MHz signal.

As shown at B of Fig. 10-11, the parallel-resonant circuit can also be used in series between two points. The parallel-resonant circuit is tuned to 10 MHz (the undesired signal) and, hence, offers a high opposition to it because of the high impedance at resonance. Since the desired signal of 5 MHz is not at resonance, a low impedance prevails and permits passage of the 5-MHz signal.

For more effective filtering of the unwanted signal, a combination of the two aforementioned methods can be used as shown at C of Fig. 10-11. The parallel combination that is in series between circuit sections No. 1 and No. 2 is tuned to 10 MHz and hence offers a high impedance to the unwanted signal, while the shunt parallel-resonant circuit is tuned to 5 MHz and thus offers a high impedance without undue signal attenuation. For the signal of 10 MHz, however, a low impedance is present and signal shunting occurs. Such a combination of the two filter-circuit arrangements is more effective than either of the methods shown at A or B.

Combinations of series- and parallel-resonant circuits can also be used for filter purposes. At C of Fig. 10-11, for instance, the 10-MHz parallel circuit could be replaced with a series-resonant circuit tuned to 5 MHz. This series circuit would have a low impedance for the desired 5-MHz signal and hence would pass it. For the undesired 10-MHz signal, however, the series circuit would present a high impedance and thus minimize its passage to the next section.

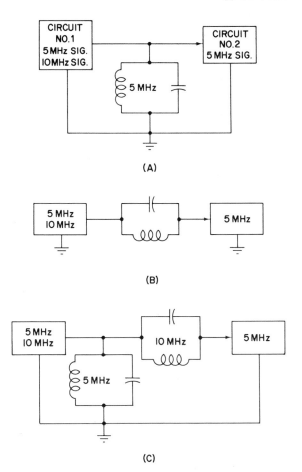

FIGURE 10-11. Parallel-resonant circuits used as filters.

10-8.   Applied Mathematics

In electronics, the lowercase Greek letter ω (omega) is used as a symbol for the *product* of the angular velocity (6.28) multiplied by the signal frequency:

$$\omega = 6.28f \qquad (10\text{-}10)$$

Thus, the symbol ω is a convenient way to express $6.28f$. Typical derivations of formulas employing this symbol are as follows: Since $X_L = X_C$ at resonance,

$$X_L - X_C = 0$$

$$\omega L - \frac{1}{\omega C} = 0 \qquad (10\text{-}11)$$

$$\omega^2 = \frac{1}{LC} \qquad \left(6.28f^2 = \frac{1}{LC}\right) \qquad (10\text{-}12)$$

$$LC = \frac{1}{\omega^2} \qquad (10\text{-}13)$$

$$f = \frac{1}{6.28\sqrt{LC}} \qquad (10\text{-}7)$$

Thus, the product of L times C is a function of $1/\omega^2$. Also ω times the inductance is, therefore, an expression for the inductive reactance. A study of the foregoing will help in solving typical problems.

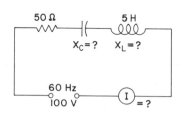

FIGURE 10-12. Calculation of un-known values in series resonance.

As an example, consider the circuit shown in Fig. 10-12. Here, a 100-V 60-Hz signal source is applied to a series circuit consisting of a 50-Ω resistor in series with an unknown capacitor and an inductance of 5 henrys. The problem is to find both the inductive and capacitive reactances at reso-nance (60 Hz), as well as the total current and the voltage drop across each component. The inductive reactance is found by multiply-ing ω by 5 henrys, which gives 1884 Ω. Since the circuit is at resonance, the capacitive reactance would be of the same value. Total current is a function of the voltage divided by the impedance and since the impedance at resonance is established by the resistance, the current is 2 A, as follows:

$$I_{\text{Total}} = \frac{E}{R} = \frac{100}{50} = 2 \text{ A}$$

Once the current value is known, the voltage across the individual components is a function of the current multiplied by the resistance or reactance. The following calculations show these values:

$$E_R = IR = 2 \times 50 = 100 \text{ V}$$

$$E_L = IX_L = 2 \times 1884 = 3768 \text{ V}$$

$$E_c = IX_C = 2 \times 1884 = 3768 \text{ V}$$

This indicates that the voltage drop across the resistor is 100 V, with 3768 V across each reactance. As previously explained, the individual voltage drops across the components in a circuit composed of L, C, and R may some-times exceed the source voltage. The *total voltage across the inductor and capacitor, however, is zero since the voltages are of opposite phase and their combined effects cancel.*

Since the inductive reactance opposes the capacitive reactance in a reso-nant circuit, the voltage and current will again be in phase so that when solving for power the apparent power becomes the true power. Because there is no phase angle, the power for the circuit shown in Fig. 10-12 would be calculated as follows:

$$P = EI = 2 \times 100 = 200 \text{ W}$$

FREQUENCY EFFECTS ON L, C, AND R

Component	Effect of Frequency	
	Increase	Decrease
Resistance (R)	None	None
Capacitance (C)	None	None
Capacitive reactance (X_C)	Lowers X_C	Raises X_C
Inductance (L)	None	None
Inductive reactance (X_L)	Raises X_L	Lowers X_L
Series capacitor-resistor combination (Z)	Lowers Z	Raises Z
Series inductance-resistor combination (Z)	Raises Z	Lowers Z
Parallel capacitor-resistor combination (Z)	Lowers Z	Raises Z
Parallel inductance-resistor combination (Z)	Raises Z	Lowers Z
Series resonance (Z)	Raises Z	Raises Z
Parallel resonance (Z)	Lowers Z	Lowers Z

Review Questions

10-1. Compare the impedance and current relationships (in terms of high and low values) with respect to parallel resonant and series resonant circuits.

10-2. Briefly explain what is meant by the Q of a resonant circuit.

10-3. Of what significance are *half-power* points on a resonant curve?

10-4. If the product of LC is known, how can the resonant frequency of the circuit be determined?

10-5. Briefly explain what effect distributive capacitance has on circuit Q.

10-6. (a) In a series-resonant circuit, what is the phase angle between voltage and current, and which leads?
(b) How do these factors differ for a parallel resonant circuit?

10-7. (a) When two inductors are overcoupled, what type of band-pass characteristics are obtained?
(b) How do these characteristics differ when coils have critical coupling?

10-8. (a) Which furnishes more signal-energy transfer across a transformer, loose coupling or critical coupling?
(b) Which provides the greatest selectivity?

10-9. Designate the *type* of series-resonant filter circuit you would employ to transfer a 40-MHz signal from one stage to another, while minimizing the transfer of a 100-MHz signal.

10-10. Designate the *type* of parallel-resonant filter circuit you would employ to transfer a 60-kHz signal to a subsequent electronic circuit, while minimizing the transfer of a 300-kHz signal.

10-11. Indicate a combination of series and parallel circuitry that will filter a 200-MHz signal from a subsequent stage while permitting the coupling of a 50-MHz signal.

10-12. Explain two methods that can be employed to tune the primary and secondary circuits of an RF transformer.

10-13. What precautions must be taken to design an effective choke coil that will isolate a signal frequency and prevent its leakage to other circuits?

10-14. Explain what the lowercase Greek letter omega (ω) represents in electronics, and give a typical example of its use in a formula.

Practical Problems

10-1. If an electronic device utilizes the resonant circuit shown in Fig. 10-13, what is the total current flow in *milliamperes*? Also, solve for the power consumed. What is the value of the inductive reactance at resonance? What is the value of inductive reactance at 180 Hz?

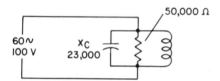

FIGURE 10-13. Illustration for Problem 10-1.

10-2. What is the Q of a series-resonant circuit if the capacitive reactance is 2000 Ω and the resistance is 10 Ω? If the resistance value is 100 Ω, what is the Q? Which resistor reduces the selectivity of the circuit the least?

10-3. In an industrial control circuit an undesired resonance effect is secured because of the presence of inductance and capacity. If LC is equal to 0.0001, what is the resonant frequency that must be eliminated?

10-4. In a test instrument, a series circuit is employed that has a resistance of 200 Ω, a capacitive reactance of 4000 Ω, and an inductive reactance of 250 Ω when used at 1000 kHz. At what frequency is this circuit resonant?

10-5. The filter network used in a shortwave transmitter is similar to that shown earlier at C of Fig. 10-11. The 5-MHz parallel-resonant circuit was replaced by a *series*-resonant circuit. To what frequency must the series-resonant circuit be tuned?

10-6. In an interference filter of a television receiver the circuitry shown earlier in Fig. 10-4 was used. A 45.75-MHz signal was to be passed but a 47.25-MHz signal was to be blocked from the output. Reproduce Fig. 10-4, showing the frequencies to which the resonant circuits are tuned.

10-7. An RF choke has an inductive reactance of 50 kΩ at 20 MHz but the distributed capacitance has a shunt reactance of 25,000 Ω. At what frequency would X_C be eight times higher than X_L?

10-8. In an industrial circuit for automation, it was found that the product of L (in microhenrys) and C (in microfarads) was 0.0253, producing an undesired resonant effect. What is the frequency of the resonant circuit thus formed?

10-9. An electronic signal filter had a series-resonant circuit in which the product of L (in microhenrys) and C (in microfarads) was 0.0704. What is the resonant frequency? What is the frequency of the signal that is applied to the load if the input to the filter consists of signals having the following frequencies (in kilohertz): 100, 300, 600, 900, 1200?

10-10. The product of L (in microhenrys) and C (in microfarads) was found to be 0.225 in a control circuit. What is the value of $(6.28 \times f)^2$, and what is the frequency of the resonant circuit thus formed?

Principles of
Electronics

II

Vacuum Tubes 11

11-1. Introduction

While resistors, capacitors, and inductors form important elements in electronic circuitry, other devices are needed for the efficient generation of signals, for rectification, detection, amplification, voltage regulation, electronic switching, and many other purposes necessary in the wide applications of modern electronics. Until the advent of transistors and other solid-state units, vacuum and gas-filled tubes provided the only reliable and convenient means for obtaining the necessary generation, amplification, and other signal-processing functions required. Sizes ranged from the extremely small types utilized in portable or high-frequency devices (handling fractional watts of power) to the huge sizes employed in radio and television transmitting stations. In the latter applications they handled thousands of watts of signal power and some required water-circulating jackets around their heat-producing anode elements.

While the solid-state components described in subsequent chapters have replaced vacuum tubes in numerous electronic applications (home receivers, phono or tape amplifiers, portable equipment, electronic organs, computers, etc.), tubes are still extensively used in transmitting equipment, photoelectric devices, television picture and oscilloscope screens, numerical display units in computers, and microwave-signal generators. Thus, the basic principles that apply to vacuum and gas-filled tubes must still be covered thoroughly in any comprehensive study of electronics.

11-2. Thermionic Emission

As mentioned in the first two chapters, a conductor has an atomic characteristic that permits the free movement of electrons since some of the latter are not so rigidly bound to the limits of the nuclear orbit that they cannot be removed from such an orbit. If such a conductor is heated, the velocity of the electrons revolving around the nuclear orbit increases, and this acceleration imparts to the electrons sufficient kinetic energy so that they leave the orbital influence of the atomic structure. When many electrons leave their atomic orbit, they can escape beyond the surface of the conductor material. The free electrons form a cloud of electrons around the conducting surface and create what is referred to as a *space charge*. Since this cloud is composed of free electrons that have left their central orbit, such electrons can be readily pulled entirely away from the influence of the conductor material, by subjecting the electrons to a positive electric charge that will attract the negative charge representative of the electrons. Thus, a continuous flow of electrons can be established from a conducting surface to some other surface having a plus charge. This method of procuring electrons by a heating process is known as *thermionic emission* (from the word *thermal*, referring to heat) and is the basic operating principle of a vacuum tube.

The famed American inventor Thomas A. Edison (1847–1931) observed this thermionic emission phenomenon during research on incandescent lamps. Edison was experimenting with methods for extending the life of the incandescent lamp filament. During such experiments, a metal plate was installed within the lamp envelope. The plate was placed near the filament, for the purpose of absorbing heat from the latter, in an effort to extend its useful life through temperature reduction. Edison connected a galvanometer in series with the positive terminal of the battery supplying electric energy to the filament. The galvanometer indicated that in the circuit that had been formed there was a current flow between the filament of the incandescent lamp and the metal plate. This is known as the *Edison effect*, and though its discoverer recorded the phenomenon (1883), he attached no particular significance to it and considered it as a mere laboratory curiosity.

Before the turn of the twentieth century the noted English electrical engineer J. A. Fleming (1847–1945) conducted extensive experiments with the so-called Edison effect and in 1905 patented the *Fleming valve*. This valve, or tube, was the forerunner of the modern vacuum tube. The original Fleming valve contained a filament wire and metal plate within a glass envelope, with most of the air pumped out to form a partial vacuum. Such a simple tube type is now referred to as a *diode*.

11-3. The Diode

Since the development of the diode by Fleming, numerous elements have been added to the device to form the many varieties of tubes currently in use. The simple plate and filament arrangement originally patented by Fleming, however still finds usage in electronics. The simple diode is shown schematically at A of Fig. 11-1, where the circle represents the glass or metal envelope

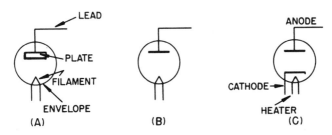

FIGURE 11-1. Diode vacuum-tube symbols.

of the vacuum tube. The plate and the lead extending from it are usually drawn at the top of the circle, as shown, while the filament, which is heated from an electric source, is at the lower part. Such a tube can also be illustrated schematically as shown at B, where the plate is a simple horizontal line.

The element within a vacuum tube that is the source of electrons is known as the *cathode*. The element within the tube which has a positive charge placed on it and which attracts and receives the electrons emitted by the cathode is known as the *anode*. Thus, the filament shown at A and B is also the cathode, while the plate is the anode. Vacuum tubes are also designed in which the filament itself is not utilized for the cathode but acts only as a heater. A separate cathode element is employed in such tubes, and the heater filament is placed sufficiently close to the cathode so that the heater will raise the temperature of the cathode to the point where it will emit electrons. This type of tube is shown at C of Fig. 11-1 and is known as an *indirectly heated* tube, as opposed to the *directly heated* tube shown at A and B. All these types are diodes since only one cathode and one anode are employed. Thus, the term *diode*, used to represent a two-element tube, refers to the cathode and anode elements, regardless of whether a directly heated or indirectly heated type is employed.

Cathodes can be made of either tungsten or thoriated tungsten. Other cathodes are oxide-coated, and still others belong to the cylindrical type of cathodes employed in cathode-ray tubes, which have a deposit of barium oxide capable of producing a large amount of free electrons.

The tungsten-type cathodes are primarily employed in heavy-duty industrial-electronic equipment, transmitting for radio and television, and other applications where tube filaments must emit a considerable amount of electrons and, thus, must be brought to a very high temperature. Tungsten filaments are operated at approximately 2500° Kelvin. (Kelvin measurements can be converted to centigrade degrees by adding 273 to the Kelvin designation.) Such temperatures range above 4000° Fahrenheit.

Thoriated tungsten cathodes are composed of pure tungsten to which some thorium oxide has been added. The addition of the thorium increases thermionic emission to a considerable degree over that of the pure tungsten filament types. Such a thoriated tungsten filament, while more efficient, is more subject to damage by overloading. Overloading will cause an evaporation of the thorium layer when the emission limits are exceeded and, in consequence, the efficiency will decline. Thoriated tungsten filaments are operated at approximatley 1900° K.

The oxide-coated filament is formed by coating the conductor with barium or strontium oxides. The conductor that forms the filament is usually a nickel-alloy wire. Oxide-coated filaments produce a high order of thermionic emission with temperatures much lower as compared with the other filament types. For temperatures as low as 1150° K, the oxide-coated filaments offer a high emission efficiency and for that reason are extensively used in applications where a high amount of filament power is not required. Thus, the oxide-coated type of emitters are generally used in low-powered electronic applications.

The two types of cathodes, the directly heated and indirectly heated, are shown at A and B of Fig. 11-2. The directly heated type is shown at A and consists of a wire filament to which the electric energy is applied (at the two terminals at the bottom). The indirectly heated type has a spiral heater

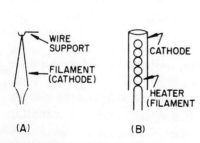

FIGURE 11-2. Filament sections of vacuum tubes.

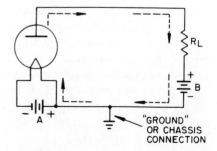

FIGURE 11-3. Current flow in diode vacuum tube circuit.

element plus a cathode metal-sleeve structure. It is coated with an emitting material such as the barium oxide mentioned earlier and the cathode structure is insulated from the heater by an asbestos sleeve.

The basic circuitry for applying potentials to a diode tube is shown in Fig. 11-3. A battery or other voltage source is applied to the filament wires to heat them to incandescence and thus liberate electrons. A series resistor is used, as shown, and the signal energy handled by the diode appears across it. Hence, the resistor is known as a *load resistor* (R_L) and circuit effects are discussed later.

The cathode-filament battery or power source (marked A in Fig. 11-3) has one side connected to the anode voltage source (marked B). The polarity of the anode battery must be such that the plate of the tube has a *positive* polarity with respect to its cathode. Thus, electrons leave the filament and are attracted by the positive anode, after which they flow through R_L and the battery and return to the filament, thus flowing in a closed-loop-type path. The direction of electron flow is shown by the dashed-line arrows.

The *ground* or *chassis* connection that is shown simply indicates that the chassis of the electronic device is used for interconnecting the current path from the bottom of one battery to the filament. Often a metal chassis is thus used for completion of circuitry without running extra wires between circuit components. In modern electronic references, the word *ground* generally indicates only a common chassis connection and not necessarily a direct earth electric path.

For Fig. 11-3 the amount of I that flows depends on the applied E and the circuit resistance. With a low value of plate voltage, very few electrons flow from the filament to the plate and through the load resistor. As the plate voltage is increased, however, there is also an increase in current, up to the point where the condition known as *saturation* occurs.

Saturation is the highest value of plate current that will flow, regardless of an increase in plate voltage. At saturation, all the electrons that are emitted by the cathode are drawn to the plate. For plate voltages below the saturation point, plate current is correspondingly lower, which indicates that not all the electrons emitted by the cathode are reaching the anode.

When electrons are emitted by a cathode structure, the loose and free electrons form a cloud around the cathode, as previously mentioned. This space-charge cloud forms a negative group of electrons that set up an electrostatic field. The latter tends to repel other electrons coming from the cathode structure. For low values of plate voltage, the anode does not have sufficient drawing power to attract all the electrons or to pull electrons away from the cathode and through the space charge. Only those electrons that initially have sufficient kinetic energy to overcome the space-charge barrier reach the plate to contribute to current flow. With an increase in anode voltage, however, the electrostatic force exerted by the rising anode potential exerts a greater influence on the electrons and, thus, has a greater effect on overcoming the space charge.

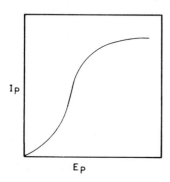

I_P

E_P

FIGURE 11-4. Graph of plate current versus plate voltage in diode.

With a fixed plate potential and an increase in cathode temperature, a curve for the plate voltage would be obtained similar to that shown in Fig. 11-4. The saturation is also reached for an increase in cathode temperature, and such *temperature saturation* is the limit at which the cathode is able to emit electrons.

An increase in plate potential above the point where current saturation occurs will increase the *velocity* of the electrons and cause them to strike the plate with a greater impact, but the amount of electrons reaching the plate will not increase.

11-4. Rectification

The diode whether a tube or solid-state type is useful as a *rectifier* and, in this application, it exhibits the ability of converting ac to dc. Thus, in such instances where it is necessary to obtain dc when only an a-c source is available, the diode unit is employed. Specific applications include the conversion of the a-c power found in homes to the necessary dc for radio and television receivers, as more fully detailed in Chapter 14.

Another application for the diode is *detection*, in which the diode is employed to obtain the audio- or television-signal components from the modulated carrier wave. This process is more fully detailed in Chapter 18.

Both the detection and rectification processes are essentially similar to their basic function. Both take advantage of the fact that electron flow in a diode tube is only in one direction, i.e., from cathode to anode. If, for instance, the *B* battery shown in Fig. 11-3 were reversed so that the battery would apply a negative potential to the plate, electron flow would cease since the negative charge on the anode would repel the negative electrons that form the space charge around the heated cathode. Under such a condition, the transit of electrons from the cathode to the plate does not occur. This principle of current flowing in only one direction through a diode is utilized for detection and rectification, and a simple circuit of this type is shown at A of Fig. 11-5. When an a-c signal is applied to terminals T_1 and T_2, a negative alternation (such as the first shaded one shown) would place a negative potential at T_1 and a positive potential at T_2; hence, electron flow

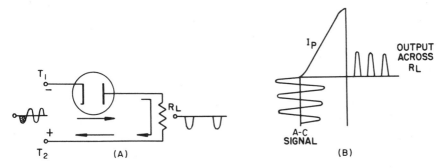

FIGURE 11-5. Rectifying principle of diode.

would occur through the vacuum tube, as indicated by the solid arrows. This flow would start at zero, just as the negative alternation of the ac does, and would reach a peak value and then decline to zero again. This rising and falling current would also flow through the resistor and cause corresponding voltages across the latter. This voltage change across the load resistor (R_L) would reproduce the first negative alternation, as shown.

The second input alternation, which is in the positive direction, would cause terminal T_1 to have a plus polarity and terminal T_2 to have a negative polarity. Under this condition, the anode of the tube would be negative and, hence, would repel rather than attract electrons. In consequence, the tube would *not* conduct during the time interval of this second alternation across T_1 and T_2. During the *third* alternation, which is again of negative polarity like the first, the proper polarities are again applied across the diode tube and conduction occurs again. Thus, a second negative alternation appears across the load resistor. Successive negative alternations will reappear across the load resistor in the form of *pulsating* dc. This is graphically shown at B, which illustrates the application of the a-c signal to the tube.

This graph represents the plate-current (I_p) rise along the vertical axis. The application of the a-c input signal is represented on the graph as extending to the point where the negative-going portion of the signal causes no current flow, while the positive-going portion of the signal will cause current flow proportionate to the emitting characteristics of the filament and the voltage of the input signal. The output across the load resistor (R_L) shows the current pulses that flow through this resistor. Such *current* variations through the load resistor will, of course, also create *voltage* variations across the load resistor, such voltage changes having waveshapes similar to the half cycles, or negative alternations. As seen, this is not pure dc but rather a pulsating dc, which occurs only at intervals. Filter circuits, however, as described in Chapter 14, are employed to smooth out the ripple and produce a dc sufficiently free from variations for application to circuits requiring dc.

11-5. The Triode

One of the truly great advances in vacuum tubes occurred in 1906, when Dr. Lee DeForest (1873–1961) the noted American inventor introduced the *grid* for the vacuum tube. The grid consists of a third element added to the basic diode and is placed between the cathode and anode. The grid, which is also known as a *control grid*, is represented schematically as shown at A of Fig. 11-6. The term *control grid* refers to the ability of the structure to control the

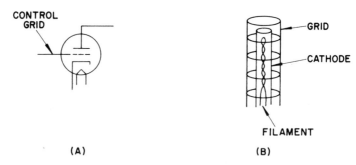

FIGURE 11-6. Tube grid and structure.

amount of electron flow through a tube. The grid is usually composed of fine mesh wires, suspended across two stiff wire supports. Within the grid-wire structure are the cathode and filament elements as shown at B.

The remarkable characteristics of the control grid are due to the fact that a relatively small voltage placed on the control grid can influence and control a considerable amount of plate-current electron flow in the tube. If a negative voltage is applied to the grid, the potential will set up an electrostatic charge around the grid wires that will repel electrons. Since the grid is placed close to the cathode structure, it has considerable influence on the amount of electrons that will flow from the cathode to the plate. If only a small negative voltage is applied to the grid, the grid charge is not sufficient to repel all the electrons leaving the cathode, and some plate current will flow. If a larger negative potential is applied to the grid structure, more electrons will be repelled. As the negative potential on the grid is increased, eventually the negative potential on the grid reaches such a value that it creates an electrostatic field of sufficient intensity to repel *all* electrons back to the cathode. Current flow within the tube then stops. This condition, where the grid potential has reached the point where it stops all current flow, is known as the tube *cutoff.* This progressive influence on decreasing current flow is shown in Fig. 11-7.

At A a small negative potential is applied to the grid, and current flow, as read by a meter, may be 20 mA (the amount depending on the type of tube and the value of the series load resistor). At B, a higher negative potential

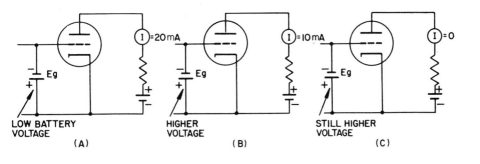

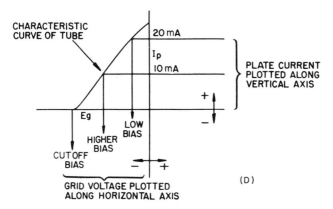

FIGURE 11-7. Effects of bias on tube conduction.

is applied to the grid and current flow has dropped to 10 mA. A still higher negative potential, as at C, repels all current and the meter now reads zero. This condition is illustrated graphically at D, where the grid potential shown at A is designated by the vertical line marked "low bias." Here, a high plate current flows (20 mA). For the grid voltage at B, a correspondingly lower plate current flows, while the condition illustrated at C indicates no current flow along the *x* axis, which represents zero on the graph. An additional increase of negative potential no longer affects current flow because the grid voltage is beyond tube cutoff.

If the grid potential is reduced to zero, maximum current (established by the plate voltage and the value of the load resistor) flows through the tube. If the voltage at the grid is made positive instead of negative, no more current will flow since tube saturation has been reached. Actually, plate current would decrease, for a plus voltage on the control grid would cause the latter to have an electrostatic field of such a polarity that it would attract rather than repel electrons. Under such a condition, the control grid draws electrons and, hence, grid current flows. Since such a current flow comes from the cathode structure, it would tend to decrease the saturation current level of the anode circuit.

In conventional circuitry, such as amplifiers for increasing weak signals to the level needed, a *fixed* negative potential is applied to the grid. This fixed voltage is known as *grid bias*, and it may be of a value between the cutoff point of the tube and zero, as shown in graph A of Fig. 11-8. The bias, since

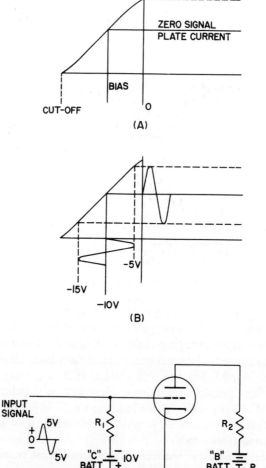

FIGURE 11-8. Graph of grid-voltage change versus plate-current change.

it is negative in polarity and repels some electrons, would lower the current flow below the saturation level and establish a fixed value of current flow referred to as the *zero-signal plate current*. When a signal such as an a-c waveform is now applied to the grid circuit, the signal voltage alternately adds and subtracts from the fixed bias value as shown at B of Fig. 11-8. The circuit representation is shown at C, where resistor R_1 is a grid resistor placed in series with the battery or other voltage source for supplying the negative grid bias. (When battery operated tube-type radios were common, the bias battery was known as the *C battery* to distinguish it from the filament *A battery* and the anode *B battery*.)

At C the resistor R_2 is the load resistor previously referred to. Often, for convenience, the filament of the vacuum tube is not shown since its function is only to heat the cathode and, hence, has no bearing on the signal-handling characteristics of the circuit. The filament symbols are often omitted from commercial schematics of electronic devices in such instances where the filament wiring is conventional and the circuit hookup would be obvious to the trained technician.

The a-c voltage, representative of the signal, is applied to the grid and cathode terminals, as shown in Fig. 11-8. The first alternation of the incoming signal, shown at B and C, is of positive polarity and, therefore, decreases the bias and causes an increase in the current through the tube. If, for instance, the bias is a negative 10 V, the positive polarity of a 5-V a-c signal would oppose 5 V of the negative *C* polarity and leave only a negative 5 V at the grid of the tube during the peak of the positive alternation. Hence, the plate current within the tube would rise and reach a peak, as shown at B. When the input signal swings in the opposite direction to reach a negative 5 V, this negative voltage adds to the existing negative 10 V of bias to produce a total bias of minus 15 V. The greater electrostatic charge now at the grid will repel more electrons so that the plate current drops to a low value, as shown at B. Since the plate-current change may be several hundred milliamperes or more for a small change in grid voltage, a much higher signal power and voltage develop in the plate circuit than is represented by the relatively low-power, low-voltage grid input signal. This control of a large amount of output power by a relatively small amount of input signal voltage is known as *amplification*. (For simplicity in basic analysis, the schematic at C does not show coupling or bypass capacitors. The latter are, however, essential to proper signal amplification and are covered in detail in Chapter 15.)

Besides having the property of amplifying audio- and radio-frequency signals, vacuum tubes can also be used to generate fundamental radio- and audio-frequency signals in circuits known as *oscillators*. Oscillators, amplifiers, and other circuits employing vacuum tubes (as well as transistors) are more fully discussed in succeeding chapters.

11-6.   Characteristics of a Triode

Information regarding the application of triode tubes, as well as the characteristics of such tubes under various voltage applications and signal intensities, can be obtained from graphs that indicate one set of conditions when two other conditions are established. That is to say, if a certain bias and plate voltage are applied to the tube, a definite amount of current will flow. If the *grid voltage* were held *constant* and the plate voltage were changed successively, different values of plate current would result.

Another procedure is to hold the *plate voltage constant*, while the bias on the control grid is varied. Here again a different value of plate current would be obtained for each change of control-grid bias value. Thus, sets of *characteristic curves* can be graphed, which give a visual indication of the relationship existing among grid voltage, plate voltage, and the resultant plate current. The circuit for doing this consists of the usual tube element voltage supplies, across which are placed variable resistors (potentiometers) so that the grid voltage and the plate voltage can be altered. Voltmeters and milliammeters to read voltage and current, respectively, are included in the circuit.

One set of curves that can be obtained is shown in Fig. 11-9, which gives the plate-current values for various grid bias voltages and plate voltages. Another set of curves, known as the plate-current grid-voltage curves, is shown in Fig. 11-10. This indicates the plate-current values for changes of grid voltage and different values of plate voltage. The plate-current plate-

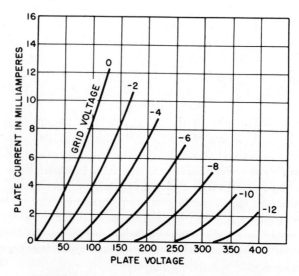

FIGURE 11-9. Plate-voltage plate-current curves.

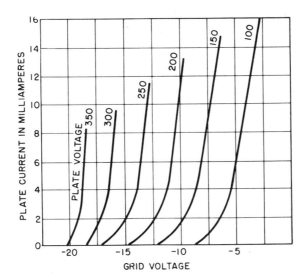

FIGURE 11-10. Plate-current grid-voltage curves.

voltage curves are obtained by maintaining a fixed-bias value and taking the current reading while changing plate voltages. When a set of points has been established and a line drawn through these points, the various current changes for one value of fixed bias will be obtained, to indicate the effect of plate-voltage changes. The plate-current grid-voltage curves are obtained by keeping the plate voltage constant and changing the grid voltage to obtain various values of plate current. This procedure establishes one plate-voltage curve. The plate voltage is then changed to a new value and held at this value, while the grid bias is again varied to obtain new plate-current readings. The manner in which such curves are utilized is indicated in the discussions that follow.

11-7. Amplification Factor

The *amplification factor* of a tube indicates the ratio of a plate-voltage change to a grid-voltage change, with the plate current held constant. This relationship may be set down as follows:

$$\mu = \frac{dE_p}{dE_g}\bigg|\ I_p \text{ constant} \qquad (11\text{-}1)$$

Here, μ (mu) is the amplification factor of a tube, and the lowercase letter d indicates a change [as an alternative the Greek capital letter delta (Δ) is often employed]. Thus, the amplification factor is an indication of the tube's ability to amplify an a-c *signal*. Hence, it represents a *dynamic* characteristic rather than a *static* characteristic, as is the case for the plate-voltage plate-current and plate-current grid-voltage curves. As seen from the formula just given, if a small voltage change on the grid produces a correspondingly large plate-voltage change, the amplification of the tube is high. The formula indicates that the plate current is held constant but this refers to the dc. During the actual amplification process, the small change of grid voltage, which produces a large change of plate voltage, would also produce a change of plate current from the value established by the dc power supply.

The amplification factor of a tube depends on the tube design and consists of such factors as element spacing and closeness of the grid to the cathode, as well as closeness of the mesh of the grid structure. When a smaller change of plate voltage occurs for a given grid-voltage change, a lower amplification factor is indicated. Thus, the ability of a tube to amplify a signal is referred to as the *amplification factor* of the tube.

The amplification factor of a tube can be ascertained from the static characteristic curves of the tube. A typical example can be provided from an inspection of Fig. 11-11. Since the amplification factor of a tube is found by dividing a change in plate voltage (with constant current) by a grid-voltage change, these variables can be obtained from the graph of Fig. 11-11. Assume

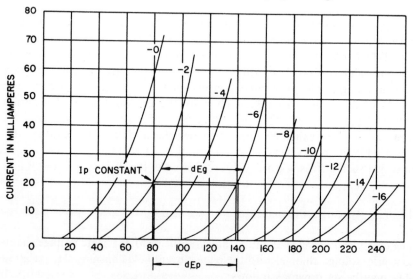

FIGURE 11-11. Graph of amplification factor.

that the plate-voltage change is that which is indicated between the two double vertical lines shown, 80 V and 140 V. This would give a change of 60 V. Using the 20 mA horizontal line as the constant current reference, the bias voltages between the plate voltage changes are ascertained. As shown in Fig. 11-11, the two bias points are 2 V and 6 V, giving a 4-V change. The amplification factor, when the voltage *changes* are substituted in the formula, is 15, as shown below:

$$\mu = \frac{140 - 80}{6 - 2} = \frac{60}{4} = 15$$

Another tube characteristic is the *plate resistance* (r_p). This indicates the opposition to the a-c signal current flow between the cathode and anode of a tube. The plate resistance is not a measurement of the plate voltage divided by the plate current in the absence of a signal but rather the opposition encountered for a *change* of plate voltage and a *change* of plate current. It is the ratio of a plate-voltage change to a plate-current change, with the grid voltage held constant. The formula is expressed as follows:

$$r_p = \frac{dE_p}{dI_p} \bigg| E_g \text{ constant} \qquad (11\text{-}2)$$

Thus, the plate resistance relates to the dynamic resistance characteristics of a tube and, for this reason, it is sometimes known as the *a-c plate resistance* or *dynamic plate resistance*. The plate resistance solved by the formula given above produces an answer in ohms.

As an example, consider again the graph of Fig. 11-11. To hold E_g constant, *one* bias line is chosen. Assume the bias line −4 V is used. A plate-voltage change is then selected and could be from 100 to 130 V. This represents a 30-V change. For the 100-V point on the −4 bias line, the current is approximately 15 mA, and for the 130-V point the current is approximately 45 mA (a current change of 30 mA, or 0.030 A). Setting these changes down gives

$$r_p = \frac{30}{0.030} = 1000 \ \Omega$$

Another characteristic of a vacuum tube is the ratio of a plate-current change to a grid-voltage change, with the plate voltage held constant. This is known as the *transconductance* (g_m) of a tube and is expressed mathematically as

$$g_m = \frac{dI_p}{dE_g} \bigg| E_p \text{ constant} \qquad (11\text{-}3)$$

The transconductance is an approximate figure of merit for the tube and indicates the amount of signal current change that is produced for a given plate-voltage change. The transconductance is also referred to on occasion as the *mutual conductance* of a tube and is a reciprocal function of the plate resistance. The formula solves for the unit quantity expressed in mhos. (It has been noted that a mho is the word *ohm* spelled backward.)

Using Fig. 11-11 again, an example of finding the transconductance is provided by choosing a constant plate voltage, such as 120. A current change can then be chosen, as from 5 to 32 mA, because these current values occur exactly where the −6 and −4 bias lines intercept the 120-V vertical line. Thus, the transconductance is

$$g_m = \frac{0.027}{2} = 0.0135 \text{ mho (or 13,500 } \mu\text{mhos)}$$

The relationships among plate resistance, transconductance, and amplification factor are

$$g_m = \frac{\mu}{r_p} \qquad \mu = g_m r_p \qquad (11\text{-}4)$$

In triode tubes, the plate resistance may vary from a low value of a few hundred ohms to several thousand ohms. The transconductance is usually designated in micromho values and may be equal to several thousand micromhos for the average triode tube. The amplification factor for triode tubes usually does not exceed 100 and in most instances ranges between 5 and 50.

Generally, in triode tubes where only the grid arrangement is altered to change the amplification factor of a tube, the plate resistance is related to the amplification factor. If the grid structure has a coarser mesh, it will have less influence on the electron flow and, hence, the amplification factor will be lowered. In such a tube, the plate resistance is also at a lower value than would be the case if the grid structure were finer. A more closely meshed grid structure enables the grid to have a greater influence on the electron flow so that the amplification factor and the plate resistance are increased.

At A of Fig. 11-12 is shown a simplified form of signal-voltage triode

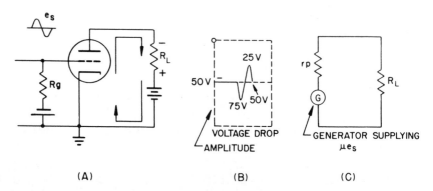

FIGURE 11-12. Amplification process.

amplifier. The input signal is applied across the grid and cathode circuit. As previously stated, the input signal will alternately increase and decrease the negative grid bias. Such an a-c signal can consist of electrically equivalent audio voltages or other signal energy that must be amplified. The *plus* alternation of the signal decreases bias and, hence, current flow *increases*. Since electron flow through the tube is in the direction shown by the arrows, the voltage drop across the load resistor will have a minus polarity at the anode side of the resistor. Therefore, a negative voltage drop exists across the load resistor. The increase in plate current flow through the anode circuit (because of the plus alternation on the grid) will also increase the voltage drop across the load resistor. Since this is a *negative-going* voltage drop, the *increase* in potential would have to be represented as a *decrease* in the negative-polarity direction, as shown at B.

In Fig. 11-12, it is assumed that the voltage drop caused by the power supply across the load resistance is 50 V. An increase in plate current may cause a 75-V negative signal to develop across the load resistor, as shown at B. When the incoming grid signal is at the midpoint between two alternations, the grid-signal voltage at that instant is zero and, in consequence, the bias on the tube is not altered from its original fixed value. Thus, the plate current returns to its normal value and the voltage drop across the load resistor also returns to its normal 50-V value. During a negative signal alternation at the grid of the tube, less current flows because the bias has been increased. Less current flow through the load resistor would decrease the negative voltage drop to a 25-V value. When the input-signal negative alternation again reaches zero, the normal grid bias brings the voltage drop across the load resistor back to 50 V, as shown at B. Thus, an a-c type of signal is reproduced in the plate circuit, though it is not a true a-c signal since it still has a d-c component. Because, however, the signal varies above and below a middle point reference line, this signal has a-c characteristics and can be converted to a pure a-c signal by coupling it to a next stage, via either a coupling capacitor or a transformer. (Only the a-c signal component will be transferred since dc does not go through either a capacitor or a transformer.)

The signal input (e_s) shown at A may have an amplitude of only a fraction of a volt and yet could cause the generation of a 25-V signal, as shown at B. (Actually, this is a 50-V signal, when considered from the peak of one alternation to the peak of the next. Insofar as the a-c component goes, however, the minus 50-V zero-signal reference line is considered as zero. Thus, as in all a-c voltage discussions, the signal has a peak value measured from zero to one alternation peak, or an *effective* value of 0.707 times the peak voltage.)

For the reasons outlined above, there is a 180°-phase difference between the input signal and the output signal of this single-stage amplifier. Thus, the amplified signal (μ_{es}) is 180° out of phase with the grid input signal. This factor is of considerable importance in many electronic circuits where the relative polarity of a signal must be ascertained. This is particularly true of video detectors and amplifiers in television receivers.

The circuit at A of Fig. 11-12 can be redrawn in a more basic form to show the electric characteristics, as at C. Here the cathode-anode sections of the tube are considered as a generator since in combination they generate an amplified version of the type of signal applied to the grid input. Because the plate and cathode circuits are instrumental in establishing a voltage drop across the load resistor, a representation of the tube as a generator for supplying the amplified signal energy is in order. Since all tubes have some internal plate resistance, the plate resistance of the tube shown at C has been designated as a series resistor, with the generator supplying the amplified signal-energy output. Obviously, current flow through the tube will develop a voltage drop across the plate resistance, as well as the load resistance. The changing current, representative of the a-c signal, will therefore cause an a-c signal-voltage drop across r_p and R_L. Thus, the full amplification factor at which a tube is rated cannot be realized across R_L since part of the signal voltage will develop across the internal resistance. If the load resistor has the same value as the plate resistor, the maximum amount of *power* will be transferred but the *signal voltage* obtained across the load resistor will not be at a maximum. As the value of the load resistor is increased over the value of the plate resistance, a proportionately greater voltage will be developed across the load resistor. In general practice (as more fully detailed later), the load resistor is never made more than three times the internal plate resistance. Such being the case, no more than approximately 75% of the rated amplification factor of the tube is realized.

11-8.   Tetrodes

The tetrode tube has four signal-active elements, as compared with three for the triode. In the tetrode, an additional grid is added, as shown in Fig. 11-13.

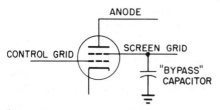

FIGURE 11-13. Tetrode tube.

This additional grid is placed between the anode and the control grid and is known as a *screen grid*. The screen grid is usually placed at *signal ground* by use of a bypass capacitor. Thus, the screen grid acts as an isolation factor between the anode (output) circuit of the amplifier and the input (grid) circuit.

The desirability for isolation comes about because of the existence of tube capacities. Such interelectrode capacities in a triode occur between the cathode and grid, between the grid and the plate, and between the plate and the cathode. Since these capacities have a measurable reactance, signal-voltage drops occur across such reactances and, thus, some of the amplified signal energy in the anode circuit is coupled to the input circuit, with the consequence that undesirable effects are created. The screen grid, however, isolates the output circuit from the input circuit by establishing an electrostatic shield between them.

The screen grid is operated at a plus d-c potential and, hence, is also instrumental in accelerating the electron flow from cathode to plate. Since the screen grid acts as an electrostatic shield between the anode and the grid (as well as between the anode and the cathode), the electron emission from the cathode is influenced only to a very slight degree by changes of plate voltage. Thus, the screen-grid element of the tube establishes a condition where the electron emission is influenced almost entirely by the control-grid and screen-grid voltages. Since the ratio of plate-voltage change with respect to plate-current change is now increased, the plate resistance of a tetrode is much higher than for a triode.

The amplification factor for a tetrode is also much higher since the ratio of a larger plate voltage to the grid-voltage change is greater. The positive screen grid diverts some of the electrons that would normally flow to the plate. These diverted electrons flow through the screen circuit and become the screen-grid current. Because some of the electrons that would normally flow to the plate are diverted, the transconductance is usually lower in a tetrode than in a triode.

If electrons strike a metal object with a high velocity, the impact of the arriving electrons will knock off other electrons from the metal plate. This condition is known as *secondary emission*. (The *primary* emission is the source of electrons from the cathode. The release of additional electrons because of the impact of the original electrons is a secondary emission process.) X Rays can be generated by the secondary emission process because they are formed by having a high-velocity electron beam strike a metallic object; the electrons that are deflected from the plate form the high-frequency X rays. In ordinary vacuum tubes, the velocity of the electrons is not sufficiently high to produce measurable amounts of X rays. In television picture tubes, however, X rays are produced when anode voltages higher than about 15,000 V are used.

Secondary emission characteristics are present in a tetrode when the screen-grid potential is higher than the plate potential. The higher *screen-grid* potential increases the velocity of the electron stream, and the lower *plate voltage* does not offer sufficient attraction to bring back the electrons that are knocked off. In consequence, such electrons that are knocked off the plate are attracted to the more highly positive screen grid. If the screen-grid voltage is held constant and the plate voltage is gradually increased, the electrons reach a higher velocity and the screen-grid current increases because of the absorption of secondary emission electrons. This characteristic is shown in Fig. 11-14, where the initial rise of current is indicated for low plate voltage. The current can rise gradually because the velocity of the electrons arriving at the plate is not great enough to dislodge secondary electrons. Once the plate voltage has reached a value equal to or higher than the screen-grid potential, the amount of secondary emission electrons is decreased at a rapid rate. In Fig. 11-14, the point where the current line levels out is where the plate

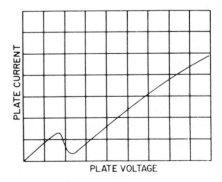

FIGURE 11-14. Tetrode characteristics.

voltage is sufficiently higher than the screen-grid voltage and, hence, attracts the secondary electrons knocked off by the electron stream. This attraction on the secondary emission electrons occurs because the plate now exerts a much greater attractive force than the screen grid. Hence, all secondary emission electrons that are dislodged are immediately returned to the anode of the tetrode tube.

The tetrode tube is still used in some electronic circuit applications, though this tube has been generally superseded by the five-element *pentode* type described next.

11-9. The Pentode

To minimize the effects of undesirable secondary emission from the plate, such as is encountered in tetrode tubes, an additional grid is added in the pentode tube. This additional grid lies between the screen grid and the anode of the tube, as shown in Fig. 11-15. The additional grid is known as the *suppressor grid* and is in the same form as the screen grid and the control grid; i.e., it is a spirally wound wire mesh affair.

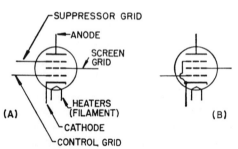

FIGURE 11-15. Pentode vacuum-tube symbols.

The suppressor grid is coarser in structure than either the screen grid or the control grid. This coarse structure minimizes the influence of the suppressor grid on the primary electron flow. The suppressor grid, however, is usually connected to the cathode. In some pentodes, the suppressor grid is connected to the cathode within the tube, while in other pentode types the suppressor-grid terminal is brought out of the tube so that it can be connected externally to the cathode or to ground, as shown at A and B of Fig. 11-15.

When the suppressor grid is thus placed at signal-ground potential, it has a negative polarity with respect to the anode. Hence, when electrons strike the anode with sufficient velocity to dislodge secondary emission electrons, the latter will be driven back to the anode because of the repelling force of the electrostatic charges established by the negative potential on the suppressor grid (like poles repel). Thus, the suppressor grid will effectively repel secondary emission electrons and prevent their reaching the screen grid, even though the screen potential may be somewhat higher than the anode potential.

The pentode tube has a higher plate resistance than triodes and tetrodes, and the amplification factor as well as the transconductance is also higher than for triodes and tetrodes.

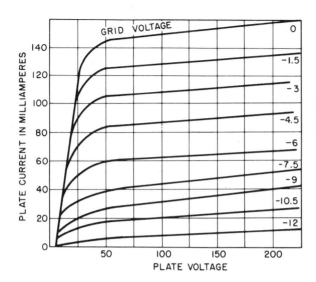

FIGURE 11-16. Pentode-tube characteristic curves.

The characteristic curves for pentodes differ from those for triodes and tetrodes, as shown in Fig. 11-16. For the plate-current plate-voltage curves of the particular tube represented, there is a sharp rise of plate current during an increase in plate voltage from zero to approximately 20 V. After that, however, a plate-voltage increase would cause little change in the plate current. During low plate voltages, the anode does not have sufficient attraction for the electrons to overcome the negative electrostatic barrier established by the cathode-connected suppressor grid. At such low anode voltages, a suppressor grid becomes a repelling factor and, thus, the screen grid, which is the first positive potential encountered by the electrons, will attract the electrons so that greater screen-grid current will flow but anode current will be low. As the anode voltage is raised, however, it will overcome the barrier established by the suppressor grid, and the increased velocity of the electrons, established by virtue of the plus screen grid, assures a virtually constant current with only a slight rise for increasing voltage. For a change of control-grid-signal voltage, however, a considerable change is established with respect to voltage amplification and, hence, the amplification factor (μ) of the tube is considerably higher.

A typical amplifier circuit for the pentode tube is shown in Fig. 11-17. The screen grid contains the usual screen voltage-dropping resistor and

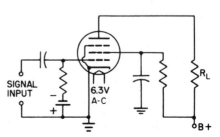

FIGURE 11-17. Pentode amplifier with fixed battery bias.

bypass capacitor, and the suppressor grid can be grounded directly to the chassis or to the cathode. As with the tetrode, the screen grid is placed at *signal ground*, to establish an electrostatic shield between the control-grid input circuit and the anode output circuit of the amplifier. The signal grounding of the screen grid also prevents signal-voltage variations across the screen voltage-dropping resistor. If signal-voltage variations were permitted across the screen-dropping resistor, the latter would share the functions of the load resistor. When the latter occurs, gain declines because in an ordinary pentode tube there is a fixed value of transconductance. This means that for a given grid input-signal-voltage change there is a corresponding plate signal-current change. Because the current flow in a tetrode or pentode tube flows to the screen grid of the tube as well as to the anode, it is evident that if the signal-current changes were also permitted to enter the screen-grid circuit, there would be a division of *signal current* within the tube, just as current divides in a parallel-resistive circuit. Since the transconductance of the tube establishes a fixed value of such signal current, a division of this current between the screen resistor and plate load resistor would mean that a maximum amount of signal current no longer flows through the load resistor because the screen resistor is sharing some of this signal current. Because the output signal is taken from across the load resistor only and not from the combination load resistor and screen resistor, any reduction in the signal current through the load resistor would alter the amplification of tube.

Pentode vacuum tubes can be employed in amplifier circuits for producing either signal-voltage amplification or power amplification. In either application, the pentode has a high degree of efficiency and can be employed for either audio or RF signal voltages. The disadvantage of the pentode, as compared to the triode, is that the pentode has a higher harmonic distortion than the triode, as more fully discussed later.

11-10. Special Tube Types

There are two basic types of pentode tubes: the sharp cutoff and the remote cutoff. The sharp cutoff tube is one in which an increase in the negative bias potential soon causes the grid to become so negative that it repels all electrons from the cathode and none reach the anode. Since the amplification factor of such a tube is fairly constant for changes of grid bias, the sharp cutoff tube is also known as the *constant mu* tube.

There are occasions when it is desirable to have a tube that does not have a sharp cutoff characteristic. When such is the case, a tube known as the *variable mu* tube is utilized. The variable-mu tube has a specially constructed control-grid structure. The pitch of the grid wires is constructed differently from an ordinary grid: the grid wires are given greater spacing around the midpoint than at the ends. Such a grid structure still permits varying the current flow by changing grid bias. As the grid bias is increased, less current flows, just as with an ordinary grid, except that at high negative potentials the grid is unable to cause a complete cessation of current flow because of the wide spacing of the grid wires. The wide spacing still permits electrons to go through the grid structure and reach the anode, at negative grid potentials that normally would cause tube cutoff. In such a tube, the amplification factor varies with each change in grid-bias voltage. Hence, the designation *variable-mu* or *remote cutoff* tube. The tube is also known as a *super-control tube.*

The variable-mu tube had been used extensively in tube-type radio, FM, and television receivers in special circuitry to vary the bias of the tube for different levels of signals that arrive at the receiver. Such bias changes are useful since they prevent overloading from strong stations and also increase the amplification factor of the tube for weak stations. (In radio receivers, the process is known as *automatic volume control;* in television receivers, the picture contrast is also controlled and here the circuit is known as *automatic gain control.* The abbreviation for automatic volume control is AVC, and for automatic gain control, AGC. These two systems are discussed more fully in subsequent chapters.)

Pentodes, like triodes and tetrodes, are designed for use as either signal-voltage amplifiers or signal-power amplifiers. An improvement over the pentode power amplifier is a tube known as the *beam power tube.* This tube type has a special internal construction that reduces the harmonic distortion below that generated in pentodes and tetrodes. (Harmonic distortion is discussed more fully in Chapter 15.) The beam-power tube also permits greater power handling with higher transconductance. This tube, however, has an amplification factor that is lower than the pentode-type tubes, though higher than triodes since most of the latter do not have an amplification factor in excess of 100. The plate resistance of the beam-power tube is also lower than that of the pentode types but higher than in triodes.

In the beam-power tube the control-grid and screen-grid wires are aligned during manufacture to cause the beam to leave the cathode in layers. Beam-forming plates are also used to concentrate the electron stream into a narrow path. The result is that the electron beam is, in effect, focused to the anode in a concentrated beam; hence the name *beam-power* tube. The concentrated beam and its increased velocity result in greater power output as well as a high degree of circuit efficiency and low harmonic distortion. Secondary emission is reduced because the beam-forming plates are also connected to the cathode, thus acting as suppressor grids. Typical beam-power tubes included the 6V6, 6L6, 6CD6, 12BQ6, and 17DQ6A.

Numerous other tube types had been developed, among them multielement types and combinations of several different tube types contained in one glass envelope. Typical special-purpose tube symbols are shown in Fig. 11-18. At A the five-grid tube was extensively used in tube radios for mixing the

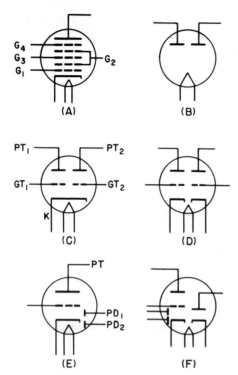

FIGURE 11-18. Types of multielement tubes.

incoming RF signal (applied to one of the grids) with a signal generated by an oscillator (and applied to another grid). The dual-grid element (G_2) shielded the two input sources and isolated the input and output sections. Such tubes were designated as *pentagrid converter* types and included the 6SA7 and 6BE6.

At B is shown the dual-diode tube used for full-wave rectification in power supplies. For home receivers, hi-fi units, etc., solid-state diodes have taken over the function. Industrial and transmitting applications utilizing the dual diode are more fully discussed later.

Two triodes in one envelope are shown at C, with a single cathode serving both tubes. For this tube, letter and number symbols are used for identifying the elements. Thus, the grid for the first triode is marked GT_1 and the associated anode for this grid is marked PT_1. Another dual triode is shown at D, except here separate cathodes are used for the individual triodes.

At E is shown a single tube containing two diodes and a high-mu triode, with the diode plates identified as PD_1 and PD_2. At F is shown a tube containing three diodes as well as a high-mu triode that had been popular in combination FM and AM tube-type radios, where the tube combined the detector for the FM signals with the detector for the AM signals, plus also embracing the first audio-amplifier tube.

Some special tubes of this type have been named *Compactrons* because they form a *compact* assembly to save space. Compactrons include the 12AL 11, which serves as a dual-pentode audio detector and output audio amplifier; the 33GY7, which is a pentode-diode; and the 17JZ8 triode-pentode, as well as others.

Other special-purpose tubes are also employed in electronics, including the Magic Eye tube in receivers, which helps the station tuning process by closing when the tuning is correct, or in test equipment. Other special-purpose tubes are the cathode-ray tubes used in television receivers and oscilloscopes. Both of the foregoing tubes use a phosphor coating that will glow when struck by an electron beam, as more fully detailed in Chapters 21 and 22.

Typical low-power transmitting tubes are shown in Fig. 11-19. Often

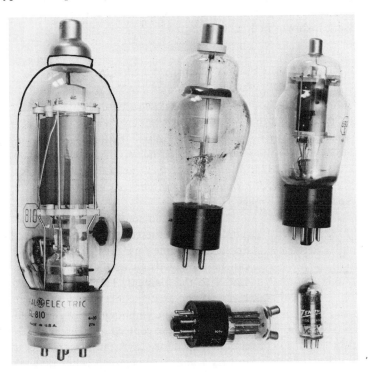

FIGURE 11-19. Typical vacuum tubes

the anode element connection is brought out at the top to isolate this high-voltage point from the grid and cathode. As shown, the grid connection may also be brought out separately for isolation purposes. Tubes often have a small pin at the sides for proper placement in the socket, or the tube prongs themselves have wider spacings at one part for placement identification. On other occasions a center post has a ribbed side for proper insertion into the socket. Tube manuals indicate the socket connections for a particular tube and give the particular tube-prong numbers involved. The socket connections

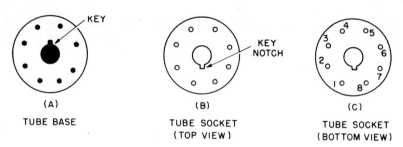

FIGURE 11-20. Octal socket keys and prong hole numbering.

shown in tube manuals are the bottom view, and the numbering system starts at the left and goes in a clockwise direction. Typical is the octal socket identification shown in Fig. 11-20.

11-11. Tube-Numbering System

Many modern tubes are numbered so that some identification is possible, though, of course, all the information pertaining to a particular tube cannot be obtained from an inspection of the number alone. The initial digit or digits indicate the voltage applied to the heater (filament) terminals. Thus, a 6L6 tube indicates that approximately 6 V (actually 6.3 V) can be applied to the filaments. A 12AT7, on the other hand, is a 12-V (actually 12.6 V) tube with respect to the filament potential.

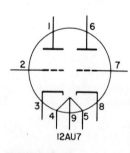

FIGURE 11-21. Tube with tapped filament.

A tapped filament arrangement is sometimes employed with the tubes designed for a-c filament applications. One such tube is the 12AU7A, which can be used with an application of either 6.3 or 12.6 V to the filament. This tube is shown in Fig. 11-21. If the heater connections No. 4 and 5 are tied together, the combination can be used in conjunction with the center tap (pin 9) for 6.3-V application. If used for 12.6 V, the latter voltage is applied to pins 4 and 5. Other such tubes that take either 6.3 or 12.6 V are the 12AV7, 12AX7, etc.

In most modern general-type tubes, one or two letters follow the initial number or numbers. Originally, these letters had specific significance but the large number of tubes that are currently employed has required the use of so many single and double letters that it is difficult to assign a specific meaning to the particular letter employed. Consequently, reference should be made to the tube manual for an indication of whether the tube is a converter, an amplifier, or a rectifier.

The numeral that follows the letters sometimes represents the number of elements within the tube and on other occasions indicates the useful elements or prongs employed. Consequently, not much information can be gained from the last number. Additional letters sometimes follow the last number, such as G or GT. The letter G following the tube symbol indicates "glass" and identifies the tube as having a standard-sized glass envelope. The letters GT indicate "glass-tiny" and show that a glass envelope is used having smaller dimensions than the G-type tube. Except for miniature types, modern tubes without the G or the GT following the last number are metal-envelope tubes.

From the foregoing, it is evident that the only correct tube-numbering identifications are the first two numbers, which indicate the approximate heater or filament voltage, and the last two letters, which refer to the size of the glass envelope.

11-12. Series-Filament Strings

The filaments of some tubes have been designed to operate at voltage sufficiently high (35, 50 V, etc.) so that a specific number of them can be used in series across the 120-V a-c line voltage. Tubes so used are also designed to draw the same current; otherwise shunting resistors must be employed. When a number of tubes are used in a series string that have a lower total voltage rating then the voltage source, a series resistor can be used as shown in Fig. 11-22. Such a resistor is usually a negative-temperature-coefficient type known

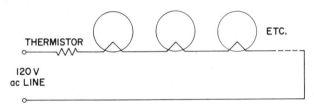

FIGURE 11-22. Series filament string.

as a *thermistor* and precedes the string as shown. The resistance of a thermistor decreases with an increase in temperature and is high when the thermistor is cold. Thus, the thermistor holds down the filament current when the voltage is applied to the string so as to minimize a voltage surge. (When tube filaments are cold, their resistance is low and, hence, an excessive amount of current may flow when the electronic device is first turned on.)

A number of electronic devices (including some television receivers) use series heater strings made up of tubes specially designed for this purpose. Such tubes permit series-filament operation without the necessity for using shunt resistors across some of the tubes to equalize currents. Some series string tubes are designed for 600-mA filament current operation, while others have a rating of 450 mA. Such tubes have identical warm-up characteristics (approximately 10 s) so that surge voltage problems are reduced to a minimum. At the same time, the thermistor-type resistor is not needed and if additional resistance must be placed in the series string to bring the total voltage drops up to the line voltage potential, ordinary resistors can be employed. Tubes of this type include the 3BZ6, 3AU6, 5T8, 10DE7, and others.

11-13.   Gas-Type Tubes

In some industrial-electronic applications, gas-filled tubes are used instead of the high-vacuum types. The gas employed is usually a vapor derived from mercury, neon, or argon. The characteristics of a gas-filled tube differ to a considerable extent from those of the high-vacuum type. With the gas-filled tube, the plate current increases from zero as the plate voltage rises from zero, in a fashion similar to that for the high-vacuum type. At a certain plate potential, however (usually above 10 V), the electrons have sufficient velocity so that when they collide with the gas atoms they will cause ionization. (Ionization is discussed in Sec. 1-9.) The ionized gas neutralizes the space charge at the cathode so that no barrier is present to limit current flow. Thus, as soon as ionization occurs, the plate-current flow reaches maximum.

The space charge is neutralized because the positive ions are attracted by the electrostatic fields of the negative cathode. Upon reaching the cathode region, the ions capture free electrons to replace those that had been lost during ionization. Thus, the space-charge electrons are constantly removed to re-form the gas atoms with a consequent neutralization of the space charge.

Once ionization occurs in a gas-filled tube, the plate resistance is extremely low because the space charge has been neutralized. Because of their characteristics, gas-filled tubes find applications in commercial power supplies (as discussed in Chapter 14) where they act as highly efficient rectifiers capable of handling high power. Such diodes are not employed for receiver purposes since the ionization process generates a high noise that must be filtered from the power supply. Also, unless precautions are taken, the danger of voltage breakdown and arcing is also greater for the hot-cathode gas diodes used for rectifiers. A typical gas-filled rectifier is shown in Fig. 11-23. The black dot within the tube circle of the symbol indicates that the tube is a gas-filled type.

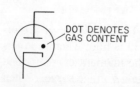

DOT DENOTES GAS CONTENT

FIGURE 11-23. Symbol for gas-filled tube.

Gas-filled tubes are also employed for voltage-regulation purposes, as described in Chapter 14. The voltage-regulator type of gas-filled tube is of the cold-cathode type, in which ionization occurs because of the electrostatic charge built up between the cold cathode and the anode. After a certain grid potential is reached, ionization occurs and the tube conducts heavily. As with the high-vacuum diode rectifiers previously discussed, the current flow is only in one direction during tube conduction and, hence, gas diodes are suitable for rectification and voltage-regulating purposes.

11-14. Thyratrons

Hot-cathode gas-type tubes are also available with a control electrode between the cathode and anode. The control electrode is similar to the control grid of an ordinary high-vacuum tube but affects current flow differently. If a high negative voltage is applied to a control electrode of a gas-filled tube, and this voltage is gradually reduced, it will be found that plate-current flow starts suddenly and immediately reaches a maximum value once the control-electrode bias voltage is insufficient to hold the tube in a nonconducting state. Once current flow has started and the tube is ionized, the control electrode loses its control on current flow. Thus, during tube conduction, the grid bias can be increased far beyond the normal cutoff value without affecting the plate-current flow. The *plate voltage* must be *lowered* below the ionizing potential before plate current will cease. When current flow stops, the control grid will again hold the tube in nonconduction as long as the high negative bias voltage is maintained in the control electrode.

The reason for this conduction behavior, which differs so radically from that of the high-vacuum tube, is the presence of the positive ions that result when current starts to flow. These ions, which have a plus polarity, are attracted to the negative grid since opposite poles attract. The positive ions surround the grid and shield its electrostatic fields. The shielding effect prevents the grid from repelling electrons and controlling current flow. As pointed out earlier, the positive ions also neutralize the space-charge region at the cathode so that the limitations normally imposed on current flow are removed.

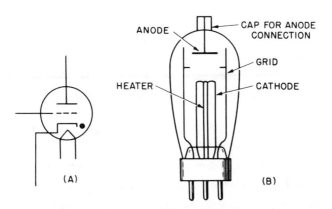

FIGURE 11-24. Thyratron tube symbol and construction.

The thyratron-type tube has internal elements as shown at A of Fig. 11-24. This tube is useful as a triggering device for electronic equipment. The large commercial thyratrons have a cylindrical control electrode, as shown at B. Even with the large types employed in commercial applications, only a small grid potential is necessary to trigger the tube into sudden and high conduction. Some thyratrons use mercury vapor, whereas others use argon or hydrogen. With hydrogen a more rapid triggering action is obtained, and the sudden current rise within the tube is useful in the formation of square waves or sharp pulse spikes having an extremely rapid rise time. The usefulness of square waves is discussed more fully later. (Additional thyratron and other gas-tube types are discussed in Chapter 14.)

Review Questions

11-1. Explain briefly the basic principles of *thermonic* emission.

11-2. Is a tube having a filament, cathode, and anode a diode or a triode? Explain your answer.

11-3. Of what significance is the ground connection shown in schematics?

11-4. Define the terms *saturation* and *space charge* in relation to vacuum tubes.

11-5. List some of the uses for diodes, triodes, and pentodes.

11-6. Briefly explain the purpose for placing a negative d-c bias potential on the grid of a vacuum tube used for amplification of signals.

11-7. Explain what information must be available for plotting the type graph shown in Fig. 11-9, which is representative of the characteristics of a typical vacuum tube.

11-8. Briefly explain the terms *amplification factor* and *dynamic characteristics*.

11-9. How does the *plate resistance* of a vacuum tube differ from d-c resistance?

11-10. What does the transconductance of a tube indicate? What is the unit value of the transconductance?

11-11. Briefly explain the purposes for the the screen grid and the suppressor grid in a vacuum tube.

11-12. Why is the amplification factor of a pentode higher than that of a triode?

11-13. What is the difference between a variable-mu tube and one having sharp cutoff characteristics?

11-14. In a 6SQ7-GT, what does the first digit represent, and what is the significance of the final two letters?

11-15. What is a *thermister*, and of what value is it in electronic circuitry?

11-16. Of what value are gas-filled tubes, and how are they identified schematically?

11-17. Briefly explain the differences between the control factors of the control grid in a vacuum tube and those of the control electrode in the thyratron gas-filled tube.

11-18. What gases are used in thyratron tubes? Does any type have an advantage over others? Explain.

Practical Problems

11-1. In a design laboratory a 6W6GT tube was tested while connected to a triode (screen grid connected to anode). With a bias of -15 V, 100 mA of plate current flowed with an anode voltage of 200. When the grid-bias voltage was changed to -30 V and the plate voltage raised to 310 V, the same plate current flowed (100 mA). What is the amplification factor of this tube?

11-2. The same tube as in Problem 11-1 was operated at -30 V bias. At 200 V on the anode, 10 mA of plate current was read. At 300 V on the anode, the plate current rose to 85 mA. What was the plate resistance?

11-3. A 12AU7A medium-mu triode was analyzed for operation in an industrial control system. At 200 V, the plate current was 5 mA with a bias of -8 V. When the bias was decreased to -4 V, the plate current rose to 15 mA. The plate voltage was held at a constant amplitude. What is the transconductance?

11-4. In the design of a vacuum tube it was found that a plate current of 45 mA flowed when the plate voltage was 100 and the bias -5 V. When the bias was changed to -10 V, the plate voltage had to be raised to 200 to have the same 45-mA plate-current flow. What was the amplification factor of this tube?

11-5. An amplifier tube for a transmitting system has the following characteristics: $I_p = 10$ mA when $E_p = 90$ V; $I_p = 2$ mA when $E_p = 40$ V. What is r_p?

11-6. What is the transconductance of a vacuum tube used in an electronic testing device if a change of two grid volts produces a change of plate current of 2 mA?

11-7. In the vacuum tube graphed in Fig. 11-11, what is the plate resistance for a voltage change from 80 to 100 V on the -2 V bias line?

11-8. For the tube graphed in Fig. 11-9, calculate the transconductance for the following operating conditions:

$$E_p = 150 \text{ V}, \qquad E_g = -4 \text{ V}, \qquad I_p = 4 \text{ mA}$$

11-9. What is the amplification factor of the tube graphed in Fig. 11-9 for the following operating conditions:

$$E_p = 150 \text{ V}, \qquad E_g = -4 \text{ V}, \qquad I_p = 4 \text{ mA}$$

11-10. Tube No. 1 at 200 V draws a plate current of 30 mA, at 2 V of bias. At 6 V of bias, the plate voltage must be raised to 600 V to establish the same plate-current flow of 30 mA. Tube No. 2 at 200 V draws a plate current of 30 mA at 3 V of bias. At 6 V of bias, however, the plate voltage must be increased to 350 V to have the same 30 mA of current flow. Which tube, No. 1 or No. 2, is a better amplifier?

Solid-State Fundamentals   12

12-1. Introduction

Solid-state diodes, triodes, and similar devices have circuit applications similar to vacuum and gas tubes but differ radically in the manner in which they function. No cathodes, grids, or anodes are present, though comparable elements exist that perform the same tasks. The absence of a filament or heater eliminates the warm-up time required in tubes and contributes to economical operation. In addition, the solid-state units are considerably smaller than tubes with the same power-handling ratings. Because of their long life and other advantages, solid-state devices are extensively employed in numerous circuits that formerly utilized gas or vacuum tubes.

The design and function of solid-state diodes and transistors are related to their crystalline structure and atomic characteristics. Hence, Sec. 1-9 should be reviewed because the principles contained in this chapter are an extension of that material. The fundamental aspects of diodes and transistors are covered in this chapter and transistor circuitry is treated in Chapter 13.

12-2. The Covalent Bond

The covalent bonding characteristics found in hydrogen and atmospheric oxygen was discussed in Chapter 1. The covalent bond also occurs in germanium and silicon, which are the building blocks for solid-state devices. Because of the crystal structures formed, the covalent bond found in carbon is closely related to the germanium and silicon bonds. In order to understand the particular binding characteristics that take place with any one of these three elements, note the outer ring valence electrons shown for each element in Fig. 12-1. In each of these elements, there are only four electrons in the outer ring.

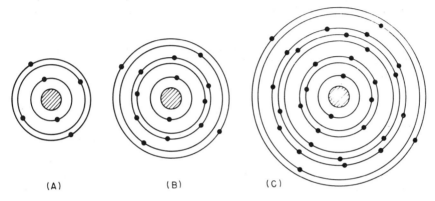

(A)   (B)   (C)

FIGURE 12-1. Atoms of carbon, silicon, and germanium.

The silicon atom uses three primary rings but only *two* subshells of the outer ring contain electrons. What would constitute the third subshell is an empty orbit. Hence, the first subshell of the outer primary shell group is completely filled with two electrons but the second subshell requires four more electrons to fill it completely. When the various silicon atoms are brought together, there is a sharing of valence electrons by the atoms and a crystal-lattice network is formed. The crystal is not in the cube arrangement as shown earlier in Chapter 1 for sodium chloride but instead is of a tetrahedron arrangement as shown in Fig. 12-2. Note that the arrangement

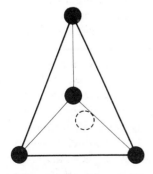

FIGURE 12-2. Formation of crystal lattice.

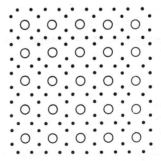

FIGURE 12-3. Crystal-lattice network.

is in the form of a four-sided structure with atoms at each corner, plus an atom at the center of the structure. As shown, the central atom (indicated by the dotted outline) is surrounded by four other atoms. Each of the surrounding atoms shares one of its outer ring electrons with the central atom. Because of the crystal-lattice network, *each* atom within the structure is surrounded by four other atoms. This can be illustrated more clearly in two-dimensional forms as shown in Fig. 12-3. In this illustration only the outer ring electrons and the nucleus of each atom are shown, for the sake of simplicity. A tight covalent bond is formed because of the sharing of valence electrons by the various atoms.

As seen from the illustration, the sharing of adjacent electrons by any particular atom means that each atom has eight electrons in the first two shells of the outer ring. This comes about because the electrons in the outer rings not only revolve around their own nucleus but also revolve around the nuclei of adjacent atoms. The electrons are still rigidly bound to any particular nucleus around which they revolve, even though shared by adjacent atoms. Because of the filling up of the two outer subshells, the entire structure presents a highly stable state of an element with a considerable degree of hardness and rigidity. Such a structure does not lend itself to the carrying of electric energy by means of electric motion because the outer ring electrons are not free but, instead, are rigidly bound in their orbital paths.

The same crystal-lattice network covalent binding described before occurs when either germanium atoms or carbon atoms are brought into close proximity. With carbon atoms, the covalent bonds produced represent the carbon crystal (diamond). Note, from Fig. 12-1, that carbon has four electrons in the second ring. Since this ring can accommodate only eight electrons (in both subshells), this means that a crystal-lattice network formed from carbon closes the entire outer shell. This condition of a completely filled outer shell forms an extremely stable and rigid bond, and this is the reason for the particular hardness and brittle characteristic of a diamond. With germanium and silicon, however, only two of the outer ring subshells are completely closed and while this forms a stable bond and also a crystal, the fact that the outer shell is not completely filled means that the silicon and germanium bonds can be broken more easily than the carbon crystal by the application of voltage, heat, or other energy.

For a clearer illustration of the foregoing, partial orbital shells of the three elements are shown in Fig. 12-4 and represent *valence-bound* atoms. Thus, the second primary ring of carbon is completely filled so that the crystal carbon is extremely hard (diamond). For silicon, however, only the first two subshells of the third primary ring are filled. Although this still forms a fairly rigid bond, the fact that the third subshell is empty means a less rigid binding than occurs with carbon. With germanium, also, only two subshells of the outer ring are filled in a valence-bound atom, and again there is a less rigid bond than occurs for carbon.

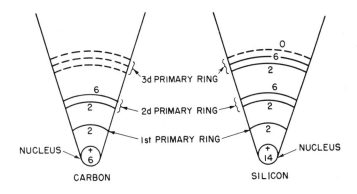

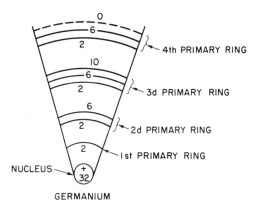

FIGURE 12-4. Valence-bound atoms of carbon, silicon, and germanium crystals.

In Fig. 12-4, it will be noted that if an effort were made to move an electron out of the outer subshell of the carbon atom, sufficient energy would have to be applied to break the valence bond and move the electron over a gap into the third primary ring. With silicon and germanium, however, only sufficient effort is necessary to bridge the space between a subshell since the third subshell of the valence ring is empty.

Electrons nearest the nucleus are at low energy levels, whereas those in the outer orbits are at higher energy levels. In a solid such as a crystal, the energy levels of individual atoms re-form and create *bands of energy levels*. The upper bands of the germanium crystal are

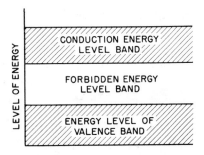

FIGURE 12-5. Energy levels of an atom in a crystal-lattice network.

shown in Fig. 12-5 and consist of the valence-band energy level, the forbidden band, and the conduction energy level band. In the valence band, the electrons are tightly bound and, hence, are not free to carry current easily. The forbidden band is the gap between the conduction band and the valence band as shown. It is this forbidden band (where electrons cannot remain) which must be crossed by electrons which are to be moved from the valence band to the conduction band. The latter is a band where the energy level of the electrons is sufficiently high to permit electron movement and current flow.

In an insulator, the forbidden band is so large that few electrons can be provided with sufficient energy to reach the conduction band. With the germanium or silicon crystal, however, the forbidden band is more narrow, and normal temperature ranges provide sufficient energy for the valence electrons to permit them to reach the conduction band. The number of electrons that will reach the conduction band depends on the width of the forbidden band, as well as on the amount of energy applied, either by temperature or electric pressure. (Light can also be a source of energy for electron movement, as described more fully for photoelectric devices.) Thus, these crystals are known as *semiconductors*. For ordinary conductors, such as copper, and silver, many free electrons are present at room temperatures and the forbidden region is extremely narrow (or nonexistent) so that the valence band and the conduction band are virtually one.

12-3.   Current Flow Versus Hole Flow

We have already learned that current flow in a conductor consists of an electron movement under the influence of energy pressure, such as electromotive force. This normal current flow can also occur in the case of the semiconductor, where the covalent bond has been broken and an electron of the atom moves into the conduction band. This factor of an electron leaving the valence band and moving into the conduction band creates a condition that is peculiar to transistors. This curious condition is the formation of *holes* that can, in fact, be considered as current carriers, just as electrons are carriers of current.

Figure 12-6 will help clarify the conception of holes in transistors. When an electron moves out of the valence band and into the conduction band, it leaves a vacancy in the atom and the latter becomes a positive ion. Since the hole creates a positive area in the subshell from which the electron is removed, the hole can be considered as having a positive charge, just as the electron has a negative charge. The nature of the hole in the atom is such that it can sustain current flow by electron and hole movement from *one valence band to another*, without the electrons moving in the conduction band. Assume, for instance, that the free electron of the atom shown

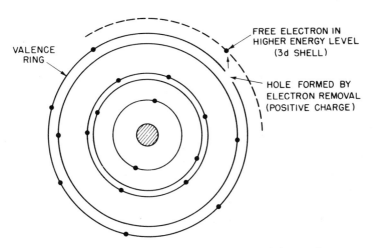

FIGURE 12-6. Silicon crystal atom with energy applied to an electron.

in Fig. 12-6 has moved on, leaving a positive ion. It is then quite possible for an electron from the valence band of an adjacent atom to break its bond and move into the hole in its neighboring atom. When this occurs, the atom that originally had the hole is now a *neutral* atom again, having regained the missing electron. The adjacent atom, however, now has a vacancy (hole) where the electron broke its bond and moved into the hole of the neighboring atom. Thus, even though electrons have moved along (current flow), the energy was confined to the valence bands because of the hole movement.

Figure 12-7 will aid in acquiring the concept of hole movement. Here each numbered circle represents an electron that is going to break its covalent bond and move to a neighboring atom at the right. Note that, at A, a hole already exists between electrons No. 5 and No. 6. As electron No. 5 moves to the right and into the hole of the adjacent atom, the hole (in effect) moves to the left and now occupies the space formerly held by electron No. 5, as at B. Next, electron No. 4 breaks its covalent bond and moves into the hole of the atom at the right, to create the condition shown at C. Thus, it is evident that *hole flow is opposite to current flow* in the semiconductor material.

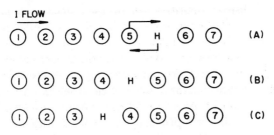

FIGURE 12-7. Current flow versus hole flow in transistor.

12-4. Impurities

So far our discussions have covered the factors relating to the pure germanium or silicon crystal, plus the concept of hole flow versus current flow. In the manufacture of a transistor, however, the pure crystal structure must be modified by the addition of another element before the transistor can perform the general functions of the vacuum tube. This modification of the crystal consists of adding a controlled amount of so-called impurities to the crystal structure. There are two such impurity types utilized, one type having atoms of three valence electrons and the other type having atoms of five valence electrons. To aid in understanding how each contributes to transistor function, each will be treated separately.

The impurities mentioned above are actually pure elements as such and are referred to as impurities only with respect to the germanium or silicon crystal structure. Considering the germanium or silicon elements as the pure or original state, the addition of any other elements labels the latter as so-called impurities. The impurities that have three valence electrons include

Acceptor (P) impurity	Atomic number	Valence electrons
Boron	5	3
Aluminum	13	3
Gallium	31	3
Indium	49	3

These impurity elements have an important function in transistors so that it is worthwhile to analyze their effect when they are combined with the pure crystal elements. Any one of the four impurities listed in the chart can be employed to form what are known as positive areas (referred to as P zones). The P-zone designation means that a zone within the crystal structure has been formed by the impurity, such a zone or area having a positive relationship with respect to the surrounding areas. Any one of the four listed impurities can be used since each has only three valence electrons, even though their atomic numbers differ. This can be proved by setting up the rings and subshells in a drawing and allocating the proper number of electrons successively from the first ring to the last subshell, in accordance with the number of electrons indicated by the atomic number. As an example, in boron, with an atomic number 5, the first primary ring is filled with its quota of two electrons, as is the first subshell of the second ring (see Fig. 12-8). Since the atomic number 5 indicates a total number

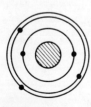

FIGURE 12-8.
Boron atom (3 valence electrons).

of five electrons, four are used up in the first ring and first subshell of the second ring. Thus, the second subshell contains only one electron. Hence, the valence ring (outer ring) contains only three electrons.

When boron (or one of the other acceptor impurities) is inserted into a crystal-lattice network, covalent bonds are again formed, as previously seen. Now, however, the impurity atom contributes only three valence electrons to the covalent bonds that are formed. This is seen in Fig. 12-9, which

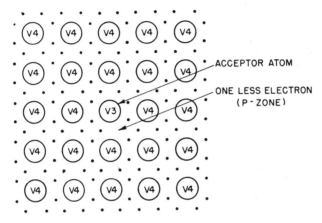

FIGURE 12-9. Crystal atoms with one impurity atom added to form *P*-zone.

again shows a crystal-lattice network which could be either germanium or silicon. For convenience, only the outer ring valance electrons are shown for each atom. Each atom is marked V3 or V4, to indicate the number of valence electrons for that particular atom.

The number of impurity atoms that are inserted is relatively small as compared to the surrounding germanium or silicon atoms. Often, impurity atoms are inserted in the proportion of one part impurity to 10^{15} germanium or silicon atoms.

As shown in Fig. 12-9, the impurity atom, with its valence ring containing only three electrons, creates a deficiency of electrons with respect to the valence bonds that are formed. Previously, it was mentioned that each atom shares its electrons with four adjacent atoms and, in consequence, each bound atom has the equivalent of eight electrons in its valence ring. With eight electrons in the valence ring of either germanium or silicon, the two subshells of the outer ring are completely filled, as mentioned earlier, and hence a rigid bond is formed. The impurity atom, however, contributes only three electrons to the valence binding process. Hence, the circulation and sharing of the neighboring electrons forms an *area* or *zone* that is deficient in one electron and creates a situation equivalent to the hole previously described. Since there is one electron less in this area, it is known as a *P* zone, indicating a *positive* area. Since this atom has only three valence electrons, it has the ability to accept another electron in order to fill completely the valence ring. Hence, this impurity atom is known as an *acceptor atom*.

As mentioned previously, the existence of a hole in a crystal structure creates a situation where a valence-bound electron from an adjacent atom can move into the hole, causing another hole to appear in the atom formation that the electron has left. Thus, hole flow can constitute current flow, in a manner different from the movement of free electrons in the conduction band. It is sometimes convenient to think of the current carriers within such a structure as the holes rather than the electrons to distinguish this type of current flow from that where free electrons in the conduction band move along to create current flow.

The impurity *atom* does not actually have a deficiency of electrons because in its normal state it only has three valence electrons. It is only when such an atom is placed within the germanium or silicon crystal network that a deficiency exists *with respect to adjacent atoms*. When an electron moves into the hole of the *P* zone and enters the valence ring of the impurity atom, the latter becomes a negative ion. Even though the entry of an electron into the orbit of the impurity now fills the outer two subshells completely, there is not so rigid a bond created as in the case of pure germanium or silicon atoms because the impurity atom has an unwanted electron in its orbit. It is, therefore, relatively easy to move this electron on. How these factors are an important aspect of transistors will be covered in greater detail after the following discussion of the other type of impurity.

The second type of impurity that is utilized in transistors consists of elements having 5 valence electrons. Elements of this type are

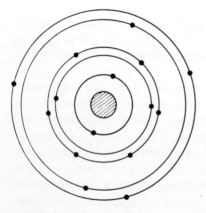

FIGURE 12-10.   Phosphorus atom (5 valence electrons.)

Donor (*N*) impurity	Atomic number	Valence electrons
Phosphorous	15	5
Arsenic	33	5
Antimony	51	5

In the impurities just listed, each valence ring contains 5 electrons, even though the atomic numbers of the impurity elements differ. Again, if the atom structure were laid out on paper and the electrons indicated by the atomic number placed in their respective orbits, it would be found that the outer ring contains 5 electrons. As an example, the element phosphorus is shown in Fig. 12-10. Phosphorus has an atomic number 15; hence, in its neutral state, there are 15 electrons in orbit. As shown, the first ring contains its quota of two electrons. The second ring has two subshells, with the first subshell containing its quota of 2 electrons and the second subshell 6 electrons, giving a total of 8. These 8 electrons plus the 2 in the first ring, total 10 electrons, leaving 5 for the next ring. The next ring (the third ring) can have three subshells but only two are employed for the remaining 5 electrons, as shown.

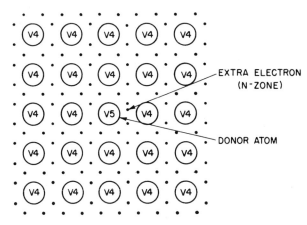

FIGURE 12-11. Crystal atoms with one impurity atom added to form *N* zone.

Figure 12-11 shows the crystal-lattice network, with each crystal atom marked V4 to indicate the normal four valence electrons. The impurity atom is designated V5, to indicate the impurity atom with its five valence electrons. Of the latter, however, only four valence electrons are used to form the covalent bond; hence, there is an extra electron that will be shared by adjacent *atoms*. Thus, the extra electron can be moved readily out of this zone or area and is not needed for binding purposes. Because the impurity injects an extra electron into the area, the impurity atom is known as a *donor atom* and forms a negative zone (*N* zone) because an excess of electrons is designated as negative, while a deficiency of electrons is designated as positive.

The extra electron contributed by the donor atom is in the nature of a free electron in the conduction band. When such donor electrons carry current, the situation is similar to that which exists in copper, silver, or other conductors. Upon the application of fields generated by electromotive force, for instance, the electrons move along in succession and constitute normal current flow in the conduction band, as opposed to the current flow that exists by virtue of hole movement. As with the acceptor impurity, the donor atom is a neutral element. Hence, when an electron moves away from the donor atom, the latter becomes a positive ion.

12-5.   *P* and *N* Junctions

Heretofore, the pure state germanium and silicon crystals were discussed, and the factors involving the formation of *P* and *N* zones were also covered. When *P* and *N* crystals are joined to form a union, the junction device can be compared to the diode vacuum tube in its general characteristics. Hence, a *PN* junction will exhibit a high resistance characteristic in one direction and a low resistance characteristic in the other direction. The basic *triode*-type transistor is formed either by joining *PNP* sections (in that order) or by combining *NPN* sections. Initially, however, it is necessary to investigate more fully the characteristics of the simple *PN* formation, as a basis for understanding the more complex *PNP* or *NPN* transistor function.

A *P*-type crystal, by itself, represents a neutral-state material because the net negative charge of the electrons in any single atom is balanced by the net positive charge of the nuclei. (Even the *P*-type impurity atom, with its three valence electrons, represents an electrically neutral atom.) If there is a hole movement at the valence band, the electrons moving from one atom to the other do not disturb the electrically neutral state since the overall charges of the total electrons and the total nuclei still balance. With the *N*-type crystal, by itself, there is also a total and net electrically neutral state since any movement of the free negative electrons will be counteracted by the positive nuclei.

When the *P*- and *N*-type crystals are combined, there is a countereffect of one material on the other. This effect can be considered to be produced by the attraction existing between the two regions of relative opposite polarity. Hence, *at the time of combining*, some electrons move from the *N* region to the *P* region and, also, holes move from the *P* region to the *N* region. A hole movement in one direction and an electron movement in the other mean, of course, that current flow exists. Such hole and electron movement, however, ceases once the *P* and *N* sections have been finally joined. (The methods employed for forming junctions of *P* and *N* zones will be discussed later.)

Because the *P* and *N* crystals were in a neutral state prior to joining, the movement of holes from the *P* zone and electrons from the *N* zone means that *neither* crystal section is now electrically neutral. Since the *P* crystal lost some holes (by electrons moving in from the *N* zone), the *P* crystal now has a predominantly *negative* charge. The *N* region, on the other hand, having given up some of its electrons, now has a net *positive* charge as compared with its former neutral state. Inasmuch as one section of the *PN* junction has a net positive charge and the other section a net negative charge, there is a *potential difference* between the two sections. This potential difference between the two sections means that a voltage exists, just as it would across the terminals of a cell or battery. The amplitude (voltage difference) between the two sections depends on the construction (in terms of the type of crystal and the type of impurities). This voltage is known as the *potential barrier* that exists between the two sections forming the junction. It is the influence of an *external* voltage on the potential barrier that creates the transistor function.

The external voltage (a battery or other power source) applied to a transistor or other solid-state device is known as a *bias* voltage. This term must not be confused with the grid bias potential applied to a vacuum tube because the transistor has no grid element as such. The various terminals of a transistor can be likened to the anode, grid, and cathode of a vacuum tube for discussion purposes, as will be shown later, but in reality the transistor does not have a grid, cathode, or anode. With the vacuum tube, as mentioned earlier, the voltage applied to the filaments is referred to as the *A* voltage. The plate-voltage supply is known as the *B* voltage or *B* power, while the grid potential is known as *C* bias. In transistors, on the other hand, the voltage is referred to as *bias*, regardless of where it is placed.

There are two types of bias potentials used for transistors. One of the bias potentials is known as *forward bias*, and this term refers to the condition established when the positive polarity of the battery is connected to the *P* region and the negative polarity of the battery is connected to the *N* region. The forward-bias connections can be shown as at A of Fig. 12-12,

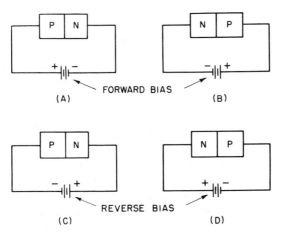

FIGURE 12-12. Forward- and reverse-bias connections.

or at B. Note that, in either case, the negative terminal of the battery is connected to the negative-zone crystal, and the positive side of the battery is connected to the positive-zone crystal. The other type of bias is shown at C and D of Fig. 12-12 and is known as *reverse bias*. Here the battery polarities are reversed with respect to the *P*- and *N*-zone polarities. In either drawing, C or D, the positive battery terminal is connected to the *N*-zone crystal, while the negative terminal of the battery is connected to the *P*-zone crystal. Each type connection (forward and reverse) has a particular influence on the *P-N* junction and contributes to transistor behavior.

As mentioned earlier, the union of the *P*- and *N*-type crystals upsets the equilibrium of the electrically neutral states of the individual crystal units, with the result that the *P* crystal represents a net negative charge and the *N* crystal, a net positive charge. With forward bias, as shown at A or B, the potential barrier of the *PN* junction is reduced because the plus potential of the battery, when applied to the *P* zone, tends to counteract the net negative charge on the *P*-zone side of the junction. On the other hand, the forward bias of the negative battery potential applied to the *N* region counteracts the net positive charge existing there. The result of the forward bias is, therefore, that the potential barrier has been reduced and current flow occurs. Electrons now move freely from the *N* region to the *P* region and hole flow occurs from the *P* region to the *N* region. Thus, forward bias means a lowering of the internal resistance of the junction and, consequently, less hindrance to current flow through the transistor and around the battery circuit.

When reverse bias is employed, as illustrated at C and D the polarities of the battery connections are such that the potential barrier is not reduced. With the negative terminal of the battery connection to the *P* region, the battery forces electrons into that region, while withdrawing them from the *N* region into the positive battery terminal. Because the negative terminal of the battery is connected to the *P* zone having a net negative status, it would seem that no current could flow since opposite poles repel. Similarly, the *N* region has a net positive charge and, again, the positive battery terminal would represent an opposite and, hence, repelling factor. The battery potential, however, is made higher than the potential barrier so that the battery or other voltage source has the energy to force some electrons to flow against the normal barrier polarities. Also, during the combining process of the two regions, not all the holes of the *P* area were filled nor did all the free electrons of the *N* region move to the *P* region. Hence, some holes in the *P* region and some electrons in the *N* region are still left for producing current flow. Because the potential barrier has not been overcome completely (as was the case with forward bias), the current flow that occurs is much less than for the forward-bias condition, even when batteries of the same voltage are used. Thus, for reverse bias, a high resistance circuit is present and the exact ohmic value depends on the design factors and processing of the junction. Thus, current flow is limited and may be only a fractional value of that flowing for forward-bias conditions.

12-6. Solid-State Diodes

From the foregoing, it is evident that *PN* junctions have some characteristics of the vacuum-tube diode. In the tube, current flows only when the anode is positive and the cathode is negative. Reversing the applied voltage polarity (plate negative and cathode positive) stops current flow. With the solid-state *PN* diode a similar (though not quite identical) condition occurs. Current flows in one direction with forward bias but very little flows with reverse bias. The difference in resistance (and hence in current flow) between the two conditions is, however, sufficiently great (particularly for silicon diodes) to permit the solid-state diode to perform virtually all the functions of the vacuum-tube diode in electronic circuitry, as discussed and illustrated more fully in later chapters.

The symbol for the solid-state diode is shown at A of Fig. 12-13. The

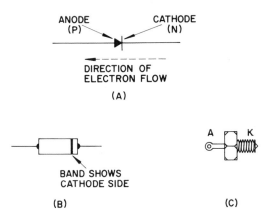

FIGURE 12-13. Diode symbol and shapes.

straight line at right angles to the triangle point represents the *cathode* toward which electrons flow and is usually marked with a positive sign on the actual unit. The triangular section is termed the *anode* (indicative of the *P* region of the junction).

Two typical diode shapes are shown at B and C and size depends on the power dissipation conditions under which the unit must operate. For detection purposes where fractional signal-potential values are found, the units have the appearance as shown at B but the diameter may be as small as $\frac{1}{16}$ in. In many of these units the cathode end is marked with a band as shown. For power-supply rectification, where high currents are involved, many diodes are shaped as shown at C, where the bolt arrangement at the cathode side permits direct mounting on a metal chassis or metal sections with heat-dissipating flanges (termed *heat sinks*) as more fully described in later chapters.

Besides detection and rectification, the solid-state diodes have a number of other uses, some of which are shown in Fig. 12-14. At A is shown a light-emitting diode where forward biasing of a specially formed *P-N* junction causes a conversion of electric energy to light. The light is produced without

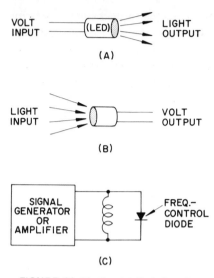

(A)

(B)

(C)

FIGURE 12-14. Special diode functions.

the thermal aspects of incandescent lamps and such diodes are useful for indicator lamps, visual readout devices for test equipment and computers and for similar applications formerly handled by pilot bulbs, neon lights, numerical display tubes, etc.

The reverse function is shown at B of Fig. 12-14, where a photodiode converts light into electric signals again by using silicon *P-N* junctions which are processed to be sensitive to light photons which strike the junction, as more fully discussed later. Such diodes are useful for sensing devices in industrial control, alarm systems, and reading devices in computer input sections.

At C is shown a diode in which the junction capacitance can be changed by applied voltage, thus forming a variable-capacitance diode of the *PN* type. Such diodes are also termed *varactors*; they replace tuning (variable) capacitors in electronic circuitry and also serve to control the frequency of oscillators. Thus the bulk and mechanical aspects of the variable capacitor are replaced by such special diodes.

Other special-purpose diodes include the *tunnel* type capable of high-speed switching and amplification; the silicon-controlled rectifier for power switching; the *zener*, useful for voltage regulation; and others designed specifically for high-frequency or microwave power applications. These diodes and their associated circuitry will be discussed and illustrated on an individual basis throughout subsequent chapters.

Review Questions

12-1. Give three advantages of solid-state devices over vacuum tubes.

12-2. Briefly explain how a crystal-lattice network is formed by silicon atoms.

12-3. Why does a crystal-lattice network formed from carbon produce an extremely stable and rigid bond?

12-4. Briefly explain what is meant by bands of energy levels, and how these relate to conductors and insulators.

12-5. How does current flow in a conductor differ from that established by the hole carriers of current?

12-6. List at least three acceptor and three donor impurities. Indicate the number of valence electrons in each impurity, and specify which form P zones and which form N zones.

12-7. What effect does a donor impurity have on the crystal-lattice network of germanium or silicon?

12-8. What effect does an acceptor impurity have on the crystal-lattice network of germanium or silicon?

12-9. In a PN junction, what occurs to make the P region have a predominantly negative charge and the N region have a predominantly positive charge?

12-10. Briefly explain what is meant by a potential barrier in a PN junction.

12-11. Briefly explain what is meant by forward and reverse bias, and illustrate these conditions by simple drawings.

12-12. Draw a schematic containing a diode, a resistor, and a battery. Indicate the cathode with K and the anode with A, and show the direction of electron flow.

12-13. For low signal-energy diodes, how is the cathode side sometimes identified?

12-14. Regarding PN junction diodes, briefly explain what effect forward bias has on the potential barrier, and also what effect reverse bias has on this barrier.

12-15. Identify three other diode types in addition to the detecting-rectifying type, and indicate their primary function.

12-16. What methods are employed to dissipate heat in diodes handling high currents?

Practical Problems

Note: The following problems constitute a review of some of the material covered in earlier chapters.

12-1. In an electronic circuit the power dissipated in a 15 kΩ resistor is 13.5 W. What is the voltage drop across the resistor?

12-2. In a control system, R_1 is in parallel with R_2; R_1 has a value of 450 kΩ and R_2 has 1.6 mA of current flow through it. If the voltage across the parallel circuit is 360, what is the total resistance of the network?

12-3. A 50 μA meter having an internal resistance of 2 kΩ is to be converted to a voltmeter with a 20-V full-scale deflection. What must be the value of the series resistor?

12-4. The current through a 2-kΩ resistor reads 50 mA (effective value). What is the peak value of the a-c voltage drop across the resistor?

12-5. In a series circuit composed of a resistor and an inductance, the apparent power was given as 60 W and the true power as 30 W. By what phase angle does the voltage lead the current in this circuit?

12-6. An inductance of 0.09 H was placed in series with a resistor of 150 kΩ.

At what time in μs after the voltage is applied to the circuit will the inductor current reach 63% of full value?

12-7. In a filter circuit an inductor with a reactance of 69 Ω was in series with a resistor of 92 Ω. What is the Z of this circuit?

12-8. In an industrial electronic graphing device it was necessary to match a circuit Z of 105,625 Ω to a scriber unit having an impedance of 25 Ω. What must be the turns ratio of the matching transformer between the circuit and the scriber?

12-9. A 0.05 μF capacitor is in series with a 200 kΩ resistor. At what time in seconds after the voltage is applied to the circuit will the voltage reach 63% of maximum amplitude?

12-10. A series circuit is composed of a 40.5 Ω resistor, a capacitor having a reactance of 24 Ω, and an inductor with a 78 Ω reactance. What is the Z of the circuit?

12-11. In a parallel circuit a resistor had 24 mA of current flow through it, an inductor had 8 mA, and an a-c meter in series with the capacitor read 26 mA. If the applied voltage is 85.8, what is the Z of the circuit? What is the total current supplied by the power source?

12-12. What is the Q of a series resonant circuit if the resistance is 14.5 kΩ, and the capacitor and inductor each has a reactance of 290 kΩ?

12-13. In a resonant circuit the LC product equals 0.00282, where L is in μH and C in μF. What is the resonant frequency in megahertz?

12-14. In a vacuum tube a change of grid voltage from 2 to 5 caused a plate-current increase of 50 mA. When the plate voltage was reduced from 200 to 50 V, however, the current decreased by 50 mA to the original value. What is the amplification factor of this tube?

12-15. In a vacuum tube the plate voltage was changed from 25 to 100 V, and the current changed from 1.5 mA to 9 mA. The grid voltage was held at a constant value. What is the plate resistance of this tube?

Transistors and Circuits 13

13-1. Introduction

Transistors and other solid-state devices, when utilized in circuits appropriate to the characteristics encountered, function in a fashion similar to vacuum tubes and gas tubes in their ability to amplify, generate, and process electric signals and switch dc and ac as required. Progressive strides in the development and improvement of solid-state units have made it possible for them to substitute for tubes in virtually all electronic circuitry. Thus, except for cathode-ray tubes or similar special-purpose types, the all solid-state radio and TV receivers are common, as are transistorized computers, test equipment, tape players, hi-fi stereo systems, etc.

Tubes are still found in special industrial circuitry, in some transmitting applications at high-power levels, and in older equipment still in usage. In these applications, as well as in the UHF and microwave regions, however, significant inroads have been made by the newer solid-state devices capable of operation in the gigahertz regions and at power levels comparable to that of tubes. As mentioned in the introduction to the previous chapter, the many advantages of solid-state devices far outweigh those of tubes in terms of lower voltages, smaller sizes, warm-up time reduction, elimination of filamental potentials, etc. Since the advent of the field-effect transistor described in this chapter, higher impedance characteristics (comparable to tubes) are also realized.

Since solid-state devices have electronic characteristics different from those of tubes, the appropriate circuitry also differs somewhat to accommodate the impedance differences. The basic transistor and other solid-state circuits are analyzed in this chapter in addition to the practical factors regarding the solid-state units themselves.

13-2.   Transistor Types

The general appearance of transistors is shown in Fig. 13-1. Those at A and

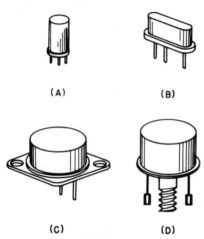

(A)   (B)

(C)   (D)

FIGURE 13-1. Transistor shapes.

B represent the lower-power types (either wired into the circuitry or plugged into miniature sockets). The larger types shown at C and D are bolted to the chassis as was the case for the power diode illustrated in C of Fig. 12-13. When transistors generate considerable heat during operation, special metal sections (heat sinks) are used as with the power diodes. Additional heat-sink factors are discussed more fully later in this chapter.

As with the solid-state diodes discussed in the previous chapter, there are many types of transistors available. Some are designed for low-frequency signal services; others are for high frequencies ranging through the microwave regions. Special types are also available for pulse signals, control circuits, and computer logic systems. Photosensitive types are also in use, as more fully covered later.

As mentioned in Chapter 12, transistors can be formed by using *P*- and *N*-region combinations to produce either *PNP* transistors or *NPN* types. In either of these, a triode (three-element) unit is formed with amplifying and signal-generating characteristics comparable to the vacuum tube. These two transistor types are shown in Fig. 13-2 but before discussing the manner in which the transistor amplifies it is helpful to become familiar with the terms assigned to the various parts of the transistor, plus the symbol designations.

At A of Fig. 13-2 the *PNP* transistor is shown; this type drawing is used for analytical discussions of transistor function and is not representative of the true symbol generally used. Instead, those at B and C are employed and on occasion these may be shown drawn within a circle. Whether B or C is used depends on convenience in circuit layout. The arrow in the symbol represents the transistor terminal which is connected to a zone called the

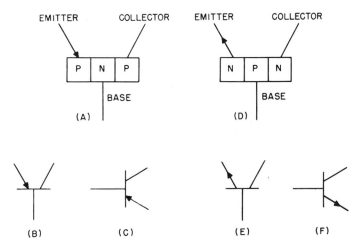

FIGURE 13-2. Transistor terminals and symbols.

emitter. If the arrow points *toward* the zone, it indicates a *P* zone; hence the transistor must be a *PNP* type. If the arrow points away from the zone, it still denotes the emitter but now also indicates that the zone in question is an *N* type. Thus, when the emitter arrow points away from the symbol as at D, E, and F, an *NPN* transistor is shown.

Both the *PNP* and the *NPN* transistors have two like zones (either *P* or *N*). The alike zones are those of the emitter and collector, with the latter represented by a slanting line, as shown in Fig. 13-2. The lead terminal from the middle zone is known as the *base*, and is represented by a straight line at right angles to the line that represents the transistor body (the horizontal line in B or E). Generally speaking, the base of the transistor can be compared to the grid of a vacuum tube. The emitter is comparable to the cathode of a vacuum tube, while the collector may be compared to the anode (plate).

13-3.   Basic Transistor Circuitry

An introductory discussion of the basic aspects of amplification was given in Chapter 11, and it was shown that the input signal (which is to be amplified) is applied between the grid and cathode circuitry of a tube. The amplified output signal was developed across the anode load resistor. While the discussion was confined only to the basic amplification circuitry, it did however introduce a process extensively used in electronics. Other amplifier types are also of importance and they will be discussed subsequently as a more solid foundation is acquired. Initially, though, it is necessary to discuss the same type of basic amplification system as it relates to the *transistor*.

A *PNP* transistor in a basic circuit is shown in Fig. 13-3. Note that the

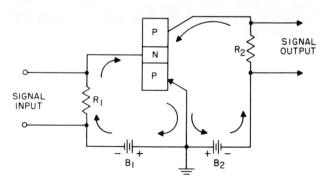

FIGURE 13-3. Basic PNP grounded-emitter circuit.

emitter-base (P and N zones) are biased in the forward direction, while the collector, with a P zone, has a negative polarity applied to it and, hence, is biased in the reverse direction. This method of biasing the input and output sections of a transistor circuit also applies to the NPN type; that is, the input is biased in the forward direction and the output (collector side) is biased in the reverse direction.

Resistor R_1 is provided for purposes of accepting the input signal, while resistor R_2 is the output resistor across which the amplified version of the signal appears. Direction of electron flow for the input and output sides is indicated by solid arrows within the drawing. As shown, electrons from the input side battery (B_1) flow from the negative side of the battery, through R_1, and to the base of the transistor (the N zone). From the latter point, it flows to the emitter side and back into the battery. For the collector side, electrons flow from battery B_2 through resistor R_2 and to the collector. For a return to the plus side of the battery B_2, electrons must flow through the center N zone and to the P zone of the emitter.

As mentioned in Chapter 12, forward bias causes current to flow rather freely, and this condition prevails at the base-emitter side of the circuit of Fig. 13-3. With electrons flowing from the base to the emitter, hole flow is in the opposite direction. Holes flowing from emitter to base, however, represent like charges and hence will repel each other and tend to diffuse and spread out, and some holes thus reach the collector area. [The base region is purposely made thinner than the emitter and plate sections so that electron (and hence current) flow in the base region has a shorter distance to travel in diffusing to the collector side.]

The holes that reach the collector area are somewhat accelerated because they have a positive charge (positive current carriers) and are attracted by the fields of the negative battery potential applied to the collector. Normally, the collector, with its reverse bias, has a high potential barrier because the reverse-bias battery did not reduce the potential barrier, as was the case with the forward bias on the emitter side. As mentioned earlier, the barrier causes the collector P area to have a net negative charge and the base N area to have a net positive charge. The electron flow from the base-emitter battery overcomes much of the potential barrier in that side of the circuit, thus influencing the N zone. The hole flow from the emitter, in diffusing toward the collector, tends to reduce the potential barrier there because the holes are "collected" by the collector and the negative charge is reduced. Hence, the collector P zone tends more and more to become an acceptor area and will thus permit an increase in electron flow from the collector battery B_2. Thus, the hole flow to the collector area constitutes a *primary* carrier of current (sometimes referred to as a *majority* or *chief* carrier of current). Hence, in the *PNP* transistor, the chief carriers of current are the holes, while electron flow, in this case, is the secondary or minority carrier of current.

From the previous paragraph, it becomes obvious that the emitter hole flow influences current flow in the collector side. Thus, the hole diffusion to the collector caused a reduction in the collector barrier potential, which, expressed in another way, means that the high resistance of the barrier section was reduced because of the effects of the hole flow set up by the current flow in the base-emitter side.

This influence on collector current flow by current and hole flow in the base-emitter side is what permits amplification to occur. Obviously, if more current were caused to flow in the base-emitter side, there would be an increase in hole flow. The increase in hole flow means that more holes reach the collector area and further reduce the potential barrier. In consequence, there would be an increase in current flow in the collector side. Conversely, if the current flow in the base-emitter side were reduced (as by a lower battery potential or a signal source), the potential barrier resistance at the collector would increase and current flow in the collector side would decrease.

The amplifying characteristics which come about by the application of a signal to the base-emitter circuit can be understood more readily by reference to A of Fig. 13-4 where one alternation of an input signal is shown. Since this is a positive-going alternation, it will develop a voltage across the input resistor, as shown, with plus toward the base and negative toward the battery. Suppose the battery voltage is 4.5 and the signal voltage at its peak is 0.5. Since the signal voltage sets up across the resistor a voltage drop that opposes in polarity the battery voltage, the net voltage between base and emitter would decrease to 4 V, from the original 4.5 V. On the collector side, the collector battery caused a *positive* d-c voltage drop to occur across the output resistor. Because the input signal decreased the voltage between base and emitter, the current flow through the input side also decreased and there were less holes reaching the collector. In consequence, the collector barrier resistance increased, and less current flowed through the output resistor. The reduction of collector current caused the voltage drop across

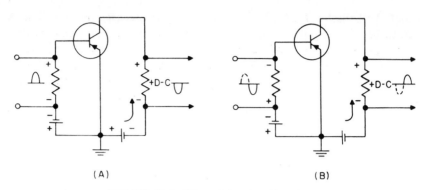

(A)   (B)

FIGURE 13-4. Effect of signal on PNP circuit.

the output resistor to decrease also, in step with the voltage change that occurred across the input resistor. Thus, an amplified version of the input signal appears at the output.

For the second alternation of the input signal, as shown at B, the negative input signal alternation sets up a voltage drop of the polarity shown, with negative toward the base and positive toward the battery. This voltage now aids the battery voltage because the voltage across the resistor is as though another battery (of lower voltage) were placed in series with the original battery. Thus, if the battery voltage is 4.5 and the signal voltage at its peak is 0.5 V, the voltage between base and emitter would reach a 5-V value. The increase in voltage between base and emitter increases current flow through the input side and, hence, there is an increase in hole diffusion to the collector. In consequence, collector barrier resistance decreases and the increased current flows through the output resistor. The result of the current increase is a rise in voltage across the output resistor, as shown. (Note that there is a phase reversal between the input and output signals, as was the case with the vacuum-tube amplifier.)

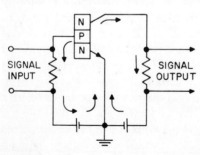

SIGNAL INPUT

SIGNAL OUTPUT

FIGURE 13-5. Basic NPN grounded-emitter circuit.

When an *NPN* transistor is used to form a basic amplifier circuit, as shown in Fig. 13-5, amplification again occurs because the collector potential barrier is again influenced by the emitter side. With the *NPN* transistor, however, current flow and hole flow are the reverse of what they were with the *PNP* type. The *NPN* transistor must still be biased in the forward direction at the base-emitter side, which means that a negative polarity must be applied to the *N*-zone emitter, and a positive polarity to the *P*-zone base, as shown in Fig. 13-5. The collector, on the other hand, must again be biased in reverse; that is, the collector must have a positive polarity applied to its *N*-zone, as shown. Direction of electron flow is indicated by the solid arrows within the drawing. Note that electrons flow toward the emitter and out of the base to the positive side of the battery. On the collector side, electrons flow toward the emitter, out of the collector and back to the positive terminal of the collector battery.

For the *NPN* transistor, the electron flow from emitter to base diffuses (like charges of electrons repel each other) and encroaches on the *N* section of the collector area, attracted by the fields set up by the positive potential of the battery applied to the collector. The collector *N* area, as mentioned earlier, has a potential barrier, with the *N* area having a positive charge because of the holes that entered the area when the *N* and *P* zones were combined. The electron flow from the emitter, in diffusing upward and being collected by the collector section, fills up the holes in the *N* area, thus contributing to the presence of free electrons and permitting a reduction of the potential barrier. Therefore, there is an increase in current through the collector side because of the influence of the electrons from the emitter section. For this reason, the electrons are the principal or primary current carriers in the *NPN* transistor, with the holes taking a secondary place and becoming minority carriers of current. If the potential difference between the base and emitter voltage is changed, there will be a change in the amount of electrons reaching the collector and hence a change in the potential barrier and collector current flow. Thus, as with the *PNP* type, amplification occurs when a signal is applied to the input because of the changes that occur in the base-emitter current.

Note that, with either the *PNP* or the *NPN* transistor, changes of base-emitter current are influential in altering the amount of current flow in the collector section. For this reason, the transistor is primarily a current-amplifying device and hence can be considered as a power amplifier. (Such amplifiers will be discussed more fully later.) Signal voltages can, of course, also be obtained at the output because such voltages would develop across the output resistor by virtue of the current flow through this resistor.

For the *NPN* transistor circuit, the signal inputs affect the collector side in a fashion similar to that described for the *PNP* transistor illustrated in Fig. 13-4. The *NPN* version is shown in Fig. 13-6. Note, however, that when a positive alternation is applied to the input resistor at A, the polarity of the voltage drop across the input resistor *aids* rather than opposes the battery voltage. Also note that electron flow through the output resistor is now from collector down to the positive terminal of the battery. Hence, there is a *negative* d-c voltage drop across the output resistor. Since the first alternation of the input signal aids the battery potential, there will be an increase in current flow through the base-emitter circuit and, hence, more electrons reach the collector to reduce the potential barrier. The collector current increases, causing an *increase* in the voltage drop across the output

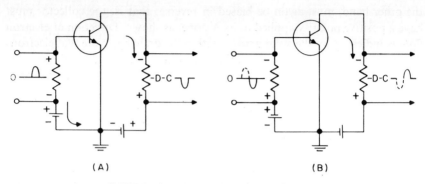

FIGURE 13-6. Effect of signal on NPN circuit.

resistor (in the *negative* direction—that is, a larger negative voltage drop). Thus, there appears at the output an amplified version of the input signal. As with the *PNP* circuit, there is a phase reversal of the signal between input

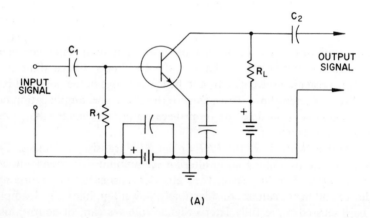

(A)

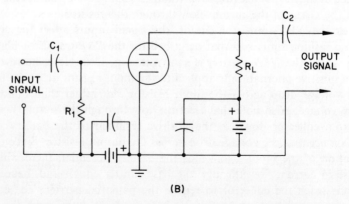

(B)

FIGURE 13-7. Comparison of NPN transistor amplifier to tube type.

and output. The second alternation of the signal will develop an opposing voltage across the input resistor, as shown at B of Fig. 13-6, with a consequent reduction of the net voltage between base and emitter. This voltage decrease causes a current decrease and fewer electrons reach the collector. Hence, the collector barrier is increased and less current flows through the output resistor in the collector circuit. The decrease in current through the output resistor causes a decrease in the *negative polarity* voltage drop across the resistor (which is equivalent to a positive-going voltage change).

Figure 13-7 shows the basic *NPN* transistor amplifier circuit at A, for comparison with its vacuum-tube counterpart at B. Coupling capacitors are used at the input and output sections to keep back the d-c part of the signal and transfer only the a-c component. Bypass capacitors are also employed across the batteries, to prevent the signal energy from suffering losses in traveling through the battery resistance. From Fig. 13-7, it is obvious that the grounded-emitter transistor amplifier compares to the grounded-cathode vacuum-tube circuit. Other types of transistor and vacuum-tube circuits will be discussed later in the section on amplifiers.

13-4.   Transistor Characteristics

Characteristic curves for transistors can be graphed in a fashion similar to those for vacuum tubes. Changes in input voltage to the basic transistor amplifier are made, while noting the changes in the base current that occurs. Voltage is applied to the collector side and the collector currents are graphed for various changes of base current. Collector voltages are changed and, again, collector currents are graphed with respect to base currents. The result is a graph of the characteristic curves of a transistor, as shown in Fig. 13-8. These curves relate to the basic ground emitter circuit discussed in this chapter.

As with the characteristic curves of vacuum tubes, information regarding a particular transistor can be obtained, including the current gain of a grounded-emitter amplifier circuit. Current gain in a grounded-emitter circuit is known as *beta*, and the symbol for it is the Greek lowercase letter β. Current gain, in such an instance, refers to the *signal* current gain that results by a change of signal current in the base side, with respect to the amplified version of the signal current in the collector side. As an illustration of how current gain (β) can be found from the characteristic curves of Fig. 13-8, assume that the collector battery voltage is held at 10 V (the battery voltage at the base side is unimportant here since only the signal current changes in the base side need to be considered). If a signal voltage is now applied to the base-emitter input of the circuit, to cause a change of base current from 60 μA to 180 μA, as shown by the vertical dotted arrow in Fig. 13-8, there will be a change of collector current from 3.25 to 7.75 mA, as shown

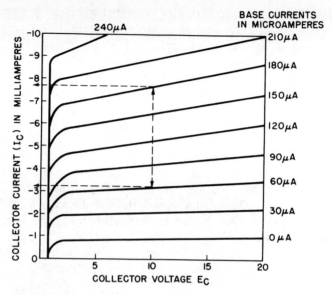

FIGURE 13-8. Typical characteristic curves of transistor in grounded-emitter circuit.

by the horizontal dotted arrows. The formula for calculating the current gain is

$$\beta = \frac{dI_c}{dI_b} \tag{13-1}$$

This formula indicates that current gain is equal to the change of collector current divided by a change of base current. The lowercase letter d preceding each current symbol in the formula is the symbol used for a change or difference (differential); as in the formula used for vacuum-tube characteristics, the Greek capital letter *delta* (Δ) is sometimes employed. When the values obtained from the chart in the foregoing example are set down, the formula indicates that the current gain for this particular transistor is 37.5, as follows:

$$\beta = \frac{dI_c}{dI_b} = \frac{7.75 \text{ mA} - 3.25 \text{ mA}}{180 \ \mu\text{A} - 60 \ \mu\text{A}}$$

$$= \frac{4.5 \text{ mA}}{120 \ \mu\text{A}} = \frac{0.0045}{0.000120} = 37.5$$

This calculation shows that the signal current in the base-emitter side is amplified 37.5 times the collector side. The relation of such current gain to power gain is covered more fully in the section on power amplifiers.

13-5. Fabrication Differences

A number of processes may be used in the fabrication of basic transistor types and some are covered in the following discussions to emphasize the fundamental differences in the end result. The earliest manufacturing method used an *N*-type crystal slab to which was fused the emitter and collector wires. The fusion process created *P* zones, as shown at A of Fig. 13-9 and

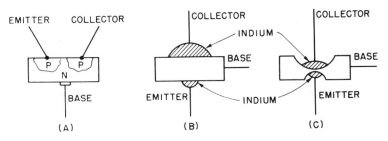

FIGURE 13-9. Transistor types.

this transistor was called the *point-contact* type. As shown, the base lead was attached to the *N* section to complete the transistor. The point-contact transistor has an extremely high noise level compared to modern transistors, hence the signal-to-noise ratio suffered in amplificator circuitry. Its only use at present is in some microwave detection procedures.

Another transistor is the junction type, which can be either *NPN* or *PNP*. Several processes are used for forming the junction transistor. One method is by growing the basic crystal structure. The process involves the placing of a pure germanium crystal plus impurities in a high-temperature furnace. The material is melted and the basic crystal seed is withdrawn and cooled; *P* regions are formed adjacent to *N* regions and, after additional processing, the junction transistor results. Another method consists in again using a basic *N*-type slab, as with the point contact type, and placing a small portion of indium on each side, as shown at B. The crystal slab and the indium deposits are heated at high temperatures, which causes the indium to melt and combine with the slab structure. The indium forms a chemical alloy with the *N*-type germanium, and *P* zones are then formed in the alloy area. Thus, a *PNP* transistor results. A larger deposit of indium is used for the collector side, as shown in Fig. 13-9. This type of transistor is sometimes known as an *alloy-junction* transistor. The *NPN* transistor types may also be manufactured by the alloy-junction process, using a *P*-type germanium crystal slab and employing an impurity element having five valence electrons, such as arsenic or antimony.

Another method utilized in the construction of transistors is the surface-barrier process. As with the alloy junction, indium is used with respect to a basic *N*-type crystal slab, as shown at C of Fig. 13-9. In this process, two jets that can shoot an electrolyte into a concentrated area on both the top and bottom of the crystal are employed. A voltage is applied across the jets so that the electrolyte can conduct current, and an etching process results when the jets shoot the electrolyte on both sides of the crystal. This etching process eats away the center sections of the crystal until only a thin layer remains. The polarity of the electrolyte current is then reversed and an indium plating process begins. During the plating process, the chemical bonds between the indium and the *N*-type slab form a surface area that becomes the *P* zone. Hence, this type of transistor is known as the surface-barrier transistor, with the potential barrier existing at the surface area of the crystal, which is chemically joined with the indium.

The most widely used transistor type is the triode, though tetrode transistors have been developed and utilized to some extent in special circuitry. For the tetrode transistor, an additional electrode is connected to the center *P* zone of an *NPN* transistor, as shown at A of Fig. 13-10. This additional

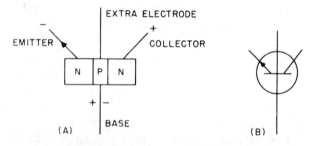

FIGURE 13-10. Tetrode transistor.

electrode provides a reduction in the zone-diffusion path link and, hence, the basic resistance is decreased. Thus, a higher efficiency is realized and operation at higher frequencies than normal is possible. The base resistance is lowered in the tetrode type because the emitter confines electron flow in the *P*-zone section to a relatively small area close to the base contact point.

For the *NPN* type shown at A, forward bias to the emitter side is applied in conventional manner; that is, a negative potential is applied to the emitter, with the base as the reference positive potential. For the collector, a positive potential is applied as shown to provide the necessary reverse bias. The additional electrode is made slightly negative with respect to the base potential. Except for the additional voltage applied to the extra electrode, circuit connections are the same as for the other transistor circuits described in this book. The symbol for the tetrode transistor is shown at B, and may be drawn with or without the enclosing circle, depending on the draftsman's particular choice.

13-6.   Field-Effect Transistors (FET)

The junction transistors previously discussed are *bipolar* types since operation depends on two types of carriers, the electron flow (current) and the positive hole flow. Two junctions were involved with the basic transistor, one with forward bias and the other with reverse bias. The so-called *field-effect* transistor (FET) differs to a considerable degree from the junction transistor since the FET is a *unipolar* device having only a single charge carrier. As with *NPN* and *PNP* transistors, the FET units are available in *N*-channel and *P*-channel types, with current carriers in the *N*-channel types consisting of electrons and the carriers for the *P*-channel types being hole flow.

The input bias to the FET is reverse bias, and a signal variation at the input influences the magnitude of the majority carriers by altering conduction between two closely spaced regions. The field-effect transistor has high input and output impedances (comparable to vacuum tubes), is suitable for low- or high-signal circuitry, lends itself to a variety of amplifying and switching circuitry (both audio and RF), and has excellent thermal stability, and fabrication is not limited to the few components from which the junction transistor is manufactured. Thus, the FET has undergone wide applications in solid-state circuitry of all types.

There are two basic types of field-effect transistors, the junction type and the so-called MOSFET (which indicates *metallic-oxide semiconductor field-effect transistor*). Basically, in the fabrication of the MOSFET, a *substrate* (foundation slab) such as an *N*-type silicon slice is used and *P*-type source and drain regions are diffused into the slab by using masking devices. A thin layer of oxide is formed over the surface and so-called *windows* are opened through an etching process directly over the *P* regions (see A of Fig. 13-11). Metallic evaporation over the surface follows and is then removed, leaving metallic contacts with the source and drain windows, plus·a metallic span over the *P* regions, separated by an oxide coating. With the junction FET, the gate is made up of a *PN* junction diffused into the channel material.

As shown in Fig. 13-11, the polarities of the slab and regions determines whether the FET is a *P*-channel or an *N*-channel type. At A the *P*-channel type is shown, with the appropriate symbol. At B the *N*-channel FET is illustrated and the arrow for the *gate* element points to the center vertical bar. As shown at C, an additional gate element can be added to the slab (but isolated from the first gate). The additional gate forms what is termed a double-gate FET for circuitry requiring additional signal inputs. At D two additional symbols are shown and these indicate the so-called depletion types, with *sub* indicating the substrate element terminal. The four-terminal and five-terminal types shown are *N*-channel units. For *P*-channel types, the arrow is in reverse direction.

A common-source (grounded source) FET amplifier circuit in basic form is shown in Fig. 13-12. This compares to the grounded-emitter-type transistor amplifier discussed earlier in this chapter. The *P* and *N* zones are identified for discussion purposes, though normally only the *G*, *D*, and *S* designations are included. Source and drain output current flow is established by battery B_2 and the path includes the load resistor and the *N* zones. In an *N*-type structure the current flow consists primarily of electrons; hence

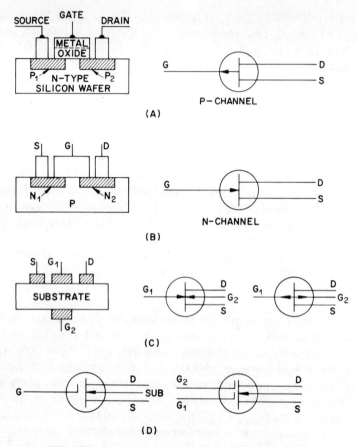

FIGURE 13-11. Field-effect transistors and symbols.

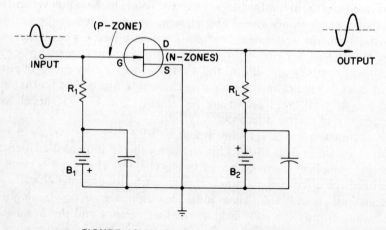

FIGURE 13-12. Common-source FET amplifier.

these are the *majority carriers* for the primary current of B_2. Battery B_1 supplies the reverse bias between the gate element and the source element (at ground potential). The input signal (if a sine wave) increases or decreases the amplitude of the reverse-bias potential supplied by B_1.

With reverse bias applied to a *PN* junction, current carriers move away from the junction, leaving depleted regions. Such *depletion regions* (lack of current carriers) now surround the junctions and extend into the *N* regions, thus providing a potential barrier of the type discussed earlier in Chapter 12. Hence, the greater the encroachment of the depletion region on the *N*-zones, the less carriers will be present in this zone for current conduction. Thus, the extent of the depletion region's movement into the *N*-zones is directly established by the amplitude of the reverse-bias potential. The signal input adds and subtracts from the reverse bias and hence moves the depletion region more and less into the *N*-zones area that serves to carry the main current. In consequence, a low amplitude signal variation produces an amplified signal output much greater than the input signal. The *N* regions can be likened to a corridor that is widened and narrowed by changes in the reverse bias, thus having a decided effect on the amount of current flow through it. The average current flow through the *N* regions is limited by the potential of B_2 and the degree of impurities that were injected into the *N* crystal.

As with the ground emitter amplifier, there is a phase reversal between the input signal and that at the output, as shown in Fig. 13-12. If the *N*-channel FET is replaced by a *P*-channel type, the polarity of each battery (B_1 and B_2) would have to be reversed. General operation would be the same, however, except that internally holes are the majority current carriers.

13-7. Circuit Variations

Other methods for connecting FET and transistors into electronic circuitry are shown in Fig. 13-13. At A the common-gate FET circuit is shown and this compares to the transistor common-base circuit shown at B. For both these circuits there is no phase reversal between the input and output signals. Since the gate at A and the base at B are at signal ground, effective isolation exists between input and output circuits and better performance and stability are realized in some applications, as discussed more fully later when practical circuitry is covered. For both circuits the input signal is impressed between ground and the input line. Batteries are bypassed to prevent signal energy from developing across them.

When the positive alternation of the input sine wave signal appears between the source and ground (gate) of the circuit at A, the bias from B_1 is reduced. This also reduces the depletion region controlling the majority current carriers between the output circuitry (gate and drain); hence more current flows through the load resistor R_L. Since the voltage-drop polarity

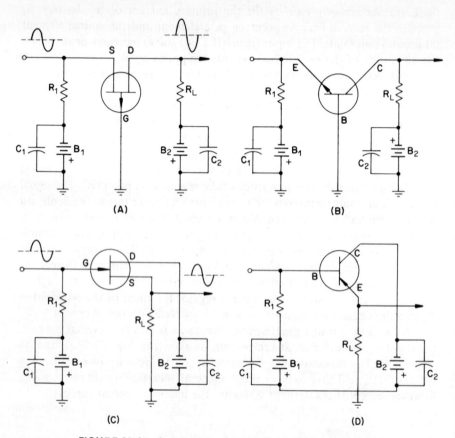

FIGURE 13-13. Common gate, base, drain, and collector circuits.

across R_L from B_2 is positive toward the drain element, a positive alternation of the output signal is produced. Thus, the output signal is in phase with the input signal. As shown later, phase differences across FET units, transistors, and tubes often is of particular significance in special circuitry and in relation to harmonics and reduction of distortion.

For the circuit at B, a positive signal alternation between the emitter and ground (base) also decreases bias (in this case the *forward* bias between emitter and base). Forward bias reduction also lowers current flow in the collector circuit; hence the voltage drop across R_L declines. Since, however, the polarity of B_2 places a negative polarity at the collector end of R_L, a reduction in the negative voltage drop across R_L raises the signal level above ground (more positive), thus developing the same phase as the input.

For both A and B circuitry, negative alternations also produce the same phase output alternation. Similar results are obtained if the *P*-channel FET at A is replaced with an *N*-channel unit or the *NPN* transistor at B is changed to a *PNP* type. In either case the polarities of B_1 and B_2 would have to be reversed.

The circuit at C of Fig. 13-13 is a *common drain* because the drain element is at signal ground because of C_2. (Without the low reactance of C_2 the drain element would be above ground by the ohmic value of the battery resistance.) Capacitor C_1 also places the bottom of R_1 at signal ground, thus in effect bypassing the battery resistance. The output is now obtained from a resistor in the *source* circuit, as shown. Such a circuit again has no phase reversal between the input signal and the output. Signal voltage gain is less than unity though signal current gain is possible. The circuit is useful for impedance-matching purposes and for obtaining a low output impedance (with a high input impedance). Since the phase of the output signal "follows" that of the input, the circuit has been termed a *source follower*. As such, it resembles the tube-type circuit where the output is obtained from across a cathode resistor (with the anode placed at signal ground). The tube type is called a *cathode follower*.

The transistorized version is shown at D, where the collector is placed at signal ground by C_2, thus making the circuit a *common-collector* type. The output is obtained from across the resistor in series with the emitter, as shown, thus forming an *emitter-follower* circuit. The signal phase at the output is again the same as the input. As with the circuits at A and B, different type FET and transistor units could be used (*P* channel for C and *NPN* for D) provided appropriate changes in polarities are made for B_1 and B_2. Otherwise, operational characteristics are virtually the same. For the circuit at D, as with the FET source follower and the tube cathode follower circuit, signal-voltage gain is less than unity, though signal-current gain can be realized.

13-8. FET and Transfer Characteristics

Characteristic curves can also be prepared for the FET units as was the case for tubes (Figs. 11-9 to 11-11) and transistors (Fig. 13-8). A typical gate-to-drain set of characteristic curves for a field-effect transistor are shown in Fig. 13-14. As with tubes and transistors, such curves provide information on the operational features of a particular FET. Note the drain-to-source (V_{DS}) voltage is plotted along the x axis and the drain current (I_D) along the y axis. The various curves represent those obtained for a specific d-c gate (V_{GS}) voltage (voltage between *gate* and *source* terminals).

An important designation in FET characteristics is the so-called *pinch-off voltage* (V_p or V_{po}). This is the point at which the gate-bias voltage V_{GS} is such that it causes the drain current to drop to zero for a specific value of V_{DS}. Such a pinch-off voltage is comparable to the plate-current cutoff condition obtained with vacuum-tube characteristics.

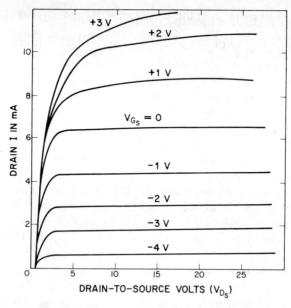

FIGURE 13-14. FET gate-to-drain characteristics.

Another important FET parameter is the *transconductance*, expressed as

$$g_m = \frac{\Delta I_d}{\Delta V_g} \qquad (13\text{-}2)$$

Thus, for the FET, transconductance is the value in *mhos* obtained by taking the ratio of a small change in drain current to a small change in gate voltage on a chart of gate-to-drain characteristics as shown in Fig. 13-14. Thus it is a figure of merit regarding the ability of the gate voltage to control the drain current. It is comparable to g_m for tubes as shown in Equation 11-3, obtained from $E_p - I_p$ curves.

FET transconductance is usually given with the gate bias at zero and a minimum is obtained at the pinch-off voltage. The FET curves (Fig. 13-14) are more closely allied to the vacuum-tube *pentode* characteristic curves (see Fig. 11-16). If g_m and the load-resistance (R_L) value are known, signal gain is given by

$$\text{Gain} = g_m \times R_L \qquad (13\text{-}3)$$

Gain and signal-power factors will be discussed more thoroughly in Chapter 16.

The g_m of a FET can be considered as the signal-transfer ratio, indicating the device's sensitivity and behavior to signal changes. The forward current-transfer ratio for the grounded-emitter circuit had already been given in Equation 13-1. For the grounded-base circuit shown at B of Fig. 13-13, the ratio is termed *alpha* (α) and is the ratio of a change of collector current

I_c to a change in emitter current I_e:

$$\alpha = \frac{dI_c}{dI_e} \tag{13-4}$$

A set of transfer-characteristic curves is shown in Fig. 13-15 that is obtained for several base current values plotted against collector current and base-to-emitter voltage. Thus we can obtain transfer curves for a particular transistor or a collector-characteristic set as shown earlier in Fig. 13-8.

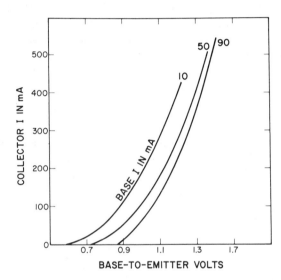

FIGURE 13-15. Typical transistor transfer characteristics.

13-9.   Negative-Resistance Units

Some solid-state devices have been designed to have negative-resistance characteristics and hence find special applications in rapid signal switching, amplification, and other uses. One such unit is the *tunnel* diode. Unlike the silicon diodes discussed earlier, the primary application for the tunnel diode is not rectification (see Chapter 14) but amplification and rapid switching. The tunnel diode can operate at much higher signal frequencies than the ordinary transistor and has switching capabilities that are faster by a ratio of over 100 to 1.

The tunnel diode is not susceptible to temperature changes to the degree found in the average transistor and hence remains stable for wide temperature variations. Also, the tunnel diode resists the adverse effects of nuclear radiation and thus finds important applications in this field. Radiation effects in semiconductor diodes and transistors normally consist of a change in the internal resistance and an increase in the noise level.

The term *tunnel diode* stems from the so-called "tunnel effect" that occurs between the *PN* junction by virtue of the extremely narrow barrier area formed by the addition of impurity elements in excess of the amount normally employed for the transistor. Hence, an electric particle reaching the barrier suddenly disappears and reappears virtually instantly at the other side of the barrier. The transfer occurs at the speed of light (186,000 m per sec), in contrast to the transistor where charges move through the barrier at comparatively slow speeds. The behavior appears as though the particle "tunnels" beneath the barrier rather than penetrating it. Hence, the tunnel diode can operate at signal frequencies extending into the gigahertz range.

The tunnel diode can function as an amplifier because of the negative-resistance characteristic. The latter is identical to that which occurs for a tetrode tube and a comparison of Fig. 11-14, with the graph at A of Fig. 13-16 will indicate the similarities. In the tetrode, the negative-resistance

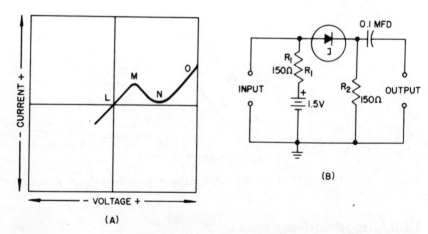

FIGURE 13-16. Tunnel diode factors.

occurs because of the secondary emission that results when the screen-grid potential exceeds the anode potential. The dislodged electrons are attracted to the screen grid and cause a decrease in plate current, with a resultant decrease in internal tube resistance. Because this occurs for an increase in plate potential, the decreased resistance is termed *negative resistance.*

The tunnel diode, as shown at A, will have a current rise when forward bias is applied. This current increase, however, from the zero point (*L*) will soon reach the peak marked *M*. An additional voltage increase will now drop current amplitude from the *M* level to *N*, as shown. As with the tetrode tube, this decrease in current represents negative resistance because of the drop of internal resistance with an increase in applied voltage. An additional voltage increase will cause a gradual rise from *N* to *O*, as with the tetrode tube.

If a fixed forward-bias potential is applied to set an operating point between M and N, amplification characteristics are obtained. An input sine wave, for instance, will alternately add and subtract from the forward-bias value and cause a comparatively high signal-current change to occur. A basic circuit application is shown at B. If the input signal is such that it increases the forward bias, current decreases and lowers the voltage drop across R_2, producing a negative-signal output change. For an input signal opposing the forward bias and thus reducing its amplitude, there will be a rise in current to produce a positive signal change at the output. Thus signal power amplification occurs.

Another solid-state device exhibiting negative-resistance characteristics is the *Unijunction* transistor illustrated in Fig. 13-17. As shown, the transistor

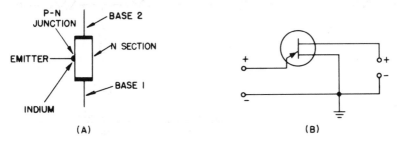

FIGURE 13-17. Unijunction transistor.

is formed from an N-type silicon slab to which is chemically bonded another element such as indium to form a PN junction. The unit has two base leads and one emitter as shown, and the basic circuit showing proper voltage polarities is at B.

When the voltage applied to the emitter is in reverse-bias form (or zero voltage), the slab acts as a conventional resistor. Because the emitter taps the crystal slab, a voltage difference exists between emitter and ground. When a forward bias is now applied to the emitter circuit, transistor characteristics are formed and the resistance of the slab between emitter and base 1 decreases, resulting in a current increase. The emitter voltage can now be decreased with a consequent current increase (negative resistance). Since a reduction of emitter potential no longer causes a decrease in base current, the unit behaves as a *thyratron* with gating characteristics. (Thyratrons are discussed fully in the next chapter.) As with other transistors, the Unijunction type can be used to form oscillator or other type circuits.

13-10. Network Parameters

In Chapter 4, resistive circuits were analyzed, using Kirchhoff, Thévenin, and Norton theorems to set up equivalent circuits. Similarly, the operational characteristics (parameters) of transistors can also be obtained by utilizing

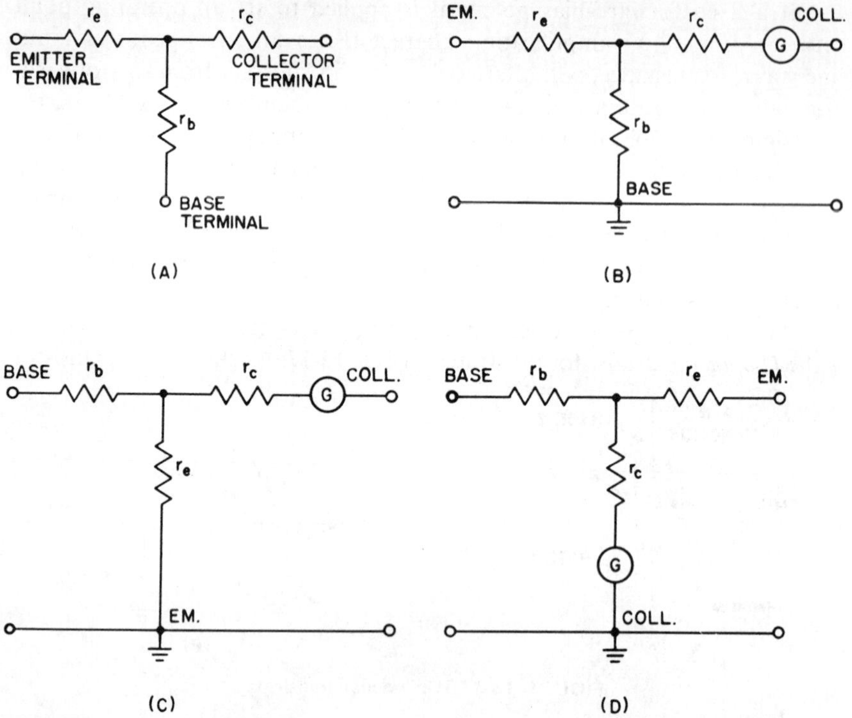

FIGURE 13-18. Equivalent transistor networks.

equivalent networks. The transistor can be considered as a resistive device represented as shown at A of Fig. 13-18. Here the emitter is indicated by a resistor r_e, the base by r_b and the collector by r_c. Together they form a T-network of resistors with ohmic values such as would be obtained by making d-c measurements. Since this T-network is a *passive* type, it is not a true representation of a transistor because such a three-resistor combination does not have amplifying characteristics. Also, the network at A is shown as a three-terminal device. Actually, however, a common input-output lead is present in a transistor circuit and hence the practical transistor must be represented as a four-terminal network.

The four-terminal equivalent network is shown at B and represents the common base (grounded base) circuit. Instead of a passive network, a generator (G) is indicated in the collector lead, making this an *active* network. Thus, the amplifying function of the transistor is indicated by the equivalent generator in the output, just as the plate and grid in a grounded-grid tube amplifier can be so shown. The grounded emitter is shown at C.

For the grounded emitter, the generator in the output line represents the emitter-collector section, as with the cathode-plate portion of the vacuum-tube circuit. Now the left arm of the T-network is shown as r_b for the equivalent base resistance that now forms the input terminal, and r_e as the representative grounded-emitter resistance. The active network for the grounded collector is shown at D.

The active resistance network characteristics of a transistor are a close approximation at low signal frequencies, or with dc. At high signal frequencies, however, it becomes an impedance network because the internal capacitances, with their decreasing reactances, are influencing factors. Input and output impedances are also affected by the ohmic value of the load resistance applied to the output, as well as the resistance of the circuit or device applied to the input. Hence, for a true evaluation of circuit parameters, the internal network values must be considered in conjunction with the external networks that are applied so that maximum signal-power transfer and maximum circuit efficiency are obtained. In the discussions that follow, we shall be concerned primarily with the resistive characteristics of the equivalent networks for a clearer understanding of the notation employed and the methods utilized.

As discussed in Chapter 4, for convenience in analyzing networks, it is expedient to consider the network as a black box wherein we have internal components of unknown values. To analyze the internal circuit, we read voltages and currents or apply test signals to the input and output terminals and evaluate their effects. The voltage and current measurements are also undertaken during the open circuit or short circuit of either the input or output terminals.

The four-terminal equivalent network of a triode transistor can be represented as a black box, as shown in Fig. 13-19. Here, V_1 is the input voltage; V_2, the output voltage; I_1, the input current; and I_2, the output current. The resistance parameters of the transistor so represented have been symbolized by

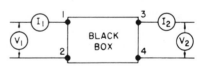

FIGURE 13-19. Equivalent black-box representation.

the so-called R parameters as well as by the more widely used H parameters. Initially we shall consider the R-type parameters since they serve as a more solid foundation for understanding the H (hybrid) types.

Because $E/I = R$, the input resistance is notated as R_{11}, to identify that this R value is obtained by using the V_1 and I_1 values. Thus the first 1 of the subscript indicates V_1 voltage and the second 1 of the subscript is the I_1 current. Thus, R_{11} is a measure of the resistance with the test voltage V_1 applied to the input terminals 1 and 2, with the output terminals 3 and 4 open; that is, $I_2 = 0$. Similarly, R_{21} indicates the forward transfer resistance, with the test voltage V_1 applied to the input terminals again and the ratio of

V_2/I_1 taken for the R_{21} value. When the test voltage is applied to the output terminals 3 and 4 and the input terminals are left open ($I_1 = 0$), we obtain

$$R_{12} = \frac{V_1}{I_2} = \text{reverse transfer resistance (feedback resistance)}$$

$$R_{22} = \frac{V_2}{I_2} = \text{output resistance}$$

From the foregoing we can derive the loop equations for the transistor network.

$$V_1 = R_{11}I_1 + R_{12}I_2 \tag{13-5}$$

$$V_2 = R_{21}I_1 + R_{22}I_2 \tag{13-6}$$

The amplifying ability of the transistor network relates to the mutual resistance (r_m), and current amplification is referred to as *alpha* (α) discussed earlier. Related to the R parameters, these are

$$r_m = R_{21} - R_{12} \tag{13-7}$$

$$\alpha = \frac{R_{21}}{R_{22}} \tag{13-8}$$

The *r*-value relationships of A in Fig. 13-16 (*common base*) become

$$r_e = R_{11} - R_{12} \tag{13-9}$$

$$r_c = R_{22} - R_{21} \tag{13-10}$$

$$r_b = R_{12} \tag{13-11}$$

The equivalent generator voltage is equal to $I_1(R_{21} - R_{12})$. The resistive parameters can also be utilized for the dynamic characteristics of the network under signal conditions. Using the lowercase d to indicate a quantity change, and holding i_c constant, we obtain

$$R_{11} = \text{slope of curve } \frac{dV_e}{dI_e}$$

$$R_{21} = \text{slope of curve } \frac{dV_c}{dI_e}$$

Holding i_e constant, produces the following:

$$R_{12} = \text{slope of curve } \frac{dV_e}{dI_c}$$

$$R_{22} = \text{slope of curve } \frac{dV_c}{dI_c}$$

For dynamic conditions where a-c signals are involved, consideration must be given to reactive components as well as resistive; hence the designations are in Z instead of R. The black box electronic circuit input would have an impedance and, with an open circuit at the output ($I_2 = 0$), we obtain

$$Z_{11} = \frac{V_1}{I_1} \qquad (I_2 = 0) \tag{13-12}$$

Again with an open circuit at the output, we obtain the forward transfer impedance comparable to R_{21}:

$$Z_{21} = \frac{V_2}{I_1} \qquad (I_2 = 0) \tag{13-13}$$

The reverse transfer impedance is obtained with the input circuit open ($I_1 = 0$) and this produces

$$Z_{12} = \frac{V_1}{I_2} \qquad (I_1 = 0) \tag{13-14}$$

The output impedance is also obtained with the input held in the open-circuit condition:

$$Z_{22} = \frac{V_2}{I_1} \qquad (I_1 = 0) \tag{13-15}$$

Now we can rewrite the loop equations given earlier (Equations 13-5 and 13-6) as follows:

$$V_1 = Z_{11}I_1 + Z_{12}I_2 \tag{13-16}$$

$$V_2 = Z_{21}I_1 + Z_{22}I_2 \tag{13-17}$$

13-11. Hybrid (*h*) Parameters

The R parameters were obtained under open-circuit conditions, representing *constant-voltage* types. As discussed in Chapter 4, however, it is often convenient to use the *constant-current* analysis, shorting out terminals as required. By combining the constant-voltage and constant-current approach, a more desirable type of parameter is obtained, one in general usage by transistor manufacturers. The combination has led to the term *hybrid* or *h parameter*, which refers to the type primarily used for obtaining operational characteristics of the *bipolar* transistors. The parameters used for the unipolar transistors (FET) are the Y types discussed later.

For the h parameters, the following notations apply for the black box concept shown in Fig. 13-19:

$$h_{11} = \frac{V_1}{I_1} \qquad \text{(an input \textit{impedance} parameter, with output terminals 3 and 4 shorted and } V_2 = 0)$$

$$h_{12} = \frac{V_1}{V_2} \qquad \text{(a reverse-transfer \textit{voltage ratio}, with input terminals 1 and 2 open and } I_1 = 0)$$

$$h_{21} = \frac{I_2}{I_1} \qquad \text{(a forward-transfer \textit{current ratio} with output terminals 3}$$

and 4 shorted and $V_2 = 0$)

$$h_{22} = \frac{I_2}{V_2} \qquad \text{(an output \textit{admittance} term with input terminals 1 and 2}$$

open and $I_1 = 0$)

Basic calculations can be used if necessary to convert R to h or h to R. Parameter R_{11}, for instance, is equal to $(h_{11}h_{22} - h_{12}h_{21})/h_{22}$. Similarly, $R_{12} = h_{12}/h_{22}$ and $R_{21} = h_{21}/h_{22}$. The equations for the R and Z parameters (Equations 13-5, 13-6, 13-16, and 13-17) involved input voltage and output voltage (V_1 and V_2). For the h parameters, however, the equations relate to input voltage and output current, as follows:

$$V_1 = H_{11}I_1 + H_{12}V_2 \qquad (13\text{-}18)$$

$$I_2 = H_{21}I_1 + H_{22}V_2 \qquad (13\text{-}19)$$

Standards have been adopted for letter subscripts for easier identification of the h parameters. The first subscript designates the characteristic, i for input, o for output, f for forward transfer, and r for reverse transfer. The second subscript designates the circuit configuration, with b for common base, e for common emitter, and c for common collector. Thus, h_{11} can be indicated as h_{ib} for the input h of a common base network. Similarly, h_{12} can be written as h_{rb} for reverse transfer in the common base network. This avoids the confusion that might result by simply using h_{11}, for instance, without designating whether it is in reference to common base, common emitter, or common collector.

At A of Fig. 13-20 is shown a basic grounded-emitter circuit. Applied to the input is the signal voltage to be amplified. This could be symbolized as e_s for *signal voltage* or as e_g for *generator voltage*, as shown in Fig. 13-20. The resistance associated with the input signal is marked R_g. The load resistance applied to the output is indicated by the standard R_L symbol. The d-c power sources E_1 and E_2 are effectively bypassed by capacitors C_1 and C_2; hence they are not part of the active parameter analysis.

The equivalent circuit for the grounded emitter is shown at B, with appropriate hybrid symbols with letter subscripts. The base current is indicated as i_b and collector current as i_c. The input voltage e_g also develops across R_g hence the actual voltage applied to the base and emitter is indicated as e_i. The output voltage across the load resistor is shown as e_o. The emitter input resistance is given as h_{ie}, and the output admittance or conductance is given as h_{oe}, measured in mhos. Note the use of the reverse-voltage transfer ratio symbol h_{re} for the emitter circuit. (In a common base this would be h_{rb}.) The forward current ratio is designated as h_{fe}, indicating the emitter designation.

The forward current-transfer ratio h_{fe} is now i_c/i_b (with $e_o = 0$), which is the same as the Equation 13-1 given earlier for signal current gain β

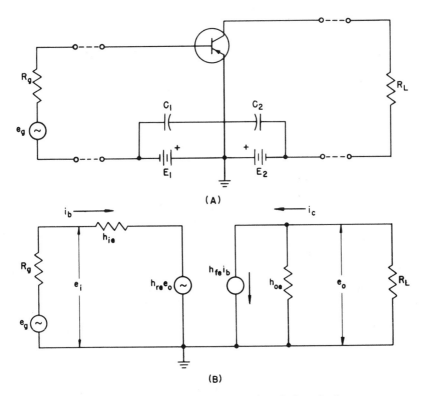

FIGURE 13-20. Grounded emitter and equivalent circuit.

(beta). Manufacturers have used A_i for signal current gain, symbolizing current amplification. The full equation is

$$A_i = \frac{i_c}{i_b} = \frac{h_{fe}}{1 + h_{oe}R_L} \tag{13-20}$$

While this equation refers to the common-emitter circuit of Fig. 13-20, it also applies to the grounded-base or grounded-collector circuits. The equation remains the same except for a change from *e* to *b* or *c* for the second subscript to suit the circuit configuration. Other equations used to analyze circuit parameters are also applicable to the three basic circuit systems of transistors, using the appropriate second subscript to denote emitter, base, or collector. The following example shows the application of Equation 13-20:

Example: The GE 2N508 transistor has a low signal designation of $E_1 = 5$ V and $I_b = 0.1$ mA. The manufacturer's *h* ratings are

$$h_{ie} = 2,800 \ \Omega$$

$$h_{oe} = 43 \ \mu\text{mhos}$$

$$h_{fe} = 110$$

$$h_{re} = 7.5 \times 10^{-4}$$

Assume a load resistance of 10,000 Ω is to be used. What is the current gain?

Solution:

$$A_i = \frac{h_{fe}}{1} + h_{oe}R_L$$

$$= \frac{110}{1 + (43 \times 10^{-6} \times 10,000)} = 77$$

The input resistance R_i can be found by the following equation:

$$R_i = h_{ie} - \frac{h_{fe}h_{re}R_L}{1 + h_{oe}R_L} \tag{13-21}$$

Using the h values given above for the 2N508 transistor, with a load resistance of 10,000 Ω, the following shows the application of Equation 13-21:

$$R_i = 2800 - \frac{110 \times 7.5 \times 10^{-4} \times 10,000}{1 + (43 \times 10^{-6} \times 10,000)} = 2224 \ \Omega$$

The signal-voltage-gain amplification of the transistor circuit is found by

$$A_e = \frac{e_o}{e_i} = \frac{1}{h_{re} - (h_{ie}/R_L)[(1 + h_{oe}R_L/h_{fe})]} \tag{13-22}$$

Again, using the values given for the 2N508 transistor as an example, the voltage amplification is found:

$$A_e = \frac{1}{0.00075 - (2800/10,000)[1 + (43 \times 10^{-6} \times 10,000)]/110} = 274.$$

The power gain of the transistor circuit is found by multiplying the signal current gain A_i by the signal-voltage gain A_e:

$$A_p = A_e A_i \tag{13-23}$$

Applying the two values previously obtained for the transistor used as an example, we find the power gain to be

$$A_p = 274 \times 77 = 21,098$$

13-12. Y Parameters

As mentioned earlier, the admittance (Y) parameters are most useful for investigating the operational characteristics of the field-effect transistor. Since $Y = 1/Z$, the following replace the h notations given earlier for the black box of Fig. 13-19:

$Y_{11} = \dfrac{I_1}{V_1}$ (an input *impedance* parameter with output terminals 3 and 4 shorted and $V_2 = 0$)

$Y_{12} = \dfrac{I_1}{V_2}$ (reverse transfer admittance with input terminals 1 and 2 shorted and $V_1 = 0$)

$Y_{21} = \dfrac{I_2}{V_1}$ (forward transfer admittance with output shorted and $V_2 = 0$)

$Y_{22} = \dfrac{I_2}{V_2}$ (output admittance with input shorted and $V_1 = 0$)

Instead of the input voltage and output current h equations (Equations 13-18 and 13-19), the FET equations for Y parameters involve input current I_1 and output current I_2:

$$I_1 = Y_{11}V_1 + Y_{12}V_2 \tag{13-24}$$
$$I_2 = Y_{21}V_1 + Y_{22}V_2 \tag{13-25}$$

As with the letter subscripts used for easier identification of the h parameters, we can use $i, o, f,$ and r for FET units to describe parameters of input, output, forward transfer, and reverse transfer. A second subscript is used to identify gate, source, or drain. Thus, the I_1 and I_2 equations (Equations 13-24 and 13-25) for *common source* circuitry, become

$$I_g = Y_{is}V_g + Y_{rs}V_d \tag{13-26}$$
$$I_d = Y_{fs}V_g + Y_{os}V_d \tag{13-27}$$

For common-gate or source-follower design, the second subscript of Equations 13-26 and 13-27 are changed accordingly. Thus, for common gate circuitry,

$$I_s = Y_{ig}V_s + Y_{rg}V_d \tag{13-28}$$
$$I_d = Y_{fg}V_s + Y_{og}V_d \tag{13-29}$$

13-13.   Lead Connections

Lead connections for specific transistor types vary to a considerable extent among manufacturers; hence no set rules can be given for guidance. Some general types are shown in Fig. 13-21, however, for reference purposes and to indicate the methods that are employed. All are bottom views of the transistors or FET units. Note the emitter-base-collector sequence of those shown at A, B, and C, with the emitter lead at the left for the type shown

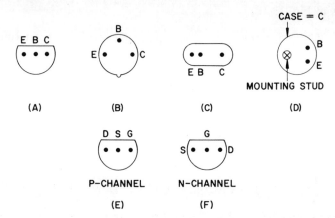

FIGURE 13-21. Examples of lead identification.

at A. For B the short flange extension identifies the sequence start. For C, the two closely spaced leads identify the emitter-base terminals. For the type at D, the mounting stud provides a means for bolting the unit to the chassis, with the case connecting to the collector. If the collector is not to be at chassis ground, an insulating washer (such as mica) is normally used between the bottom of the transistor and the chassis. A silicon lubricant is available for providing a better thermal bond to the chassis for heat-dissipation purposes. At E and F the FET lead identification is shown, with the *P*-channel type at E and the *N*-channel type at F.

While the lead sequences shown in Fig. 13-21 have been widely used, many transistors with the same lead spacings may have other sequences. For that at A, for instance, some types have been used with a sequence of emitter-collector-base. Similarly, for that at B, a base-emitter-collector sequence has been used, though generally the emitter-base-collector is favored. Consequently, a transistor reference manual should be consulted for positive identification.

As shown in Fig. 13-22, most of the larger transistors that handle higher power are designed to be bolted to the chassis for heat-dissipating purposes.

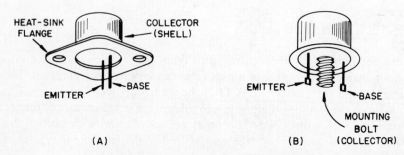

FIGURE 13-22. High-power transistor leads.

Often copper or aluminum flangers or metal sections are used as heat sinks to help dissipate heat and prevent damage to the transistor.

The identifying (numbering) systems used for transistors also vary considerably among manufacturers, even though the characteristics of the transistors are similar. Transistor numbering systems usually include letters as well as numbers and hence appear as 2SC984, 2N212, and 2N1069. Again, reference should be made to the manufacturer's specifications or transistor manuals for an identification of the characteristics of a particular type.

Because of the danger of heat damage, most solid-state devices should be soldered in a minimum of time and with a minimum of heat. This is particularly important when the soldering is made close to the transistor body. Heat-dissipating clamps are available and are clipped on the lead between the transistor and the point of soldering to cut down on the heat generated. The tips of thin-nosed pliers can also be used for this purpose.

13-14. Integrated Circuits

Electronic signals perform many tasks, including sound reproduction from speakers, formation of images on television screens and display tubes, computation and data processing in computers, and the initiation of physical movements in industrial control applications. For signals to perform such tasks they must have sufficient electric power acquired by amplification; hence voltage and current levels are highest in the output circuits, and solid-state units and associated components are larger and have higher ratings to handle increased power levels.

Prior to final amplification and usage, however, signals are usually processed in a variety of ways to provide them with the characteristics necessary in computers, control systems, etc. Such processing and signal modification can be done at pico-power levels; hence miniature circuitry is ideal to save space and power and to increase efficiency. Because of the rapid strides in solid-state manufacturing procedures, the hardware bulk of virtually all electronic circuitry can now be reduced to an extraordinary degree by a process known as *microminiaturization*. Thus, dozens of units, including diodes, transistors, FET's, resistors, and associated wiring can be assembled into a silicon chip or a so-called *integrated circuit*, often limited in size only by the necessity for bringing out connecting leads.

The advent of the printed circuit also contributed to miniaturization and mass production since it permitted the assembly of compact circuits and eliminated the more bulky interconnecting wires. This also reduced the undesired effects of stray capacitance and lead inductance. Thus, many circuit sections of TV or radio receivers and other electronic equipment consisted of thin plastic sheets with the interconnections between circuit components consisting of electroplated segments as shown in Fig. 13-23.

FIGURE 13-23. Typical printed
circuit.

Over the years many processes have been used for microminiaturization, including electroplating, electrodepositing of material on ceramic substrates, photoengraving and etching procedures, and vacuum deposition. In the latter, the active solid-state material is placed in a vacuum chamber with the substrate and heated for vaporization deposits of thin film onto the substrate. Often the silicon slice was heated (up to 1000°C) in a vapor-type atmosphere containing the impurities to be imbedded into the silicon slice. This initial process is termed *deposition* and it was followed by placing the silicon slice into another heat chamber where higher temperatures caused the surface film to diffuse into the silicon slice. This is the *diffusion* procedure.

The diffusion process, when applied directly to transistor manufacture, produced a rugged solid-state device known as the *silicon planer transistor*, which was the initial step toward the production of integrated circuits. These transistors are formed by the diffusion of both the base and emitter sections by aluminum depositing, thus contributing greatly to mechanical stability and strength.

Next in the chronological progress toward today's microminiaturization was the development of the *epitaxial* technique wherein a thin layer of *N*-type material is grown on a *P*-type silicon chip. In epitaxial form the *N*-type material becomes an integral part of the crystal structure. This process has been instrumental in obtaining high-quality power transistors with excellent linearity and improved stability for changes of temperature.

Thus, by diffusion, masking techniques, etching processes, and vacuum deposition, it has been possible to form integrated circuits containing solid-state units and associated components, complete with interconnecting wiring. The tiny integrated circuit has number-identified leads brought out as shown in the example given in Fig. 13-24. Here the IC contains four *NPN* transistors, an FET, a diode, and 10 resistors. Some IC units would, of course, contain fewer components, while others would have many more.

Integrated-circuit units take many forms, with some having a circular shape and others having a square or rectangular appearance. Some, as shown at A of Fig. 13-25, have prongs that are designed to fit into holes in the printed-circuit board for soldering to the metallic conducting strips. (The circular 10-prong connecting section is apparent in Fig. 13-23.) For the IC at B, a rectangular form is used, with connecting segments as shown. Such

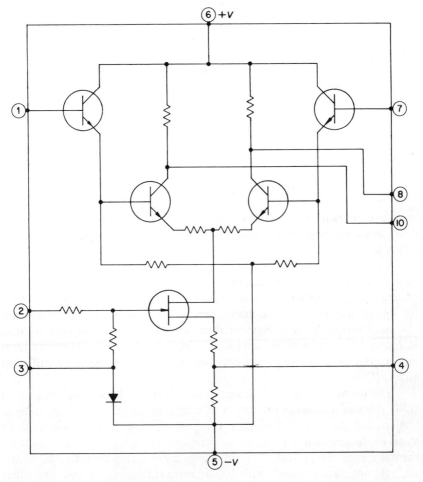

FIGURE 13-24. Examples of IC internal circuitry.

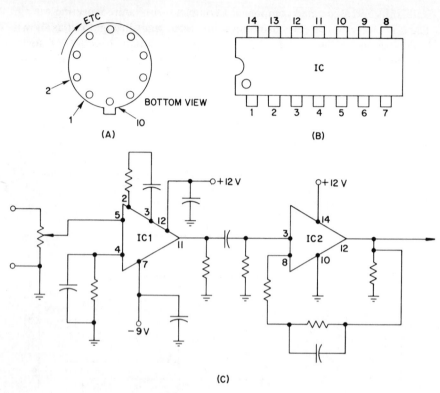

FIGURE 13-25. IC terminals and symbols.

units may measure less than a half inch in length, depending on the number of circuits contained therein and the necessity for establishing firm soldering contacts.

Schematically, integrated circuits are often shown in triangular form as at C of Fig. 13-25. On occasion they may be represented as a square or oblong. Only those terminals are usually numbered that are actively used for the particular electronic device formed. If more than one IC is used, the usual procedure is to number them in consecutive order such as IC1, IC2, and IC3. If one IC is used for two circuit sections, the triangular form may appear twice but represent only one IC, with each section identified as IC1A, IC1B.

The terms medium-scale integration (MSI) and large-scale integration (LSI) refer to maximum utilization of chip area to achieve a high concentration and density of solid-state components. The practice has been to stack oxide-dielectric layers and interconnecting metallic layers to increase the number of components that can occupy a given chip area. Consequently, such IC units are available with 100 or more transistors, diodes, and other components contained within a single IC section. No definite standards have been established regarding the number of components above which the unit can be termed MSI or LSI. The term LSI appears to be in greater usage than MSI, with IC referring to the smaller units and LSI to the very large (number of component) units.

Review Questions

13-1. In a transistor, what terminal is identified by the lead with the arrow, and what is the significance of the arrowhead pointing toward the symbol or away from it? Explain briefly.

13-2. Briefly explain how amplification occurs in a transistor.

13-3. Briefly explain the difference between triode and tetrode transistors. Use simple drawings to illustrate the explanation.

13-4. Define the terms *bipolar* and *unipolar* in relation to solid-state devices.

13-5. Briefly explain the difference between an *N*-channel FET and a *P*-channel type, and indicate in which current carriers are holes and in which type the carriers are electrons.

13-6. Compare the *gate, drain*, and *source* terminals of an FET with the comparable terminals of a transistor.

13-7. Explain what effect the *depletion region* has in a MOSFET.

13-8. Indicate which FET basic circuits have a phase reversal between input and output signals, and which do not have a phase change.

13-9. (a) Which FET circuit compares to the emitter follower?
(b) Which FET circuit compares to the common-base circuit?

13-10. Define the term *pinch-off* voltage.

13-11. Briefly explain the difference between Equations 13-1 and 13-4.

13-12. Describe *negative resistance* as it applies to the tunnel diode and the unijunction transistor.

13-13. Give some of the particular advantages of the tunnel diode over that of an ordinary silicon diode.

13-14. Explain briefly why a tunnel diode has amplification characteristics.

13-15. What is the difference between an *active* and a *passive* network?

13-16. What are the basic differences between the R and h parameters?

13-17. How is the transistor parameter h_{22} obtained?

13-18. How may h_{11} and h_{12} be written with letter subscripts?

13-19. To what transistor equation does h_{fe} relate?

13-20. Briefly define the FET parameters Y_{12} and Y_{22}.

13-21. Define the subscripts in the equation $I_g = Y_{is}V_g + Y_{rs}V_d$.

13-22. Briefly explain what measures are taken to dissipate heat in solid-state units?

13-23. Explain briefly what the following mean: IC, MSI, and LSI.

13-24. Briefly define the terms *deposition* and *diffusion* with respect to solid-state processing.

13-25. Briefly explain the *epitaxial* process.

13-26. What is the basic manufacturing process for the silicon planer transistor?

Practical Problems

13-1. In a communications system a grounded-emitter circuit was tested and it was found that a change in the transistor base I from 50 μA to 175 μA produced a collector I change from 4 to 8 mA. What is the current (β) gain?

13-2. In a common-emitter transistor signal amplifier circuit the dI_b is 100 μA and the dI_c is 2.5 μA. What is β?

13-3. (a) In the design of an FET circuit for an FM receiver the g_m was found to be 2500 μmhos. If a load resistor of 12 kΩ were used, what would be the signal gain?
(b) If R_L were increased to 50 kΩ, what would the gain be?

13-4. In testing a ground-base transistor circuit for use as an RF amplifier, the dI_c was 10 mA and the dI_e was 12 mA. What is α?

13-5. A commercial transistor has a low-signal designation of $E_1 = 5.5$ V and $I_b = 0.125$ mA. The manufacturer's h ratings are $h_{ie} = 2500\ \Omega$, $h_{ne} = 50$ μmhos, $h_{fe} = 100$, $h_{re} = 6 \times 10^{-4}$, and the recommended $R_L = 10,000\ \Omega$. What is the current gain (A_i)?

13-6. For the h values given in Problem 13-5, what is the value of the input resistance (R_i)?

13-7. For the h values given in Problem 13-5, what is the signal-voltage amplification (A_e)?

13-8. For the h values given in Problem 13-5, what is the power gain (A_p)?

13-9. For the h values in Problem 13-5, what would be the A_i if the load resistance were changed to 5000 Ω.

13-10. A transistor has the same h values as in Problem 13-5 except for $h_{fe} = 125$. With a 5000-Ω load resistance, what is A_i?

Power Supplies 14

14-1. Introduction

Solid-state devices require dc for operation, and the voltage as well as current values vary considerably for different units. Similarly, the various electronic components used in transmitting and industrial-electronic practices have a wide range of electric power requirements, particularly if vacuum or gas tubes are used. In the latter case, dc is needed for anode and screen voltages, though ac may be used on occasion for heaters.

Transistors, in contrast to tubes, require a less complicated voltage supply. Portable electronic units utilize one or more batteries as the power source, though in permanent of semipermanent installations power supplies or generators are employed. Since dc is necessary for solid-state units and tubes, the power-supply units must convert the ac in the power mains to dc. The most convenient method for doing this is to rectify the ac of the power mains and use a filter system to produce dc, as mentioned earlier. Since the power-main voltages supplied to the home range between 110 to 220 V ac, it is necessary to alter the voltage and current properties of such ac to meet the requirements of transmitters, receivers, recording devices, and other electronic units.

Some solid-state units require dc potentials below 12 V, while some power transistors may need 60 V or more. Tube filaments having no separate cathode, for instance, need dc in many applications since ac would cause

273

fluctuations in any signal voltages handled by the circuit, resulting in 60-Hz hum. Such hum would be audible in speaker systems or appear as interfering bars on television screens. For cathode-type tubes, however, ac may be used on the filaments, though the voltages must be reduced from the power-main levels to that required (6.3 or 12.6 V, etc.).

Power supplies must also be capable of supplying the amount of current required, without undue overheating. Tube anodes and screen-grid elements, for instance, may need several hundred volts but current requirements may be in milliamperes. Filaments of tubes, however, or power transistors may draw several amperes of current, depending on the needs established by the design. Thus, voltages must be stepped up or down as needed, with the capability of furnishing the required current. A common method is to utilize a transformer (referred to as a *power transformer*) as discussed in this chapter. Voltage regulation, voltage multiplication, filter factors, power-supply types, and other related matters are also covered in this chapter.

14-2. Half-Wave Power Supplies

The basic principle of rectification was discussed in the early portion of Chapter 11 and illustrated in Fig. 11-5. Such a rectifier system, which utilizes every other alternation of the a-c cycle for producing dc, is known as *half-wave* power supply because it only employs one-half of the a-c cycle for production of dc.

A typical commercial half-wave rectifying system, using a vacuum tube, is shown at A of Fig. 14-1. In the primary winding, the slanted arrow indicates a switch for turning the device on and off. The top secondary winding provides the necessary heater current for the rectifier tube. The secondary winding connected to the anode of the rectifier and ground provides a much higher voltage than the other windings since this is the source of the ultimate dc for the anodes and screen grids of the tubes. In contrast to filament windings where low voltage and high currents are necessary, the anode rectifier winding provides a high voltage with low current. For this reason, the high-voltage secondary is wound with much thinner wire, such as No. 28, 30, 32, or 34, while filament secondaries would be wound with No. 16, 18, or 20 wire.

The high-voltage winding can have as many turns as required for the amount of voltage needed. In common practice, the voltage step-up is usually of the order of 200 or 300. One lead of the high-voltage winding connects to the plate, as shown, while the other lead connects to ground. (It must be remembered that ground connections are simply made for the purpose of *providing interconnection between other similar voltage points.* In most transformer-type power supplies the negative section of the power supply is referred to as the *ground potential.*)

The cathode element of the rectifier is connected to a filter capacitor

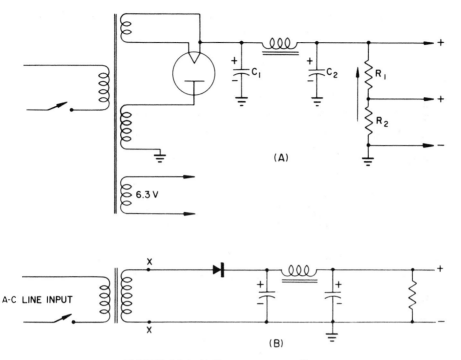

FIGURE 14-1. Half-wave power supplies.

(C_1) and to a filter choke, as shown at A. An additional filter capacitor is employed (C_2) for smoothing out the a-c ripple. Across the output terminals of the power supply, two resistors, R_1 and R_2, are used. The output voltage develops across these resistors. In some electronic devices, the resistors may be dispensed with, and the positive and negative terminals are connected directly to the circuits. The resistor network is known as a *bleeder* section since it imposes a slight current drain on the power supply and thus helps to stabilize the output voltage. The voltage-regulating function of the bleeder resistor is explained in greater detail later in this chapter. The bleeder network can also be used as a source of lower voltages, by employing several resistors in series or by using a single resistor with several taps. Thus, the center positive terminal shown at the output of circuit A in Fig. 14-1 provides a lower voltage than that obtained from the top positive terminal.

An additional secondary winding is employed (6.3 V or higher, as required) for furnishing heater power to the tubes of the electronic device. Where a large number of tubes is used, additional secondary windings may be provided. Additional windings are, of course, necessary when other values of voltages are required, besides the voltage furnished by the single 6.3-V secondary.

When a positive alternation appears at the rectifier anode, electrons flow from the ground side of the high-voltage winding and along the chassis of the power supply to the bottom of the capacitors, as well as to the bottom of the resistor R_2 in the bleeder. Since the filter capacitors have a zero charge initially, capacitor C_1 across the rectifier tube charges to a peak value of the a-c alternation. Since the peak value is calculated on the basis of 1.41 times the effective voltage, capacitor C_1 charges to a peak value of 282 V if the secondary is delivering a voltage of approximately 200. (The peak value will be reduced in proportion to the resistance of the rectifier. Pulses involve a calculation based on the average value, as mentioned earlier.) A charge is also placed across capacitor C_2 but such a charge has a value less than the peak voltage because the filter choke is in series with capacitor C_2 and reduces the voltage. Once the filter capacitor has been charged, current flow will occur through R_2 and R_1 and will return to the cathode of the rectifier.

During a negative alternation of the a-c voltage across the high-voltage secondary, the plate side of the winding will be negative and the cathode side positive. During this time, the tube cannot conduct. The filter capacitors are charged, however, and seek a path for discharge. Such a discharge path is provided by the bleeder network consisting of resistors R_1 and R_2. The capacitors start to discharge through the bleeder network in the direction shown by the arrow. When the supply is furnishing power to a load such as an audio amplifier or other electronic device, the discharge of the capacitor furnishes the power during the time the tube is not conducting. Eventually, of course, the capacitors would discharge completely but while the power supply is in operation, the capacitors are fed constantly by each positive alternation, which occurs once every sixtieth of a second. During the charging cycle of the capacitors, the time constant of the circuit is short since the tube is conducting and offers very little resistance and the capacitors are virtually across the voltage source. When the tube does not conduct, however, the time constant of the circuit is long because the capacitors must discharge through the bleeder (they cannot discharge through the vacuum tube because the latter is in a nonconducting condition). Thus, very little of the charge leaks off before a new charge is placed on the capacitors. If the resistors R_1 and R_2 were made smaller, the time constant would be shortened and the capacitors would discharge more rapidly during the intervals when the tube is not conducting. In normal operation, the bleeder resistance is made fairly high (usually between 25,000 and 50,000 Ω).

The bleeder resistor network represents a load on the power supply since some energy is consumed as the electrons flow through the resistors. When other resistors (or transistor and tube circuits) are added across the output of the power supply, additional current is consumed from the power supply (i.e., the load on the power supply is increased). Adding more circuits or resistors means that additional current is drawn from the power supply. An increase in current indicates a decrease in the value of resistance across the supply and this is virtually what occurs since each circuit or additional resistor placed across the bleeder network would reduce the resistance value across the plus and minus terminals. As more and more energy is drawn from the power supply, the decline of voltage across the bleeder becomes more pronounced. As extra current is drawn from the filter capacitors, the latter discharge to a lower level between alternations and will not charge to as high a level during voltage peaks. Thus, the output of the power supply declines as the load increases (as more current is drawn). This change of voltage output with a change of current consumption is referred to as *regulation* and will be more fully described later in this chapter.

By employing additional filtering, such as increasing the value of the filter capacitors, the sharp decline of capacitor charges can be minimized so that the output voltage will be substantially ripple-free and become sufficiently smooth for application to electronic circuitry. The filter choke also minimizes ripple because it tends to diminish the peak amplitude of the voltage through its opposition to sudden changes of voltage. The filter choke can also be considered as a series inductance that provides a high order of inductive reactance to any ripple component. By similar analysis, the filter capacitors become a low capacitive reactance shunt for any ripple component of the output d-c voltage.

While the rectifier can be a tube with a separate cathode, a rectifier with a directly heated cathode element is often employed. In such an instance, the filament of the rectifier also acts as a cathode, as explained earlier. There is virtually no difference in performance since one leg of the filament will now carry the dc rectified by the tube. The necessity for a cathode, however, arises when a single filament winding of the transformer must also furnish heater current for other tubes besides the rectified tube. In such an instance, a separate cathode is necessary to keep the high voltages out of the filament circuits of other tubes.

A typical half-wave power supply using a solid-state diode is shown at B of Fig. 14-1. In earlier devices of this type, the selenium rectifier was employed. This consisted of metal disks treated with selenium to pass current in one direction. The disks were stacked, as shown at A of Fig. 14-2, to conform to the voltage and curent requirements. The selenium rectifiers have generally been replaced by the silicon diodes, which are solid-state units operating on the *P-N* and *N-P* transistor principles discussed in the preceding two chapters. The silicon diodes are smaller than the selenium type handling the same current (and with the same voltage ratings) and also have a higher forward-to-back resistance ratio. The resistance in one direction is usually in excess of 2000 times that in the other direction and thus very little current flow is possible in the reverse direction. With the selenium types, the much lower resistance ratio permitted some current flow in the reverse direction.

Silicon rectifiers come in various sizes, from the small units shown at B and C of Fig. 14-2 to the larger types shown at D and E. Those for electronic equipment in the home (tape recorders, receivers, etc.) range from current

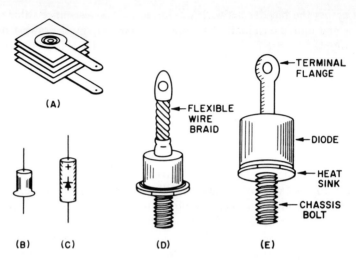

FIGURE 14-2. Silicon diodes.

ratings of 50 to approximately 800 mA. For industrial applications the current ratings are in amperes. As shown at D and E, a threaded terminal is provided to permit the unit to be bolted to the chassis, with the flange acting as a _heat sink_. With the flange bolted flat against the chassis, the heat is absorbed by the chassis and drained away from the rectifier, in similar fashion to some of the larger power transistors.

For the circuit shown at B of Fig. 14-1, a transformer is used that steps up the voltage from the line, though the power supply is often connected directly to the line at the terminals marked X when high voltage outputs are not necessary. The operation of this power supply is similar to that of the simple half-wave type discussed earlier. During positive alternations of the input voltage, the rectifier conducts and creates pulsating dc that is filtered so that a sufficiently smooth dc appears at the output terminals.

14-3. Full-Wave Power Supplies

A full-wave power supply is one that uses both alternations of the a-c cycle and hence has several advantages over the half-wave type. Since filtering of ripple is easier, smaller-value filter capacitors can be used. In addition, the full-wave supply has better regulation (maintenance of output voltage) for changes in load.

As shown in Fig. 14-3, two diodes are connected across a tapped secondary transformer winding. The secondary voltage is approximately twice that obtained at the output of the supply, as opposed to the half-wave type

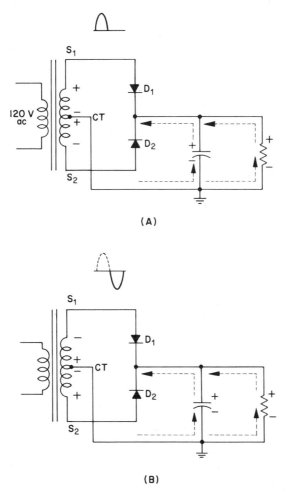

FIGURE 14-3. Electron flow in full-wave supply.

where only a single secondary winding (untapped) is required. As shown, one lead of the secondary connects to diode D_1 and the other to D_2. The diode cathodes are connected to a common terminal that will have a positive polarity during rectification. The center tap (CT) is the negative return to the bottom of the filter capacitors and output resistor and consequently this terminal is usually grounded.

The two illustrations of Fig. 14-3 show the manner in which this circuit functions. When the switch is closed, assume that initially a positive alternation appears across the secondary winding as at A. This causes the voltage polarity at the terminal applied to D_1 to be positive, while the polarity at the lower diode D_2 (fed by the bottom of the transformer) will be negative.

The center tap of the secondary winding has a polarity *that depends on its relationship to either the upper-winding terminal S_1 or the bottom terminal S_2.* If S_1 is positive, the center tap will be negative with respect to it. When S_2 is negative, the center tap will be positive with respect to it. This is a common polarity relationship, as described in greater detail in earlier chapters.

Under the condition illustrated at A of Fig. 14-3, diode D_1 will conduct because of the forward-bias type voltage present across it. Thus, for the positive alternation as shown, D_1 anode is positive and the center tap is negative in relation to it. Hence, electrons leave the center tap and travel toward the bottom of the resistor and capacitor in their return circuit to the positive terminal, as shown by the arrows. During the electron-flow process, the filter capacitor is charged as in the case for the half-wave rectifier discussed earlier. Electron flow through the resistor is also in the direction shown by the arrows. Diode D_2 is unable to conduct because of the reverse bias on it (negative anode and positive cathode).

At the second alternation of the ac across the secondary, the conditions prevail as shown at B. Here diode D_2 now conducts since its anode is positive with respect to its cathode. Again, the cathode is connected to the center tap (via the capacitor and resistor, as with D_1). Now since diode D_1 has a minus anode and a positive cathode, it does not conduct.

Electron flow in the power supply will again be *in the same direction as for the positive alternation* shown at A. Again the capacitor charges toward the peak value of the voltage and discharges across the resistor in the direction shown by the arrows. Thus the capacitor receives a charge at a rate that is twice that for half-wave rectification. As shown at A of Fig. 14-4, the pulsating dc that results has a ripple frequency of 120 Hz instead of 60 Hz. As shown at B, the discharge time between the peaks of the pulsating dc is only half that encountered in half-wave rectification. Because of the reduced interval between peaks, the filter capacitors in such supplies do not have as long a discharge time and hence are more capable of maintaining a relatively high charge. At the same time the higher ripple frequency creates a lower reactance in the filter capacitor, thus shunting ripple components more effectively. If a filter choke is used, a higher inductive reactace prevails and hence the choke is more effective in limiting voltage variations.

Two more complete versions of the full-wave power supply are shown

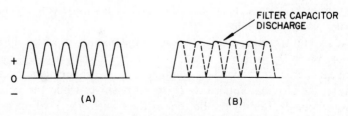

FIGURE 14-4. Function of filter capacitor.

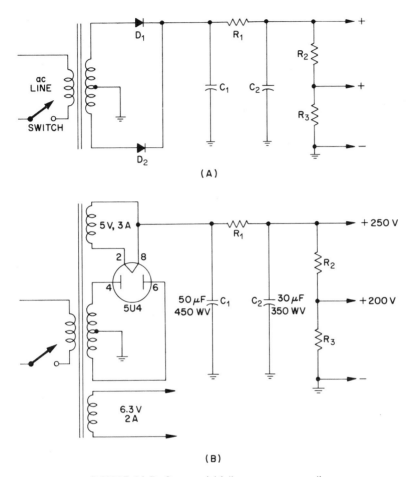

FIGURE 14-5. Commercial full-wave power supplies.

in Fig. 14-5. Instead of using a continuous line to interconnect negative points, ground connections are shown and this schematic shortcut is often used in commercial drawings. It is also indicative that the chassis is utilized as a return circuit for the electron flow from the negative-polarity source.

For both circuits A and B a series resistor replaces the filter choke mentioned earlier. With load circuitry that imposes a steady drain on the supply, resistor R_1 can replace the filter choke for decreasing circuitry bulk. If ripple increases, the capacitance of the filter capacitors can be raised. At the output, voltage-divider resistors (R_2 and R_3) are used so that more than one potential is available. Additional resistors could be used for obtaining additional lower voltages.

A typical tube-type supply is shown at B, using a dual-diode tube such as the 5U4 with pin terminals indentified by numbers. Here a separate low-voltage winding is required for the filament of the rectifier tube. An additional secondary winding is provided furnishing 6.3-V with available current values up to 2 amp, for furnishing heater potentials to tube-type circuitry.

It must be remembered that ampere ratings indicate the *maximum permissible* current that can be drawn from a particular winding and it does not indicate the amount of current that would flow in a particular system. Thus, the amount of current that is delivered by the 6.3-V winding will depend on the load imposed on it by the heaters of tubes connected across the 6.3-V source. As more and more tube heaters shunt this winding, additional current is drawn. If more current is drawn than specified for the particular transformer, the winding will overheat and may burn out. (If more voltage is impressed across the heaters of a tube than called for by the tube, filament burnout also occurs because of the increase of current above that needed.)

Full-wave rectification can also be obtained without using a center tap in the high-voltage secondary of the power transformer. To accomplish this, however, additional rectifier units are employed. A typical circuit is shown in Fig. 14-6 and this is known as a *bridge rectifier*. Pulsating dc is provided across the output and must be filtered with the conventional filter circuits shown earlier.

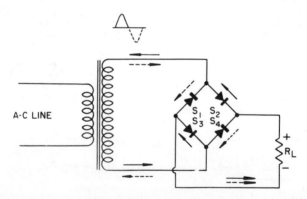

FIGURE 14-6. Full-wave bridge rectifier.

Operation of the bridge rectifier of Fig. 14-6 is dependent on the manner in which the rectifiers are connected with respect to their polarities. When a positive alternation appears at the secondary, as shown by the first alternation (in solid lines), the top of the secondary is positive and the bottom of the secondary is negative. Under these conditions, the electron flow is as shown by the solid arrows. From the bottom of the secondary terminal (negative), electrons flow through the silicon rectifier S_3 to the junction of S_1 and S_3. Since this electron flow is in the opposite direction to what S_1 will conduct, the electrons cannot enter S_1 but, instead, will flow to the bottom of resistor R_L and through this resistor to the junction of S_2 and S_4. The electrons then flow through S_2 to its return circuit that is the positive terminal of the secondary.

When the second alternation appears (shown by the dotted outline), the top of the secondary has a negative polarity and the bottom of the secondary has a positive polarity. Electrons now flow from the top negative terminal through the circuit, as shown by the dotted arrows.

14-4. Voltage Doubling

The transformer in a power supply can be dispensed with, and greater voltage than present in the a-c line can still be procured. The voltage increase is obtained by circuits using voltage-doubling or voltage-tripling systems. A typical transformerless power supply using the voltage-doubling principle is shown at A of Fig. 14-7. The voltage-doubling circuit furnishes pulsating dc, which must be filtered for smooth dc using a filter choke and capacitor, as previously detailed.

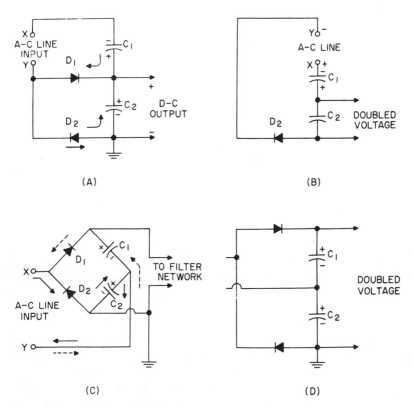

FIGURE 14-7. Voltage doubling.

Reference to the basic voltage-doubling circuit shown at A will aid in understanding how this circuit functions. Initially, assume that the line voltage applied to the terminals X and Y is an alternation that applies a negative polarity to the X terminal and a positive polarity to the Y terminal. Under this condition, electrons will flow from the X terminal and will charge capacitor C_1 with a polarity as shown. The completion of the circuit is in such a manner that electron flow is in the direction shown by the arrow. During this time, however, the diode rectifier D_2 does not permit current to flow since it is wired into the circuit in opposite fashion to D_1.

During the next alternation of the line voltage, the X input terminal would have a positive polarity and the Y input terminal would have a negative polarity. Under this condition, diode D_1 is in a nonconducting state. Diode D_2, however, now conducts and permits electrons to flow through it in the direction shown by the arrow beside D_2. This electron flow from the negative Y terminal must complete its circuit back to the positive X terminal. Because C_1 has been charged by the previous alternation, however, the charge on C_1 *will combine with the a-c line voltage* and hence double the line voltage will appear across capacitor C_2.

An inspection of the rearranged circuit shown in B will help illustrate the manner in which the a-c line voltage adds to the charged capacitor. Here you will note that the a-c line polarity at this instant is negative for the Y terminal and positive for the X terminal. This voltage source is in series with C_1. The latter has stored the energy from the previous alternation and its polarity is such that its energy adds to that of the a-c line energy in charging capacitor C_2.

During the third alternation, the X terminal is again negative and capacitor C_1 is recharged. During the fourth alternation, the line voltage and the charge on C_1 again add together to double the voltage impressed across capacitor C_2. Since the doubled voltage that is impressed across C_2 occurs for every *other* alternation, a 60-Hz hum ripple is present and this ripple must be filtered out by the usual combination of filter capacitors and filter chokes.

Another method for voltage doubling is by use of the circuit shown in C, where a symmetrical arrangement provides a ripple frequency of 120 Hz instead of a 60-Hz ripple frequency. Hence, this circuit is equivalent to full-wave rectification and while a ripple filter network is still necessary, the filtering requirements are less demanding than for the 60-Hz ripple frequency.

If the a-c line alternation is such that the input terminal X is negative and the input terminal Y is positive, electrons will flow from the X terminal in the direction shown by the solid arrows. This path of electrons from the X terminal includes the diode rectifier D_2 and capacitor C_2; then it returns to the positive terminal Y. During this current flow, capacitor C_2 is charged with a polarity as shown. During the next alternation, when the X terminal is positive and the Y terminal is negative, electrons will flow as shown by the dotted arrows. Hence, electrons flow from the Y terminal toward capacitor C_1 and charge the latter to the peak of the line voltage. Electron flow continues through diode rectifier D_1 to the positive X terminal. From the foregoing, it is evident that capacitors C_1 and C_2 are charged to the peak values of the a-c line voltage during successive alternations. The d-c output voltage is taken from across the two capacitors and hence will be approximately double that of the line voltage because the two capacitors are effectively in series, with their polarities adding to increase total voltage. While most schematic representations of the circuit are as shown in C, the circuit has been redrawn in D to indicate more clearly how the output voltage is secured from across the two charged series capacitors C_1 and C_2.

14-5. Voltage Tripling

Diodes can also be used to triple the line voltage by using a suitable circuit arrangement as shown at A of Fig. 14-8. When an alternation of the a-c input voltage is positive, terminal T_1 will be positive and T_2 negative. Now electrons will leave the lower terminal (T_2) and will flow through the rectifier circuit to T_1. Since the diode rectifier (D_1) will conduct under such polarity conditions, capacitor C_1 will charge to the peak line voltage, with a polarity as shown. (Actually the charge will be somewhat less than the peak voltage because of the resistance of D_1 and the resultant voltage drop across the

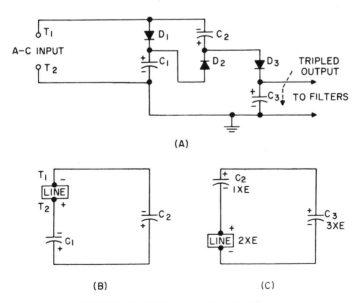

FIGURE 14-8. Voltage tripling principles.

latter. For simplicity, however, reference will be made to the peak-voltage charge.)

During the next alternation when T_1 is negative and T_2 is positive, the rectifier D_1 will not conduct but D_2 will now conduct because the proper polarity is present for it. Electrons will leave terminal T_1 and flow through D_2. It will be noted that the return path for D_2 is *via C_1 to terminal T_2*. Thus, the charge that is already across C_1 will add to the line voltage and place across C_2 a charge that is proportional to the line voltage added to the potential already across C_1.

This is more clearly shown at B of Fig. 14-8, where the circuit that charges C_2 has been rearranged to illustrate the principle involved. Note that when the alternation is such that T_1 is negative and T_2 is positive, the charge on C_1 has the same polarity relationship. Consequently, the circuit can be considered as consisting of two generators (the line voltage plus the charge across C_1). This situation is similar to that of placing two batteries in series to obtain double the voltage. Because capacitor C_2 is across these two voltage sources, the potential that appears across C_2 will be equal to the sum of the voltage across C_1 plus the a-c line voltage.

During the next alternation, when terminal T_1 is positive and terminal T_2 is negative, the charge across C_1 is replenished but, at the same time, rectifier D_3 conducts and charges capacitor C_3 (see A of Fig. 14-8). When D_3 conducts, the charge across C_3 will be composed of the line voltage *plus* the existing charge across C_2. Since the charge across C_2 represents the *doubled voltage*, the latter adds to the line voltage and the result is a triple voltage across C_3. This is shown at C, where the circuit has been redrawn for simplicity, with the rectifier omitted since it is in a conducting state and can be considered a closed circuit. As seen from this drawing, the line voltage represents one potential source in series with the potential source across C_2. The *combination* of these two voltages appears across C_3. Since C_2 has *double* the voltage and this is added to the existing line voltage, the voltage across C_3 will be three times the normal line voltage.

In the circuit shown at A, the lowest voltage can be obtained from across C_1. Twice the voltage can be obtained from across C_2, though in this instance the negative potential would not be at ground. A voltage output of three times the line voltage can be obtained from across C_3, though in all instances additional filtering would be necessary to reduce the ripple component to an acceptable minimum for application to the circuits of receivers and other electronic devices.

14-6.   Filter Factors

The filter sections that follow the rectifier should smooth the ripple sufficiently so it will not appreciably cause interference with signals being processed by the electronic gear connected to the power supply. Overfiltering is usually avoided since it increases the manufacturing cost. Power supplies for home entertainment systems usually are not so complex as those used in industry where precise and well-regulated voltage values may be necessary. Thus the use of single or double filter chokes is more common in industrial systems than receivers or recorders in the home, as are other components as more fully described later.

Figure 14-9 shows four more filter types besides those shown earlier in Fig. 14-1 and 14-5. At A of 14-9 is shown the *choke input* filter. This type uses an iron-core choke ranging from a few to approximately 25 henrys, depending on whether the supply feeds solid-state circuitry or tube-type gear. Such a choke coil is sometimes referred to as a *smoothing* choke. A single filter capacitor is sufficient for some power-supply applications, though a higher capacitance rating is necessary than would be the case if additional filter sections are used.

The choke input filter type provides better power-supply regulation than the capacitor input type, as more fully detailed later in this chapter.

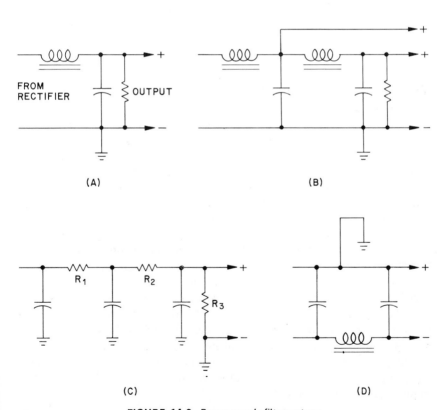

FIGURE 14-9. Power-supply filter systems.

The resistor shown at the output of the filter section is the bleeder resistor that also improves regulation since there is a constant current flow through this resistor. Such a resistor can also be used as a voltage divider, as explained earlier in this chapter.

A more elaborate choke input filter section is shown at B. This circuit has greater ability to smooth the ripple components of the rectified ac. The first inductance shown will limit the peak values of voltage because of its characteristics in opposing a current change such as results during pulsating dc. The filter choke also presents a high inductive reactance to the ripple components of the rectified waveform. Similarly, the filter capacitor presents a low-shunt reactance for ripple components. From the practical standpoint, however, the filter capacitors charge to peak values of voltage and tend to maintain a constant output voltage, free from fluctuations, because of the storage characteristics of the capacitors and their ability to oppose voltage changes. With the double section filter shown at B, the output voltage is substantially free from ripple and suitable for circuits in which the hum factor must be kept at a low level. If such a device is used for a multicircuit amplifier system, for instance, the output voltage can be applied to the early stages where a minimum of ripple is desired. For the power-output amplifier stages, where amplification is low and a slight ripple is not a serious factor, the voltage can be obtained from the junction of the two filter chokes. This arrangement also provides a higher level of voltage than is secured from the other output because the voltage drop across the d-c resistance of the second filter choke reduces the output voltage.

The capacitor input-type filter was illustrated in Fig. 14-1 using a choke and also in Fig. 14-5 where a series resistor replaced the filter choke. A more effective filtering system is shown at C of Fig. 14-9 and in some industrial applications resistors R_1 and R_2 may be replaced with filter chokes to decrease the voltage drops that occur across resistors. Capacitor input filters provide higher output voltage than choke input filters, though regulation is somewhat poorer. Higher output voltage occurs because the first filter capacitor charges to the peak value of the rectified signal waveform. With a choke input, the capacitor following it does not charge to as high a peak value because voltage has been lowered and smoothed by the input-smoothing filter choke.

For bias supplies, where a negative potential with respect to ground is desired, the filter section shown at D is often employed. Here the positive terminal of the power supply is placed at ground potential and the filter choke or series resistor is placed in the negative lead, as shown. This procedure prevents isolation of the ground portions of the power supply, which would otherwise occur if the resistor or choke were to remain in the positive lead. With bias supplies, any of the other combinations of choke or capacitor input filters can be employed.

14-7.   Regulation

The amount of voltage obtainable across the output terminals of a power supply depends on the voltage produced by the secondary winding less the voltage drops that occur across the rectifier series resistors, or filter chokes. Because the amplitude of the voltage that drops across the rectifier and other components depends on the amount of current flowing through these units, the voltage output of a power supply is also affected by the amount of current drawn from the unit. When a minimum of current is drawn from the power supply, the output voltage will be near its maximum value since the voltage drops across the rectifier and resistors or chokes will be at a minimum. When more current is drawn from the power supply, the voltage drops across the power supply, units increase and, in consequence, the output voltage is reduced. This variation of voltage output with respect to the amount of current drawn from the power supply is known as *voltage regulation*, as mentioned previously.

When the load circuit is such that its current drain from the power supply varies, voltage regulation will be poor if a capacitor input filter is used. Poorer regulation occurs because even though the capacitor at the input of the filter section will charge to the peak voltage of the rectified ac, such a peak value is impressed across the capacitor at only short time intervals because the voltage peaks of the pulsating dc have a short duration. Thus, as more current is drawn from the power supply, the charge on the filter capacitors is drained off more rapidly than it can be replaced by the current peaks of the pulsating dc. With the choke input filter, however, the peaks have been reduced to a lower level and, in consequence, there is less of a decline from the relatively high-peak value to the average value that results by virtue of the choke input filter.

The percentage of voltage regulation of a power supply can be expressed by the formula

$$\% \text{ voltage regulation} = \frac{\text{no-load } E - \text{full-load } E}{\text{full-load } E} \times 100 \qquad (14\text{-}1)$$

Equation 14-1 takes into consideration the proportions of voltage increase or decrease with a change of load on a power supply. Such a load could consist of the radio, TV receiver, amplifier, or other device attached to it, in terms of the amount of current drawn. Actually, the load on the power supply also includes any bleeder resistor network placed across it since such a bleeder also consumes power. Thus, if a solid-state audio amplifier's power supply has an output of 40 V without load but decreases to 30 V under load, the voltage regulation would be

$$\frac{40 - 30}{30} \times 100 = \frac{10}{30} \times 100 = 0.3 \times 100 = 30\%$$

The percentage value thus obtained indicates the amount of regulation. The greater the difference between the full-load and no-load voltages, the poorer the regulation of the power supply under test. Power-supply regulation is particularly important with electronic devices that have varying loads, that is, when the current drawn from the power supply varies constantly. Since good regulation improves efficiency and performance of both RF and audio amplifiers, some voltage regulation method will be found in a great number of well-designed electronic systems.

Voltage regulation can be improved in a number of ways, and the method employed depends on the type of equipment, the degree of regulation required, and the most economical procedure indicated for a particular electronic device. In some high-voltage industrial systems, choke input filters comprise one regulation design, and sometimes this is improved on by use of a *swinging choke* that is particularly designed to produce a considerable change in its inductance value with a change in the current flow through the choke inductance. The swinging choke is also known as a *saturable reactor* because it has a change of reactance for changes of its magnetic density when operated around the saturation levels of the hysteresis loop. The most efficient type of swinging choke utilizes a core material that produces the nearly rectangular hysteresis loop shown earlier in Fig. 6-7, at B. As discussed in Chapter 6, the permeability of core material is altered by varying the intensity of the magnetizing force. A change in permeability will also change the inductance and hence the inductive reactance. The swinging choke has a high-value inductance when there is little current flowing through it. With a low value of current flow the core is below the saturation level; hence permeability and inductance values are high. In some swinging chokes, the high value of the inductance is approximately 22 to 25 henrys.

With a high-value inductance, the impedance of the choke is also high, and a larger voltage drop occurs across it. Because of the latter factor, the output voltage of the power supply is lower. As more current is drawn by the load circuit, the voltage output would normally tend to decrease. When more current flows through the swinging choke, however, the magnetizing force causes saturation (or near saturation) and the inductance value may drop to a level between 5 and 10 henrys. When the inductance value drops, the series impedance decreases and the voltage drop across the choke becomes less. Hence, the output voltage of the power supply tends to rise and compensate for the drop that otherwise occurs because of the increased current drawn by the load circuit. The swinging choke is usually used as the input choke to the type of filter that was shown at B of Fig. 14-9. The swinging choke is followed by a second choke, of the ordinary smoothing variety.

Another method for improving regulation is to employ gaseous rectifiers, such as the mercury-vapor types. Because of the ionization that occurs within such tubes, the internal resistance is very low and, in consequence, there is a minimum of voltage drop across them for either high or low values of current flow. Gaseous rectifiers are employed frequently in commercial power supplies where a high degree of voltage regulation is essential in industrial control applications, critical circuits in transmitters, or other such units. Such rectifiers, however, are not suitable for receivers because of the high-frequency noise that they develop during ionization, as mentioned in Chapter 11.

Improved regulation can also be provided by use of solid-state voltage regulators or gaseous voltage regulator units. Typical regulator tubes of this type include the VR-105-30, VR-150-40, etc. The first number refers to the terminal voltage and the second number is the maximum current that the tube is permitted to pass. Tubes of this type as well as the solid-state units maintain the voltage of the power supply at a close constant level, even though the load on the power supply is such that the current varies considerably.

A widely used solid-state voltage regulator is the silicon-junction *zener* diode. It owes its regulating characteristics to the unusual breakdown (*zener*) region graphed in Fig. 14-10. The unit is available in numerous voltage ratings as well as a variety of wattage ranges to accommodate a wide span of voltage-regulating needs. When forward-bias voltage is applied to the diode, it behaves much like an ordinary silicon (or other solid-state) diode, though with lower internal resistance and greater current-passing characteristics as the applied voltage is raised.

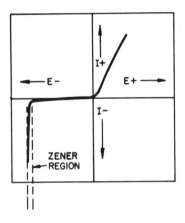

When a reverse voltage of low amplitude is applied, the internal resistance of the unit is high and only fractional amounts

FIGURE 14-10. Zener characteristics.

of current flow through it. As the reverse voltage is gradually increased, only a slight increase occurs in the internal conduction. Once, however, the reverse voltage reaches a certain amplitude, the internal resistance suddenly drops to a very low value and current shoots up to a significant amount, the value of which depends on the type zener used and the circuit that it regulates. The sudden current increase during the reverse-bias breakdown voltage is shown in Fig. 14-10.

Despite the sudden high amplitude current flow within the unit, the voltage drop across the zener diode remains practically the same as before the breakdown, making the device useful for voltage regulating. As the load draws less current, supply voltage tends to rise and this increases the current flow through the zener, bringing the voltage back to that specified. Similarly, if the load draws more current, supply voltage would drop and so would current through the zener. Consequently, voltage regulation again occurs and a constant voltage is maintained within the limits of the diode's zener region span. With zeners of over 1-watt ratings, heat dissipation may be a factor and heat sinks may have to be used as was the case for silicon rectifiers or power-amplifier transistors. Zeners develop most heat during times when the load draws the least current. With proper voltage adjustments of the supply, however, the zener will maintain the voltage for which it is rated.

The reverse-voltage breakdown is not damaging to the zener diode as it is for other diodes (solid-state or tube). The breakdown occurs when a certain reverse-voltage amplitude is reached (sufficient to penetrate the diode's internal semiconductor barrier), causing the unit to become a conductor in the reverse direction. After removal of the reverse voltage, the internal barrier region re-forms itself without damage.

The breakdown point can be closely estimated during design and manufacture by control of the internal resistivity of the silicon structure. Hence, the zener point can be set at several volts or at several hundred volts as needed. In the ordinary silicon rectifiers *not designed* for voltage regulation, the breakdown point is set sufficiently high so as to be beyond the *peak reverse voltage* (discussed later) at which the unit is rated.

A typical zener-diode voltage regulation circuit is shown at A of Fig. 14-11. This circuit is similar to the type used with voltage-regulator tubes, except that the anode of the latter was attached to the junction of R_1 and the regulated output terminal. For the zener *regulator* shown, the value of the reverse-current limiting resistor R_1 must be chosen to hold the diode within the zener region for all fluctuations of load. Thus the wattage of R_1 must also be sufficient to prevent overheating during the zener and load current flow.

The zener diodes can also be used for voltage regulation of ac as shown at B of Fig. 14-11. Here two zeners are placed *back to back* so that each half

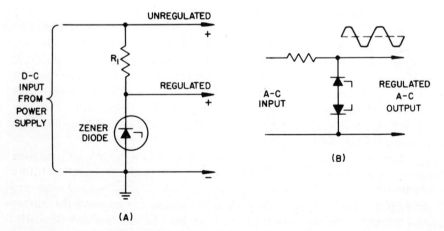

FIGURE 14-11. Zener-diode voltage regulators.

of the a-c cycle is under control. Two separate zener diodes can be employed, or special double-ended diodes manufactured specifically for this purpose can be used. With separate diodes, each should have identical operating characteristics as the other. As shown, the output waveform is clipped slightly at the peak amplitudes.

14-8.   Filter Ratings and Values

The various components in the filter section of a power supply must have the proper ratings, both in terms of the permissible voltage that can be employed on them and in terms of their value in capacity, resistance, or inductance. The power rating is also important in preventing burnout.

If a filter choke is used, it should have a value high enough to provide smooth and effective filtering. Also, the wires that make up the winding of a choke must have sufficient insulation and be of adequate size to carry the current consumed by the load without undue overheating. Thus, a choke coil would not only be rated in 5 or 10 henrys but also at 100 or 150 mA. The latter ratings would indicate the maximum permissible current flow through the filter choke. An increase in either the inductance or the rated value of current is advisable to provide better smoothing action as well as longer life.

Most filter capacitors are rated in their capacity value in microfarads, and their rating also usually includes peak voltage and working voltage. The peak-voltage rating is the maximum voltage that the capacitor can be subjected to during the initial warm-up of the receiver or other load. During such a time, the output from the power supply increases and the filter capacitors would have a higher voltage than normal impressed on them. Thus, the peak voltage that occurs before load warm-up must not exceed the rated peak value of the filter capacitor. When the load is drawing a normal amount of current and the voltage from the rectifiers has dropped to the value normal during operation, the voltage existing across the filter capacitors should not exceed the working voltage (WV) marked on the capacitor.

As a general rule the higher capacity filter capacitors are preferable to the lower ones, for smoother filter action. There are occasions, however, when an increase in the value of the capacitor used as the input to the filter network may not be advisable because of inverse peak-voltage effects described more fully later. In a tube-type supply, for instance, a filter capacitor having a value in excess of 30 μF should not be employed for the input filter unless the specifications for the rectifier tube are checked initially to make sure a higher value of capacity can safely be employed. For filter capacitors that follow a filter choke, however, the higher values can be used safely for the production of a more ripple-free d-c output and, in consequence, minimize the possibility of hum modulation of the receiver or other electronic device that is fed by the power supply.

The input filter capacitor should have a voltage rating that is above the peak value of the pulsating dc applied across it. Thus, if the transformer secondary furnishes 40 V of ac (rms), the peak value of such a voltage would be 1.41 times the rms value, or 56 V. Thus, the input filter capacitor should have a *working voltage rating* in excess of 56 V to minimize the danger of internal arcing and capacitor damage. A value of 60 V or more is preferred.

When a single capacitor of sufficiently high-voltage rating is unavailable, two filter capacitors can be wired in series to provide a voltage rating that is substantially higher than the peak value of the voltage applied across the particular capacitor. Thus, if two high-voltage capacitors are placed in series and each has a 450-V rating, the combination will withstand 900 V. Since considerably less voltage will actually be impressed across the two in combination, they will not be subjected to overload and will have a much longer life. When two such capacitors are utilized in series as the input capacitor of the power-supply filter, each should have the same value, and preferably the same d-c resistance, as the other, as measured on an ohmmeter. This matching of the two ensures the equal distribution of voltage across each. It must be remembered, however, that if an 8-μF input filter capacity is desired, *two* 16-μF filter capacitors must be employed in series. The series circuit of two 16-μF capacitors will result in a total capacity of 8 μF.

14-9. PRV Factors

As shown earlier in Fig. 14-10 for the zener diode, when a reverse-bias voltage is impressed across a solid-state diode, only a fractional amount of current flows. For the silicon diode rectifier, as with the zener, a sufficient increase in the reverse bias will reach the so-called zener region, where a significant reverse-current increase occurs (also called *avalanche*). An additional voltage increase, or an operating temperature rise, may cause a thermal runaway condition and the excessive currents cause diode rectifier burnout. The so-called peak-reverse voltage (PRV) factors are important in power-supply design since high reverse voltages are present during intervals of time when the diode is in a nonconducting state. Another precaution which must be observed is to minimize the voltage-surge conditions which occur during abrupt voltage changes and by reverse recovery transients of the diode. The amplitude of surge voltage peaks can be lowered by shunting the diode rectifier with a capacitor (C_3 at A of Fig. 14-12). Reverse-voltage breakdown can be avoided by using a rectifier having a PRV rating well above expected voltage peaks. Thus, a PRV several times that of the power supply's output voltage should be considered.

Both soild-state diodes and tube types are also rated in their forward-bias voltage (causing conduction) and the maximum current that can flow without damage. For tubes, the reverse voltage existing between the filament (cathode) and the anode during the time there is no conduction is referred to as the *peak inverse voltage*. The discussions that follow apply equally to the PIV and PRV factors.

In a simple half-wave rectifier circuit without any filter section, the peak reverse voltage would be the product of the rms value multiplied by 1.41. Thus, if the transformer delivers 40 volts, the peak reverse voltage would be $40 \times 1.41 = 56.4$ V. When, however, the necessary filter sections are added to the basic rectifier circuit, the peak reverse voltages change considerably because of the voltage charges that appear across the filter capacitors.

At A of Fig. 14-12 is shown a typical half-wave rectifier system with a capacitor input filter. Each time the rectifier diode (solid-state or tube) conducts, a charge is placed across capacitors C_1 and C_2, as described previously in this chapter. Capacitor C_1, however, receives the full charge of the rectified voltage. As shown at A, if the secondary of the transformer delivers 30 volts rms, capacitor C_1 is also charged to this voltage value with a polarity as indicated in the drawing. During the time the diode is in a nonconducting state, a negative potential appears at the top of the secondary winding L_2 and a positive potential appears at the lower end of this winding as at B. During the time the diode is in this nonconducting state, the charge on the

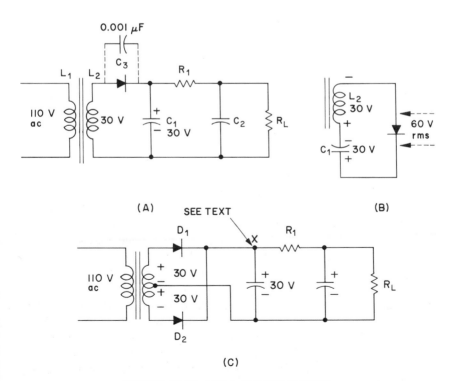

FIGURE 14-12. Peak reverse voltage factors.

input capacitor is added to the voltage produced by the secondary winding. This is shown at B, where the secondary L_2, the capacitor C_1, and the rectifier have been redrawn for simplicity. Note that the polarity of the voltage across the secondary and the polarity of the charge appearing across the capacitor are aiding and not opposing each other. In consequence, the secondary of the transformer can be considered as one generator in series with the capacitor, which represents another generator. Thus, the voltage impressed across the rectifier is the sum of these two voltages. Hence, the peak inverse voltage that appears across the rectifier during nonconduction is 84.6 V as the following calculation indicates:

$$\text{Peak inverse voltage} = 30 + 30 = 60 \text{ V (rms)}$$
$$= 60 \times 1.41 = 84.6 \text{ peak inverse volts}$$

From the foregoing, it is evident that with a capacitor input filter the rectifier unit must be able to withstand 84.6 volts in the reverse direction to normal current flow and for safety should have a peak reverse voltage rating of 100 V or more. In the gas-filled rectifier tubes used in some industrial-electronic applications the danger of arcing is much greater and the specifications for the gas-tube rectifier regarding its peak inverse-voltage rating must be referred to before using such a tube. In most instances the gas-filled rectifiers should be used with choke input filters since the input choke reduces the amplitude of the voltage that is applied to the capacitor following the filter choke.

At C of Fig. 14-12 is shown a full-wave power supply. Here the peak inverse voltage is of a value twice that produced by one-half of the secondary winding. Thus, if each half of the secondary winding delivers 30 V as shown, the peak reverse voltage is 84.6 V. This is the amount of voltage which appears across the rectifier which is in a nonconducting state. For instance, at the instant the voltage polarities are as shown at C of Fig. 14-12, the rectifier D_1 is conducting because its anode is positive, while rectifier D_2 is not conducting because its anode is negative. It is across D_2 that the peak inverse voltage appears at this instant. During the next alternation of the ac that appears across the secondary, D_2 conducts, and the peak inverse voltage appears across the cathode-anode of D_1.

During the time one diode is conducting, its internal resistance is low. (A gas-rectifier has a particularly low internal resistance during conduction because of the ionization that occurs.) For the full-wave rectifier shown in Fig. 14-12, when D_1 is conducting, the cathode will be at a potential that is almost that of the top of the secondary winding because the low internal drop between cathode and anode would create only a small potential difference in the voltage across cathode and plate. Hence, the point marked X in the drawing would have a high peak inverse-voltage value *regardless of the filter type in the full-wave power supply.* With the point marked X common with the cathode of the rectifier having the same potential as the top of the secondary winding, there will be a 60 V rms difference between the cathode and the anode of the nonconducting diode. This condition would prevail even if only a load resistor were present without the filter network.

14-10. Power-Supply Bleeders

As mentioned earlier, a resistor (or several resistors in series) may be placed across the output of a power supply, as shown in Fig. 14-5. Such a resistor (or resistors) is known as a *bleeder* because the latter maintains a constant current drain on the power supply and, by thus "bleeding" off some of the current, a slight though constant load is imposed on the power supply to help stabilize it and to improve regulation.

The bleeder should be designed to consume between 5 and 10% of the current drawn by the load (the load in this instance referring to the radio circuits, amplifier circuits, or other systems that are furnished power from the supply). Regulation is improved somewhat when 10% current is drawn by the bleeder instead of a lower value but the higher bleeder current requires a larger wattage bleeder resistor, and the additional current drawn by the bleeder must not impose an undue load on the power-supply transformer secondary windings. A bleeder should not be used if the power supply is already loaded to its full current capabilities.

If the load for the supply at A of Fig. 14-13 draws 200 mA, and the bleeder is to consume 5% of such current (10 mA), the value of the bleeder

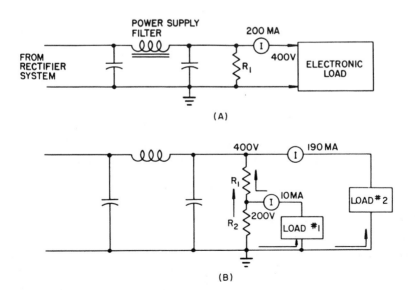

FIGURE 14-13. Load distribution in power supplies.

resistor would be

$$R = \frac{E}{I} = \frac{400}{0.01} = 40 \text{ k}\Omega$$

The wattage in power consumed by the bleeder would be

$$P = EI = 400 \times 0.01 = 4 \text{ W}$$

Since the bleeder consumes 4 W of power, it is necessary to employ a resistor having a sufficiently high wattage rating to withstand the heat generated. Hence, to minimize overheating and to provide a margin of safety, a resistor having a wattage rating of about 10 W is preferable over lower wattage values.

When two or more series resistors are used in the bleeder section to supply intermediate voltages (in addition to the maximum voltage), the bleeder section is also referred to as a *voltage divider*. Such a voltage-divider bleeder is shown at B of Fig. 14-13 and consists of R_1 and R_2. For this power supply, load No. 2 draws 190 mA at 400 V, while load No. 1 draws 10 mA at 200 V.

Assume that resistor R_2 has 100 mA of current flowing through it. As shown at B, resistor R_2 is shunted by load No. 1, which also draws 10 mA. Hence, resistor R_2 is in parallel with load No. 1 and the combination will draw a total of 20 mA through resistor R_1 since the electron flow for both R_2 and load No. 1 goes through this resistor, as shown by the arrows in the schematic.

Since there is a 200-V drop across R_1 and 20 mA of current flow through it, by Ohm's law the resistance value of R_1 must be 10 kΩ. Resistor R_2 also has 200 V across it, and with 10 mA of current flowing through it the resistance value must be 20 kΩ. Load No. 1 must also be 20 kΩ since it also has 200 V across it and draws 10 mA. The total resistance of R_2 and load No. 1 must therefore be 10 kΩ. It is obvious that both R_1 and the combination of R_2 and load No. 1 must have equal resistance values, in order to divide the 400 V in half.

Voltage dividers can employ several resistors in series to obtain intermediate values of voltage division, as required. Each load that shunts a section of the voltage divider draws additional current, and such additional current will flow through the resistors above it in similar fashion to the two loads illustrated at B.

14-11.   Thyratrons and SCR Units

The thyratron tube described earlier in Chapter 11 is also a rectifier but unlike the diode rectifiers it contains a grid for initiating current flow. Once current flows through the tube, however, the grid loses control and cannot stop conduction unless the anode voltage is removed. The basic circuit shown in Fig. 14-14 will help demonstrate the thyratron characteristics. Here when the switch (SW) is closed, the battery potential is applied between cathode and anode (in series with the load resistance R_L). Since the grid is negative, however, no current flows through the thyratron. If the grid voltage is now reduced by moving the variable arm of R_1 toward the cathode, current will flow through the tube and ionization occurs. If the

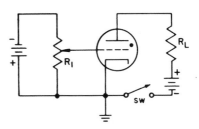

FIGURE 14-14. Circuit illustrating thyratron characteristics.

negative bias is now applied again to the grid, it will be ineffectual and current flow continues because of the ionic space charge that is set up around the grid as described in Chapter 11. In order to stop conduction, it is necessary to open the switch in the anode circuit. When current flow and ionization stop, the switch may again be closed and the negative potential at the grid will again hold the tube in the nonconduction state.

The firing time of thyratrons is related to the amplitude of the anode voltage, the gas pressure within the tube, and the type of gas it contains. The internal structure also affects firing time because it determines the grid sensitivity. In some thyratrons, the grid can be slightly negative for firing, while in others (with more shielding between anode and cathode) it is necessary to reduce the bias to zero or to a small positive value.

The rapidity with which ionization occurs is known as *ionization time* and in many thyratrons it is less than 5 μsec. The *deionization time* (the minimum interval of time for neutralization of the ionic space charge surrounding the grid after current flow has been interrupted) in many thyratrons averages 500 μsec, though some have a deionization time of less than 100 μsec. When deionization time is long, usage of the device is limited to power sources having a frequency of 60 Hz.

The presence of the grid in the thyratron takes it out of the class of an ordinary rectifier and permits it to be used for the control of the amount of power applied to a load. Hence, thyratron devices (including the solid-state type discussed later) find wide application in industrial control, automation, and in other areas where the amount of power to a load must be varied. With the thyratron, a small grid potential of negligible power can control thousands of watts of power.

Since the thyratron is a rectifier, it lends itself to operation with ac at both the grid and anode circuits. With ac, the control of the power applied to the load is determined by the relative *phase* of the grid signal versus the anode signal. Because the grid control is lost once the tube is conducting, the use of an a-c signal at the anode permits the grid to regain control. This is so because successive alternations of ac to the anode causes conduction to be interrupted, regardless of the polarity of the grid signal.

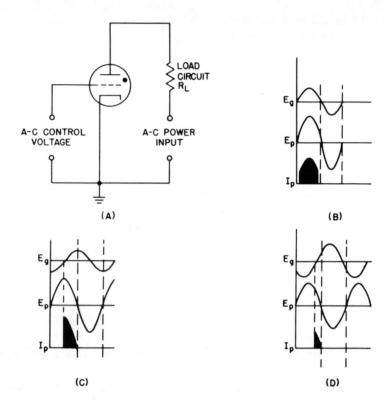

FIGURE 14-15. Thyratron control characteristics.

Operation with an a-c signal is illustrated in Fig. 14-15. At A is shown the basic circuit, with the power source for the load applied to the anode circuit below the load resistance R_L. If the signal at the grid is in phase with the a-c power input signal as shown at B, the thyratron will conduct for successive positive alternations at the anode, as shown. When the grid signal E_g is positive-going at the same time as the plate voltage E_p, plate current I_p starts to flow as soon as the anode potential reaches a value that will cause ionization. When the grid signal is negative-going, the phase coincides with E_p and conduction ceases. Thus, the tube acts as a half-wave rectifier in a fashion similar to the others described earlier.

If there is a 90° phase difference between E_g and E_p as shown at C, the tube will not start conducting until the grid bias is reduced sufficiently to permit conduction, even though the anode is positive initially. When the anode potential drops to the point where ionization ceases, conduction stops even though the grid waveform is still positive. With a greater shift of E_g as shown at D, the interval of conduction is still shorter, and in consequence less power is applied to the load circuit. Thus, by shifting the phase of the grid potential, a considerable variation of output power is possible. The

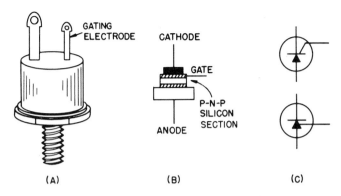

FIGURE 14-16. Silicon-controlled rectifier.

pulsating dc produced can be filtered, if necessary, for ripple reduction.

A solid-state counterpart of the vacuum-tube thyratron is the *silicon-controlled rectifier* shown at A of Fig. 14-16. This device combines some of the characteristics of both the solid-state diode and the transistor, as shown at B. (Symbols that have been used for this device are shown at C.) The threaded terminal is the anode and this is bolted to the chassis, with the flange acting as a heat sink as with the silicon diodes previously described. The controlled rectifiers come in various sizes as required, and units that are approximately 0.5 in. wide and 1.5 in. high handle currents up to 15 A at 400 V.

While the solid-state thyratrons have similar characteristics to the gas-tube types, they offer a number of advantages. The tube types of the larger sizes require forced-air or water circulation for cooling purposes; they require power for heater (filament) operation: and they need to have the filaments brough up in temperature before anode voltages are applied. The silicon-controlled types need no warm-up, their life span is much longer than the tube types; there is no filament deterioration; and they have lower internal resistance during conduction. Switching is also more rapid than with the gas-tube types (often less than 12 μsec).

The silicon-controlled rectifier can be used as a straightforward rectifier without employing a gating electrode voltage, as shown at A of Fig. 14-17. If the reverse breakdown voltage is not reached by the peak swings of the negative a-c alternations, the output will be in the form of pulses occurring at the line a-c rate as shown. When the pulsating dc (or filtered dc) is to be controlled by switching it to the load circuit at specific intervals as with the tube thyratrons, a trigger voltage is applied between the gate electrode and the cathode, as shown at B. The trigger voltage must be applied to coincide with a positive alternation of the a-c input signal, just as with the tube thyratron. By applying timed (or phased) signals to the gate, conduction can be set for the precise intervals required.

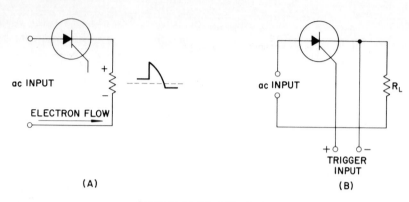

(A)   (B)

FIGURE 14-17. SCR circuitry.

14-12. Ignitron Rectifier

Another widely used gas-filled rectifier for industrial power requirements is the *ignitron*, which has the basic construction shown at A of Fig. 14-18. The tube contains a pool of mercury, an anode, and an *ignitor* element. The latter is a pointed tip of silicon carbide or boron carbide that dips into the mercury as shown. The ignitor tip is rough surfaced so that when a voltage is applied across the ignitor and mercury cathode, small flash points will occur to initiate current flow and ionization. Without any voltage applied to the ignitor, conduction would not result because of the cold-cathode type of rectifier. Thus, with a positive voltage at the anode *and* ignitor, the arcing at the ignitor starts the emission process.

The basic circuit is shown in Fig. 14-18. The diode connected in series with the ignitor lead prevents the application of a reverse polarity across ignitor and cathode. A reverse voltage (even though of low amplitude) can damage the ignitor by causing electron flow from ignitor to cathode. Resistor

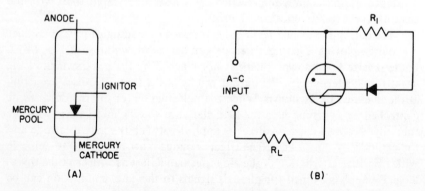

(A)   (B)

FIGURE 14-18. Ignitron and basic circuit.

R_1 drops the voltage to that required for creating the initial arc at the ignitor. The a-c supply power is in series with the load circuit, as shown, and with the mercury pool.

During the time when the a-c voltage is negative at the upper terminal and positive at the lower terminal, no conduction occurs because both the ignitron anode and the diode anode have negative potentials applied to them and hence are inoperative. When the upper terminal is positive with respect to the lower terminal, the diode conducts and initiates the emission and ionization process, permitting full conduction. Thus, the load receives rectified ac (pulsating dc) that again can be filtered to reduce the ripple component if required.

The ignitron has a number of advantages over the gas-tube thyratron. No filament excitation must be maintained; the ignitron mercury pool has unlimited life and will withstand high overloads without damage, and the anode can be placed in closer proximity to the mercury pool for lower internal resistance and increased efficiency. As with the other thyratrons, the ignitron can be fired at any point on the cycle, thus having the same advantages of applied-load power control. The disadvantages are bulk, cost, and the necessity for vertical operation because of the mercury-pool structure.

As with other rectifiers, the ignitron can be operated for full-wave rectification as shown in Fig. 14-19. The a-c supply power is applied to the two ignitrons from across the input transformer composed of L_1 *and* L_2 During the time the upper portion of L_2 is positive, ignitron I_1 conducts electrons in the direction shown by the arrow at the load resistance R_L. At this time the bottom of L_2 is negative and ignitron I_2 is in a nonconducting state. When the polarity reverses across L_2, the lower ignitron conducts in the same direction through the load circuit as was the case with the upper ignitron. The ignitors obtain their control signal from across the transformer composed of L_3 and L_4, as shown. By phasing this control signal, the full-wave alternations can be triggered at any part of the cycle desired.

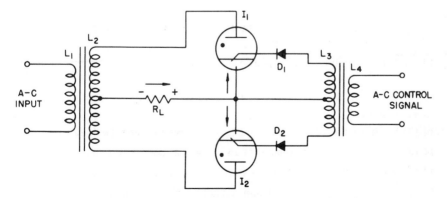

FIGURE 14-19. Operating the ignitron for full-wave rectification.

Review Questions

14-1. What are some of the characteristics that a power supply must have for efficient and adequate performance?

14-2. Briefly explain the basic operational principles of a half-wave power supply.

14-3. Briefly explain why the output voltage of a half-wave supply tends to decline when an additional load is impressed across the supply.

14-4. (a) What is the advantage of a full-wave power supply over a half-wave type?
(b) Why is less filtering needed for the full-wave type?

14-5. Briefly explain the function of a filter network.

14-6. Why is the output voltage higher for a capacitor input filter than for a choke input filter?

14-7. Briefly explain why a bridge rectifier system does not require a center-tapped transformer.

14-8. Describe one method for doubling the a-c line voltage during rectification, without using a power transformer.

14-9. Briefly illustrate and explain one method for achieving voltage tripling.

14-10. Explain what is meant by "regulation" and by "percentage of voltage regulation."

14-11. Briefly explain the advantages and function of a swinging choke.

14-12. Briefly explain how the regulation of a power supply can be improved by employing a zener diode.

14-13. (a) What factors must be considered when replacing the filter choke in a power supply?
(b) What factors must be considered when replacing the filter capacitors of a power supply?

14-14. (a) What is meant by the *peak reverse voltage* encountered in a power supply?
(b) What is meant by the bleeder network of a power supply? Explain its purpose and function.

14-15. Briefly explain how a thyratron can be used to control the amount of power applied to a load circuit.

14-16. What are the advantages of the solid-state thyratrons over the gas-tube types?

14-17. What are the advantages of the ignitron over the gas-tube thyratron?

14-18. What are the disadvantages of an ignitron as compared to a thyratron?

14-19. Why is a diode used in series with the ignitron?

14-20. Redraw Fig. 14-19, and include a choke input filter circuit.

Practical Problems

14-1. In a power-supply type such as shown in Fig. 14-5, the first filter capacitor (following the rectifier) is to have a value of 8 μF. The transformer supplies 300 V rms on each side of the center tap to the rectifier. Assuming no loss in the rectifier diodes, how can the following capacitors (in series or parallel combinations) be used to obtain the proper capacity value and safety with respect to the working voltage? Explain the reasoning behind your conclusion.

Number available	Capacity (μF)	Working voltage
2	16	150
2	8	300
2	16	450

14-2. In a capacity input power supply, the bleeder draws 20 mA and the electronic load 180 mA. Which of the following filter chokes should be used? Explain the reasons behind the choice made.

No. 1	5 H	200 mA
No. 2	10 H	150 mA
No. 3	15 H	250 mA
No. 4	20 H	100 mA
No. 5	20 H	180 mA

14-3. In Fig. 14-13, assume that load No. 2 draws 160 mA at 400 V, and load No. 1 draws 40 mA at 200 V. Resistor R_2 has 10 mA flowing through it. What must be the value in ohms of R_1 and R_2? What is the value in ohms of load No. 1?

14-4. A 50-μF, 60-working-volt capacity is required for a stereo cassette power supply. A number of 50 μF capacitors are available but each one is only rated at 30 working volts. How can a number of these available capacitors be combined to obtain a total value of 50 μF at 60 working volts? Sketch how these capacitors would be connected together.

14-5. What is the peak reverse voltage of a capacity input half-wave power supply if the secondary delivers 75 V rms?

14-6. What is the approximate inductive reactance of a choke coil of 20 henrys rated at 200 mA and used in a full-wave power supply connected to a line voltage source of 110 V, at 60 Hz?

14-7. What would be the approximate inductive reactance of the same choke in Problem 14-6 if it were used in a half-wave power supply from the same a-c line?

14-8. In an industrial power supply it was noticed that the output voltage dropped from 1000 to 800 V after the load was applied. What is the percentage of voltage regulation?

14-9. In a laboratory power supply it was found that the rated regulation was 33%, with a load current of 10 mA. After installing a new voltage-regulator diode, it was found that the no-load voltage of 2000 dropped to 1600 V with a 100 -mA load. What was the percentage of regulation with the new regulator diode? Was this an improvement in regulation?

14-10. In designing a 400-V power supply for a 5000-Ω load circuit, it was decided to include a bleeder resistor that would have a current flow through it equal to 10% of the load current to improve regulation. What must be the value of the bleeder resistor?

14-11. In a power supply for a solid-state amplifier, the bleeder current was 0.2 A and the load drew 2 A. The series filter resistor was 10 Ω.
(a) What power is dissipated in the filter resistor (which is in series with the bleeder and the paralleled load)?
(b) By how much does the filter resistor decrease output voltage?

14-12. (a) For Problem 14-11, if there is a 50-V drop across the bleeder, what power is dissipated in it?
(b) What is the resistance value of the bleeder?
(c) Would two 500-Ω parallel resistors, each rated at 7 W, suffice?

Basic Amplifier Factors 15

15-1. Introduction

Virtually all circuits found in various branches of electronics are concerned with the handling, generation, modification, or utilization of some sort of electric signal voltage or power. When such signals are first generated or obtained from circuits that handle them initially, the signal amplitude is usually insufficient to perform the functions intended; hence it must be brought up to the level required. Circuits that thus increase the voltage or power of signals are known as *amplifiers*. The design of a particular amplifier is dictated by the nature of the signal to be handled in terms of whether it has d-c or a-c characteristics, whether it is of low frequency or high frequency, and whether its voltage amplitude or power must be increased. A photocell used for automatic lighting control, for instance, need only indicate d-c changes; hence if amplification is required, the type amplifier would differ from that used to amplify the type of signals received by an antenna system.

In transmitting systems, special generators produce high-frequency signals so the latter can be used to "carry" lower-frequency signal information such as audio, picture information, or the pulse and square-wave signals described later. The high-frequency signals are necessary because low-frequency signals cannot be sent any appreciable distance, also described more fully later. The high-frequency signals in transmitters must also be amplified so that they are brought up to the proper power level for sending out over the air. At the receiver, the high-frequency signals reaching the antennas are too weak for immediate usage and again amplifiers must be employed. Similarly, specific amplifier types are required to handle the signals employed in the electronic circuits used in industrial control systems, radar networks, computers, and other commercial gear. Also the type amplifier used to increase the signal level from a microphone (which produces a signal having a-c characteristics) differs from those used to handle other signal types.

From the foregoing it is evident that signal amplification is an important area of electronics, and the various aspects and circuit characteristics must be evaluated for complete understanding. Hence, two chapters are devoted to this topic, with this chapter covering basic factors, level controls, bias, coupling, distortion, and the general aspects of amplifier design. Power amplification and related discussions are covered in Chapter 16.

15-2. Types of Amplifiers

Amplifiers fall into a number of categories, and the two fundamental types are voltage amplifiers and power amplifiers. The voltage amplifiers are designed to increase the voltage amplitude of a signal that may have negligible power. The power amplifiers either convert the voltage-type signal to one with a high-level energy component or amplify a power signal an additional amount. Besides the designations of voltage and power amplification, reference must also be made to whether an RF type signal is handled or whether a low-frequency (such as an audio) signal is involved. Thus, a voltage amplifier could also be a radio-frequency amplifier or an audio amplifier. Similarly, a power amplifier could be a radio-frequency type or an audio (or other low-frequency signal) type.

The term *radio frequency* does not imply that the signals must be ordinary radio-type signals commonly received by home radios. Radar, television, frequency modulation, and other such devices handle high-frequency signals other than only radio-frequency signals but the term radio-frequency signal (abbreviated *RF* signal) from long usage is still applied to such signals.

The characteristics of an amplifier in relation to its bias and design also place it in another specific category. Thus, if an amplifier is biased on the linear portion of its curve and if the input signal is kept within certain amplitude bounds, the amplifier is designated as a *Class A amplifier*. (The specific characteristics and linear operation are discussed in detail later in this chapter.) The Class A amplifier could be designed specifically to handle RF signals or built primarily for amplifying audio or other low-frequency waveforms. Also, it may be intended to function as a voltage amplifier or the circuit may be so made up as to perform as a power amplifier. Other amplifier types include the Class AB_1, AB_2, B, and C. These are described in the next chapter. The characteristics and circuitry of the Class A amplifiers are covered in detail in this chapter.

Low-frequency signals such as procured from photocells, microphones, tape playback heads, phonograph pickup devices, and other units described later are usually applied to a low-frequency amplifier of the Class A type initially. If the signal-amplitude level is still insufficient, additional amplifier stages are employed to increase the signal amplitude to that required for application to relays, control units, loudspeakers, recording devices, or modulators. Often, the amplification system consists of one or more stages of signal-voltage amplification so that the signal voltages can be increased to a sufficient level for application to *power* amplifiers since signal power rather than signal voltage is required to operate the output devices mentioned. If, for instance, audio amplification is involved, the sequence may be as shown in block-diagram form in Fig. 15-1. Here two stages of audio amplification increase the weak electric signals obtained from the input device and in turn apply a relatively high signal voltage to the final audio-amplifier circuits, for conversion of these signals into power signals.

FIGURE 15-1. Block diagram of audio amplifier.

15-3. Level Controls

The amplitude of the signals applied to the input of an amplifier is usually regulated by a variable resistor known as a *gain control*, which if audio signals are involved, is also called a *volume control*. The gain control is usually located in the first amplifier stage following the signal source. If, for instance, the system is designed for audio amplification, a typical volume control circuit would be as shown at A of Fig. 15-2.

In some cases, the volume control potentiometer (R_1 in Fig. 15-2) has an additional terminal, as described earlier in Chapter 3 and shown in Fig. 3-3. This extra terminal is utilized for *bass compensation*, by using a capacitor (C_1 in Fig. 15-2) from the additional terminal to ground. Bass compensation is employed so that at low volume levels some of the high-frequency signals are diminished in amplitude. When high-frequency audio signals are diminished, the low-frequency signals are comparatively greater in amplitude than the high-frequency signals and hence sound louder. The reason for accenting low frequencies at low volume levels is because the ear is less sensitive to bass notes when the latter are heard at low volume levels. Thus, when

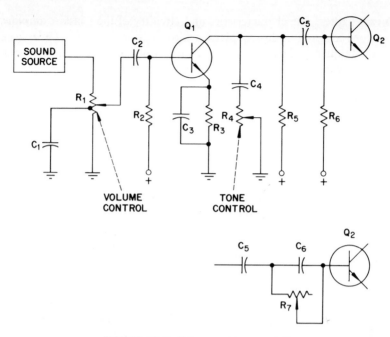

FIGURE 15-2. Volume and tone controls.

the volume control is reduced so that music is reproduced softly, the bass tones will still be audible because of the bass compensation circuit.

Capacitor C_1 has a shunting effect for higher frequencies since it provides a lower-shunt reactance for them. When the volume control movable arm is turned down so that it is opposite the extra terminal or below it, the shunting effect of capacitor C_1 is greatest.

Potentiometer R_4 and capacitor C_4 form a *tone-control* network. When the potentiometer is regulated to have minimum resistance, there will be a maximum shunting effect for higher frequencies because of the low capacitive reactance of C_4 for high frequencies. Thus, the diminished higher frequencies are not heard as much but the undiminished lower frequencies will appear more prominent. Thus, such a tone-control circuit causes the lower-frequency tones to appear louder in sound than the higher-frequency tones. When R_4 is adjusted for maximum resistance, the impedance of the network is high and the frequency response is not altered to any appreciable degree. The tone control operates at any volume levels, while bass compensation operates only at low volume levels.

In addition to the bass tone control shown at A of Fig. 15-2, a *treble* tone control can also be used, as shown at B. Here, an additional capacitor C_6 has been added in series to the coupling capacitor C_5. A variable resistor R_7 shunts C_6 and when this resistor is adjusted for zero resistance, C_6 is bypassed and not effective. As the resistance of R_7 is increased, however, C_6 becomes more effective as an increasing reactance for lower-frequency signals. With low-frequency signals diminished, the effect is to increase the high-frequency (treble) signals. Even with C_5 and C_6 in series, there is a decreasing capacitive reactance for signals of higher frequencies.

Level controls are used in television receivers for contrast and brightness control functions as shown in Fig. 15-3. Here, resistor R_1 is adjusted to pick up the desired signal-voltage amplitude for application to the base-input circuit of the video output amplifier transistor. Since this transistor amplifies the picture signal, R_1 regulates the degree of contrast and is comparable to the volume control in radios.

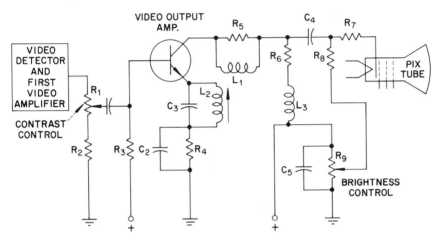

FIGURE 15-3. Contrast and brightness controls.

As also shown in Fig. 15-3, the brightness of the receiver is adjusted by R_9, which adjusts the voltage level applied to the cathode of the picture tube. As the cathode is made more positive in polarity than the grid, the grid becomes relatively more negative, thus decreasing current flow within the picture tube and hence decreasing brilliance. If, however, R_9 is adjusted for a decreased positive voltage on the cathode, the grid voltage would become less negative and more current would flow within the tube, thus increasing picture brilliance.

Additional controls are present in a color receiver, as shown in Fig. 15-4, where the color control and the tint control circuitry are illustrated. As detailed more fully in Chapter 19, the black and white signals in a color receiver are channeled through different circuitry and are applied to separate elements of the color picture tube. Thus, as shown in Fig. 15-4, the level of color intensity can be set by the color control R_1, which establishes the amplitude of the color signals applied to the input of the second chroma

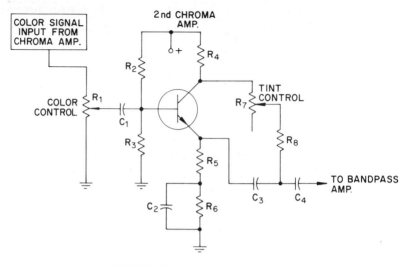

FIGURE 15-4. Color and tint controls.

amplifier. At the output is the so-called *tint* control that is adjusted for proper color rendition (such as the flesh tones).

15-4. Bias Factors

Bias refers to the voltage established between transistor elements or tube elements for the purpose of establishing the operating point for the type of amplifier desired. Basic bias factors relating to transistors have already been covered in Chapters 12 and 13. For these units both the potentials applied to the input and output sections are referred to as bias, while in vacuum-tube circuitry the bias usually refers to the voltages applied between the grid element and the cathode, as discussed more fully later in this section.

For transistor amplifiers of the Class A variety, the bias applied between the base element and the emitter element conforms to the polarity designations of the *P* and *N* zones and is known as *forward bias*, as detailed in Chapter 12. Hence, for the *PNP* transistor shown at A of Fig. 15-5, the bias polarity is negative to the base element and positive to the emitter. For the *NPN* transistor circuit shown at B, the forward-bias polarity is positive to the base and negative to the emitter. Both, however, are in the forward-bias direction as required for Class A operation. Similarly, the reverse-bias polarities for the two circuits differ as shown so as to apply the necessary polarity to the collector opposite to the zone designation of either *N* or *P*. (See also Fig. 13-13, common gate, base, drain, and collector circuits.)

Note that a resistor and capacitor (R_2 and C_1) are in series with the emitter lead. This combination stabilizes the transistor circuit and reduces the

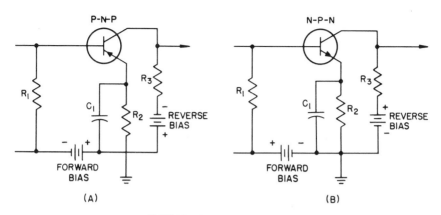

FIGURE 15-5. Transistor bias.

danger of thermal damage as more fully described for the transistor RF amplifier covered later in this chapter.

In a high-quality amplifier where the signal levels are to be raised with a minimum of distortion, bias is essential for both the solid-state devices as well as tubes. If no forward bias were provided, collector current in a transistor would be near or at zero, while absence of tube bias would cause full plate-current flow within the tube. In either case, the output currents could not alternately increase and decrease, as required during signal amplification. With proper bias, output currents during zero-signal input are held at a level below extreme values of either saturation or zero and hence can vary both above and below the zero-signal level during amplification.

The bias for tubes is the negative potential that exists at the grid in relation to the cathode. The bias circuitry discussions that follow apply equally as well to vacuum tubes used for amplification, signal generation, and signal modification as they do to tubes such as the cathode-ray types for oscilloscopes and television receivers.

Referring again to Fig. 11-8, when a positive-going *signal* is applied to the grid, the positive polarity of the *signal* has the effect of momentarily reducing the bias because the positive-going signal opposes the existing negative bias. The temporary reduction in bias will also temporarily increase plate-current flow. Similarly, a negative-going signal will decrease plate-current flow below that set by the bias. These variations in plate current will alternately increase and decrease the current through the plate resistor, above and below the idling current set by the bias. The current variations in the plate resistor cause voltage variations and such voltage variations constitute the amplified signal. The latter has a frequency and waveshape that conforms to the grid signal, as mentioned in Chapter 11.

One method for applying bias to the grid of a vacuum tube is shown at A of Fig. 15-6. Here, a battery is applied to the bottom of the grid resistor

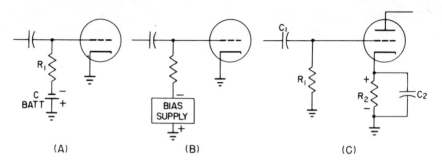

FIGURE 15-6. Various bias methods.

R_1 so that the negative terminal of the battery is toward the grid, while the positive terminal is toward the chassis (ground). Since the cathode is also at ground potential, it can be considered as connected to the positive terminal of the bias battery.

Since no current flows in the grid circuit, there is no voltage drop across R_1. Consequently full battery potential exists between grid and cathode and if a 4-V battery is utilized, the grid will be negative with respect to the cathode by 4 V. A similar bias method would be the substitution of a power supply for the battery. The use of a power supply for bias is advantageous in high-power transmitters and, in such cases, the power supply replaces the battery, as shown at B.

Another method for obtaining bias is the use of a resistor in the cathode circuit as shown at C. Since electron flow from the anode supply of the tube must be through the cathode circuit to the plate, such current flow through the cathode resistor (R_2) will cause a voltage drop across it, with a polarity as shown. If, for instance, the voltage across the resistor is 3 V, the cathode will be positive with respect to the grid by 3 V since the grid leak R_1 is connected to ground, like the cathode resistor R_2. Again (in Class A amplification), no grid current flows so that no voltage drop occurs across R_1. For the latter reason, a 3-V drop across R_2 makes the grid negative with respect to the cathode by 3 V. Such a voltage can be measured by placing a voltmeter across R_2 or by placing the voltmeter between the grid and cathode terminals of the tube. The bias voltage cannot be read by placing the voltmeter across R_1 because no current flows in this resistor and hence no voltage difference exists across it.

Capacitor C_2 is placed across the cathode resistor R_2 to minimize signal-voltage variations across the latter. This capacitor is necessary in order to realize the full benefits of the bias voltage developed across R_2, and to prevent such a bias from varying at a rapid rate. Without capacitor C_2, the voltage across R_2 would vary since the plate-current amplitude varies at a rate determined by the input signal to the grid of the tube. The capacitor filters this voltage variation from across R_2 and thus prevents a bias variation.

If capacitor C_2 of Fig. 15-6 (or C_1 in Fig. 15-5) were omitted, the circuit would be degenerative and would have reduced gain. This gain reduction comes about because the changing bias that occurs across R_2 (due to the changing plate-current amplitude) would act inversely to the signal at the grid. If, for instance, the grid signal is positive-going, it would cause the plate current to increase. This increase in plate current through R_2 would cause an increase in the voltage drop across the latter resistor. This larger voltage drop would make the cathode more positive than the grid and thus the grid would become more negative (a bias increase). The increase in negative grid bias reduces some of the current flow through the tube and the current reduction counteracts some of the increase in current flow established by the positive-going grid signal. Degeneration, of course, also occurs for a negative-going grid signal. The negative grid signal causes an increase in the negative bias, which reduces the current flow in the tube and consequently reduces the current through R_2. The reduced current through R_2 decreases the voltage drop across it and thus reduces the negative bias at the grid of the tube. This reduction in bias causes the plate current to increase somewhat, which counteracts some of the plate-current decrease caused by the negative-going grid signal. Thus, this inverse function causes a decline in the amplification of the signal.

The amount of filtering that cathode capacitor C_2 accomplishes depends on its capacity. A fairly large capacity must be used in order to have an equivalent low-shunting reactance across the cathode resistor R_2. Because the capacitive reactance shunts the cathode resistor, a lower value of capacitive reactance will cause a decrease in the impedance represented by the resistor-capacitor combination. The lower the impedance, the smaller the signal voltage drop that occurs across it and hence the less such voltage variations will affect the bias. The cathode resistor must have a value necessary for the amount of bias required and hence cannot be reduced in its ohmic value. Also, if the circuit is analyzed on the basis of the capacitive reactance bypassing signal-voltage variations because of its low reactance, it is obvious that C_2 will not provide the same degree of filtering for lower frequencies that it does for higher frequencies. If, for instance, the circuit at C of Fig. 15-6 is used to amplify low-frequency signals (such as audio) and the capacitor had a value of 0.1 μF, the capacitive reactance would be slightly over 1.5 kΩ at 1 kHz. If the cathode resistor R_2 is 1 kΩ, the capacitor's shunting effect is only partially effective. At 2 kHz, however, the capacitive reactance would be less than 800 Ω and hence provide a more pronounced shunting effect across the 1-kΩ resistor. Since capacitive reactance decreases for higher-frequency signals, these signals are filtered to a much greater degree than the lower. The more filtering of the signal variations in the cathode, the less the resultant degeneration. Hence, the size of the capacitor C_2 should be of such a value that the reactance at the lowest signal frequency handled is appreciably less than the ohmic value of the cathode resistor it shunts.

In some instances the cathode capacitor is deliberately omitted in order to improve frequency response, even though this results in some sacrifices in signal gain in favor of a good low-frequency signal amplification. (This factor also applies to the C_1-R_2 combination in Fig. 15-5.)

Still another method for obtaining bias is the use of resistors across the power supply (or battery) as shown at A of Fig. 15-7. For purposes of discussion, assume the power-supply output potential is 310 V. Instead of grounding the bottom of R_2, as was done in the bleeder networks shown earlier, that section of the resistive network is not grounded but the ground

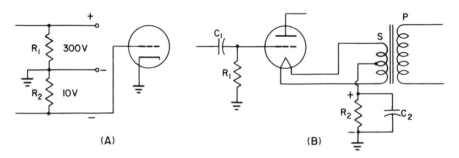

FIGURE 15-7. Additional bias methods.

terminal is placed at the junction of the two resistors, as shown. Thus, the positive terminal of the power supply will furnish 300 V with respect to the grounded center section, which is of a negative polarity with respect to the positive terminal. The negative bias for the tube is obtained from the bottom section of the lower bleeder resistor R_2 because this point is more negative than the ground terminal. (Actually, the bottom of R_2 is negative with respect to the top of R_2 and thus the grounded negative terminal is positive with respect to the lower end of R_2.) If the voltage drop across R_2 is 10 V, the grid is negative by 10 V with respect to the cathode since the cathode is also grounded and, therefore, common with the grounded terminal of the power supply.

When vacuum tubes employ the directly heated principle, as at B, the bias can be established by utilizing the center tap of the transformer secondary, as shown. Here, a resistor (R_2) is placed in series with the transformer center tap and ground and shunted by C_2, in similar fashion to the bias method shown at C of Fig. 15-6. In the circuit shown at B of Fig. 15-7, the power supply would be connected in conventional fashion with positive to the anode and negative to ground. Hence the plate electron flow from the power supply must, of necessity, be through resistor R_2, to reach the filaments and plate. This path establishes a voltage drop across R_2, with a polarity as indicated, and thus it will make the grid more negative with respect to the cathode (filament) of the tube.

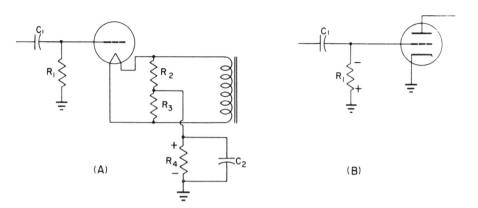

FIGURE 15-8. Bias in filament and self bias.

In instances where the directly heated tube is fed by a transformer that does not have a center tap, the method shown at A of Fig. 15-8 is employed. Here, two resistors (R_2 and R_3) are placed across the filament winding of the transformer. These resistors have a low ohmic value (approximately 30 Ω *each* for a 6-V filament section). The two resistors establish an electric center at their junction. At this junction, the bias resistor R_4 is added, as shown, with the usual bypass capacitor C_2. Thus, the voltage drop across R_4 will give the cathode (filament) a positive potential with respect to the grid and hence the grid will be negative with respect to the cathode. Since R_2 and R_3 have low ohmic values, they will not have an appreciable effect on the bias because the cathode resistor would be much larger in value. Resistors R_2 and R_3 also act as a hum-reducing circuit, by placing the ground connection (via C_2) at the electric center of the transformer. A single potentiometer can replace resistors R_2 and R_3, with the variable arm of the potentiometer attached to the top of R_4. The potentiometer circuit provides a means for balancing the system to minimize any hum that may be present in the amplifier.

Yet another form of bias is the contact potential bias that is obtained from electrons striking the grid wires in a vacuum tube. A circuit of this type is shown at B of Fig. 15-8. At first glance, it would seem that this tube is unable to develop a bias. The grid resistor R_1 has a high value, however, and if the bias requirements are low, as is the case with high-mu tubes, enough bias will be developed by virtue of the electron bombardment of the grid structure. The random electrons which strike the grid wires will establish a voltage across R_1 which has a low potential value. The fractional volt, however, is often sufficient for bias purposes for certain tubes having low bias requirements.

Another method for obtaining bias is that found in certain circuits such as the oscillators, limiters, and Class C amplifiers discussed later. In such

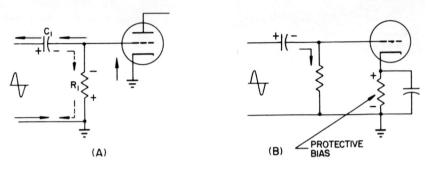

FIGURE 15-9. Generation of self bias.

devices, the grid signal has a high value and drives the grid of the tube posi-tive during positive signal peaks. When the grid is positive, electrons flow from cathode to grid and this flow consists of the signal-energy component. Thus, for a brief interval of time, the grid and cathode of the amplifier tube may be considered as the anode and cathode of a diode rectifier. The resultant electron flow is in the direction shown by the solid arrows at A of Fig. 15-9, and the flow path continues through the cathode-grid section of the tube and on to capacitor C_1. The electron flow toward capacitor C_1 will cause an accumulation of electrons on that side, giving it a negative potential.

The piling up of electrons on one side of the grid capacitor will cause a repulsion of the electrons on the other side of the capacitor and, in con-sequence, the electrons will flow away from the left side of capacitor C_1, giving that side a positive potential. If the positive peak of the signal voltage is 10 V, for instance, the capacitor will charge to this approximate value. When the signal declines below its 10-V peak, the tube will no longer conduct because capacitor C_1 holds a 10-V negative charge at the grid and since this is a higher negative potential than the declining positive-going grid signal, the grid is predominately negative so that no current will flow between cathode and grid. Hence, no additional charge of energy is placed across the coupling capacitor C_1. During this time (and while the signal is going through its negative alternation), the capacitor C_1 will discharge through the grid resistor R_1, in the direction shown by the dashed arrows. This dis-charge establishes a negative bias potential across the grid resistor R_1, with a polarity that is negative at the grid and positive at ground (cathode), as shown.

Since the grid resistor is of a high value and since the tube does not conduct until the next positive signal alternation, the capacitor does not discharge fully but rather maintains a fairly constant bias potential between grid and cathode. Before the capacitor charge can decline to an appreciable extent, another positive signal alternation arrives at the grid and will again cause grid conduction. In consequence, the positive grid voltage will again recharge capacitor C_1 to its full value, to repeat the initial process.

During the time when the grid of the tube conducts, the grid circuit may be considered as having a short time constant (short RC). This short time constant permits a virtually full charge to appear across C_1. During the time the grid of the tube is not conducting, however, there is a long time constant established and, therefore, the rate of discharge for C_1 is much lower than its charging rate. At the time the tube conducts, the signal energy flows through the tube rather than through R_1 because during the time the tube conducts it has a very low impedance and hence is in virtual shunt across the grid leak R_1. Thus, current flow takes the easiest path, which is through the vacuum tube, since the latter has the lowest resistance. Resistor R_1 and the cathode-grid sections of the vacuum tube during conduction may be considered as two parallel resistors, one of which has a high ohmic value and the other has a very low ohmic value. During the signal peaks, current flows through the low ohmic value resistor (the tube) in greater proportion than through the higher resistor, composed of R_1. When the input grid signal value is less than the positive peaks, the resistance represented by the vacuum tube can be considered as having been removed from the circuit because during no-current flow conditions the tube resistance is infinitely high. The latter condition leaves only resistor R_1 as a discharge path.

The method shown at A of Fig. 15-9 has the advantage of being able to develop a bias that is beyond cutoff, as required for certain types of circuits, which will be described later. Cutoff bias cannot be secured by a cathode resistor alone because if the point is reached where the bias is at cutoff, no current flows through the cathode resistor. In the absence of such current flow, no bias would be developed and hence the current flow would again be at a maximum. Thus, the cathode resistor cannot be used to bias the tube beyond the cutoff point, no matter how high a value of cathode resistor is employed.

The disadvantage with the type of bias obtained by driving the grid positive is that when there is no signal input present to the grid of the tube, an excessive amount of current flows through the tube. For this reason, circuits employing this bias method often utilize an additional source of bias, by use of a cathode resistor and bypass capacitor. Such an additional bias is for protection against excessive current flow. The protective bias circuit in the cathode is shown at B. The protective bias need not have a high value but only a value sufficient to limit the current flow that would result during the absence of the input signal when no cutoff bias is developed.

15-5. Coupling Methods

There are several methods for interconnecting the various amplifier stages. One of these is the *resistance-coupled* or *resistance-capacitance-coupled* method as it is also called. A typical system of this type is shown at A of

Fig. 15-10. Here, C_1 is the coupling capacitor that links the signal energy from a previous stage (or from a microphone, playback head, etc.) to the base input of the first amplifier transistor Q_1. Resistor R_1 supplies the necessary forward bias to the base (in relation to the emitter) and the input-signal voltage also develops across this resistor. The combination of C_2 and R_2 form the stabilizing network previously discussed.

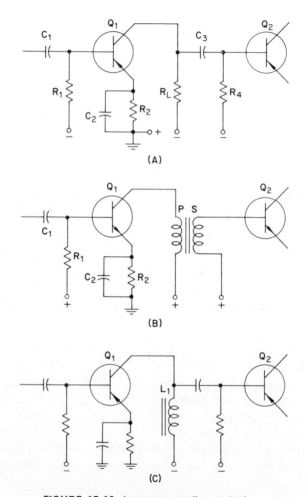

FIGURE 15-10. Interstage coupling methods.

The amplified signal energy develops across the collector resistor, known as a *load resistor* (R_L) and, from the latter, the signal is transferred to the next amplifier stage with another coupling capacitor (C_3). The advantage of this coupling method is economy since no costly transformers or other inductances are employed. The disadvantage is that the coupling capacitors limit low-frequency response because they have a rising capacitive reactance for signals of lower frequencies.

Another coupling method for audio-signal amplifiers is shown at B of Fig. 15-10. This system utilizes a transformer to couple the energy between the first transistor Q_1 and the second Q_2. While such coupling is more costly than the resistance-coupled method, an increase in signal voltage can be obtained across the transformer when the number of turns of the secondary are greater than those of the primary. A good-quality transformer is needed, however, to minimize losses at low- and high-frequency signal regions (because of shunting effects of capacitance between wire turns and layers and because of the varying reactance of the transformer inductors for frequency changes).

Another method for interstage coupling is shown at C of Fig. 15-10. Here an iron-core inductance is utilized in place of the resistor or transformer and it is across this inductor that the signal voltages are developed. This energy is then coupled to the next stage, using a conventional coupling capacitor and base resistor, as shown. The additional bulk and cost of the coil do not warrant its inclusion in modern circuits and, for this reason, it is rarely encountered in audio amplification systems, though it is used in some RF circuits.

As shown at B and C of Fig. 15-10, combinations of coupling methods can be employed where deemed convenient or dictated by design factors. At B, for instance, *R-C* coupling is used at the input of Q_1 and transformer coupling is used between stages. At C, however, L_1 is used in conjunction with the *R-C* coupling at the input.

One stage can be coupled to another stage without the use of a coupling capacitor or transformer, by direct coupling, if circuit-design factors permit its usage. This method is sometimes used in pulse and square-wave amplifiers, as well as in high-fidelity audio systems and in picture signal amplifiers of television receivers because it does not suffer the disadvantages of either capacity or transformer coupling. In capacity coupling, the reactance may be fairly high for lower frequencies, unless a large-sized capacitor is employed. A large capacitor, however, may have excessive leakage and might shunt signals to ground if placed too near the chassis because of capacity effects between the capacitor and the chassis. Also, as mentioned earlier, a transformer has a changing reactance for different frequencies, in addition to distributed capacities, all of which affect the frequency response. With direct coupling, all frequencies are transferred from one stage to another without some being diminished with respect to others.

In the design of direct-coupled circuitry particular attention is paid in establishing proper potentials of transistor elements. A typical example of a direct-coupled audio amplifier is shown in Fig. 15-11, where *NPN* transistors are used. For Q_1 a positive dc is applied to the base and since the potential is 2.5 V, the base is positive with respect to the emitter, thus satisfying the forward-bias requirements for the *NPN* unit. The collector of Q_1 has 10 V applied; hence it is also positive with respect to the emitter. Since, however, the collector is an *N* zone, this forms the required reverse bias between collector and emitter.

In direct coupling the collector is connected to the base, thus also applying a positive 10-V potential to the base. Because the emitter of Q_2 has an 8-V potential, however, the base is positive with respect to the emitter, thus supplying forward bias. For the reverse bias between the collector and emitter of Q_2, the collector has 70 V applied; hence it is also positive with respect to the 8-V emitter.

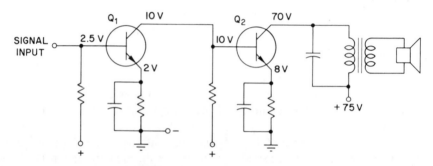

FIGURE 15-11. Direct coupling.

An FET amplifier with direct coupling is shown at A of Fig. 15-12. Here, the gate of Q_1 has a potential of 2 V but the source is 3 V, thus making the gate negative with respect to the source. Since the drain element, however, has 6 V applied, it is positive with respect to the source element, thus satisfying the bias requirements for the *N*-channel field-effect transistor. Since the drain voltage of Q_1 is also present at the gate of Q_2, the source voltage for Q_2 is increased to 8 V so it has a positive potential with respect to the gate. Since the drain element has 22 V present, it is positive with respect to the lower 8 V at the source of Q_2.

A three-stage transistor circuit combining several design features is shown at B of Fig. 15-12. Here the input signal is obtained from the variable arm of the volume control (as against ground potential) and coupled to the base input of Q_1 by the series capacitor. A positive voltage is applied to the base (1.5 V for this circuit) and the lower emitter voltage establishes the necessary forward bias for the *NPN* transistor. The 7 V at the collector of Q_1 makes it positive with respect to the emitter, thus providing the necessary reverse bias. Note the capacitor to ground at the collector. This capacitor has a sufficiently large capacitance so it has a low reactance for the signals

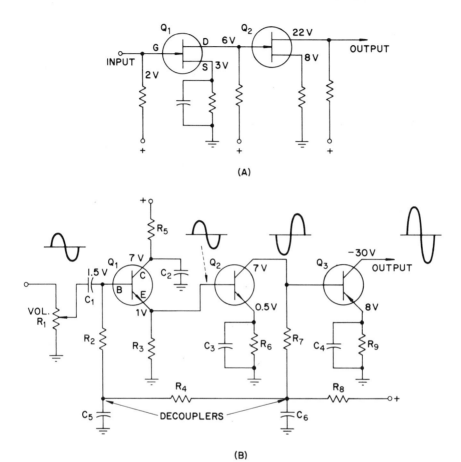

FIGURE 15-12. FET direct coupling and emitter-follower.

and hence effectively shunts them to ground. Thus, instead of a collector-output circuit, the signal is obtained from the emitter and coupled directly to the base of the second transistor (Q_2). Such a circuit is termed an *emitter follower* since the phase of the amplified signal follows that of the input, as discussed earlier. (This is comparable to the *cathode follower* using vacuum tubes, where the output is obtained from the cathode resistor.) Since, in the emitter follower, the collector is at signal ground, this design has also been called a *grounded-collector* circuit.

The emitter-follower circuit is useful for impedance-matching purposes (it has a low output impedance) and its step-down characteristics are obtained without the disadvantages of the transformer. Since there is no signal-phase inversion, it can also serve as an isolation stage to minimize interaction between two critical circuits. There is no *signal-voltage* gain, though *signal-*

current gain can be obtained. Since sufficient signal current is obtained to drive other circuits, it is also widely used in computer circuitry, industrial-control networks, and in receivers as well.

Since the signal must be obtained from the emitter of Q_1, the emitter resistor does not have a bypass capacitor as is the case for Q_2 and Q_3, where the resistor-capacitor network increases thermal stability as discussed earlier. For Q_2, the 1-V potential at the emitter is also present at the base of Q_2. Since, however, only 0.5 V is present at the emitter of Q_2, the base is positive with respect to the emitter. The 7-V potential at the collector of Q_2 provides the necessary positive-potential reverse bias for the *NPN* transistor.

Direct coupling is again used between Q_2 and Q_3 but now a *PNP* transistor is used for Q_3. When an *NPN* transistor is directly coupled to a *PNP* type (sometimes termed a *complementing* circuit), the change in voltage-polarity relationships between the two transistors must be considered. Thus, the 7 V at the collector of Q_2 is also present at the base of Q_3 but since this element must be negative with respect to the emitter (forward bias for the *PNP* type), the emitter of Q_3 has a higher potential to make it positive with respect to the base. For the collector of Q_3, the reverse-bias requirements are met by applying -30 V to the collector, making it negative with respect to the emitter by -22 V.

15-6.   Decoupling Networks

At B of Fig. 15-12, C_5 with R_4 and C_6 with R_8 form *decoupling networks* which help isolate stages from the common coupling which would occur because of the same power supply feeding the various stages. While resistors R_4 and R_8 also drop d-c potentials to that required, any *signal* voltages that appear across them will be shunted because of the low reactance of capacitors C_5 and C_6. Decouplers are extensively used in both audio and RF amplifiers for signal isolation purposes and hence minimize the tendency for either regeneration or degeneration, depending on the phase of the signal energy that is coupled between the stages.

Decouplers are also used in tube-type amplifiers, as shown in Fig. 15-13, where a pentode tube is employed. Here R_3 is the screen-voltage dropping resistor that sets the proper level of potential needed. Capacitor C_3 is a signal bypass, thus preventing transfer of any signal energy that may develop in the screen grid. Thus, signal energy is returned to the ground circuit (and back to the tube section via the cathode ground). Resistor R_5 and capacitor C_4 decouple the load-resistance section and confine signal-energy variations to R_4 only. The decoupling resistor R_5 is usually much lower in value than R_4, and R_5 consists of approximately 5000 to 10,000 Ω. Capacitor C_4 is made sufficiently large so as to have a low reactance for the lowest frequencies handled by the amplifier.

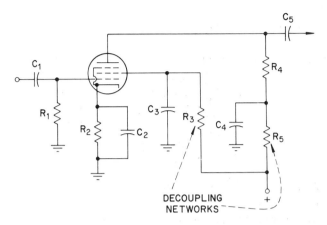

FIGURE 15-13. Decoupling networks in pentode-type amplifier.

The decoupling network can also be employed to increase low-frequency response in an amplifier, by proper choice of capacitor C_4. As this capacitor has a fairly high reactance for low frequencies, it will develop a signal-voltage drop across the impedance formed by C_4 and R_5 for lower frequencies. Thus, for low audio frequencies, the total load-resistance value is increased since it is composed of R_4 in combination with the decoupling network. The effectively increased R_L represents two signal-voltage drop components in series. Because the total ohmic value is increased, the signal energy developed across the combination of resistor R_L and the decoupling network will be increased for the lower frequencies. At the higher audio frequencies, C_4 has a low reactance and acts as an effective shunt preventing any signal voltages from developing across R_5. Thus, the low-frequency signal components are boosted and their potential is raised in proportion to the higher audio-frequency signal components.

The same requirements apply when the amplifier is used for pulse signals, square waves, and other types of waveforms discussed later. Not only do such special signals have low-frequency components that must be retained in their progress through an amplifier but high-frequency signal components as well that must not be diminished. In audio amplification the circuit must also be capable of handling the *dynamic* range of sound, that is, the soft and loud passages, without overloading on the latter and distorting them. Obviously, then, an amplifier does not have the sole function of increasing signal level alone but must amplify without undue loss of any part of the frequency range of the signals applied to the circuits and without distorting or otherwise degrading the quality of the original signals, whether such are audio signals, pulses, or other type waveforms. A better understanding of the necessity for good amplifier design will be gained by considering the characteristics of sound, composite signal waveforms, and distortion factors relating to amplifying systems.

15-7.   Amplifier Characteristics

When the air pressure is increased and decreased at a certain rate, the air pressure variations strike the eardrum and produce a sensation of sound for the listener. Among individuals, the sensitivity of the ear for the various sound frequencies differs to some extent. Younger people can hear sounds having frequencies as high as 20 kHz, while older people experience a decline in their hearing at the upper audible frequency range and, consequently, may hear sounds having frequencies only to 15 kHz or, in later life, only to 10 kHz.

The frequencies of speech sounds, as well as the frequencies of all the fundamental tones of musical instruments, are below 5 kHz; hence, one might wonder about the usefulness of sounds having frequencies extending to 10 or 20 kHz. The primary use for such high-frequency sounds, however, is that they have the ability for lending certain characteristics to fundamental-frequency tones that provides them with distinguishing features. Such features permit recognition of such tones as emanating from a specific source as compared to other similar tones having identical fundamental frequencies.

When a clarinet and a violin play the note A in the middle musical register, a tone having a fundamental frequency of 440 Hz is generated by each instrument. Despite the fact that each is generating a fundamental-frequency tone of 440 Hz, however, the ear immediately recognizes one note as being the tone derived from a violin and the other from a clarinet. This difference, which also characterizes the individuality of various other instruments, is due to the *harmonic content* of the fundamental tone generated by each musical instrument. Thus, a musical instrument not only produces a fundamental tone but also other tones that have higher (though related) frequencies than the fundamental-tone frequency.

The harmonic-frequency tones are usually lower in amplitude than the fundamental and have decreasing amplitudes for higher harmonic frequencies. The amplitude of the various harmonic-frequency tones, however, plus the particular harmonic-frequency tones generated by the musical instrument, combine to produce the identifying tonal characteristics of that instrument.

How a fundamental-frequency signal can combine with others to form a composite-frequency signal is shown in Fig. 15-14. At A, a single alternation of the fundamental-frequency signal is shown, and, for purposes of discussion, assume it represents a 500-Hz signal. Another signal is shown at B and this one has a frequency three times that of the one at A. Thus, the signal at B represents a third harmonic with a frequency of 1500 Hz. When these two signals appear at the input of an amplifier, the circuit receives a signal that is the additive sum of the two; that is, at points where both signals

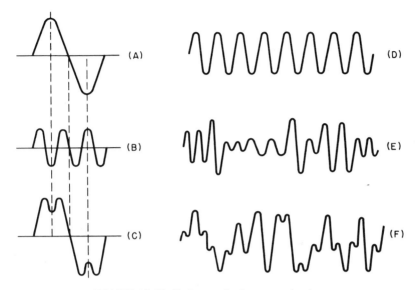

FIGURE 15-14. Various audio-frequency signals.

(at A and B) are in phase, the resultant signal (at C) that appears at the amplifier input will have an increased amplitude. Where there is a phase difference between the signals (A and B), the resultant signal (C) will have a correspondingly lower amplitude. Thus, when a fundamental frequency signal of 500 Hz is combined with another signal having a higher frequency, a resultant signal occurs, as shown at C. This waveform is now complex, rather than simple, since it contains signals having other frequencies as well as the fundamental frequency. A musical tone would contain many other tones of higher-order frequencies, making the resultant waveform that appears at the input of the amplifier even more complex.

To illustrate the foregoing discussion by showing successive cycles of the a-c signal waveform, a pure fundamental frequency is shown at D. Here, each cycle is identical to the others, both in amplitude and in frequency. In E various fundamental frequencies are shown and the overall waveform begins to have a more complex appearance. Not only are cycles of various frequencies present but amplitude changes occur as well. Thus, the waveform at E could represent the injection into the amplifier of various fundamental-frequency tones, some having higher volume levels than others. In F the type of waveform produced by music or speech reproduction is shown. Here there are not only various fundamental-frequency tones and amplitudes present but harmonics as well. During speech or music, the waveform would be undergoing a continuous change as different tones and volume levels are introduced into the input of the amplifier.

Because of the harmonics just mentioned (or *overtones*, as they are sometimes called), it is necessary for audio-amplifying systems employed in electronics to be able to handle the wide span of frequencies required for true reproduction of the original sound. The rendering of the harmonic-frequency tones, as well as the fundamental, during music reproduction, adds reality to the reproduced music and gives the illusion that the musical instrument is actually *present* in the room. This characteristic of a good amplifier is known as *presence* to define reproduction that closely resembles the original.

The typical frequency span of various sounds, as well as the approximate levels of sound by decibel comparisons (see the discussion on decibels in Sec. 5-6), is given in the following tables:

TABLE 15-1. Typical Frequency Span of Various Sounds

Type of Sound	Approximate Frequency Span (in Hz)
Desirable range for good speech intelligibility	300 to 4,000
Audibility range (normal hearing, young person)	16 to 20,000
Piano	26 to 4,000
Baritone	100 to 375
Tenor	125 to 475
Soprano	225 to 675
Cello	64 to 650
Violin	192 to 3,000
Piccolo	512 to 4,600
Harmonics of sound	32 to 20,000

TABLE 15-2. Approximate Levels of Sound by Decibel Comparison

Type of Sound	Relative Intensity (in dB)
Reference level	0
Threshold of average hearing	10
Soft whisper; faint rustle of leaves	20
Normal whisper; average sound in home	30
Faint speech; softly playing radio	40
Muted string instrument; softly spoken words (at a distance of 3 ft)	50
Normal conversation level; radio at average loudness	60
Group conversation; orchestra slightly below average volume	70
Average orchestral voume; very loud radio	80
Loud orchestral volume; brass band	90
Noise of low-flying airplane; noisy machine shop	100
Roar of overhead jet-propelled plane; loud brass band close by	110
Nearby airplane roar; beginning of hearing discomfort	120
Threshold of pain from abnormally loud sounds	130

15-8.   Harmonic Distortion

An amplifier should reproduce at its output a signal that is an exact duplicate
of the input signal, in all respects except amplitude. Thus, the amplifier
should build up the signal that is applied to its input without adding other
signal frequencies to it or otherwise distorting the original signal. Transistors
(and tubes), however, are not perfectly linear devices. That is to say, their
characteristic curves exhibit curvatures and hence do not consist of straight
lines. If the characteristic curves were perfectly straight lines, no distortion
would be introduced because the output waveshape of the amplified signal
would be the same as the input-signal waveshape. How curvature of transistor
(or tube) characteristics introduces distortion can be more easily understood
by referring to Fig. 15-15 where a nonlinear characteristic curve is shown
at A and a straight-line (linear) curve is shown at B. For purposes of emphasis, the curvature for the nonlinear characteristic shown at A has been confined to the lower portion of the curve. Also, the following discussions apply
to both transistor or tube characteristics.

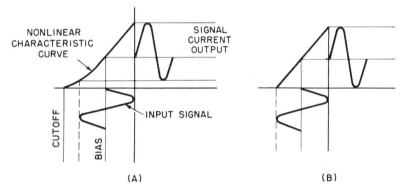

FIGURE 15-15. Distortion of output signal.

The input signal (a pure sine wave in this instance) can be considered
as being applied to the base input of an *NPN* transistor or to the grid input
circuit of a tube-type amplifier. As the input signal alternately adds or subtracts from the bias potential, there are corresponding changes in the output-
signal current, representative of the amplified output energy flowing through
the load resistance.

As shown at A, the output-signal current has a waveshape that is not
representative of the input signal. The change in output-signal current is
greater for the first alternation than for the second and hence some harmonic
distortion exists. When either alternation is greater in amplitude than the
other compared to the original signal, it is termed *amplitude distortion*.

If the transfer characteristics are perfectly linear as shown at B, then a variation of the input signal causes an output-current change proportional to the input which faithfully follows the amplitude changes which occur at the input. The result is a distortion-free output signal.

To minimize distortion, transistors or tubes can be operated on their most linear portion of the characteristic curve by setting the bias value at the approximate center of the linear portion and holding the input-signal amplitude changes to a sufficiently low value so that the amplitude excursions do not go into the curvature regions of the curve. When the transistor or tube is operated in this fashion, it is known as *Class A*. The bias setting permits output-current flow with or without an input signal to the transistor or tube. A representative operating point for a triode tube is shown in Fig. 15-16.

In Fig. 15-17 a triode characteristic curve is shown for comparison with the pentode curve shown at B. Because of the steeper slope of the pentode characteristic curve, a larger plate-current change is obtained for a relatively smaller input grid signal. In consequence, pentodes or beam-power tubes are capable of much greater signal amplification than triodes.

Triode tubes develop harmonic distortion that has primarily an even harmonic content. Thus, triode tubes generate a second harmonic distortion, as well as additional harmonics. The second harmonic will be the most dominant, while the higher order of even harmonics will be progressively lower in amplitude than the second harmonic. Pentode tubes produce a har-

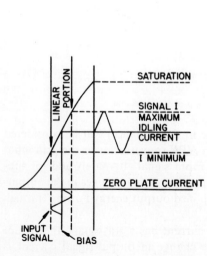

FIGURE 15-16. Class A operation.

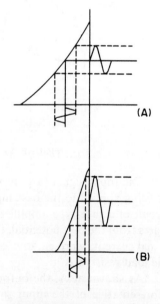

FIGURE 15-17. Linear operation—triode versus pentode characteristics.

monic distortion primarily composed of odd harmonics. In pentodes, the third harmonic is the most dominant, with successively higher odd harmonics having a progressively lower amplitude.

A good amplifier is designed so that such a distortion will be at a minimum. An acceptable level for distortion is 5%. Distortion below 5% is usually not noticeable to the average ear, though in high-fidelity applications the distortion is often reduced to less than 1%. In some applications, distortion as high as 8 or 10% can be tolerated, where the primary purpose is speech reproduction or where the *quality* of music that is reproduced is not too important a factor. In audio or RF *voltage* amplifiers, Class A operation is usually employed, in contrast to some of the other amplifier types, such as Class AB_2 or B, as more fully described for power amplifiers in the next chapter.

The harmonic distortion that occurs when an amplifier is operated in the nonlinear portion of the characteristic curve of a transistor or tube means that harmonic-frequency signals are *generated within* the tube or transistor. Because such signals were not present in the original signal applied to the amplifier, these harmonic signals are referred to as *distortion* and are considered undesirable. We must, however, distinguish between *harmonic distortion* and the *harmonic signals* normally present in complex waveforms such as musical tones and square waves. The harmonic signals present in musical tones and other complex waveforms usually encountered are applied (with the fundamental tone) to the input of the amplifier and, hence, both the fundamental tone and its inherent harmonic content should be amplified to the same degree. If other harmonic signals are generated within the amplifier, they represent an undesired addition to those signals being amplified because neither the generation of even or odd harmonics is desirable in amplifier circuitry.

To emphasize the difference between harmonic distortion and the natural harmonic content of musical or other audible tones, see Fig. 15-18. In A, a pure 400-Hz sine wave is applied to the input circuit of the amplifier. If the output signal contains not only the original fundamental 400-Hz signal but also a 1200-Hz signal, the output signal is considered to contain harmonic distortion. (The additional 1200-Hz signal represents a third harmonic component of the original signal. The 1200-Hz signal, however, was not present in the original signal but was generated within the amplifier and added to the 400-Hz signal.)

If, on the other hand, the output had consisted of only a 400-Hz audio-amplified tone, the output signal would have been undistorted. The output signal shown in A is considered to be distorted only because it contains signals that were not present at the input of the amplifier. The 400-Hz output, plus the 1200-Hz output, could be an undistorted output under other input conditions, as shown at B. Here two input signals to the grid of the amplifier are indicated. One is a 400-Hz tone and the other is a 1200-Hz tone. These

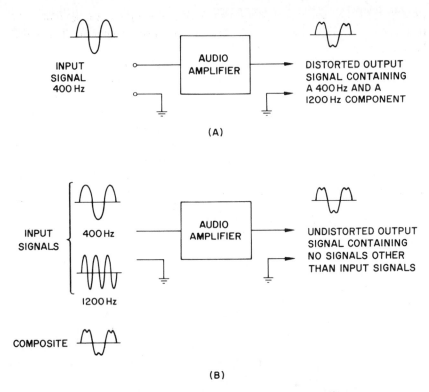

FIGURE 15-18. Harmonic distortion in an audio amplifier.

two frequencies could represent *the two fundamental tones generated by two musical instruments.* Since the input-signal voltage must either increase or decrease, the two signals will combine in additive fashion mentioned earlier. Hence, the output will be an amplified version of this combined input signal, and the output would contain both the 400-Hz audio tone and the 1200-Hz audio tone. In this instance, the output signal *is not distorted* since no signal frequencies are present except the signal frequencies originally introduced into the circuit.

15-9. Frequency Distortion

There are various factors which may prevent an amplifier from having a flat frequency response over the desired range of frequencies which are to be handled. In a transformer-coupled amplifier, both the primary and secondary windings exhibit various reactances to signals of different frequencies. At the higher audio frequencies, the reactances of the transformer are high and, in consequence, larger signal-voltage drops occur across the transformer. At low audio frequencies, however, the transformer reactances are also low and, hence, the low-frequency signals are not amplified to the same degree as the higher signal frequencies. In addition, the transformer windings have considerable *distributed capacitances*, which occur between individual turns of wire as well as between layers of the wire turns making up the primary and secondary. Hence, higher signal frequencies cause the distributed capacities to have low capacitive reactances, which tend to shunt signals having high frequencies. The result of the foregoing is a loss of low frequencies because of inductive reactances in the transformer and a loss of high frequencies because of the distributed capacitances. This is known as *frequency distortion*. To minimize such losses, it becomes necessary to employ transformers having larger cores with characteristics such that permeability is increased considerably. The increase in permeability causes an increase in inductance, permitting fewer wire turns to produce the required inductance and thus lessening the total distributed capacity.

If a coupling capacitor is employed instead of a transformer, a variable factor is again introduced with respect to amplification. The higher signal frequencies find a low reactance and thus encounter little opposition in reaching the next amplifier stage. The lower-frequency signals, however, find a higher reactance and hence are diminished (attenuated) in transferring across the coupling capacitor. If the coupling capacitor is increased in size (larger capacity), the reactances for lower frequencies will be decreased but the larger physical size of the capacitor may set up shunt capacities to the chassis (or nearby components) and may thus tend to shunt some of the higher-frequency signals. Even though the stray capacity existing from the body of a capacitor to the chassis may be small, the higher-frequency signals may find a sufficiently low reactance so that an appreciable portion of the high-frequency signal energy is lost. Even undue lengths of wire or printed-circuit conductors that connect a coupling capacitor from the output circuit of one stage to the input circuit of the following stage may introduce losses. Long leads tend to increase inductive effects because even a short length of wire has some inductive characteristics. The inductance of the wire length may be small but high-frequency signals find a greater inductive reactance in the wire length than do low-frequency signals; hence the high-frequency signals suffer some attenuation. Also, any wire-connecting circuits may present some shunt capacity to the chassis or other nearby components. If such wires carry signals, the stray capacity will provide a shunt reactance that will cause signal losses.

Signal losses also occur because of the interelement capacitances of solid-state devices and the interelectrode capacitances of vacuum tubes. Such capacitances exist among base and emitter, collector base, etc., and in tubes between the grids, cathodes, and anodes. For a minimum signal loss at high-frequency operation, transistors and tubes must be selected that were specifically designed to have low capacitances at the upper frequency ranges.

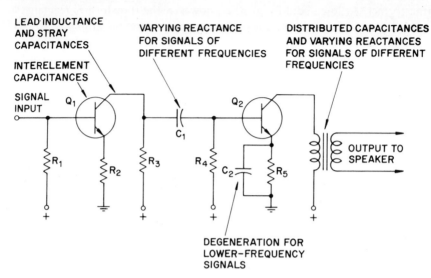

FIGURE 15-19. Loss factors in an amplifier.

Tube miniaturization and integrated solid-state circuitry have done much to extend efficiency of circuit operation at high frequencies.

The various factors that can contribute to the loss of signals of certain frequencies are shown in Fig. 15-19, and this summarizes the principles outlined in the foregoing discussions. In a well-designed audio amplifier, where all necessary signal-loss precautions have been observed, it is possible to obtain a substantially flat response for signals ranging in frequencies from approximately 30 Hz to well over the hearing limit of 20 kHz. The frequency response of an amplifier can be ascertained by applying various signals to the input, starting from the lowest frequency and progressively increasing the frequency of the signals in steps. Each signal applied to the input of the amplifier is held at a constant amplitude. The output signals from the amplifier are then measured for amplitude by a calibrated test instrument, as shown at A of Fig. 15-20. The results, in the form of a graph of the frequency response, then appear as shown at B.

The overall gain that can be realized from an amplifier is dependent on the gain factor of the transistor (or the amplification factor of the tube) as well as on the value of the load resistor used and the output impedance of the transistor or tube. As the signal frequency is increased, upper limits will be reached in terms of signal transfer efficiency. For a common-base circuit, the cutoff point for signal frequency is considered when the *alpha* (α) value drops to 0.707 of the value obtained at 1 kHz. For the common-emitter circuit, the frequency cutoff is reached when beta (β) drops to 0.707 of the 1 kHz value.

Vacuum tubes, because they are physically larger than the solid-state

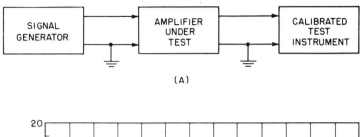

(A)

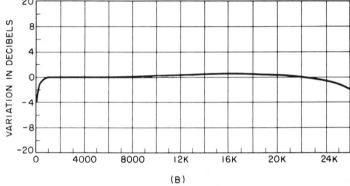

(B)

FIGURE 15-20. Measuring audio-frequency response of an amplifier.

devices handling comparable signal levels, are more prone to attenuation factors because of capacitances within the structure. In triode tubes, for instance, the input capacitance existing between grid and cathode will shunt high-frequency components of a complex wave but, in addition, the grid-plate capacitance also has a pronounced effect on input capacitance. This comes about because the amplified signal that appears in the anode circuit will be coupled back to the input circuit via the grid-plate interelectrode capacitance. Since the amplified signal at the output is 180° out of phase with the input signal, it cancels some of the input signal in proportion to the amplitude of the signal fed back. The end effect is similar to what it would be if the grid-to-cathode capacitance *were increased*; this characteristic is known as the *Miller effect*. The tube itself could have only a few picofarads of input capacity but the Miller effect could increase this to an equivalent of 100 pF or more and would be the primary factor in the shunting effect of higher-frequency signal components. With greater amplifier gain, the Miller effect is correspondingly increased and input capacitance lowered.

Since input capacitance shunts higher-frequency signals, it can also affect the higher-frequency components of a complex waveform applied to the input. This is shown in Fig. 15-21. Here the grid-plate capacitance C_1 provides the Miller effect and has the end result of increasing the input capacitance represented by C_2. If a waveform (as shown) were applied (containing a fundamental and a third-harmonic component), the higher-

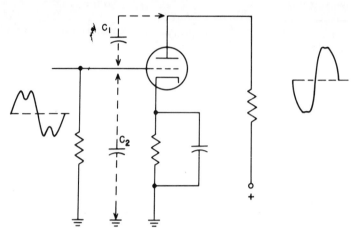

FIGURE 15-21. Miller effect and frequency distortion.

frequency portion would find a greater shunting effect due to the input capacitance, resulting in a decrease in the high-frequency portion of the amplified signal. Thus, the output-signal waveform no longer duplicates the input, and frequency distortion is the result.

15-10.   Phase Distortion

An input resistor such as R_1 of Fig. 15-22 in conjunction with the coupling capacitor C_1 forms a voltage divider for the applied input signals. Thus, the base-emitter input actually obtains the signal from across R_1 only as shown. Because the coupling capacitor C_1 has a low reactance for signals of higher frequencies, there will not be so much signal-voltage drop across it

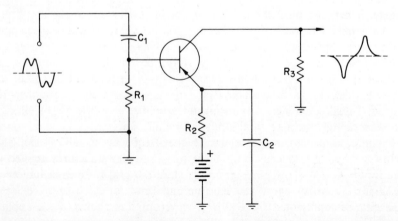

FIGURE 15-22. Phase distortion.

for the higher-frequency signals (or components of the signal) as there would be for signals of lower frequency, again causing some frequency distortion. At lower frequencies the reactance may be appreciable and the input-signal voltage developing across the capacitor causes a signal decrease across the resistor R_1. Since the signal applied to the base input depends on the voltage drop across R_1, there is an obvious decrease in the base signal over that which would prevail if the reactance of C_1 were only a fractional value of the impedance formed by R_1. [An *impedance* is present because the base input has a base-emitter capacitance (or cathode-grid capacitance in a tube) and this shunts the input resistor R_1.]

Thus, the shunting interelement capacitance of the transistor (or interelectrode capacitance of a tube) plus the varying reactance of C_1, combine to cause a phase shift between high- and low-frequency signals as well as attenuating signals of certain frequencies, as mentioned earlier. When higher-frequency signals (or the high-frequency components of a complex wave) undergo a phase shift with respect to the lower-frequency signals, the result is *phase distortion* of the amplified signal, as shown in the output waveform in Fig. 15-22. This can be understood more readily by reference to Fig. 15-14 shown earlier. If the third-harmonic component shown at B were shifted (either to the left or right) in comparison to the fundamental signal at A, the output waveform would no longer resemble that at C but may have the phase-distorted waveshape as shown at the output of Fig. 15-22. This distortion is particularly objectionable in amplifiers handling visual-type signals such as in radar, television, and oscilloscopes. Phase shifting will cause a shift of picture information and a blurring of images.

It must be noted that there is a 180° phase shift between the input and the amplified output signals of the common-emitter circuit of Fig. 15-22. This is not considered phase distortion since it is an inherent characteristic of the grounded-emitter circuit and the original phase can be obtained by passing the signals through another similar circuit. In television receivers, signals must have proper phase at the picture tube so a rising amplitude causes a grid-bias increase and hence a decrease in beam current. These factors are covered more thoroughly in Chapter 19.

15-11. Signal-Voltage Gain

In the grounded-emitter transistor circuit (also termed *common emitter*), the input impedance (base emitter) may range from a low value of 50 Ω to approximately 4 kΩ, depending on type. Output impedances (collector emitter) may range from 75 Ω to about 50 kΩ. For tubes, however, the input resistance (grid-cathode) of a tube is high and may be well over a megohm. The plate resistance of a triode vacuum tube may range between a few thousand ohms and 15 kΩ, depending on tube design. (See Equation 11-2.)

Thus, in a vacuum tube, there is an impedance or resistance decrease across the tube, while in a transistor there is an increase.

The output resistance of a transistor or tube has a bearing on amplification because the inherent resistance of the unit acts as though it were in series with the load resistance, as shown in Fig. 15-23. Here the generator (G) is considered as consisting of the output section of the transistor or tube (the collector emitter in the common-emitter circuit or the anode-cathode of the conventional tube circuit). The plate resistance of the tube is indicated as R_p, though this would also comprise the output resistance of the solid-state devices. This resistor is shown as external to the generator to illustrate its *series* relationship with respect to the load resistor (R_L). The input resistance of the following transistor or tube effectively shunts the load resistance (particularly at higher frequencies where the coupling-capacitor reactance is low). Its effect must be considered where its value is near that of the load resistance. For tubes and solid-state FET units, however, the much higher input resistance of the following stage (in shunt) does not materially lower the effective value of R_L. Consequently, the value of R_L and the output resistance of the transistor or tube need only be considered in setting up equations for calculating gain.

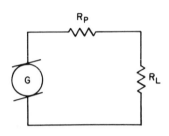

FIGURE 15-23. Equivalent circuit of an amplifier.

Equations for signal gain in solid-state circuitry have been given in Chapter 13, and reference should be made to Equations 13-1, 13-3, 13-4, 13-22, and 13-23. For tubes, the signal-voltage gain may be found by taking the ratio of the signal output voltage to the signal input voltage, as affected by the amplification factor and the series resistances of R_p and R_L:

$$\text{Voltage gain} = \frac{\mu R_L}{R_p + R_L} \qquad (15\text{-}1)$$

Often, the load resistor is chosen to have a value some two or three times that of the plate resistance, for an increased signal-voltage drop across the load resistor. (A value of load resistor substantially higher than three times the plate resistance may result in signal distortion and hence is usually avoided.) Since the chosen load resistance is usually of a value no higher than three times the plate resistance, the maximum gain that can be realized is only three-fourths of the rated amplification of the tube. Thus, when the load resistor is three times the plate resistance, one-fourth of the signal develops across the internal plate resistance of the tube, and the other three-fourths across the load resistance. (Impedance matching, for maximum *power* transfer, is discussed in the next chapter.)

Pentode tubes have a much higher amplification factor than triodes, as well as a much higher plate resistance. The latter, for pentodes, may range from several hundred thousand ohms to over a megohm. In consequence, the load resistor is not made larger than the plate resistance because of the current limitation that would result. Instead, the load resistor is made substantially lower than the plate resistance; hence it is convenient to base the calculation for voltage amplification on the transconductance (g_m) of the tube, as follows (see also Equation 13-3):

$$\text{Voltage gain} = g_m R_L \tag{15-2}$$

This is derived from the previous formula for voltage gain but since R_L is much smaller than R_p, the addition of R_L to the plate resistance value in the denominator of the equation will alter the value of the denominator very little, hence the equation may be written as

$$\frac{\mu R_L}{R_p} \tag{15-3}$$

Since μ/R_p represents the transconductance (g_m) of a tube, the formula finally can be expressed as $g_m R_L$, as shown in Equation 15-2.

15-12. RF Amplification

As mentioned in the introduction to this chapter, RF signals encountered in both transmitters and receivers must be amplified to bring them to the levels necessary for proper transmission and reception of radio and television. As with audio amplification, RF amplifiers must be designed to handle not just a single-frequency signal but a group of signals having various frequencies. In transmitters, the basic RF signal (known as the *carrier*) must be modified by a process known as *modulation* in order for the carrier to convey the audio or video signals, as more fully detailed in a subsequent chapter. This modulation process generates additional signal frequencies immediately above and below the carrier frequency, such additional frequencies being termed *side bands*. Thus, the RF amplifier must be capable of amplifying to the same degree both the carrier and the side bands involved.

How much space a station may occupy in the frequency spectrum is determined by the Federal Communications Commission (FCC), which also allocates the particular carrier frequency which a station must employ as well as the power which may be used in transmission. In standard AM radio, for instance, the allocated spectrum is from 550 kHz to approximately 1600 kHz. (See Fig. 15-24.) Each station occupies approximately 10 kHz. (The reason for the approximation is that the total space occupied by a station may vary because the number of side bands will vary during broadcasting, as more fully explained later.) Because a number of stations occupy the broadcast spectrum, the RF amplifiers that are employed must have selective

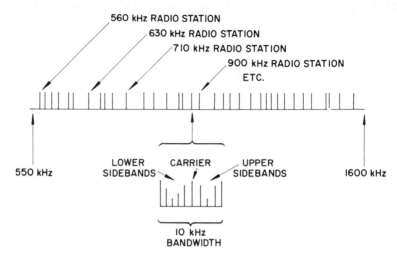

FIGURE 15-24. AM radio station spectrum.

characteristics. Selectivity means that the amplifier must be capable of selecting and amplifying a band of frequencies for a particular station, while rejecting nearby signals of adjacent stations.

The selectivity of an amplifier must be chosen to accommodate the type of transmission to be handled. In AM broadcasting, the selectivity of an RF amplifier must be such that it ranges over approximately 10 kHz as previously mentioned. In FM (frequency modulation), however, an RF selectivity (known as *band pass*) of 150 kHz is necessary. As compared to AM, FM is actually a wide-band type of selectivity. Wide-band selectivity is also necessary in television transmission and reception, where the tuner amplifiers must have a selectivity ranging to 6 MHz (6000 kHz). To obtain selectivity in an amplifier so that it will select the desired station, while rejecting unwanted stations on either side of the desired one, tuned resonant circuits must be employed. The Q of such circuits is regulated by introducing a certain amount of resistance so that the band pass will be sufficient to accommodate the carrier and the side-band signals. Selectivity factors were described at greater length in Chapter 10 and the resonant theory section of that chapter should be reviewed to refresh the reader's memory of the fundamentals that apply to this topic.

At Fig. 15-25 a basic common-emitter RF amplifier is shown. Inductances L_1 and L_2 form a transformer arrangement in which L_1 couples the signal energy from the previous stage to L_2. Both L_1 and L_2 have shunting capacitors (C_1 and C_2) that are variable so that both the L_1 and L_2 sections can be tuned to resonance. Signal energy is inductively coupled from L_1 to L_2, and the high-impedance circuit of L_2 and C_2 (parallel resonance) develops a high signal-potential component which is applied to the base-emitter input.

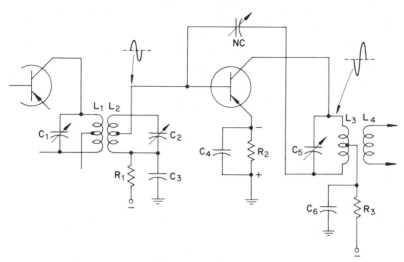

FIGURE 15-25. Common-emitter RF amplifier.

Capacitor C_3 places the bottom of the resonant circuit at signal ground, thus making it common to the emitter terminal since C_4 also has a low reactance for signals handled by the amplifier. Inductor L_2 is tapped to provide for an impedance match between the resonant circuit and the base input. Since the base-input impedance is much lower than that of the resonant circuit, the tap supplies the low-impedance coupling point.

When a triode transistor (or vacuum tube) is used as an RF amplifier as in Fig. 15-25 with a tuned circuit at both the input and output sections, the entire circuit would oscillate; that is, it would generate a signal of its own instead of simply amplifying an incoming signal. While oscillators as such are important in electronics, they consist of special circuits, as described in Chapter 19. An amplifier, on the other hand, should not exhibit oscillatory characteristics.

Oscillations occur in the triode RF amplifier because the interelement capacitances in a solid-state device (or the interelectrode capacitances in a tube) act to couple the amplified signal energy at the collector (or anode) back to the base input (or grid of a tube). This *feedback* converts the amplifier into an oscillator, and the signals which are consequently generated are undesired and cause interference with the input signals which are being amplified.

To prevent undesired oscillations in an RF triode amplifier, it is necessary to neutralize the circuit, as shown in Fig. 15-25, by using a neutralizing capacitor (NC). Note that the collector voltage applied via R_3 is to a tap on the inductor L_3; hence a signal can be obtained from the *bottom* of this inductance, such a signal being 180° out of phase with the collector signal at the top of the inductance. Thus, the capacitor NC couples a portion of the signal from the collector side to the base side, of proper phase to neutral-ize the feedback causing oscillations. The amount of signal fed to the base can be regulated by adjustment of NC since this is a variable capacitor. Hence the reactance of NC is set so the signal energy fed back is of the same amplitude as the undesired signal coupled by the interelement capacitances of the transistor.

For solid-state circuitry, the word *unilateralization* has been widely used as a common designation for neutralization. Thus, a *unilateral circuit* is considered as one having only a single-direction path for the signal to be amplified. Consequently, if an input signal is applied, an amplified replica appears at the output. If, however, a signal is applied (or is present) at the output circuitry, no path in reverse to the input is present and hence the signal is unable to reach the input side.

Strictly speaking, unilaterilization should be considered as the process by which an external feedback system is designed and adjusted for canceling both *resistive* and *capacitive* internal coupling between the input and output circuitry of a transistor, plus other reactive internal coupling. Neutralization, on the other hand, cancels only *reactive* internal feedback between output and input.

Resistor R_2 and capacitor C_4 in the emitter circuit are to increase circuit stability and minimize a thermal runaway condition where transistor currents may increase as internal heat rises. Electron flow through R_2 from collector to emitter to ground establishes a voltage drop across R_2 that opposes the positive potential present at the emitter. Assume, for instance, that the for-ward bias between the base and emitter would be 6 V if the emitter were grounded. With R_2 in the circuit, however, a drop of 2 V changes forward bias to 4 V, thus decreasing conduction through the transistor. (As the emit-ter is made less positive, the base, in turn, is also made less negative. If emitter potential is increased to a higher positive level, the negative base potential also rises since forward bias is the relationship of potentials and polarities between base and emitter.)

Thus, during operation, if currents tend to rise above normal, there is an increased drop across R_2 that decreases forward bias and thus lowers the current flow, tending to normalize it. Capacitor C_4 shunts R_2 and, because of its low reactance for signals of resonant frequency, prevents signal-voltage variations across R_2.

A typical RF grounded-base amplifier (also termed *common base*) is shown in Fig. 15-26. (Reference should also be made to Fig. 13-13B for com-parable low-frequency circuits.) For Fig. 15-26, input and output transform-ers are again used as with the common-emitter type of Fig. 15-25. Instead of variable-capacitor tuning, adjustable cores are used (tuning slugs) with the arrows between windings denoting adjustable cores. Variable-capacitor tuning could, of course, also be employed as required.

The forward-bias positive potential is applied to the emitter of the *PNP*

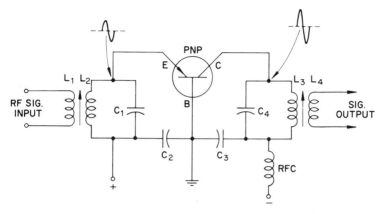

FIGURE 15-26. Grounded-base RF amplifier.

transistor through L_2, with C_2 placing the bottom of the resonant circuit at signal ground by virtue of the low reactance of C_2 for the resonant-frequency signals. The reverse bias (negative to the collector) is also applied through a resonant-circuit inductor (L_3). Again, a capacitor (C_3) provides the necessary low reactance to place the bottom of this resonant circuit at signal ground. (For an *NPN* transistor, battery polarities would be the reverse of those shown.)

Compared to the common-emitter amplifier, the common base is less affected by low-frequency oscillations since it has less gain at frequencies below the resonant frequency than the common emitter. Also, the signal frequencies above the resonant frequency extend to higher regions before they tend to decline, thus facilitating wide-band operation when this is required. Also, with the common-base RF amplifier, the output circuit is isolated from the input by the grounded-base element; hence there is rarely a need for neutralization as with the common-emitter RF amplifier.

An RF amplifier designed around a field-effect transistor (FET) is shown in Fig. 15-27. To illustrate alternate methods for applying voltages, the applied potentials shunt the resonant circuits; hence power-supply current does not flow through them. Such design is termed *shunt feed* in contrast to the *series feed* shown for the amplifiers in Fig. 15-25 and 15-26. For series feed the supply current flows through the resonant-circuit inductor (or load resistor, if one is used). Shunt feed could, of course, also be used for the previously discussed amplifiers, as could series feed for that shown in Fig. 15-27.

A radio-frequency choke (RFC) is in series with the supply potential and the gate of the FET. This RFC (L_4) provides a high reactance for the resonant-frequency signals and prevents their leakage to the power-supply system. Capacitor C_2 blocks the dc and prevents the supply potentials from being grounded through the low resistance of inductor L_2. A similar arrange-

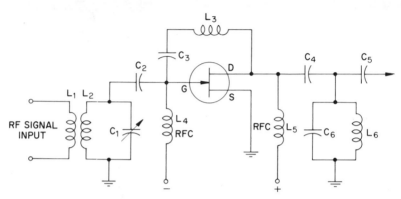

FIGURE 15-27. RF amplifier.

ment exists at the drain side where C_4 blocks the dc applied to the bottom of RFC L_5.

Unilateralization is provided by inductor L_3 and capacitor C_3. The latter also prevents direct coupling of dc between the gate and drain elements of the FET. The degree of feedback is regulated by the values selected for the series inductor and capacitor. As these elements are tuned closer to the resonant frequency, the impedance of the series circuit drops and hence more RF energy is fed back for neutralization purposes.

Tube-type RF amplifiers are shown in Fig. 15-28. As with the common-emitter triode-transistor RF amplifier, neutralization is necessary. Thus, a similar feedback arrangement is used and the neutralizing capacitor (NC) again regulates the degree of degeneration introduced, as with the circuit shown in Fig. 15-25. Resonant-circuit input and output systems are utilized, as with the transistor RF amplifiers. Since filament voltages are often designated as *A* potentials, the anode voltages are referred to as the *B* voltages (and bias supplies as *C* voltages). The *B*+ feed point shown is applied to a decoupling network where an RFC is used instead of a resistor.

A typical pentode RF amplifier is shown at B of Fig. 15-28. Here the input-signal energy from the previous stage is again coupled from L_1 to L_3 and consequently developed across the parallel-resonant circuit composed of L_2 and C_2. Capacitor C_3 places the bottom of this resonant circuit at ground potential for the signal, while R_1 performs the function of the grid resistor. A conventional cathode resistor R_2 and capacitor C_4 furnish the necessary bias for the tube. If this amplifier is utilized in broad-band work such as television, an additional resistance may shunt the resonant circuit, as shown by resistor R_3. Use of such a shunting resistor will broaden the characteristics of the parallel resonant circuit and will lower the Q sufficiently so that the circuit is enabled to cover the requisite band of frequencies. [When the Q is lowered, the selectivity is also decreased but when several

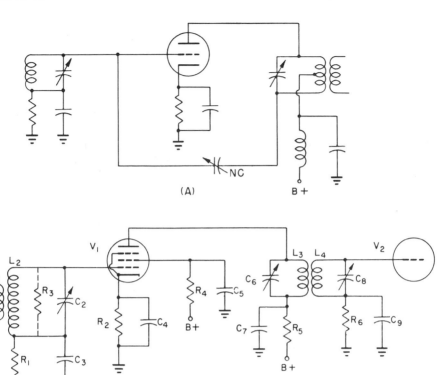

FIGURE 15-28. Tube-type RF amplifiers.

RF stages are employed in succession (*cascade*), the desired selectivity for good rejection of unwanted signals can be reestablished.]

The screen voltage for tube V_1 is applied via the voltage drop in resistor R_4, with C_5 having a bypass effect on signal-voltage variations at the screen. In the plate circuit, another resonant circuit is formed, using capacitor C_6 and inductance L_3. This circuit is fed from the B voltage supply through the decoupling network composed of R_5 and C_7. Capacitor C_7 also establishes the bottom of the plate resonant circuit at ground potential for the signal. If capacitor C_7 has its ground side connected near cathode capacitor C_4, a direct return path for a completely closed anode signal circuit is provided, with a short span of circuitry. The decoupler also minimizes interaction between successive RF amplifier stages, in similar fashion to that employed for the audio-frequency amplifier. The energy developed across the resonant circuit composed of C_6 and L_3 is coupled to the grid resonant circuit of the next stage through inductive coupling, by L_3 and L_4. In the grid circuit of the next stage, a similar arrangement is found, which is virtually identical to the input grid circuit of the previous stage.

If the pentode RF amplifier is used at very high frequencies, the inter-electrode capacities, even though small in a pentode, may still have sufficiently low reactance at such high frequencies to provide some coupling between input and output circuits. For this reason, a pentode tube may exhibit the tendency to oscillate, when utilized at high RF frequencies, and some form of neutralization may be necessary. One method, often utilized, is to use a screen-grid bypass capacitor that is smaller in value than normally employed so that it will not be fully effective as a bypass capacitor. The inadequate bypass capacitor introduces some screen-grid degeneration, which will then minimize the tendency toward oscillation in the amplifier circuit.

The grounded-grid RF amplifier is shown in Fig. 15-29. This circuit is the tube counterpart of the grounded-base transistor amplifier shown earlier in Fig. 15-26. For the circuit of Fig. 15-29 the input signal is developed across the high-impedance resonant circuit consisting of L_2 and C_2, between cathode and ground. (Resistor R_1 is for bias purposes, and capacitor C_3 places the bottom of the cathode resonant circuit at signal ground.)

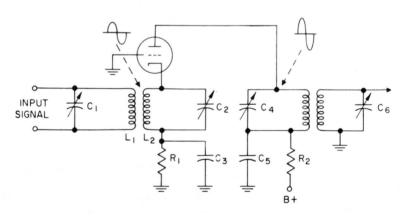

FIGURE 15-29. Grounded-grid amplifier.

While it appears as though the signal is applied to the cathode circuit only, it is actually present between cathode and grid. Assume, for instance, that a positive alternation appears across the cathode circuit. The positive alternation will increase the cathode voltage drop and an increase in cathode potential means an increase in the negative *grid* potential. Hence, current flow through the tube decreases and plate voltage rises. A negative-going alternation at the cathode circuit decreases the cathode potential and hence lowers grid bias. Plate current now increases and plate voltage drops, developing a negative-going signal at the output. From this analysis, it can be seen that the phase of the output signal is the same as that of the input signal, a characteristic of the grounded-grid amplifier.

The fact that the grid is grounded minimizes oscillations in the triode amplifier because the grounded grid acts as an electrostatic shield between the grid-cathode and the grid-plate circuits, just as the screen grid (at signal ground by virtue of a bypass capacitor) isolates the input and output circuits of pentode amplifiers.

The output circuit of the grounded-grid amplifier is conventional, as shown in Fig. 15-29. Resonant circuits are employed and may be tuned by variable capacitors C_4 and C_6, as shown, or by movable-core tuning slugs within the coils.

15-13. Load Lines

In earlier chapters, descriptions were given for obtaining static characteristics and curves for vacuum tubes and transistors. In Sec. 11-6, triode characteristics were given and curves illustrated in Fig. 11-9, 11-10, and 11-11, while pentode characteristic curves were shown in Fig. 11-16. Transistor characteristics were covered in Sec. 13-4, and characteristic curves for the grounded-emitter circuit were shown in Fig. 13-8. Gate-to-drain characteristics of the FET unit were illustrated in Fig. 13-14.

While such characteristic curves indicate the amount of output current that flows through the transistor or tube for a given bias and anode or collector voltages, they do not indicate the *dynamic characteristics* of the tube or transistor, that is, the characteristics that are established when a signal is applied to the base input (or grid input of a tube). During signal input, the output current varies by an amount established by the amplitude of the input signal and the signal gain of the tube or transistor.

In order to illustrate the conditions that occur when the tube or transistor is operating with a signal input and is amplifying, a *load line* must be drawn. Such a load line is represented by a diagonal line drawn through the characteristic curves as shown in Fig. 15-30. (This is for a triode tube and the graph contains grid-voltage curves. Load-line factors for transistor characteristic curves will be covered later in this chapter.)

With a load line such as shown in Fig. 15-30, the dynamic operating conditions of the tube can be calculated and such information can be gathered as power output, plate efficiency, and percentage of harmonic distortion, as well as the value of the load resistor itself.

For a load line illustrated in Fig. 15-30, a load resistance of approximately 5342 Ω is indicated. This value can be calculated by assuming a maximum signal swing and then ascertaining the maximum and minimum values of plate current and plate voltage. For the tube illustrated, assume a normal bias of minus 20 V. The maximum signal swing would presumably start at this bias value of minus 20 V and swing to zero, then back to 20 V, and on up to minus 40 V of bias. Assuming such a signal swing, the ohmic value of

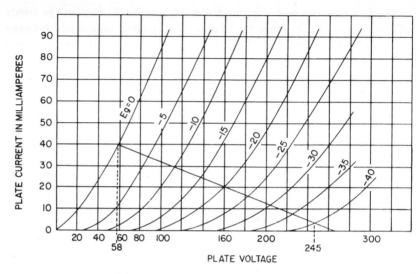

FIGURE 15-30. Load line for a triode tube.

the load resistance can be found from the following formula:

$$R_L = \frac{(E_{\max} - E_{\min})}{(I_{\max} - I_{\min})} \tag{15-4}$$

Thus, the maximum voltage swing would be approximately 245 V, based on the intersection of the minus 40-V bias line with the load line. Minimum voltage swing would be approximately 58 V, established by the load-line termination at zero bias. When the signal swings to minus 40, the current would be approximately 5 mA (0.005 A) and, at zero bias, the current would be approximately 40 mA (0.04 A). When the difference in voltage (187 V) and the difference in current (0.035 A) are used according to Equation 15-4, the ohmic value of the load resistance will be found. (The accuracy of the results are determined by how carefully the current and voltage values are established from the graph.)

When no load line is given on a set of tube characteristics, one can be drawn in arbitrarily and calculations can be made from it. A horizontal load line will give the least distortion, and also the least output. If the load line is sloped toward the vertical position, the output and distortion will increase. As the load line is slanted toward a vertical position, the line enters the nonlinear portion of the static characteristic curves that corresponds to the region of high distortion. (The subject of linearity versus distortion is discussed more fully in the next chapter.) Thus, a higher value of load resistance will, within certain limits, decrease distortion below that which would prevail for a lower value of load resistance. With a lower value of load resistance, however, the frequency response limits of the amplifier are extended since there is less shunting effect of the interelectrode capacity of the tube.

As the load resistance approaches the interelectrode capacity values, the latter will have a greater shunting effect. Since the higher-frequency components of the signals lower interelectrode capacitive *reactance*, the upper-frequency response is decreased with a large value of load resistance. For this reason, a low value of load resistance is used in video amplifiers, where it is necessary to pass frequencies up to 4 MHz. For audio work, however, where frequencies above 20 kHz are rarely encountered, the larger value of load resistance will give good frequency response with greater output.

The larger the load resistance, the more horizontal the load line becomes. Thus, several load lines can be drawn in a static set of curves and calculations will then indicate which one is preferable from the standpoint of signal-amplitude output with a minimum of distortion. In most instances, a compromise will have to be made between the distortion and signal output levels.

If a representative load line is desired without drawing several arbitrarily, two points can be established on the static set of characteristic curves that will permit the load line to be drawn. One such place is the *operating point*, which can be calculated by use of the following equation:

$$\text{Zero signal bias} = \frac{-(0.68 \times E_b)}{\mu} \tag{15-5}$$

In this equation, E_b is the chosen value of d-c plate voltage at which the tube is to be operated. This is derived from the fact that cutoff is normally equal to E_b/μ and bias is proportional to Equation 15-5, which establishes the bias so that operation is on the linear part of the characteristic curve. The constant 0.68 was established by RCA engineers in a number of tests conducted on many tubes. The formula is an approximation and gives an average bias for the general tubes encountered. Average plate current, however, should always be chosen so that it does not exceed the plate dissipation of the tube.

One point for drawing the load line has now been established, the zero-signal bias. If the desired value of load resistance is known beforehand, the following formula can be used to establish the place where the load line intersects the zero plate-current axis:

$$E_b + I_b R_L \tag{15-6}$$

In this formula, the operating plate voltage is given as E_b, and I_b is the operating plate current.

As the limit of the signal swing is at zero, the load line ends at zero bias, establishing the third point. This gives us $E_{\min}$ and $E_{\max}$ automatically, for if the bias changes from zero to 40 V, then the extent of signal swing would also indicate the minimum and maximum voltage and current points.

The tube-type amplifier circuits described in this chapter are of the signal-voltage type. In a signal-voltage amplifier, the primary purpose is to raise the signal voltage (not power); hence the load resistor is made fairly large in audio applications or parallel-resonant circuits with a high impedance are used in RF amplifiers. While power output is low, it can be solved for if necessary, once the load line has been drawn. The formula used is given here for reference, though it is more applicable to the power-amplifier circuits discussed in the next chapter.

$$\text{Power output} = \frac{(E_{max} - E_{min})(I_{max} - I_{min})}{8} \qquad (15\text{-}7)$$

The power-output formula is based on the fact that the alternating components of voltage and current have peak amplitudes that are one-half of the total swings of plate potential. The power delivered to the load resistor is equal to one-half the product of the peak a-c signal voltage and the peak signal current as in Equation 15-7. In d-c circuits, the power in watts is equal to the product of voltage times current, and in resistive a-c signal circuits, such as amplifiers, the same calculations apply. Therefore, if the extent of the signal swings are multiplied and the result again multiplied by a factor that takes into consideration the effective values (as required in a-c calculations), the result will be the average power delivered to the load resistor, as derived from the Equation 15-7 for power output.

The plate efficiency of the tube can also be calculated on the basis of the load line, using the following equation:

$$\text{Plate efficiency} = \frac{(E_{max} - E_{min})(I_{max} - I_{min})}{8E_p I_b} \times 100(\%) \qquad (15\text{-}8)$$

The second harmonic distortion can be solved by either of the following equations:

$$\frac{(I_{max} + I_{min})/2 - I_b}{I_{max} - I_{min}} \times 100 \qquad (15\text{-}9)$$

$$\frac{(I_{max} + I_{min}) - 2I_b}{2(I_{max} - I_{min})} \times 100 \qquad (15\text{-}10)$$

In drawing load lines, the operating bias may be furnished, in which case one point is already established for the load line. If the value of the load resistance is also known, the second point can be established by the equation previously given: $E_b + I_b R_L$.

Load-line factors for transistors have many of the aspects of tube load lines, though the characteristic curves as shown in Fig. 15-31 closely resemble the pentode curves instead of triode curves. (See Fig. 11-16.) As with tubes, if such curves are not available, they can be plotted by reading collector current for an applied collector voltage, for a fixed-base current. After various collector values have been applied, the base current is changed and a new set of values obtained.

Note the dashed-line curve in Fig. 15-31. This represents the constant power dissipation line (in watts) for the collector side as specified for a particular transistor. Thus, for the transistor graphed in Fig. 15-31, the 48 mW

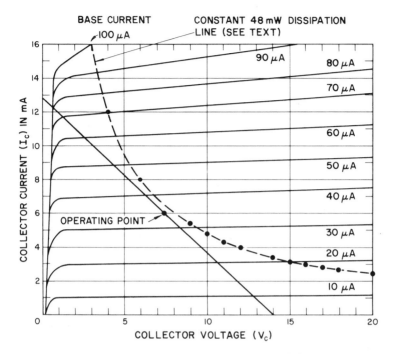

FIGURE 15-31. Transistor load line and collector dissipation line.

is the maximum energy dissipation before overload occurs. This dissipation curve is plotted along points established by the product I_cV_c. Thus, at any point where the dashed curve intersects with abscissas and ordinates, the current-voltage product equals 48 mW for this transistor.

Thus, to stay within the dissipation rating of this transistor, the load line must be drawn in the graph area below and to the left of the dissipation line. Maximum power gain of the transistor is obtained when the load line is drawn tangent to the dissipation line. The load impedance (or resistance) may again be calculated by Equation 15-4, which is actually a computation of the *slope* of the load line:

$$R_L = \frac{dV_c}{dI_c} \qquad (15\text{-}11)$$

The operating point has been selected so a linear signal swing for Class A operation is available above and below the operating point. The point is at 7.5 V for V_c, 6 mA for I_c, and 35 μA for I_b as shown in Fig. 15-31. At the operating point, $I_cV_c = 0.006 \times 7.5 = 45$ mW, which is below the maximum permitted according to the power dissipation line. Similarly, no other portion of the load line touches or extends to the right or above the dissipation line.

For the load resistance, using Equation 15-11, $dV = 0$ to 14 and $dI = 0$

to 12.8. Thus, we obtain

$$\frac{14}{12.8} \times 10^{-3} = 1093.7 \ \Omega$$

For Class A operation, the signal swing along the load line would not encroach on the base-current curvatures at the left. The operating point as well as the load-line slope can be altered to provide for a different value of load resistance, greater signal swing, etc. In Fig. 15-32, for instance, two load lines are shown for comparison purposes. Though both use the same operating point, each could have a different operating point if desired.

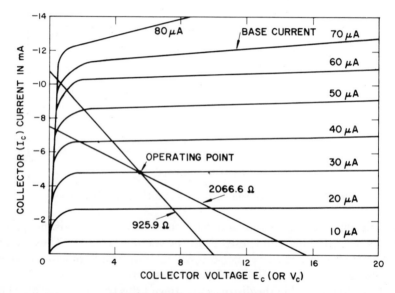

FIGURE 15-32. Load-line comparisons.

The steeper load line has a $dV = 0$ to 10 and a $dI = 0$ to 10.8:

$$R_L = \frac{10}{10.8} \times 10^{-3} = 925.9 \ \Omega$$

For the other load line, however, a greater ohmic value is obtained:

$$R_L = \frac{15.5}{7.5} \times 10^{-3} = 2066.6 \ \Omega$$

Power output can again be found by using Equation 15-7 as for tube amplifiers. As shown earlier in Fig. 13-14, characteristic curves for the FET also resemble those of the pentode tube, and load-line factors as discussed for the transistor again apply. A drain-dissipation line replaces the collector-dissipation line, and again this indicates the maximum wattage permitted for a specific transistor type.

Review Questions

15-1. Explain the essential differences between a *voltage amplifier* and a *power amplifier*.

15-2. List some of the signal types handled by the Class A amplifier.

15-3. Briefly explain the operational principles of tone and level controls.

15-4. Compare the current-flow conditions that prevail when no forward bias is applied to a transistor and no grid bias is applied to a vacuum tube.

15-5. Why does degeneration occur when a bypass capacitor is omitted across an emitter or cathode resistor?

15-6. Draw two schematics, one showing forward and reverse bias to an *NPN* transistor and the other showing grid bias applied to a tube from the power supply.

15-7. (a) What is the advantage of *RC* coupling over transformer coupling?
(b) What is the advantage of direct coupling over *RC* or transformer coupling?

15-8. (a) Give at least two useful purposes for the emitter-follower circuit.
(b) To what tube-type circuit does the emitter follower compare?

15-9. Briefly explain why decoupling circuits are extensively used in all types of electronic circuitry.

15-10. Briefly explain what factors contribute to the individual characteristics of musical tones that distinguish the musical instrument even though the same signal frequency is being produced.

15-11. Why is it necessary for an audio amplifier to have a frequency response beyond the frequency range of the fundamental tones to be amplified?

15-12. Explain briefly what causes harmonic distortion in an amplifier.

15-13. What are the basic differences between frequency distortion and phase distortion?

15-14. Compare the approximate input and output impedances of transistors and tubes.

15-15. The voltage gain of a tube can be expressed as $g_m R_L$. To what solid-state devices does this formula also apply? Explain why.

15-16. Briefly explain what method is used to prevent an RF amplifier from oscillating.

15-17. Why is the common-base RF amplifier essentially a *unilateral circuit*?

15-18. Explain one method for minimizing the tendency of thermal runaway in a transistor.

15-19. Illustrate schematically a shunt-fed and a series-fed circuit.

15-20. Define the term *load line*, and discuss its usefulness.

15-21. Of what significance is the *constant power dissipation line* in a set of transistor curves?

15-22. What is the effect of a large-value load resistance on frequency response of an amplifier?

15-23. If a load line inclines toward the horizontal, would the load resistance be higher or lower than with a load line that inclines toward the vertical?

15-24. Briefly explain the derivation for Equation 15-7, which solves for power output for either a transistor amplifier or a tube amplifier.

15-25. What is the relationship of a transistor-type amplifier's load line to the dissipation line in terms of maximum power gain?

15-26. For Class A amplification, what determines the location of the operating point on the load line?

15-27. (a) Is a comparable collector-dissipation line also found in FET characteristic curves?

(b) What type vacuum tube has curves similar to that of the field-effect transistor?

Practical Problems

15-1. A pentode audio amplifier has a 15-kΩ load resistor and the g_m is rated at 4000 μmhos. What is the voltage gain?

15-2. A triode in a signal-voltage amplifier has a mu of 100 and an R_L of 80 kΩ. If R_L is 160 kΩ, what is the signal-voltage gain?

15-3. If the load resistance for the circuit of Problem 15-2 were doubled, what would be the new signal-voltage gain for that circuit?

15-4. In a laboratory test of a pentode amplifier, it was found the tube has a g_m of 9800 μmhos and a load resistor of 100 kΩ. What is the signal-voltage gain?

15-5. The E_p-I_p curves for a triode tube indicate the following values for a given signal swing:

$$E_{min} = 200 \text{ V}, \qquad E_{min} = 25 \text{ V}$$
$$I_{max} = 40 \text{ mA}, \qquad I_{min} = 5 \text{ mA}$$

What is the value of the load resistance?

15-6. What is the approximate zero-signal bias for a triode operated at 200 V and having a mu of 50?

15-7. Where does the load line for a triode amplifier intersect the zero plate-current axis if the operating current is 20 mA, the operating voltage is 160, and the load resistance value is 5340 Ω?

15-8. The characteristic curves for an amplifier indicated the following values for a given signal swing:

$$E_{max} = 240 \text{ V}, \qquad E_{min} = 40 \text{ V}$$
$$I_{max} = 40 \text{ mA}, \qquad I_{min} = 6 \text{ mA}$$

What is the power output for this amplifier?

15-9. A transistor common-emitter load line extends from the 8-mA point on the vertical axis to the 16-V point on the horizontal axis. What is the ohmic value of R_L?

15-10. In a common-emitter amplifier a sine-wave signal input causes a collector voltage change of 5 to 30 V and a collector current change of 0.5 A to 50 mA. What is the signal-power output?

15-11. In a FET amplifier, the operating point is at $8V$ on the load line and the no-signal collector current is 10 mA. The drain-dissipation line is given as 90 mW. By how many watts is the operating point below the drain-dissipation line?

Power Amplifiers 16

16-1. Introduction

Power amplifiers are designed to furnish *signal energy* to a load rather than *signal voltage* as with the small-signal amplifiers discussed in the preceding chapter. The load resistance (R_L) fed by a power amplifier may convert the signal energy to a physical movement as is the case with a loudspeaker or a recording stylus that plots curves electronically. Such devices are known as *transducers* because they convert one form of energy (electrical) to another form (mechanical). On the other hand, the load may utilize the signal energy without reconverting it, as is the case when a power amplifier feeds another stage that requires power input to the next circuit. The power output may also be applied to an antenna system for transmission or to tape recording heads, industrial control devices, and other similar units. Thus, power amplifiers are called on to handle signal energy from a few milliwatts to many hundred kilowatts. The signal energy may be of the RF type or the audio type.

Small-signal amplifiers, as discussed in the previous chapter, are usually Class A, although large-signal power amplifiers are employed in Class A, AB_1, AB_2, B, and C applications. Of the foregoing, Class C amplifiers can only be used for RF power amplification, though for audio amplification all the other types from A to B can be employed.

In comparison with small-signal amplifiers, where signal currents are

356

low, power amplifiers have relatively higher signal currents. In small-signal amplifiers, the load resistor is usually of a high value to obtain a large signal-voltage drop, while in power amplifiers the load impedance and the circuit arrangement are such as to permit a high signal current.

A power amplifier will transfer a maximum amount of signal energy to the load (a loudspeaker or another amplifier stage) when the output imped-ance of the power amplifier matches the load impedance. If the load resistance or impedance is lower than the circuit impedance that generates the signal, less power output will be developed. Less power is also developed for a mis-match where the load impedance is higher than the source impedance.

In small-signal amplifiers, an impedance match is usually not important and, in many instances, the load resistance is two or three times higher than the output resistance. The reason for this lies in the fact that a maximum trans-fer of power is not desired or needed because signal-*voltage* amplification is the primary goal, as mentioned earlier.

In power amplification, *impedance matching* is more closely adhered to, in order to provide for a maximum signal-power transfer. It is only when distortion is to be minimized that a slight mismatch is tolerated. Because impedance matching is often misunderstood and because it is an important aspect of the general subject of power amplifiers, a more thorough discussion of impedance matching follows.

16-2. Impedance Matching

As mentioned in the preceding chapter, a transistor or a vacuum tube can be considered as a generator since their output circuits generate an amplified version of the signal applied to the input. Thus, if an analysis is made of how a load resistor is to be applied to the output of a transistor or vacuum tube, we must consider the transistor or vacuum tube as a generator having an internal resistance, as discussed earlier.

To simplify the discussion, low values of resistance will be employed to illustrate the application of Ohm's law. Figure 16-1 thus shows a generator (G) with an internal resistance of 8 Ω, which is represented as a series resistor (R_G), as shown at A of Fig. 16-1. The generator has an output of 48 V and since the load resistor (R_L) of 8 Ω is in series with R_G, total resistance is 16 Ω. Thus, 3 A of current flows through the network. The power consumed by the load resistance is 72 W. This is clearly evident from simple Ohm's law calculations, as follows:

$$\text{Current } (I) \text{ through network} = \frac{E}{R} = \frac{E}{R_G + R_L} = \frac{48}{16} = 3 \text{ A}$$

$$\text{Voltage } (E) \text{ across } R_L = I \times R_L = 3 \times 8 = 24 \text{ V}$$

$$\text{Power dissipated in } R_L = E_{R_L} \times I = 24 \times 3 = 72 \text{ W}$$

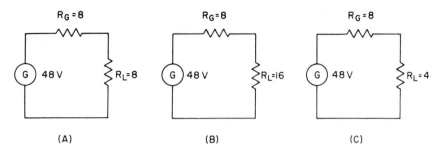

FIGURE 16-1. Impedance matching.

As long as the generator voltage remains at 48 V and its internal resistance is 8 Ω, no other value of load resistance will be furnished as much power as one having an 8-Ω value. When the load resistance has a value either higher or lower, less power will be developed in it. At B of Fig. 16-1 the load resistance has been changed to 16 Ω to illustrate how the power in the load decreases when its resistance is other than 8 Ω. With the 16-Ω load resistance, the power in the load drops to 64 W, as indicated by the following calculation:

$$I = \frac{E}{R} = \frac{E}{R_G + R_L} = \frac{48}{24} = 2 \text{ A}$$

$$E \text{ across } R_L = I \times R_L = 2 \times 16 = 32 \text{ V}$$

$$\text{Power in } R_L = E_{R_L} \times I = 32 \times 2 = 64 \text{ W}$$

It will be noticed that, in this instance, the voltage across the load resistor has increased and, if an output *voltage* were desired instead of an output power, the increase in the value of the load resistance would be preferable. Since, however, we are concerned with the power available across the load resistor, the power delivered from the generator to the load is of primary importance.

The opposite condition from B is shown at C, where the load resistance has now been reduced to 4 Ω. By Ohm's law calculation, the power dissipation in the load resistor is again below the 72 W obtained when the internal resistance of the generator matches the load resistance. This is indicated by the following calculation:

$$I = \frac{E}{R} = \frac{E}{R_G + R_L} = \frac{48}{12} = 4 \text{ A}$$

$$E \text{ across } R_L = I \times R_L = 4 \times 4 = 16 \text{ V}$$

$$\text{Power in } R_L = E_{R_L} \times I = 16 \times 4 = 64 \text{ W}$$

It will be noted that the current through the load resistor at C of Fig. 16-1 is larger than when the load resistor matches the internal resistor of the generator, as shown at A. The lower value of load resistor at C, however, results in a lower voltage drop and a corresponding decrease in power output. As the value of the load resistor is decreased or increased below or above that which constitutes a match with the internal resistance of the generator, the resultant power in the load resistor will always be less than the maximum obtained when the impedances are matched. (For simplicity, these examples assume that the generator voltage remains constant with varying current drains. This is rarely the case, however, unless voltage regulators are employed. In any event, impedance matching procures a maximum power transfer.)

Previously, it was mentioned that the load line indicates the dynamic operating characteristics of the tube or transistor. The load line itself does not necessarily have the same ohmic value as the output impedance of the transistor or tube because the slope of the load line is chosen to obtain the best results, in terms of minimum distortion and a good range of frequency response. Thus, the load-resistor value is chosen by performance requirements and may be somewhat higher or lower than the amplifier's output resistance. When the load resistance is not the same as the amplifier's output resistance, the maximum power output of the amplifier is not realized, though the output may be *near* the maximum. The slight decrease in power output is sacrificed in the interests of better performance.

In power-output amplifiers, an actual resistor is not used for the load because a high resistive value would decrease current flow too much and result in considerable power reduction. Instead, the actual load is the input impedance to another stage, or the *voice coil* of a loudspeaker. To minimize excessive inductive reactance variations in the voice coil for different frequencies, voice-coil impedances are kept quite low, ranging from a few to about 50 Ω, with 4 Ω and 8 Ω common. Because the voice coil must transform audio signal power into audible sounds, the voice coil is the actual load resistance and, hence, must be matched to the calculated load resistance to prevent power loss, as more fully explained in the following discussion.

16-3. Audio-Power Amplifiers

Figure 16-2 shows a typical audio output-power amplifier, such as employed in low-power phonograph amplifiers or in the audio output systems of some radio or television receivers. Capacitor C_1 couples the signal from a prior small-signal amplifier output to the base input of the power amplifier. The collector-emitter sections of the transistor circuit have now brought the signal level to that needed for application to a loudspeaker. Some coupling device is needed, however, to achieve impedance matching and power transfer since the few turns of the speaker's voice coil provides an impedance of only several ohms.

Since a maximum transfer of power is desired, a transformer is utilized, as shown, to match the calculated load resistance to the voice coil. The

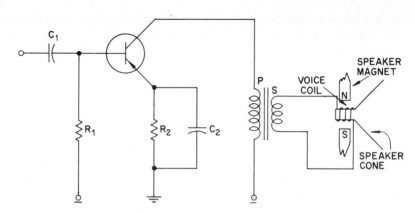

FIGURE 16-2. Transformer coupling to voice coil.

primary of the transformer, even though it occupies the same position as that of the load resistor in voltage amplifiers, does not constitute the load resistance for the audio output-power amplifier. Actually, the load resistance is the loudspeaker *voice coil*, and the primary and secondary of the transformer only serve as an impedance adjusting device so that the relatively low ohmic value of the voice-coil impedance can be made to appear equal to the higher value of the load resistance chosen for the particular power-amplifier transistor or tube used.

Because a transformer can be employed for stepping up or stepping down voltages, as explained in Chapter 8, an impedance transfer can also be obtained. If the number of turns in the secondary is less than the number of primary turns, a high impedance in the primary can be stepped down to appear as a low impedance to the secondary. Conversely, a low impedance in the secondary will appear as a high impedance in the primary circuit, by virtue of the step-up function obtained with the transformer.

The step-up or step-down ratio is not directly proportional to the number of turns in the primary and the secondary. That is to say, if there are 400 turns in the primary and 100 turns in the secondary, the impedance ratio will not be four to one. Since the primary and secondary windings are wound one on top of the other, there is a cross-transfer of impedance consisting of capacity, resistance, and inductance. In consequence, the turns ratio of the transformer (for matching two dissimilar impedances) must be calculated on the basis of the following formula, which was also given in Chapter 8:

$$\text{Turns ratio} = \sqrt{\frac{Z_1}{Z_2}} \qquad (16\text{-}1)$$

Thus, if the larger impedance is designated as Z_1 and if this quantity is divided by the smaller impedance Z_2, the square root of the ratio would indicate the number of turns necessary to obtain an impedance match in the transformer.

Suppose, for instance, that a vacuum tube is employed that requires a load resistance of 10,000 Ω for best performance. The amplifier is to drive a loudspeaker having a voice-coil impedance of 5 Ω. The calculation for finding the turns ratio would then be

$$\sqrt{\frac{10,000}{5}} = \sqrt{2000} = 45$$

The foregoing calculation indicates that the transformer must have a turns ratio of 45 to 1 in order to match the relatively high load resistance, as determined by the load line, to the low voice-coil impedance.

The actual number of turns in the primary and secondary is not indicated by the formula but only the turns ratio. Thus, if there are 225 turns in the primary, there would have to be 5 turns of wire in the secondary. The relative impedances will be maintained as a match, as long as the proper turns ratio prevails.

The transformer is assumed to have virtually no internal resistance. If the resistance of the primary winding is appreciable, some power will be consumed in the primary. This is undesirable since such power consumption in the transformer is wasted. As much as possible of the power generated within the transistor or tube should develop across the voice-coil impedance so that the loudspeaker cone will be actuated at a higher efficiency. To obtain a minimum of d-c resistance, the primary winding can be made of fairly large wire. Actually, the wire must have a diameter sufficiently large to carry the collector or plate current without undue overheating. An overheating of the transformer indicates that the internal resistance is consuming considerable power and such power, whether audio signals or dc from the power supply, is a waste of energy. When the primary has such a low resistance that it is a negligible factor, the *signal power* circulating in the primary is transferred into the secondary with the greatest efficiency. A well-designed transformer has an efficiency of well over 95%.

The primary of the output transformer *is not* the actual load resistance; neither is the secondary. This point needs emphasizing since many students often mistakenly assume that the primary of the transformer is the load resistance because current from the power supply flows through it to the transistor or tube, in similar fashion to the load resistor in a voltage amplifier. It must be understood, however, that neither a pure inductance nor a pure capacitor consumes electric energy; hence neither the primary nor the secondary, with its negligible resistance values, will consume any appreciable audio power. Rather, the audio-signal power is applied to the voice coil and the resultant signal current sets up varying magnetic fields. The latter aid or oppose the fields of the speaker magnet, causing the loudspeaker diaphragm cone to vibrate and produce audible sounds. Thus, the electric energy of the power amplifier is converted into acoustical energy.

If the transformer has an appreciable distributed capacity, some of the high-frequency audio-signal components will encounter a shunting effect and hence will be attenuated. For this reason, the transformer design should be such that a minimum of distributive capacities are present. For the same reason, the wire size should be held at a minimum because a larger wire size contributes to capacity effects between adjacent turns, as well as between adjacent layers of the transformer.

Another factor in transformer efficiency is the quality and quantity of the laminated core. A larger core increases permeability and hence the number of turns in the primary and secondary can be reduced while still holding the required amount of inductance constant. The reduced number of turns means less distributive capacity, as well as a decrease in the internal resistance of the transformer. Consequently, less signal energy is shunted by distributive capacities and less signal energy is wasted because of the series d-c resistance of the primary winding. The d-c resistance will consume audio-signal-energy power (which is ac), as well as d-c power from the power supply.

Another factor to be considered is the variation of inductive reactance with changes of signal frequency as mentioned earlier. A low audio-signal frequency will decrease the reactance of the primary and hence less signal voltage develops across it than is the case for higher signal frequencies. This factor tends to attenuate the lower-frequency audio-signal components and upsets uniform response. Good transformer design, however, helps keep such variations to a minimum, as well as reducing the higher audio-signal losses by virtue of their being shunted due to distributed capacitances within the transformer. While transformers are not ideal except from the overall efficiency standpoint, they serve useful purposes in electronic applications and hence are widely used.

In transistor circuits, both bypass capacitors and coupling capacitors are chosen to have much larger values than such capacitors have in vacuum-tube circuits. With the relatively low input impedances of transistors and the low battery resistances, the capacitive reactances of bypass and coupling capacitors must be much lower than comparable units in vacuum-tube circuits. Hence, it is common to find bypass and coupling capacitor values from several to well over 10 μF, in some instances. With the low voltages encountered in transistor radios or other low-power equipment, the size of the capacitors is kept down by using low-voltage units. As a comparison, coupling capacitors in vacuum-tube amplifier circuits may range from 0.01 to 0.25 μF.

16-4. Push-Pull

Two power-amplifier transistors or tubes may be utilized in a symmetrical circuit arrangement known as *push-pull* to produce a greater signal-power output. One method for doing this is to employ an interstage transformer between the small-signal amplifier output and the two push-pull transistors.

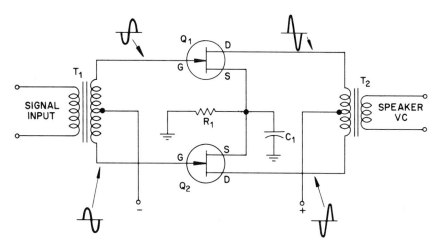

FIGURE 16-3. FET push-pull amplifier.

A typical circuit is shown in Fig. 16-3, where the FET units are used, though push-pull transformer circuitry is also applicable to tubes and transistors, as shown later.

Push-pull operation requires a 180° phase difference between the signals applied to the input element of the transistors or tubes used, as shown for the gate inputs of Fig. 16-3. There are several methods for obtaining such out-of-phase signals, and one of them is the transformer T_1 arrangement as shown. The secondary of the transformer has a center tap that effectively divides the signal energy into two signals, each of which is out of phase with the other. (A resistive circuitry of phase inversion can also be used, as discussed later.)

For the circuitry of Fig. 16-3, the source elements are connected to a common resistor R_1 and to ground. Capacitor C_1 prevents signal energy from developing across R_1. An output transformer T_2 with a center tap is also used and a positive voltage applied to the center tap to satisfy the drain-voltage requirements of the N-channel FET.

When a push-pull system is operated as Class A, idling current flows through each transistor (or tube) with or without signal input, as is the case with the single-ended power-amplifier circuitry discussed earlier. The idling currents through the output-transformer primary will be a steady-state condition in the absence of an input signal. (Because there is no current *change* through the transformer primary, no voltage is induced into the secondary of the output transformer.)

The amplified output signal from each transistor is also out of phase with its input signal; hence the signal developed at the drain element of Q_1 is out of phase with the signal at the drain of Q_2. Thus, while current through one-half the primary winding of T_2 is increasing, the current through the other half of the primary is decreasing. Consequently, as one drain side of the transformer becomes more positive, the other drain side becomes more negative. This results in the formation of a high-voltage signal, consisting also of the combined current changes. The total changing field in the primary now induces a voltage across the secondary of the output transformer, and the resultant current flow represents signal power from both transistors. Thus, the combined negative-positive signal *alternations* in the primary produce a high-energy single alternation in the secondary. Similarly, the next out-of-phase signal alternations at the primary induce a single high-energy alternation at the secondary.

The opposing current changes in the push-pull transformer tend to reduce transformer core saturation and, at the same time, the changing fields created by the current changes induce in the secondary the sum of the audio power developed by each transistor. In many cases a marked reduction of harmonic distortion results, particularly of the even-harmonic order. Thus, it is possible that the audio-power output from a push-pull amplifier may be *more than twice* the power developed by a single transistor or tube because of the reduction of harmonic distortion. This is particularly true of triode tube-type modulators (audio amplifiers) or the RF push-pull tube-type amplifiers discussed later, wherein the triode characteristics tend to generate even-harmonic-type distortion. Pentodes produce a high degree of *odd-harmonic* distortion and hence their usage in push-pull does not result in a marked reduction of distortion. If, of course, the push-pull circuitry is more carefully designed than the single-tube (or transistor) circuitry and better components are utilized (or operation is confined to a more linear portion of the load line), harmonic distortion of all types will be minimized.

If the input signal to a transistor or tube is a sine wave, where each alternation has the same amplitude as the other (though opposite in polarity), the amplified signal should also have identical alternation amplitudes (though higher than the input-signal amplitude). Thus, if one alternation of the output signal is greater than the other, even-harmonic distortion results. With the balanced circuit arrangement of push-pull, however, the distortion is canceled out.

This harmonic reduction principle can be more clearly understood by reference to Fig. 16-4, where the diagram at A shows the distortion produced by one transistor or tube and the graph at B indicates the signals obtained from the other push-pull transistor or tube. Thus, for the particular characteristics shown, each transistor or tube has an idling current of 20 mA under no-signal conditions. (For convenience, reference will be made to tubes in the following discussions, though the principles also apply to solid-state devices exhibiting nonlinear characteristics.)

Note that the two tubes have the same nonlinear characteristics because, for a given signal, the current will rise more in the positive direction than it will decrease in the negative direction. Thus, V_1 shown at A may have a plate-

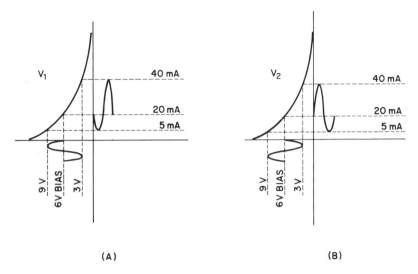

FIGURE 16-4. Distortion in individual tubes.

current decrease of only 15 mA (from 20 to 5 mA when a negative signal voltage appears at the grid). For a positive-going signal alternation of the same amplitude on the grid, however, the plate current rises by 20 mA (from 20 mA to 40 mA). These unequal plate-current changes would introduce severe second harmonic distortion because of the uneven amplitude of the output sine-wave alternations produced by this tube.

When the *input* signal has a constant amplitude for each alternation, the output alternations of the amplified signal energy should also have identical alternation amplitudes. In push-pull, however, the current for the two tubes would combine to form the total current. Thus, if the characteristics of the second tube, V_2 shown at B of 16-4, are similar to those of V_1, shown at A, the *overall* effect would be diminished by the push-pull circuit arrangement. Therefore, when current flow in one tube *decreases* by 15 mA, it *increases* by 20 mA in the other tube, giving a total current *change* of 35 mA. At the second alternation of the input signal, the first tube would increase in current by 20 mA, while the second tube would decrease by 15 mA since the decrease in each tube is less than the current increase. Thus, the result would again be a total change of 35 mA and both alternations of the output-signal waveform would be equal. Under this condition, the second and even-harmonic components would have been eliminated or diminished. Push-pull does not, however, eliminate all harmonic distortion. Odd harmonic distortion is relatively unaffected, though all even harmonic components are diminished in amplitude. Such reduction of distortion, however, is proportional to the balance in circuit and tube design. If the transformer is not perfectly center-tapped or if one tube draws more current than the other (or has different emission characteristics than the other), the unbalance that occurs would affect the harmonic reduction characteristics of push-pull design.

For resistor R_1 of Fig. 16-3 (or for a common cathode resistor in tube-type push-pull), signal *variations* are at a minimum whether or not the bypass capacitor C_1 is used. The reduction of signal variations across R_1 is due to the symmetrical aspects of push-pull, where changes of signal current through each push-pull transistor or tube will be equal but opposite in polarity. In Class A operation, the idling current to each transistor or tube depends on the type transistor or tube employed. Assume, however, that 20 mA flows through each transistor with no signal input. Thus, 40 mA of current flows through R_1. For any given input-signal polarity, current through one transistor increases and current through the other decreases. With a symmetrical circuit, the current decrease in one will be proportional to the current increase through the other; hence the current through R_1 remains unchanged. Capacitor C_1 is only necessary when an unbalanced circuit causes voltage variations across R_1.

For the FET circuit of Fig. 16-3, *P*-channel transistors could be used with an appropriate change of power-supply potentials. Similarly, *PNP* or *NPN* transistors could be used in push-pull. A typical circuit using *PNP* transistors is shown at A of Fig. 16-5. Here two transformers are again used in conventional grounded-emitter circuitry.

Transformer T_1 is for phase-inversion purposes and applies a signal to the base input of the upper push-pull transistor, as well as to the base input of the lower transistor. The signals applied to the base inputs are out of phase with each other. A d-c voltage of negative polarity is applied to both base circuits, via the tap on the secondary of T_1. Resistor R_1 provides the necessary voltage drop for the base-emitter circuits, and capacitor C_1 acts as a signal-voltage bypass for R_1. Each emitter of the push-pull stage transistors is placed at ground potential (positive), in accordance with forward-bias requirements. The collectors of the push-pull transistors are at negative potential (reverse bias), which is applied via the center tap of the primary of T_2. As with vacuum tubes, the total power output depends on the type of transistors used, their relative efficiency, and the amplitude of the input signals applied.

Another version of a transistorized push-pull output-power amplifier is shown at B of Fig. 16-5. Here, no *output* transformer is required as with the version shown at A. For the upper *push-pull* transistor T_2, the necessary bias for the emitter (negative for the *NPN* type) is obtained from battery (or other power source) B_2 and applied through the speaker voice coil. The positive potential for the T_2 collector is applied by the ground connection from the bottom of battery B_2, thus completing the reverse-bias polarities for the output side. The forward bias for the emitter of T_2 is obtained from

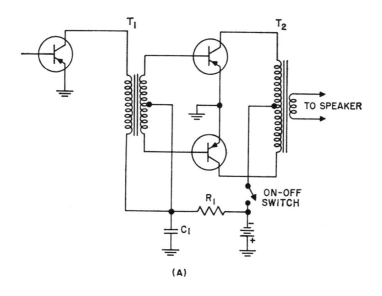

(A)

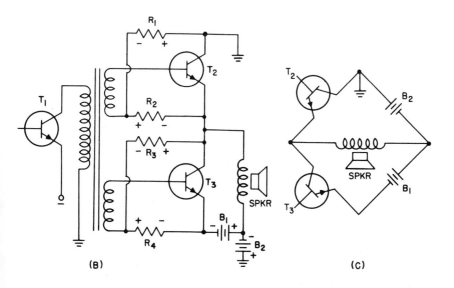

(B)   (C)

FIGURE 16-5. (A) Transistor push-pull power amplifier. (B) and (C) Transistorized output and equivalent circuit.

the voltage-divider resistors R_1 and R_2. A careful study of the circuit will indicate that resistors R_1 and R_2 are in shunt with the negative and positive battery potentials of B_2, via the ground path of T_2. Hence, the polarity of the voltage drops across these two resistors is as shown at B, establishing the base of T_2 as positive with respect to the emitter in accordance with the forward-bias requirements of the *NPN* transistor.

The forward bias for transistor T_3 is also obtained by voltage division, using resistors R_3 and R_4, which shunt battery B_1. Hence, the voltage drop across R_4 applies a positive potential to the base of T_3 and a minus potential to the emitter (with respect to the base). The negative-bias potential for the emitter of T_3 is procured from battery B_1 and the necessary positive potential for the reverse bias of the collector of T_3 is applied through the speaker inductor.

The interstage coupling transformer has a split secondary winding, as shown, to provide out-of-phase signal voltages. Thus, the base of each transistor is fed with a signal that is out of phase with the signal at the other transistor base, in conventional push-pull design, just as was the case with the tapped secondary winding previously discussed.

Note that the collector-emitter sections of T_2 and T_3 are actually in series across the split battery potential source. The circuit design, however, is such that the two transistors are in parallel with the load (the speaker coil). This becomes more apparent by an inspection of C of Fig. 16-5, where the equivalent bridge circuit is shown. If the two push-pull transistors are well matched, each will have a voltage drop across it equal to that of the other. Also, if both power sources (or batteries) have the same voltage and internal R, the network bridge will be balanced and *no* direct *current* will flow through the speaker inductor. When, however, an audio signal appears across the interstage transformer, one transistor base receives a positive signal alternation and the other transistor base receives a negative signal alternation. Hence, one transistor will conduct more than the other, causing an unequal current flow and an unbalance of the bridge network. This unbalance will cause audio-signal voltages to appear across the speaker inductance (voice coil) and audio-signal currents to flow through it, producing the sound output.

With this transistorized push-pull system the speaker load impedance required to match the transistor push-pull circuit is only a fractional value of what is normally required in conventional vacuum-tube push-pull stages. Since the transistor circuitry has a low impedance, a good match ·can be obtained with low-ohm voice coils. If such a circuit were attempted with conventional push-pull design, a center-tapped voice coil would be required. Because only one-half of such a voice coil would receive signal energy at any time interval, the efficiency would be lowered and design and production problems would be increased.

16-5.   Phase Inversion

While a transformer functions satisfactorily as a device for obtaining the necessary phase difference between the input signals of the push-pull transistors
or tubes, transformers are costly when properly designed to minimize signal
losses. Transformers also contribute bulk to the amplifier system, besides
producing stray magnetic fields. The latter require shielding precautions to
minimize spurious signal pickup by nearby circuit components. For these
reasons circuit design is often used that dispenses with the interstage push-
pull transformer in favor of some other means for obtaining the necessary
phase reversal of the signal.

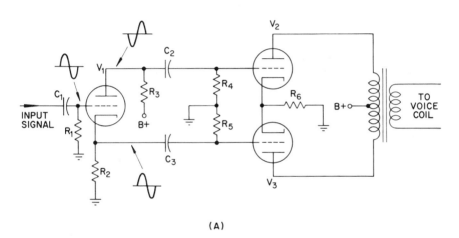

(A)

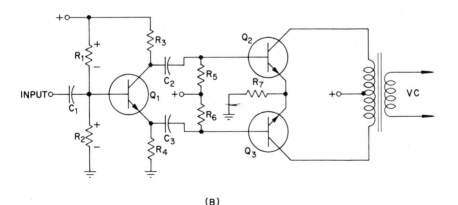

(B)

FIGURE 16-6. Split-load phase inversion.

One method is to use the *split-load phase inversion* system shown in Fig. 16-6. For the tube type shown at A, the output signal is essentially divided between two load resistors, R_2 and R_3. Thus, if the grid input signal is of the polarity shown in Fig. 16-6, the first alternation (positive) will cause an increase in the current flow through the cathode resistor since a positive alternation applied at the grid has the effect of decreasing bias. A bias *decrease* causes an *increase* in plate current and, hence, there will be a proportionate voltage *increase* across the cathode resistor. Such a voltage change across the cathode resistor represents an *in-phase* signal condition with respect to the input signal, as opposed to the *out-of-phase* signal developed across the load resistor, R_3. For the circuit of Fig. 16-6, cathode resistor R_2 and load resistor R_3 are made equal in value so that each push-pull tube will have grid signals of equal, though opposite polarity, values. Coupling capacitors C_2 and C_3, which transfer the signal voltages to the respective grids of the push-pull amplifier tubes, are also chosen to have equal values. Also, grid resistors R_4 and R_5 have equal values.

The unbypassed cathode resistor R_2 causes considerable degeneration and, consequently, the gain of the stage associated with the phase-inverter vacuum tube V_1 is adversely affected. In most instances, the gain for the phase-inverter tube V_1 is less than unity, which means that the tube has the prime function of phase inversion and not of amplification. A bypass capacitor cannot be employed across cathode resistor R_2 because it would nullify the signal-voltage changes that occur across the cathode resistor and, hence, no signal would be obtained from this resistor for application to the grid of V_3.

Thus the *cathode-follower* principle is combined with the anode load-resistor output method and since the cathode signal *follows* in phase that of the grid, we obtain the contrasting signal with respect to the phase-inverted one at the anode. A similar circuit design prevails at B of Fig. 16-6, where the emitter-follower circuit principle is combined with the collector load-resistor output, thus again effectively splitting the load resistor into the dual arrangement. (See also B of Fig. 15-12 and the related discussions on *follower* circuitry.)

For the circuit at B, resistors R_1 and R_2 form a voltage divider across the power supply and thus furnish the necessary forward bias between base and emitter. With voltage-drop polarities as shown, resistor R_2 makes the base more positive than the emitter, satisfying the *NPN* bias requirements at the input. As with the tube type at A, resistors R_3 and R_4 are of equal value, with equal-value dc through them and equal-amplitude signal voltages across them. The phase relationships are identical to those described for the circuit at A and conventional *RC* coupling is used for transferring the split-phase signals from Q_1 to the input-base circuits of Q_2 and Q_3.

A paraphase circuitry for achieving the necessary phase inversion is shown in Fig. 16-7, where a separate transistor (Q_2) is used specifically for obtaining the necessary 180° phase shift. (This circuit is also applicable to

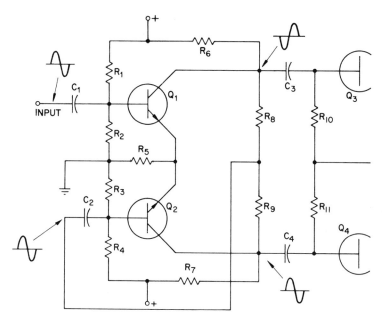

FIGURE 16-7. Paraphase circuitry.

vacuum-tube type push-pull circuits.) Transistor Q_1 is the conventional interstage small-signal amplifier, with the input signal applied to the base circuit
via C_1. Resistors R_1, R_2, R_3, and R_4 are voltage dividers for obtaining the
necessary forward bias as was the case for the circuit of Fig. 16-6. The load
resistors across which the amplified signals develop are R_6 and R_7. The two
additional resistors R_8 and R_9 span the two collectors as shown and it is
from their junction that the signal for the base input of the inverter is obtained. Coupling capacitor C_2 has the same value as C_1 and a balanced circuitry
is maintained by using matched transistors, matched load resistors, and
identical values for R_8 and R_9.

Since the signal phase across the common-emitter amplifier inverts 180°
between input and output, the signal which develops across R_9 must be 180°
out of phase with the signal which appears at the base input of Q_2. Similarly,
the signal that develops across R_8 is out of phase with that at the base input
of Q_1. At the junction of resistors R_8 and R_9, a signal of proper polarity for
application to the base input of the phase-inverter Q_2 is *automatically present.*
This must be so because the circuit must achieve a balance in order to function at all. If no signal appears at the base input of Q_2, for instance, there
would be no signal developed across R_9. Consequently, the signal that is
derived from R_8 and fed to the input of Q_2 would have a proper phase to
start the system into symmetrical operation. Thus, an output would be
developed across R_9 that would tend to oppose the signal developed across
R_8. Complete cancellation, however, cannot occur. Hence, a circuit is established that is virtually self-adjusting, and the signal that appears at the input

of Q_2 will always be approximately of the same amplitude as the input signal applied to Q_1. The system is self-balancing and since the input circuits to the push-pull transistors are not disturbed by loading effects that occur for some other phase-inversion systems, the input impedance is not affected. (Some loading effects occur when the input signal for Q_2 is obtained by tapping R_{10}, the base-input resistor of Q_3, or the grid resistor of a push-pull tube.)

Since resistors R_8 and R_9 shunt the load resistors R_6 and R_7, voltages from the power supply as well as signal voltages develop across them. Since, however, the center tap at the junction of R_8 and R_9 is coupled (via C_2) to the base of Q_2 and not to signal ground, these resistors do not function as the actual load resistors. Resistors R_6 and R_7, however, are at signal ground at the power-supply source because of filter capacitors (or bypass capacitors, when used). Hence the signals from across the load resistor R_6 are applied to the base emitter of Q_3, while the signals developed across R_7 are applied to the base-emitter circuits of Q_4.

The complementary phase-inversion system is illustrated in basic form in Fig. 16-8. Here the phase inversion occurs because the output signals from Q_1 are applied to the base inputs of two common-emitter circuits consisting of one *PNP* and one *NPN* transistor. Resistors R_3, R_4, R_5, and R_6

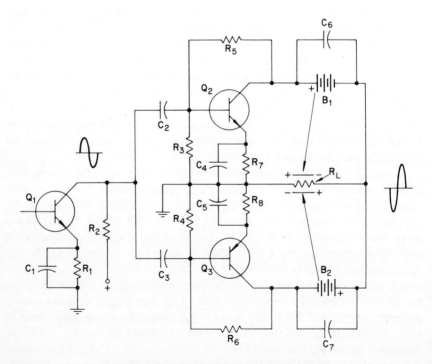

FIGURE 16-8. Complementary phase inversion.

are for bias-setting purposes, and the *RC* networks in the emitters are for stability purposes as explained for earlier circuits.

Note that battery B_1 (or power supply) applies the necessary positive potential for reverse bias to the *NPN* transistor Q_2. A separate source, B_2, supplies the necessary negative potential to the collector of the *PNP* transistor Q_3. A common load resistor R_L is used, and this load could be the primary of an output transformer or the voice coil of the speaker.

In the absence of an input signal, the idling currents through R_L are equal and opposite in polarity; hence they cancel. (Current flow, however, still exists around the loop formed by the B_1 and B_2 batteries in series with the collector-emitter circuits.) When a *positive* alternation of an *input* signal appears at Q_2, it aids forward bias; hence conduction through Q_2 increases. For the positive-signal alternation at the input of Q_3, however, the result is a decrease in the forward bias and a lowering of conduction through Q_3. When conduction through Q_2 increases, the voltage drop across R_L increases in the negative direction, resulting in a voltage change representative of one alternation of the output signal. At the same time the decreased conduction through Q_3 decreases the positive signal-voltage drop across R_L, thus additionally raising the amplitude of the negative output-signal alternation.

For the negative alternation of the input signal, the forward bias for Q_2 is decreased and conduction declines, thus lowering the negative-polarity voltage drop across R_L. For Q_3, however, the negative input signal increases forward bias and raises conduction. Thus, the positive-polarity change that occurs across R_L represents the second alternation of the output signal. Thus, the basic circuit function is typically push-pull and the normal phase inversion occurs across the common-emitter circuits as for a single-ended output stage.

Reference should be made to B of Fig. 15-12 and the related discussions on complementary circuitry. Commercial push-pull circuits may combine the direct-coupling principles with complementary circuitry and use the output system shown earlier in Fig. 15-5. Others may use some transformer coupling and achieve phase inversion with the paraphase method.

16-6. Inverse Feedback

Inverse feedback in an amplifier is a method for feeding a portion of the amplified signal back to the input of the amplifier or to some other previous stage, in such a manner that the signal that is fed back is opposite in phase to the signal at the circuit to which the feedback is applied. Two basic types of feedback are utilized, one being the *negative-voltage type*, and the other the *negative-current type*.

When a signal voltage is fed back to a previous stage out of phase, degeneration occurs because the out-of-phase signal will cancel a portion of the input signal, in proportion to the amount of signal fed back. Hence, the process is often referred to as *negative feedback*. Voltage feedback reduces circuit impedance and also decreases the gain of the amplifying system. On the other hand, however, it reduces harmonic distortion, as well as noises generated within transistors or tubes. For these reasons, inverse feedback is used extensively in electronic circuitry.

Feedback principles apply equally to tube and transistor circuitry. Initially, tube-type applications are discussed, with transistor systems covered later. As shown in Fig. 16-9, the feedback may be obtained from the secon-

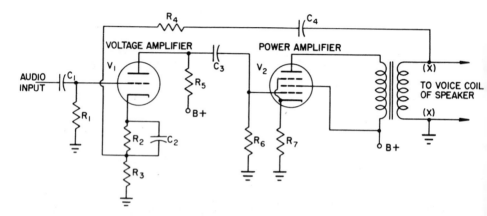

FIGURE 16-9. Inverse feedback in tube-type circuitry.

dary winding of the output transformer and applied to the cathode circuit of a previous stage. Feedback, in such an instance, could also be from the plate of V_2 to a previous stage. Since a negative, or out-of-phase, signal voltage is desired, the places marked X at the output-transformer secondary must be reversed if positive feedback should occur. The signal voltage which is fed back from the secondary of the output transformer is applied across resistor R_3 which applies the signal to the cathode circuit of V_1. The amount of signal voltage that is fed back is determined by the values of resistors R_3 and R_4. Capacitor C_4 is a coupling capacitor that prevents any d-c components in the cathode circuit from being shunted by the low d-c resistance of the output-transformer secondary.

Inverse feedback reduces harmonic distortion generated within vacuum tubes or transistors in proportion to the amount of signal voltage that is fed back. For instance, if a pure sine wave is applied to the grid of V_1 and duly amplified in the anode circuit, signal distortion may occur because of nonlinear vacuum-tube characteristics. The distortion would become part of the amplified signal at the output of V_1 and thus would be applied to the next amplifier stage, where the original distortion would add to any new distortion that is generated. The accumulated distortion finally reaches the loudspeaker and becomes audible.

The manner in which inverse feedback reduces harmonic distortion will be evident by tracing the process involved. Initially, assume a 5-V, 2-kHz signal is applied to the input of the amplifier. If the feedback circuit is such that 0.5 V of the amplified signal is fed back, the latter will cancel 0.5 V of the input signal. Such cancellation will reduce the output of this stage since less signal is now present at the input. Normally, the reduced gain is not a detriment because if the amplifier has reserve power, the volume control can be turned up to bring the gain to the level desired.

The 0.5-V signal that is fed back not only contains the 2-kHz original signal but also any harmonic distortion signal present at the output of the amplifier. The distortion signal, however, being higher in frequency than the 2-kHz signal, will not be canceled at the feedback point across R_3 because of the frequency difference between the feedback distortion signal and the input signal. Thus, if the distortion signal that is fed back is 0.1 V, there would be 4.6 V of the input signal left after the feedback component had been applied to the input system across R_3, plus 0.1 V of distortion signal. Now, the 4.6-V original signal plus the 0.1-V distortion signal are amplified by V_1, and the distortion component introduced by the feedback will be inverted in phase when it appears in the output across load resistor R_5. The amplified distortion, now out of phase with the particular distortion developed by V_1, cancels out a portion of the *distortion developed by the tube*. The amount of cancellation that occurs depends on how nearly the negative feedback signal amplitude matches the amplitude of the distortion developed in the tube. If the amplitude of the signal that is fed back is made high, the reduction in harmonic distortion is increased. As mentioned, however, the greater the feedback, the higher the degeneration and gain loss that occurs. Any tube noises or other spurious signals which are developed within the tube and are not a part of the applied signal will also be reduced in proportion to the amount of negative-voltage feedback which is utilized.

The reduction of plate resistance that occurs at the output tube when using inverse feedback has some advantages. The lower plate resistance tends to dampen, and thus diminish, the resonance effects of loudspeaker systems because any spurious signal developed at the loudspeaker voice coil and fed back to the anode circuit would find a shunting effect, due to the lowered impedance.

Inverse feedback also broadens the frequency response of an amplifier. If, for instance, a particular amplifier does not amplify the lower-frequency signals as much as other frequency signals, there would be reduced output for such signals and hence the amplitude of the feedback voltage of such signals would also be lower than other signals. Consequently, less inverse feedback degeneration occurs for the lower-frequency signals, and thus the overall gain of the amplifier would tend to flatten out.

The resistive values of R_3 and R_4 determine the amount of inverse feedback. Experimentally, R_3 and R_4 can be varied until harmonic distortion has been reduced to the degree where the output power has not dropped to a value too low for usefulness. Coupling capacitor C_4 should have a value sufficiently large so that it will have a low reactance for the low-frequency signals handled by the amplifier.

To express the mathematical relationships of feedback systems, the lowercase Greek letter *beta* (β) is used to indicate the amplitude of the signal voltage fed back. The symbol is preceded by a minus sign when the feedback is inverse or negative (the type with which we are presently concerned). With oscillators, where positive feedback is used for regeneration, the symbol is preceded by a plus sign. Thus, the symbol β indicates the decimal equivalent of the percentage of output signal voltage that is fed back. Another symbol, A', designates the *signal voltage amplification with feedback*. The A without the prime sign indicates signal-voltage amplification *without feedback*.

The product $A\beta$ is used to indicate the *feedback factor*. Thus, $1 - A\beta$ is a measure of the feedback amplitude. In equation form, the signal-voltage amplification with feedback is

$$A' = \frac{A}{1 - A\beta} \quad \text{or} \quad \frac{A}{1 + A\beta} \qquad (16\text{-}2)$$

where

A' = the signal-voltage amplification *with* feedback

A = the signal-voltage amplification *without* feedback

β = the decimal equivalent of percentage of output signal voltage fed back.

When the feedback factor $-A\beta$ is much greater than 1, the signal-voltage gain is independent of A and the equation for signal-voltage amplification with feedback becomes

$$A' = -\frac{1}{\beta} \qquad (16\text{-}3)$$

Because inverse feedback reduces distortion, it is also convenient to calculate the amount of distortion that is present with feedback. D' indicates the distortion of the output signal voltage with feedback, and D shows the distortion of the output signal without feedback. Thus, the equation is

$$D' = \frac{D}{1 - A\beta} \qquad (16\text{-}4)$$

Thus, both the gain and distortion is reduced to the degree established by the amplitude of the feedback $1 - A\beta$. If, for instance, the latter were 2.5 and the gain without feedback were 30, the gain would be reduced to 12:

$$A' = \frac{30}{2.5} = 12$$

If the distortion had been 5% before feedback, it would be reduced to 2%:

$$D' = \frac{5}{2.5} = 2$$

Negative-*current* feedback is also shown in Fig. 16-9 and is represented by the unbypassed cathode resistor R_7 of the power-amplifier tube V_2. Negative-current inverse feedback obtains the feedback signal voltage from an unbypassed cathode resistor. The amplitude of the feedback signal is thus proportional to the signal-current flow through the cathode resistor. Negative-current feedback increases the plate impedance of the amplifier tube, as opposed to the decrease in plate impedance for voltage feedback. As with the latter system, however, gain is decreased because of degeneration.

The function of negative-current feedback depends on the inverse signal voltage with respect to grid bias that occurs across the cathode resistor. Without a signal input to the amplifier stage, normal bias develops across R_7, by virtue of the current flow through the resistor. The cathode becomes more positive with respect to ground and since the grid is connected to ground through the grid resistor, the grid is negative with respect to the cathode, as previously mentioned.

Resistor R_7, not being bypassed, will develop signal voltages across it whenever signals are applied to the grid of the tube. If, for instance, a positive alternation of signal is applied to the grid, plate current will increase. The increase in plate current through R_7 will also increase the voltage drop across this cathode resistor. The higher voltage at the cathode will *increase* the negative grid bias and cause plate current to decrease. Thus, the grid signal and the signal-current change in the cathode resistor act inversely to each other. Where the positive grid signal causes a plate current increase, the changing voltage across the cathode resistor causes a plate-current decrease. With the negative-alternation input signal, degeneration also occurs because the negative grid signal decreases plate-current flow. A decrease in current through R_7 lowers the cathode voltage with respect to the grid potential; hence plate current increases because bias is decreased.

As with negative-voltage feedback, the frequency response of the amplifier with negative-current feedback increases, thus permitting the amplifier to handle a wider range of frequencies. Omission of the cathode bypass capacitor increases low-frequency response because of the absence of the increased reactance at lower frequencies. If the bypass capacitor has too high a reactance, some negative-current feedback occurs for lower frequencies. Without the capacitor, the cathode circuit is not instrumental in lowering the gain for signals having lower than normal frequencies. The value of resistor R_7 determines the degree of negative-current feedback. If the ohmic value of R_7 is increased to a sufficient degree, the amplifier stage will have a gain less than unity because of the excessive degeneration that results.

The application to transistor circuitry of the two inverse feedback systems just described is shown in Fig. 16-10. Voltage feedback is coupled from the output of the second transistor stage to the emitter circuit of the first stage by resistor R_8 and capacitor C_4. Electron flow in the emitter circuit is in the direction shown by the arrow at R_2. The voltage drop across R_2

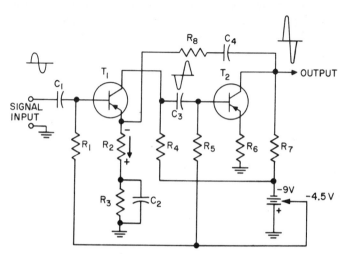

FIGURE 16-10. Inverse feedback systems in transistor amplifiers.

opposes to a certain degree the normal positive charge of the emitter versus the base of transistor T_1. Thus, instead of the emitter having 4.5 V, it would have only 4 V if 0.5 V develops across R_2. If a signal, as shown, is applied to the input of the amplifier, the first positive alternation will also oppose the normal forward bias because the positive alternation of the signal will lower the negative base potential. Thus, current flow in the base-emitter circuit is reduced. The inverse-feedback signal, however, places a positive alternation across R_2 and thus decreases the opposing voltage developed across the latter resistor. In consequence, base-emitter current increases. Thus, where the input signal alternation tends to decrease base-emitter current, the feedback voltage tends to increase base-emitter current, providing an inverse function. Reduction of harmonic distortion, increase of frequency response, and all other factors discussed for the vacuum-tube counterpart of voltage feedback also apply to the transistor circuits shown in Fig. 16-10.

The absence of the bypass capacitor across resistor R_6 also causes degeneration, and this is a current feedback similar to the condition set up for the vacuum tube in Fig. 16-9, where R_7 was not bypassed. Again, there is a reduction of gain, a change of circuit impedance, as well as a reduction of harmonic distortion, just as with the vacuum-tube negative-current feedback system.

Use of *NPN* transistors instead of the *PNP* transistors shown in Fig. 16-10 does not alter the feedback circuitry. With *NPN* transistors, the only change would be that of battery-supply potentials. There would still be the same phase reversal of the signal across each grounded-emitter circuit and the same feedback principles would apply.

16-7.   Classes of Amplifiers

The amplifiers previously discussed have been Class A. These are suitable for a variety of signals, both in the audio and RF ranges. As shown earlier in Fig. 15-16, Class A operation, such an amplifier is so biased that operation is on the linear portion of the characteristic curve of the tube or transistor used. The signal applied to the input has an amplitude that does not reach current cutoff or saturation levels on the characteristic curve. A Class A amplifier has a minimum of harmonic distortion as compared to the other amplifier types described next. Efficiency (conversion of d-c power to signal power), however, is relatively low, depends on the bias and signal amplitudes, and may range between 10 and 15%.

Class A amplifiers can be resistance-, impedance-, or transformer-coupled types. Such amplifiers find extensive applications as power-output amplifiers, as previously illustrated, though they also find application as voltage (small-signal) amplifiers in the *RF* stages of television, radio, and other receivers. Class A amplification may also be used in a single-ended arrangement or in push-pull as previously described.

Another type amplifier is Class AB_1, with characteristics as shown in the upper left drawing of Fig. 16-11. This type is used when more output power is needed than can be obtained from Class A. For Class AB_1 the bias is usually set so the no-signal idling current permits the input signal to swing the collector (or anode) current to the cutoff point and to the saturation level as shown.

Class AB_1 is used in audio power-output amplifiers, where the slightly higher distortion level is outweighed by the increased efficiency and power output. The subscript "1" had its origin in tube circuitry where it indicated that the input signal's amplitude was held just below the point where the grid would be driven into the positive region. Efficiency for the Class AB_1 amplifier ranges between 20 and 35%, depending on exact bias setting and other design factors.

Another amplifier is the Class AB_2, which has operational characteristics as shown in the upper right drawing of Fig. 16-11. The bias is set on the lower portion of the curve and the input signal has sufficient amplitude to drive the output current into the cutoff region as well as into the saturation area as shown. Thus, output power and efficiency are increased over the AB_1 type, though distortion also increases as shown by the output signal. Distortion is high because of operation on the nonlinear portion of the curve and the high-level drive of the input signal (which causes clipping of the output-signal alternations).

Again, the subscript ("2" in this instance) has its derivation from early tube-type Class AB_2 amplifiers where it designated that some grid-current

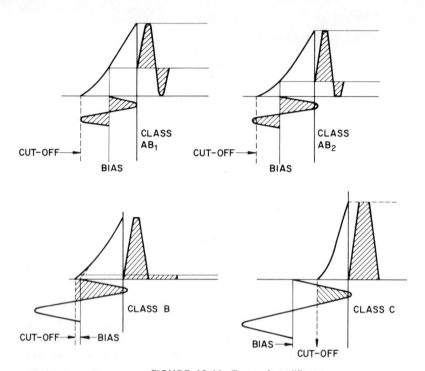

FIGURE 16-11. Types of amplifiers.

flow occurs because the positive peaks of the input signal had sufficient amplitude to overcome bias and drive the grid positive. Thus, the input signal had to deliver power since grid current flowed. Push-pull operation is preferable with Class AB$_2$ for increasing power output and reducing distortion. Efficiency ranges between 35 and 50%, again depending on bias and signal-input drive factors.

Schematically, the circuits for Class AB$_1$ and Class AB$_2$ resemble those for Class A. The differences present are in terms of higher-power d-c supplies, bias differences, and the amplitudes of input signals. Thus, an inspection of the amplifier schematic may not necessarily indicate whether the amplifier is Class A, AB$_1$, or AB$_2$, unless the design features are analyzed with respect to the amount of bias being used and whether or not higher output power is available than would be obtained in Class A operation.

Another type of amplifier is that known as Class B. The Class B amplifier can be designed to handle either audio- or radio-frequency signals, and a typical Class B *audio amplifier* in a push-pull arrangement is shown in Fig. 16-12. The Class B amplifier is biased approximately at cutoff, as shown in the lower-left drawing in Fig. 16-11. Because one transistor only reproduces about one-half of the signal, *two are necessary in order to reproduce the full signal information* in audio work. In a push-pull arrangement, one

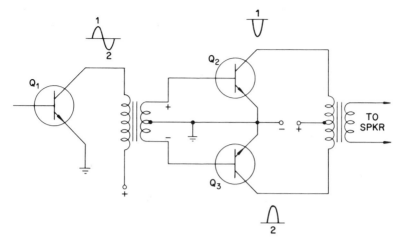

FIGURE 16-12. Transistorized Class B amplifier.

transistor produces the amplified positive half of the output cycle, while the other transistor produces the amplified negative half of the output cycle.

The bias for Class B is often set at what is referred to as the *projected cutoff*, as shown in Fig. 16-11. The projected cutoff is the bias point that is found when the linear portion of the curve is projected downward to meet the horizontal line of the graph. The dotted-line section of the curve in Fig. 16-11 illustrates the projection of the linear portion of the curve to get projected cutoff. Because the projected cutoff is slightly below the actual cutoff point, a low value of idling current flows through the transistor (or tube) when no signal is applied to the input. With an input signal, however, the output current varies as shown in Fig. 16-12. The input signal will develop half alternations in the collector circuit that are produced by operation on the linear portion of the characteristic curve. Despite projected cutoff operation, however, signal distortion is still much higher than for Class A, although, in a carefully balanced circuit, distortion can be held at a level comparable to that obtained from Class AB_2 operation. To obtain good balance, the transistors in the Class B push-pull arrangement must, therefore, be closely matched to hold harmonic distortion at a minimum. It is also essential that the center taps of the transformers be at the actual electrical center of the inductance for a balanced system.

For Fig. 16-12, note that the base-emitter leads of each transistor are at the same potential. Thus, instead of the usual forward-bias arrangement at the input of the transistor, zero potential is used. With the base and emitter at the same potential (zero), there is no forward bias to overcome the potential barriers within the transistor and collector current drops to a value almost at cutoff, the latter being equivalent to the projected cutoff bias used for vacuum-tube Class B operation. (As with vacuum tubes in Class B, if the collector were operated at zero current flow, the distortion would increase because of operation on the nonlinear portion of the transistor's curve.)

If the first alternation of the input signal is positive, as shown, the top of the transformer secondary is at positive potential, while the bottom is at negative potential. The positive potential at the base of the upper transistor establishes the equivalent of a forward bias (positive to the base and negative to the emitter, in an *NPN* transistor). Hence, amplification occurs for the first alternation in the top transistor. For the lower transistor, however, the negative potential at the base, and the relatively positive potential to the emitter, create a reverse-bias condition that will cut off collector current completely (similar to the signal extending into the cutoff region of a vacuum tube). When the second alternation (negative) of the input signal arrives, the lower transistor has an amplified output alternation, while the upper transistor is now at cutoff. Thus (as with vacuum-tube Class B systems), each transistor contributes one-half of the amplified signal output cycle. For *PNP* transistors, circuit operation is identical and the supply battery polarities are merely reversed.

The Class B amplifier is characterized by high efficiency, which ranges between 60 and 70%, but with a correspondingly higher distortion as compared to other types of audio amplifiers.

16-8.   RF Power Amplifiers

Power amplifiers for handling RF signals usually consist of Class B and Class C types and are used primarily in radio, television, FM, or other broadcast transmitters. The Class C type (discussed later) is generally used for amplification of an RF pure carrier signal. The Class B, on the other hand, is usually employed when it is desired to amplify a modulated RF-type signal in transmitting systems. The Class B amplifier for RF can be employed either in push-pull or as a single-ended stage. It will function as a single-ended stage because of the energy interchange effect between the capacitor and inductance of the resonance circuit. This energy exchange (flywheel effect) supplies the missing alternation for the individual pulses of collector (or plate current) that flow in Class B. The pulses of output current occur at a repetition rate corresponding to the frequency of the input signal, and such plate-current pulses supply continuous RF energy to the resonant circuit composed of an inductance and capacitor. The interchange of energy between the capacitor and inductance of the parallel-resonant circuit forms a complete sine-wave type of RF signal, even though only supplied with timed pulses of power.

Figure 16-13 illustrates a typical single-stage Class B tube-type RF amplifier, though a power transistor could, of course, also be used. The input signal develops across the grid and cathode sections, as shown. A bias supply,

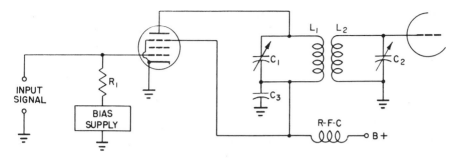

FIGURE 16-13. Class B R-F amplifier.

consisting of a battery or power supply, furnishes the necessary negative bias near the cutoff point. The amplified signal energy develops across the parallel-resonance circuit composed of capacitor C_1 and inductance L_1. Capacitor C_3 is a return signal path from the resonant plate circuit to the cathode circuit. Thus, capacitor C_3 prevents signal energy from developing across the power supply. In series with the B voltage is an RF choke (RFC) that because of its high reactance for the RF signals acts as an opposition for signal energy and reduces leakage of the latter to the power supply. By transformer coupling of L_1 and L_2, the energy in the plate resonant circuit is transferred to the grid resonant circuit of the next stage. The plate resonant circuit could also be coupled to an antenna system if the Class B stage is the final RF amplifier of a transmitter. Since a pentode is used, neutralization should not be necessary.

The Class C amplifier, in contrast to the other types previously discussed, is only an RF power amplifier and cannot be employed for audio-power amplification. Hence, the Class C circuit is used only in RF power amplifiers in transmitters. Bias for Class C tube operation is set *beyond* the grid-voltage cutoff point of the tube (usually from two to three times cutoff). For large-signal power transistors in a common-emitter circuit, sufficient reverse bias is applied between base and emitter to attain the desired bias setting beyond cutoff. Signal input must be capable of furnishing power, as with Class AB_2 and Class B amplifiers. The signal input, therefore, must have sufficient amplitude to drive the grid positive at alternate positive peaks. In a Class C amplifier, the grid-current flow during the positive alternation peak of the input signal will charge the grid capacitor, which in turn will discharge across the grid leak, as previously discussed in Chapter 15 and illustrated in Fig. 15-9. Thus, the Class C amplifier is capable of furnishing its own bias.

Since bias is set beyond the cutoff point, collector or plate current flows for only a portion of the applied input signal (less than the duration of one alternation). This factor is illustrated at the lower right in Fig. 16-11. The shaded portion of the positive alternation of the input signal is the only section causing current flow because the remainder of the input signal is beyond the cutoff bias point. Hence, collector or anode current flows for only a portion of each positive alternation of the input signal, as shown. The remainder of the input signal (beyond the cutoff point) is not reproduced by amplification, and thus the Class C amplifier is not useful for audio amplification, even if a push-pull circuit is employed. If a Class C circuit were utilized for audio, severe distortion would result because some of the audio components would be missing from the amplified signal developing across the load impedance or resistance. Thus, Class C is useful only for RF amplification, where only the *frequency* and *amplitude* of the signal are involved and not the *waveshape*.

A typical Class C tube-type RF single-ended amplifier is shown at A of Fig. 16-14. The energy from the previous stage is coupled to the input of the Class C amplifier via capacitor C_1. The input signal energy is known as *excitation*. The self-bias that develops by virtue of the grid being driven positive will cause a current flow through R_1, which can be read by a milliammeter placed in series, as shown. Since there would be no bias developed in the

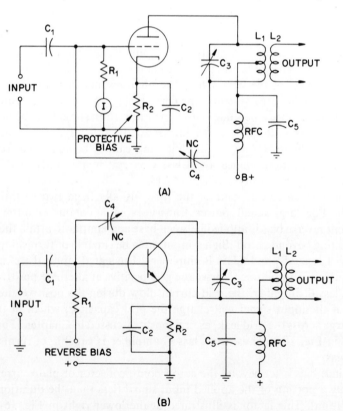

FIGURE 16-14. Triode Class C amplifiers.

absence of an input signal, a cathode resistor and capacitor are also employed for protective bias, as shown. As with the Class B amplifier previously discussed, the amplified signal energy develops across the anode resonant circuit composed of C_3 and L_1. Because a triode tube is employed here, neutralization is necessary, as discussed previously. Adjustments are made to the neutralizing capacitor (NC), consisting of C_4.

The high efficiency of the Class C amplifier is due to the short duration of the plate-current pulse. The current surges in the plate circuit pulse the resonant circuit (sometimes referred to as the "tank") into a flywheel type of oscillation. Thus, when the pulse appears across the tank circuit shown in Fig. 16-14, it charges capacitor C_3 to its full value and this, in turn, discharges across L_1. The collapsing field of L_1, after C_3 has been exhausted, produces a back-emf that recharges capacitor C_3. This flywheel motion reproduces a sine wave in the tank circuit, despite the momentary pulses of energy that are present in the anode circuit.

Since electron flow is from cathode to plate, and thus down through the tank circuit to the power supply, the pulse of energy developed across the tank circuit is in a negative-going direction. During the time of maximum current flow, the impedance of the tube is low, while the impedance of the tank circuit is high for the resonant frequency. This means that most of the energy from the power supply is developed across the tank circuit during a time when tube impedance is low and consuming little energy. In consequence, a high order of efficiency is obtained in the Class C amplifier, and a well-designed unit may have an efficiency rating in excess of 90%.

At B of Fig. 16-14 a transistorized Class C amplifier is shown. As with the circuit at A, the input signal is coupled via capacitor C_1 to the base and via ground to the emitter. For positive alternations of the input signal, the reverse bias is overcome and the high-amplitude forward bias developed by the input signal causes the transistor to conduct heavily, thus pulsing the resonant circuit composed of C_3 and L_1 into its flywheel-effect resonant condition, with an interchange of energy between C_3 and L_1 as described for the circuit at A. For a negative-alternation input, the reverse bias is increased additionally, driving the transistor farther into the cutoff region, as with the tube-type amplifier.

Stabilization is obtained by resistor-capacitor combination C_2 and R_2, and the RFC presents a high-inductive reactance for signal energy, thus effectively preventing its coupling to the power supply or other stages. Similarly, C_5 is a direct return to the emitter of the RF energy, thus placing the bottom of the collector tank circuit at *signal* ground. As with the triode tube, neutralization is necessary and the coupling of the neutralizing capacitor is similar to the arrangement at A.

While the circuits of Fig. 16-14 are essentially similar, some differences are present because of the inherent differences between tubes and transistors.

Also, if a *PNP* transistor were used at B, polarities would have to be reversed with a positive potential applied to the base for reverse bias.

Tuned resonant circuits are highly selective, as mentioned earlier. Selectivity, however, can be regulated by the degree of resistance introduced into the resonant circuit. Resistance will lower *Q* and decrease selectivity. Thus, a Class C amplifier can be designed to handle a signal of only one particular frequency. Signals of other frequencies will be diminished because of the flywheel-effect selectivity of the tuned resonant circuit. With resistance added to the circuit, the amplifier can be designed so that its selectivity is somewhat broadened; that is, the amplifier will handle a narrow band of signals having frequencies grouped around the center resonant frequency.

In the input circuit for A of Fig. 16-14, the value of capacitor C_1 and resistor R_1 must be chosen so that the *R-C* constant is proper for the frequency of the applied input signal. Capacitor C_1 must be capable of charging to the peaks of the rectified energy when the grid runs positive, while resistor R_1 must be capable of dissipating the energy contained within C_1 in sufficient time to prevent grid saturation. At the same time, resistance R_1 should not be so low in value that it dissipates an excessive amount of the input-signal energy. If too much energy is drawn from the previous circuit, excessive loading effects on the previous circuit may upset the normal function of the latter.

The tuned resonant circuit at the output of each amplifier of Fig. 16-14 is adjusted for the frequency of the input signal by adjustment of capacitor C_3. When resonance is achieved for the circuit at A, however, a maximum grid signal will flow, and the largest value of bias is developed across resistor R_1. For this reason, resonance indication while tuning can be obtained by observing the grid-current flow with a d-c ammeter or by measuring the bias voltage drop that appears across resistor R_1. The protective bias furnished by C_2 and R_2 may have a value anywhere below cutoff. The larger the resistor, the greater the voltage drop, and the more negative the grid will be with respect to the cathode. A value of protective bias is chosen that will provide sufficient bias for protection against excessive plate current during the absence of the input signal, without wasting an excessive amount of power from the supply.

When a pentode tube is employed at A, no neutralization is necessary, provided that the tube is of such a design that interelectrode capacities have a high reactance for the frequencies handled by the amplifier. When a triode tube is utilized, however, neutralization is necessary to prevent feedback (regeneration) and consequent oscillation. If the Class C amplifier (transistor or tube type) oscillates because of feedback, it will generate a frequency of its own that is independent of the applied frequency of the input signal. Hence, during oscillation, the usefulness of the Class C stage as an amplifier is lost.

When the tank circuit is not tuned to resonance, maximum collector or plate current will flow because of the low impedance presented by the tank circuit on each side of the resonant frequency. With high-power Class C amplifiers, excessive currents can flow unless some protective measures are taken to limit the plate-current flow during the tuning process. Without a load, the Q of the tank circuit is very high and excessive voltages are present because of the high impedance. Under normal load conditions, the Q drops to a value as low as 10 or 15.

16-9. Load-Line Factors for Power Amplifiers

With a *resistance* coupled to the amplifier stage, the supply voltage indication on a load-line drawing is the *right-hand termination* of the load line at the zero collector-current line (or plate-current line in a tube). For the characteristic curves shown in Fig. 16-15, the supply voltage would be -28 V if this were a resistance-coupled (small-signal) amplifier. The signal-voltage swing across the load resistor for a signal input can then reach the maximum value of -28 V, since this is the highest potential available from the power supply.

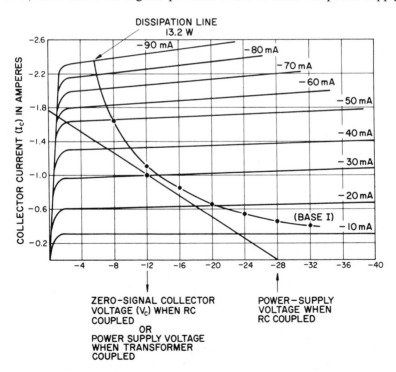

FIGURE 16-15. Dynamic load-line factors for PNP power transistor (RC vs. transformer coupled).

The operating voltage would be -12 V, since this is the drop across the load resistance without signal input.

With an output transformer-coupled amplifier, or loudspeaker load coupling in power-amplifier circuitry, the actual supply voltage would only be -12 V for this transistor, the same as the operating voltage for the dynamic load line. The function, insofar as the dynamic characteristic is concerned with input signal, however, is essentially the same as with an actual load resistor in series with the power supply. In a transformer-coupled system, the actual load is that which is applied to the secondary of the transformer, such as a loudspeaker voice coil. The function, in terms of collector-current swing, comes about because of the action of the inductance through which the collector current of the power amplifier must flow.

Thus, if the input signal to the base swings in a positive direction (less forward bias) sufficiently to reduce the collector current to a small value, the sudden change in current through the inductance induces a voltage across this inductance, *which adds to the collector-supply potential.* The effect is to increase the *instantaneous* collector potential to a value considerably above the operating potential and hence above the collector-supply potential.

When the input signal swings in the opposite direction and increases forward bias, a sudden current change again occurs in the transformer primary inductance, and a voltage is induced across the primary inductance that is opposite to the previous change. Thus, the collector-voltage signal swing reaches values above and below the normal operating potential.

As long as the average d-c collector current does not change when the signal is applied to the base input circuit, the same dynamic characteristics hold for resistance loads as for the transformer-coupled loads, and reference should be made to Chapter 15 for representative equations and processes for calculations (see also Fig. 15-31 and 15-32).

Note the constant power-dissipation line (see Chapter 15) is indicated as 13.2 W and that the load line is close to it for obtaining maximum power output. Using either Equation 15-4 or 15-11, the load line has a value of 15.9 Ω and hence is sufficiently low to permit direct coupling to a voice coil of 16 Ω.

$$R_L = \frac{dV_c}{dI_c} = \frac{(0 \text{ to } 28)}{(0 \text{ to } 1.76)} = 15.9 \text{ Ω}$$

The 1.76 is an approximate value but sufficiently close for practical purposes. Thus, the load resistance can also be considered to be 16 Ω for all practical purposes.

The actual signal-power output depends on the magnitude of the input signal and how nearly it causes a full swing of collector current. If, for instance, the collector voltage changes from 4 V to 20 V during the presence of an input signal, Equation 15-7 produces the following:

$$\frac{(20-4)(1.5-0.5)}{8} = \frac{16}{8} = 2 \text{ W}$$

If, however, sufficient input-signal drive were present to cause the collector voltage to change from 0 to 24 V, the power output also increases:

$$\frac{(24 - 0)(1.76 - 0.25)}{8} = \frac{36.24}{8} = 4.53 \text{ W}$$

Distortion should be calculated on the basis of the maximum signal swing since low values of input signals cause operation on a more linear portion of the characteristic curves and is not a true indication of what occurs at higher signal levels. Similarly, the distortion should be calculated on the basis of a pure sine-wave input signal, where it causes a base-current change of equal amplitude on each side of the base-current operating point.

Thus, for Fig. 16-15, with an equal change of I_b above and below the operating point, we can obtain values from 10 to 30 mA (at the operation point I_o) and from 30 to 50 mA for the signal swing in the opposite direction. We can use Equation 15-9 for solving second-harmonic distortion, substituting symbol I_o (operating collector current) for I_b (operating plate current):

$$\frac{(I_{\max} + I_{\min})/2 - I_o}{I_{\max} - I_{\min}} \tag{16-5}$$

Thus, at 10 mA I_b the I_c minimum is 0.3 amp, while at 50 mA I_b the I_c maximum is 1.62 (approximately). Setting these in Equation 16-5, we obtain

$$\frac{(1.62 + 0.3)/2 - 1}{1.62 - 0.3} = \frac{0.04}{1.32} \times 100 = 3\%$$

The percentage of collector efficiency can be found by

$$P_{\text{eff}} = \frac{\text{signal-power output}}{\text{d-c input power}} \times 100 \tag{16-6}$$

Since variations in signal amplitudes would alter loading effects on the amplifier, Equation 16-6 should be applied only to constant-amplitude sine-wave-type test signals that present a constant load. For Fig. 16-15 we have:

$$\frac{4.5 \text{ W}}{(1 \times 12)} = \frac{4.5 \text{ W}}{12 \text{ W}} \times 100 = 37\%$$

Triode-tube load-line characteristics were illustrated in Fig. 15-30, and comparative pentode characteristic curves are shown in Fig. 16-16, with a representative load line drawn in (see also Fig. 11-16). Note the resemblance to junction transistor and FET curves. Since pentode and beam-power tubes generate a greater amount of third-harmonic distortion, calculations for ascertaining the percentage of such distortion should be considered for a full indication of general characteristics. (Transformer-coupled loads versus resistance loads coupled to the output without a transformer have the same characteristics and supply-voltage factors discussed for Fig. 16-15.)

A typical circuit for the tube from which the curves of Fig. 16-16 were obtained is shown in Fig. 16-17. Here, a beam-power tube is shown in the output-power amplifier stage of an audio system. The cathode resistor is

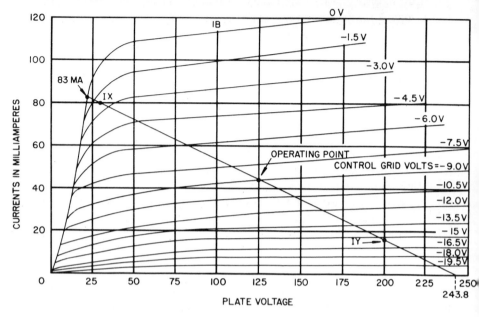

FIGURE 16-16. Pentode-load line.

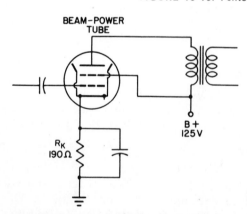

FIGURE 16-17. Beam-power output audio amplifier.

listed as having a value of 190 Ω and the power supply voltage is 125 V. Since the screen grid is also connected to the same voltage source, 125 V also appears at the screen. Assume that, for this tube, 3.3 mA of screen current are flowing and that the recommended load resistance is 2700 Ω. The operating current is 44 mA for the anode.

If the load line shown in Fig. 16-16 were not drawn in, two points on the characteristic curves could be ascertained from the information given above. Initially, the bias value can be calculated and this will determine the operating point on the 125-V vertical line. Since 44 mA of current are flowing through the cathode resistor for the anode and 3.3 mA are also flowing through the cathode resistor for the screen grid, the total current flow through the cathode resistor is 0.0473 A. When this current value is multiplied by the value of the cathode resistor (190 Ω), the bias value is found to be 9 V. This establishes the operating point shown in Fig. 16-16. The place where the load line meets the zero current line at the right can be solved for as follows by using Equation 15-6:

$$E_p + I_p R_L = 125 + 44 \times 2700 = 243.8 \text{ V}$$

This establishes the second point, and the load line can now be drawn in as shown.

The second harmonic distortion can now be calculated as follows from Equation 15-10:

$$\frac{I_{max} + I_{min} - 2I_b}{2(I_{max} - I_{min})} = \frac{83 + 8 - (2 \times 47)}{2(83 - 8)} = \frac{3}{150} = 0.02$$

$$0.02 \times 100 = 2\%$$

To solve for *third*-harmonic distortion, the following formula must be employed:

$$\frac{I_{max} - I_{min} - 1.41(I_X - I_Y)}{I_{max} - I_{min} + 1.41(I_X - I_Y)} \times 100 \qquad (16\text{-}7)$$

In the foregoing, the minimum values of current swing must be subtracted from the maximum values and the values of I_X and I_Y ascertained. The latter two values are based on the full signal swing. Since the bias is minus 9 V, the full signal swing would be from this value to zero bias and from minus 9 V to twice this value, or 18 V. Calculations of third-harmonic distortion, however, must be based on the rms values of each peak alternation of the full signal swing. The value for I_Y gives the rms value of the negative signal swing at the grid and is found as follows:

$$0.707 \times 9 \text{ V} = 6.3$$

$$6.3 + 9 = 15.3 \text{ V}$$

The foregoing 15.3 V bias value indicates the rms value of one alternation of the signal swing. When this bias value is marked on the load line, as shown in Fig. 16-16, it indicates that approximately 16 mA flows at this instantaneous value of signal. The other required point of I_X is found by subtracting the rms value of 6.3 from the 9-V bias value, as follows:

$$9 - 6.3 = 2.7 \text{ V}$$

This 2.7 V establishes the I_X point on the load line, as shown. This point is approximately 79 mA. Now that the two plate-current values for I_X and I_Y have been found, they can be set into the formula for calculating the third harmonic distortion, as follows:

$$\frac{83 - 8 - 1.41(79 - 16)}{83 - 8 + 1.41(79 - 16)} = \frac{75 - 88.8}{75 + 88.8} = \frac{13.8}{163.8} \times 100 = 8.4\%$$

In transformer-coupled push-pull circuitry, one transistor (or tube) is connected across only one-half of the primary of the output transformer as shown earlier. Hence, the load impedance that is reflected to the transistor from across the transformer will only be one-quarter of the total impedance reflected into the total primary winding. This is so because the reflected impedance varies as the square of the turns ratio. Thus, in push-pull amplifiers, the load resistance must be one-half the value which would be used when only one transistor or tube is used in order to obtain at least double the power output which would be obtained from a single transistor of the same type. As with single-ended circuitry, the maximum dissipation rating for each transistor must be observed and the load line selected that does not dissipate more power than permitted for each transistor.

Review Questions

16-1. What are the basic differences between power amplifiers and small-signal amplifiers?

16-2. Under what conditions is there a maximum transfer of signal power between the output of an amplifier and its load?

16-3. In a power-output transformer feeding a speaker, which has more turns of wire, the primary or the secondary? Why is this so?

16-4. Why is it important to have a minimum of d-c resistance in the windings of an audio-output transformer?

16-5. List some of the advantages of using push-pull circuitry.

16-6. Briefly explain the necessity for a phase-inversion system in push-pull.

16-7. Briefly explain how push-pull circuitry minimizes harmonic distortion.

16-8. Briefly explain why transistors in push-pull can feed a speaker directly without use of an output transformer.

16-9. Briefly explain what is meant by a *split-load* phase-inversion process.

16-10. Describe the phase-inversion process for push-pull that is self-balancing and self-adjusting.

16-11. Briefly describe the phase-inversion system based on transistor complementary principles.

16-12. What advantages are obtained from voltage-type negative feedback? Explain briefly.

16-13. Explain how current-type inverse feedback can be obtained in transistors and tubes, and what effects this system has on signal gain and reduction of distortion.

16-14. (a) Can a Class B audio amplifier be used in a single-ended stage?
(b) Can a push-pull Class C amplifier be employed for audio amplification?
(c) Can a Class B amplifier be used in a single-ended stage for RF amplification?

(d) Can a single-ended Class C amplifier be employed for RF amplification?

16-15. Explain the approximate efficiency of the following amplifiers: Class A, Class B, Class C.

16-16. Briefly explain how the collector voltage can swing above the power-supply voltage in a transformer-coupled output audio-power amplifier.

16-17. Briefly explain how the characteristic curves of pentode or beam-power tubes differ from those of triode tubes.

16-18. Give the formula for finding the third-harmonic distortion in pentode tubes.

16-19. Briefly explain what is meant by percentage of collector efficiency.

16-20. Briefly explain how the reflected load resistance is altered in a transformer-coupled push-pull power amplifier (either transistor or tube).

16-21. In single-ended or push-pull power circuitry, of what importance is the power dissipation line of the characteristic curves? Explain briefly.

16-22. In what harmonic distortion analysis must we consider the rms value of each peak alternation of the full signal swing?

Practical Problems

16-1. In a solid-state public-address amplifier it was necessary to match an output impedance of 800 Ω to a speaker voice coil of 8 Ω. If a matching transformer is used, what must be the turns ratio?

16-2. The dynamic characteristic curves for a transistorized amplifier indicated $dI_c = 650$ mA and $dV_c = 26$ V. What is the ohmic value of the load resistance indicated by the load line?

16-3. In an experimental amplifier in a testing lab it was found that I_c at minimum was 0.6 A and I_c at maximum was 1.8 A. I_o equalled 1.1 A. What is the second-harmonic distortion?

16-4. Each solid-state audio amplifier in a stereo cassette system had a signal output power of 16 W. $I_o = 1.5$ A and $V_c = 18$ V in a transformer-coupled circuit. What is the percentage of collector efficiency?

16-5. For Fig. 16-15 what is the R_L if the load line touches the 1.6-A point at the left and the -32-V point at the bottom of the set of curves?

16-6. Assume the R_L in Problem 16-5 is to be matched to a 5-ohm speaker voice coil. If the secondary of the output transformer has 16 turns of wire, how many turns must the primary have?

16-7. For Problem 16-5, assuming a full signal swing from V_c of 0 to -32 V, what is the signal-power output?

16-8. If, for Problem 16-7, the d-c input power is 12 watts, what is the percentage of collector efficiency?

16-9. For a special-purpose pentode-tube amplifier, $I_{max} = 84$ mA, $I_{min} = 8$ mA, and $I_b = 45$ mA. What is the second-harmonic distortion?

16-10. A pair of 6V6 tubes in the modulator of a communications system had a common cathode resistor. Current values were 70 mA (combined I_p for both tubes) anode current and 5 mA (combined I_{sc} for both tubes) screen current. What must be the value of the cathode resistor if grid bias is to be -15 V?

Oscillator Circuits   17

17-1. Introduction to Resonant-Circuit Types

Oscillators are generators of signals and hence are used extensively to originate the various low-frequency, intermediate-frequency, and high-frequency signals required in the operation of electronic equipment. In radio and television receivers, for example, oscillators are used in the tuning stages, as more fully detailed in Chapter 19. In transmitters, oscillators are used to generate the fundamental signal that is to be sent for many miles to various receivers. Tape recorders also use oscillators for erasure of recorded tape, and electronic organs employ oscillators for generating the fundamental musical tones. Oscillators are also widely used in radar, electronic computers, and other electronic devices.

Some oscillators are employed for the generation of audio-frequency *sine-wave* signals, while others are used to generate signals having square or rectangular shapes. Oscillators are also designed to produce RF sine-wave signals ranging from a few hundred kilohertz to well above several thousand megahertz. Some oscillators employ resonant circuits, while others use resistance and capacitance combinations for the generation of signals having specific frequencies. Certain oscillators have provisions for varying the frequency of the output signal by manual adjustment (variable-frequency oscillators) and others have a fixed frequency that cannot be changed readily without circuit modifications (fixed-frequency oscillators). Generally, all oscillators have low signal output and must be followed by amplifiers to raise the signal level to that desired for practical applications.

17-2.   Flywheel Sine-Wave Generator

The sine-wave type of signals usually employ resonant circuits as frequency-*determining* devices and include such types as the Hartley, Colpitts, and others described in this section. To understand how such oscillators function, it is necessary to investigate more fully the characteristics of a resonant circuit with respect to the *flywheel effect*, mentioned with reference to the RF amplifiers in the last chapter.

When a capacitor is connected across an inductance, as shown at A of Fig. 17-1, the combination becomes a basic generator of a-c waveforms.

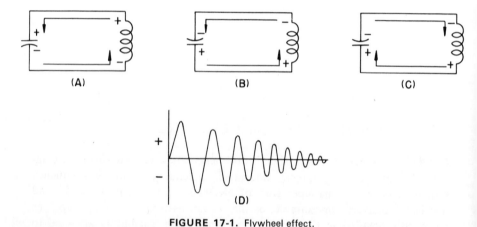

FIGURE 17-1. Flywheel effect.

If the capacitor is charged with a polarity as shown, and the *charging source is removed*, the capacitor will immediately start to discharge through the inductance and the electron flow will be in the direction indicated by the arrows. During such discharge, current flows through the inductance, setting up a magnetic field. The potential across the capacitor gradually declines and when the capacitor has completely discharged its energy across the coil, the fields of the coil will collapse and produce a back-electromotive force (voltage). According to Lenz's law, the current flow from the coil will be in the same direction as the charging current, and the back-electromotive force will cause a polarity change across the inductance, as indicated at B since the source of electrons will now be at the top of the coil and will charge the capacitor with the polarity opposite to the initial charge.

After the collapsing field has recharged the capacitor, the capacitor then discharges across the coil again, as shown at C. Thus, the reversal in the direction of the electron flow in the circuit will create a voltage that is of sine-wave form. The circulating energy that constantly reverses direction is known as the *flywheel effect*.

If there were no resistance in the circuit to consume the energy, the interchange of energy between coil and capacitor would continue indefinitely. This is similar to the momentum of a pendulum. After energy has been imparted to the pendulum to start it swinging, it would keep up such motion for an indefinite period if the energy were not consumed by the resistance of the air or the friction in the bearing or pivot by which the pendulum is suspended. Since some friction is always present, however, the pendulum eventually slows down and so does the energy circulating in the resonant circuit shown in Fig. 17-1. Some resistance is always present in the coil and, in consequence, the amplitude of the generated waveform gradually declines as the energy is consumed by the resistive component of the circuit. Thus, a parallel resonant circuit will have a gradual decline of its waveform, as shown at D. This is known as a *damped wave*.

The frequency generated by the parallel-resonant circuit depends on the inductive value of the coil and on the capacitive value of the shunt capacitor. If a larger value of capacity is used, it will take longer for the capacitor to charge fully and also longer to discharge. Hence, the *frequency* of the sine-wave would be reduced. The frequency can also be lowered by increasing the inductance. With a larger value of inductance, it would take longer for the capacitor to discharge through the coil because of the greater number of opposing fields that are produced. If either the capacitor or the coil is made smaller or if both are reduced in size simultaneously, the charge and discharge time is reduced and consequently a signal of a higher frequency is generated. The frequency that is generated depends on the resonant-circuit conditions previously detailed, and the frequency would be the one at which the inductive reactance of the coil equals the capacitive reactance of the capacitor. The formula for the resonant frequency of such a circuit is

$$f_r = \frac{1}{2\pi\sqrt{LC}} \tag{17-1}$$

This formula gives the resonant frequency when the inductance in henrys and the capacity in farads are known. The farad, however, is too large a value and in electronic work the microfarad is employed. Hence, a more workable formula is one that utilizes the microfarad value such as shown below. By also changing henrys to microhenrys, the answer will be the frequency in kilohertz:

$$f_r = \frac{159}{\sqrt{LC}} \tag{17-2}$$

A simple capacitor and inductance combination, as described, will generate the basic waveform with the desired frequency but the small amount of energy contained in such a circuit cannot be utilized practically. As soon as the energy is taken from the circuit, the fields collapse and the circuit stops oscillating. Thus, it is necessary to furnish power to the circuit so that it can deliver a continuous amount of energy. The most practical way to do this is to furnish the circuit with d-c power and permit it to convert this power to the necessary a-c energy. This can be done by employing appropriate transistor or tube-type circuits. These circuits are so designed that they supply energy at the proper time intervals to replenish the losses incurred as the a-c energy is removed from the circuit by the load. This function can be understood by inspection of the previously discussed Fig. 17-1. At A, any energy that would be applied to the circuit must have a polarity conforming to the existing polarity *at the instant of application.* If the replenishing device applied energy in opposite polarity, it would oppose the energy in the circuit and would cause a collapse of the field and of the circulating flywheel energy. The device that furnishes energy over a period of time to the circuit shown at A must apply that polarity in the proper time interval. A transistor or tube-type oscillator will perform this timing function and will pulse or supply the resonant-circuit energy of proper polarity and in sufficient amounts to maintain a constant amplitude output from the oscillator.

17-3. Feedback Oscillator

One of the first types of oscillators developed was the *feedback* type, in which part of the signal energy at the anode of a tube was coupled back to the grid circuit in proper phase for regeneration. A transistorized version of such an oscillator is shown in Fig. 17-2. Here the feedback inductance L_1 is used to couple some of the collector's signal energy back to the emitter in a common collector circuit. The feedback coil is sometimes referred to as a *tickler* coil and the degree of coupling between it and L_2 is adjusted for maximum

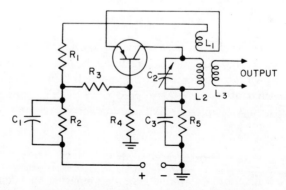

FIGURE 17-2. Transistor R-F feedback oscillator.

oscillating amplitude. Capacitor C_2 tunes the oscillator to produce the desired signal frequency.

When power is first applied to the oscillator (forward and reverse bias potentials), the initial rise of current in the circuitry provides a sufficient *change* of energy in L_2 for coupling purposes to L_1 and hence some energy is applied to the emitter circuit. This energy change, in turn, is amplified and appears across the resonant circuit at a higher amplitude. Any energy change, even though d-c in character, is sufficient to cause the flywheel effect described earlier; hence sine-wave signal energy is developed in the resonant circuit composed of C_2 and L_2. These signals are again coupled to L_1 and applied to the emitter, until finally the emitter signals have sufficient amplitude to drive the emitter alternately into the cutoff region and to the current-saturation level, at which time a state of equilibrium is reached.

Now the d-c energy of the power supply is continuously converted to a sine-wave signal energy having a frequency dependent on circuit constants of L and C. Inductor L_3 forms the necessary output secondary winding for obtaining the signals generated within the oscillator. The network R_5 and C_3 is a decoupling circuit wherein the low reactance of C_3 to the signal couples it back to the base circuit (ground) and minimizes coupling to the power supply. Positive polarity for the emitter (forward bias for a *PNP* transistor) is supplied via R_1 and R_2 through the tickler coil. The unbypassed R_4 provides some circuit stabilization for thermal changes.

17-4.   Hartley Oscillator

A typical Hartley oscillator is shown in Fig. 17-3. Such a circuit has been extensively used as the local oscillator in broadcast receiver applications as well as in certain FM transmitter applications. The resonant circuit of the Hartley oscillator is composed of the usually parallel inductance and capacitance, though in the Hartley oscillator the inductance is tapped by the emitter

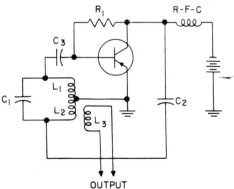

OUTPUT

FIGURE 17-3. Grounded-emitter Hartley Oscillator.

circuit (which is at ground potential). (For tubes, the tap would be by the cathode circuit.) Because of the tap, the resonant-circuit inductance is divided into two sections, composed of L_1 forming the base-emitter circuit and L_2 comprising the emitter-collector circuit.

As shown in Fig. 17-3, the power-supply potentials are shunt-fed to the collector by using an isolating radio-frequency choke (RFC) coil. Thus, the power-supply current shunts rather than flows through the resonant-circuit inductance. The isolating capacitor C_2 prevents the supply voltage at the collector from being shorted to ground through L_2. The amplified signal energy developed in the collector circuit and inductor L_2 is inductively coupled to the base section because of the autotransformer characteristics of the tapped coil. This inductive coupling thus sustains oscillations since the proper in-phase feedback network is present. Capacitor C_1 forms the other component of the resonant circuit and can be variable type for changing the frequency of the signal being generated.

Capacitor C_3 prevents grounding of the negative potentials supplied the base and collector by the power supply or battery. Inductor L_3 forms the output secondary in a transformer circuit with L_2 as the primary. Though capacitor C_1 and inductors L_1 and L_2 are the primary frequency-determining components, the final signal frequency is also influenced by the lumped inductances of the circuit, as well as by interelement capacitances within the transistor and stray capacitances of circuit wiring.

17-5.　Colpitts Oscillator

The Colpitts oscillator is shown in Fig. 17-4 and basically it resembles the Hartley. Instead of a tapped inductance, however, two capacitors are placed

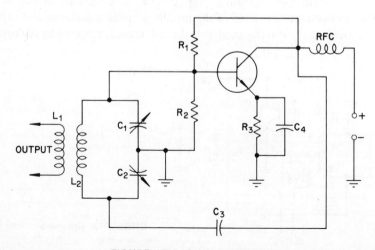

FIGURE 17-4. Colpitts oscillator.

across a common inductance (L_2) and the center of the two capacitors is tapped, again forming a resonant circuit. As with the Hartley, the inductance L_2 forms part of the base-emitter circuit, in conjunction with C_1. Capacitor C_2, however, with L_2 is part of the collector-emitter circuit as shown.

The supply energy is again shunt-fed, using the RFC for isolation of the RF signal energy. Capacitor C_3 prevents coupling the d-c potentials at the collector directly to the base (via L_2). The forward bias developed across R_1 and R_2 and applied to the base is isolated from ground by the two tuning capacitors C_1 and C_2.

Each tuning capacitor across L_2 has the movable sections grounded. Hence the grounded rotors have the same potential as the negative terminal of the supply and the oscillator's resonant circuit is effectively divided into two sections, but with coupling between the two, thus feeding back a portion of the amplified output signal to sustain oscillations, as was the case with the Hartley oscillator.

17-6. TPTG Oscillator

The input and output sections can be tuned independently of each other (without deliberately coupling the circuits together) to form what is known as a tuned-plate tuned-grid oscillator in the tube-type version or the tuned-collector tuned-base oscillator in the transistor version. Both operate in similar fashion and, for contrast to the transistor types previously discussed, the tube version (TPTG) is discussed. A typical circuit is shown in Fig. 17-5. In this type oscillator the coupling between the output and input circuits is established by the interelectrode capacity (or interelement capacitances in transistors), as well as capacitor C_4, which places the bottom of the anode resonant circuit at the same ground potential as the grid resonant circuit, thus establishing the necessary second link between input and output circuitry.

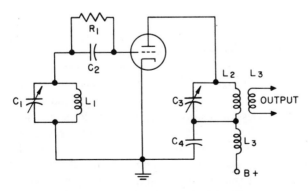

FIGURE 17-5. Tuned-plate tuned-grid oscillator.

Capacitor C_4 establishes the bottom of a plate resonant circuit at signal ground and thus maintains the RF path from plate to cathode. This is necessary to complete the connection from the resonant circuit to the oscillator tube. An RF choke is placed in series with the resonant circuit and the power supply, for isolation purposes. As with RF chokes previously mentioned, a high reactance is present for the RF energy, and this minimizes leakage to the power supply.

The frequency of this oscillator is influenced to a considerable degree by the interelectrode capacities and circuit capacitors. Also, one of the resonant circuits must be tuned off frequency for oscillations to occur. This can be understood by assuming that the grid resonant circuit composed of C_1 and L_1 is tuned to resonance. At resonance, the inductive reactance of the L_1 would have a value equal to the capacitive reactance of C_1 and since the two reactances have opposite characteristics, the respective reactance would, in effect, cancel out and the circuit would be a resistive one. Thus, the grid circuit is primarily resistive, with some capacitive reactances contributed by interelectrode capacities and the external capacitor C_2. The latter, however, has a very low reactance for the frequency of the oscillator.

If the plate resonant circuit is also tuned to resonance, it too becomes primarily resistive since the effective reactance is again canceled. In lumping together the various components that now exist in the oscillator circuit, it would be found that only resistance and capacities are present, the capacities being contributed by the interelectrode and stray capacities. Since the inductances have been effectively canceled out, oscillations cannot occur because no flywheel effect can be created when there is no interchange of energy between the inductance and the capacity.

In order to create oscillations, the resonant circuit in the plate side of the oscillator can be tuned slightly above the frequency that the oscillator is to produce. When this is done, the inductive reactance of L_2 decreases and the capacitive reactance of C_3 increases. This comes about because a resonant circuit, when tuned above the generated frequency, creates a condition equivalent to having a resonant circuit of a certain frequency and applying a lower frequency across it. When a frequency lower than the resonant frequency is impressed on a circuit, the capacitive reactance increases and the inductive reactance decreases.

The reduced inductive reactance now decreases the high impedance that formerly existed in the parallel resonant circuit. The decreased impedance is created because the inductive reactance of the coil is now lower. Hence, the signal energy finds less opposition in the coil and flows through it. The circuit becomes primarily inductive since it is the inductive reactance that has been decreased, and creates a lower opposition to signal-energy flow. Inasmuch as the anode circuit is now primarily inductive, interchange of energy between circuit capacity and the inductive anode circuit is possible. Consequently, the oscillator can generate a signal because the lumped constants consist of inductance, as well as resistance and capacity.

The function of the tuned-plate tuned-grid oscillator is closely duplicated by the crystal-controlled oscillator described next. In the crystal oscillator, only a single frequency can be obtained because of the fixed-frequency characteristics of the crystal. In the tuned-plate tuned-grid oscillator, however, the resonant circuit in the grid can be varied, as well as the resonant circuit in the plate; hence, the signal output can be changed for frequency by readjustments of the grid and plate capacitors. The desired frequency, however, is only obtained under the conditions previously detailed, with respect to tuning one of the circuits to resonance and the other above resonance.

If a pentode tube is used instead of a triode, difficulty in establishing oscillation may be encountered because of the smaller-value interelectrode capacities in a pentode. For low-frequency operation, an additional capacitor must be placed between the grid and plate terminals of the tube. For higher-frequency operation, however, the decrease in the interelectrode capacities' reactances eliminates the need for the external capacitor.

17-7. Crystal Oscillator

When the grid resonant circuit of the tuned-plate tuned-grid oscillator previously discussed is replaced by a crystal, an oscillator is formed that has a high order of frequency stability. For this reason, crystal oscillators are used extensively to control the carrier frequency of a radio, FM, or television transmitter. They are also used in other electronic applications where a high order of frequency stability is needed in a signal generator.

The type of crystal utilized is the piezoelectric quartz crystal, which has the property of vibrating at a frequency which depends on the thickness of the unit and the type of cut employed. Figure 17-6 illustrates how some typical crystal slabs are derived from the raw crystal.

In transmitting, or in other applications where accurate frequency is required, the crystal is placed in a small enclosure and is subjected to a controlled temperature. The enclosure is referred to as a *crystal oven,* and a resistance-strip heating element is utilized to maintain a constant crystal temperature for stability purposes. With most crystal types, a perceptible frequency difference results when the crystal temperature varies.

The crystal is held by two plates to form the contacting elements for each crystal surface. The two plates are then mounted in a crystal holder of low-loss plastic material. This holder usually has two terminal prongs so that the crystal can be plugged into the oscillator socket provided for it as shown in Fig. 17-7.

The two plates that hold the crystal form a capacitor since the plates are metal and the crystal acts as the dielectric of the capacity. Because of the capacity characteristics thus formed, the crystal also assumes the function

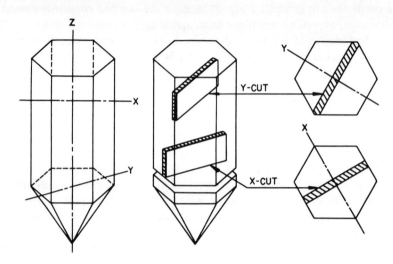

FIGURE 17-6. Piezo-quartz crystal structure.

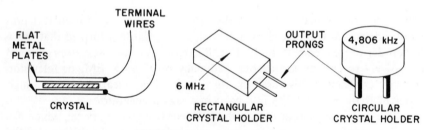

FIGURE 17-7. Crystal holders.

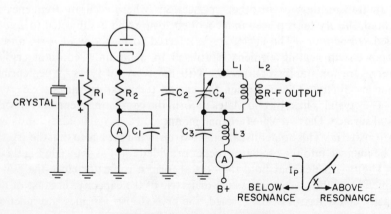

FIGURE 17-8. Crystal oscillator.

of the grid capacitor in conjunction with a grid leak. This is shown in Fig. 17-8, which illustrates the basic crystal tube-type oscillator circuit.

In similar fashion to the tuned-grid oscillator, the signal in the plate tank circuit of the crystal oscillator is coupled to the grid circuit through interelectrode capacities, to maintain oscillation. When triodes are used in crystal osillators, the interelectrode capacity is sufficient for coupling purposes. When a higher amplitude of output power is desired, however, pentodes are used and, frequently, these have insufficient interelectrode capacities to produce oscillation as mentioned earlier for the tuned-plate tuned-grid oscillator. For this reason, an additional capacitor is usually connected between plate and grid to establish the necessary degree of coupling.

The peaks of grid signal drive the grid positive, and the latter draws current. This, in turn, places a charge across the capacity represented by the crystal and, during signal intervals other than positive peaks, the charged grid capacitor will discharge across the grid leak resistor and thus establish a bias as indicated by the arrow at R_1 in Fig. 17-8. In most instances, however, a small amount of additional bias is provided by a cathode resistor (R_2), to act as a protection, should the crystal fail to operate. In the absence of oscillations, no bias would be developed at the grid leak and currents would become abnormal within the tube. The cathode resistor provides sufficient *protective bias* to limit currents below excessive values.

The capacitor C_2 acts as a bypass across R_2 to prevent signal variations across the latter, which would result in degeneration and lowered output. A milliammeter can be placed in the cathode circuit to read plate-current flow. This is useful when tuning the plate circuit to resonance. Capacitor C_1 shunts signal energy across the meter and prevents damage to the d-c instrument.

Capacitor C_3 places the bottom of the plate resonant circuit at ground potential, while L_3 is an RF choke that provides a high reactance to signal energy and prevents it from getting to the power supply.

Tuning C_4 establishes the optimum oscillation point. As shown by the plate-current waveform beside the plate-current meter, plate current is high when the circuit is off resonance. As the circuit is tuned toward the resonant frequency there is a sharp drop in plate current and then a gradual rise as C_4 tunes to a higher frequency. The most stable operation occurs within the limits X and Y of the slope of the plate-current curve. An attempt to operate the oscillator right at the dip of the plate-current curve usually results in instability and loss of oscillation.

Since the piezoelectric quartz crystal that is used operates at a frequency dependent on the type of cut and the thickness, it forms a resonance circuit that could be replaced by physical components of inductance, capacity, and resistance, in similar fashion to the tuned-plate tuned-grid oscillator previously discussed. The inductance of the crystal can be considered as an electronic equivalent of the mass that determines vibration, while the capacity is contributed by the holding plates plus the mechanical compliance of the quartz. The mechanical friction set up during crystal vibration is equivalent to the resistive component of a resonant circuit.

Because of the foregoing, the frequency of a crystal oscillator can be changed to a limited degree by introducing components of either inductance or capacity, or both, across the crystal. In some commercial applications, a small variable capacitor is placed across the crystal for limited tuning purposes so that the frequency of the crystal can be adjusted closer to that desired, without the necessity for having to grind the crystal thickness to close tolerances.

When used in transmitting, the crystal oscillator is followed by successive stages of Class C radio-frequency power amplifiers. These build up the power produced by the crystal oscillator, until the desired carrier power is secured at which the station is to operate.

As with all oscillators, plate-current or collector-current flow in the output resonant circuit (the "tank") is in the form of pulses. Because of the flywheel effect in the resonant circuit of the plate, however, the signal energy that is produced is a sine wave. Since a resonant circuit is highly selective, it will develop the greatest signal energy for the resonant frequency, while rejecting frequencies below and above the resonant frequency. Since an oscillator often generates a high order of harmonic frequencies, the resonant circuit will reject such undesired frequencies while favoring the desired frequency. Subsequent stages, which also contain resonant circuits, add to the degree of selectivity desired and further suppress spurious harmonics generated by the oscillator.

A crystal oscillator using a *PNP* transistor is shown in Fig. 17-9. A grounded-base circuit is used and, as mentioned earlier, there is no phase reversal between the signal at the emitter and that at the collector. Thus, the signal currents in the emitter are in phase with the signal currents in the collector and all that is necessary to produce oscillations is to couple some of the signal energy from the collector directly back to the emitter side. For the circuit shown in Fig. 17-9, the piezo-quartz crystal is employed as the resonant feedback loop from collector to emitter. Not only does the crystal form the feedback loop but, because of its resonant characteristics, it also establishes the oscillator frequency and

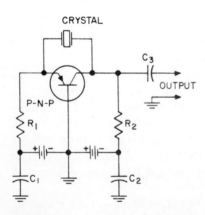

FIGURE 17-9. Crystal-controlled transistor oscillator.

maintains it with good stability. Capacitors C_1 and C_2 are for battery bypass purposes.

17-8.   Electron-Coupled Oscillator

A tube type of oscillator that is much more stable than the variable-frequency types discussed earlier is the electron-coupled oscillator. A well-designed oscillator of the latter type has a stability that almost equals that of the crystal oscillator previously discussed.

Basically, the electron-coupled oscillator employs a variable-frequency oscillator, such as the Hartley previously described or any other similar type. A special circuit arrangement is employed, however, wherein the output circuit (the load) is isolated from the primary oscillator circuit. Hence, variations in the amount of signal energy that might be drawn from the oscillator (load variations) and might normally have an undue influence on frequency stability have a negligible effect on the electron-coupled oscillator.

A typical electron-coupled oscillator is shown in Fig. 17-10, using the basic Hartley oscillator previously described. The essential difference between this oscillator and the conventional Hartley is that the *screen grid* of the pentode tube is used as the anode for the Hartley oscillator. Grids of vacuum

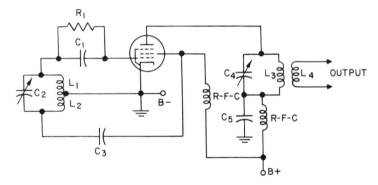

FIGURE 17-10. Electron-coupled oscillator.

tubes can be employed as anodes if a plus potential is applied to them. Since, however, the grid wires do not have the larger surface area of plates, they are unable to handle as much power as a plate element can. But the lower power generated can be increased to the amount desired by subsequent Class C amplifiers.

For the electron-coupled oscillator shown in Fig. 17-10, the Hartley circuit is that portion of the tube consisting of the cathode, the control grid, and the screen grid. Thus, the inductance split into two sections, L_1 and L_2, in conjunction with capacitor C_2, forms the parallel-resonant circuit for the

Hartley oscillator. The plate of the pentode is coupled to another resonant circuit that is independent of the resonant circuit of the Hartley oscillator. Hence, the plate resonant circuit does not contribute toward establishing the resonant frequency of the oscillator.

The Hartley oscillator section of the electron-coupled oscillator generates a signal having a frequency established by the resonant circuit, as mentioned. Thus, the signal voltages are present at the grid and signal current flows in the screen grid, the actual anode of the Hartley oscillator section. Consequently, the current that flows from cathode to the screen grid varies in accordance with the signal-amplitude changes. *Plate* current for the pentode, however, also flows from cathode to anode; hence the signal-current variations between cathode and screen grid influence the plate-current flow. When the grid signal swing cuts off current flow to the screen grid (the anode of the Hartley section), plate current also ceases. Thus, the plate resonant circuit composed of C_4 and L_3 is pulsed in a timing that coincides with the frequency established by the Hartley oscillator section. For the foregoing reasons, the output resonant circuit is *electron-coupled* to the oscillator section and loading effects are held at a minimum.

A load resistor could also be substituted for the resonant circuit of the plate. The resonant circuit in the plate, however, improves the output selectivity and thus ensures better harmonic-frequency rejection. Since all oscillators generate signals rich in harmonics, additional resonant circuits help select the desired frequency and reject the undesired harmonic components. In some applications where a higher frequency is needed than generated by the oscillator, advantage can be taken of the harmonic frequencies to produce frequency multiplication as described in Chapter 24.

17-9.   Relaxation (RC) Oscillators

In addition to the resonant-circuit oscillators, several nonresonant types are also widely used. Primary applications include the generation of low-frequency signals from approximately 30 Hz to 20 kHz in the audio-frequency spectrum. Occasionally signals having frequencies as high as 500 kHz are generated by the resistance-capacitance types. These nonresonant oscillators are known as *relaxation* types and the signal frequency is related to the values of circuit resistance and capacitance.

Instead of generating sine-wave signals, the relaxation oscillators produce special waveshapes that are easily converted to the desired shape (sawtooth, pulse, etc.) by special circuits added to the output of these oscillators. These added components modify the signal obtained from the relaxation oscillators to form the particular waveshape needed. The additional circuits include discharge systems, clippers, differentiators, and others as described later in this chapter.

The relaxation-type oscillators have circuit characteristics that permit simple synchronization of the generated frequency by an external signal of a different frequency. Thus, exact frequency control of the *RC* oscillator is possible in such applications as vertical and horizontal sweep signal synchronization in television, pulse-rate timing in radar, and control circuitry in industry.

Instead of resonant circuits, relaxation oscillators use combinations of capacitance and resistance to form charge and discharge functions in circuitry for alternately blocking electron flow through the transistors or tubes that make up the system. The two basic types of relaxation oscillators are the *blocking oscillator* and the *multivibrator*, both of which are described next.

17-10. Blocking Oscillator

Transistor and tube-type blocking oscillators are shown in Fig. 17-11. A *PNP* transistor is shown for the type at A, though an *NPN* or an FET could also be used by applying proper-polarity potentials as required. This oscillator uses the regenerative feedback principle for producing oscillations; hence it is necessary to invert the phase of signal appearing at the collector. For the circuit at A, the transformer secondary winding is used to feed a signal back to the base in phase with the existing signal.

The high degree of feedback causes the base of the transistor to be driven

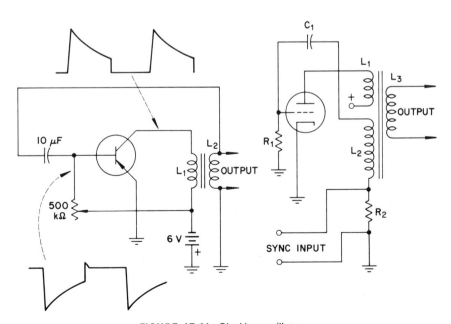

FIGURE 17-11. Blocking oscillators.

beyond the cutoff region for a time dependent on the time constant (RC) of the circuit. A charge is developed across the capacitor in the base circuit that opposes the normal reverse bias (negative with respect to the emitter for a *PNP* transistor). This charge gradually leaks off and conduction again occurs, repeating the cycle.

As shown at A, the waveforms that occur at the base and collector are not sinusoidal in waveshape and hence are rich in harmonics. Since the frequency of the generated signals depends on the circuit capacitance and resistance, the internal characteristics of the transistor plus those of the transformer have a bearing. Thus, the exact frequency of the signals generated is difficult to predict mathematically. Frequency can be changed by adjustment of the variable resistor in the base circuit, by changing the capacitance of the base capacitor and by shunting the L_1 (primary) of the transformer with a capacitor. If oscillations do not occur, the L_2 terminals may have to be reversed to assure regenerative rather than degenerative feedback.

Similar functions prevail for the tube-type blocking oscillator shown at B of Fig. 17-11. The high-amplitude feedback signal alternately blocks the grid by charging capacitor C_1 in such a direction that a negative potential appears at the grid and drives it beyond the cutoff point. Now conduction stops and the capacitor discharges through R_1 until grid bias is reduced sufficiently to again permit conduction. Thus, the conduction-nonconduction cycle is repeated.

For the circuit at B, a separate winding (called a *tertiary*) is used for feedback instead of tapping the output winding. This procedure minimizes loading effects to some extent and hence minimizes changes in signal frequency caused by the characteristics of the applied load. Such a tertiary winding could also be used for the circuit shown at A.

Synchronization (locking in the oscillator frequency by an external signal) is introduced into the grid circuit at B by resistor R_2. This procedure is also applicable to the transistor circuit at A (in the base circuit). Since such relaxation oscillators (including the multivibrator type discussed later) undergo periodic blocking of conduction, the insertion of a pulse signal at a time slightly prior to the normal start of conduction can lock in the frequency of the RC oscillator.

This synchronization principle is shown in Fig. 17-12. At A is the original signal produced by the relaxation oscillator and present at the base of a transistor (or grid of a tube). For a tube, the waveform represents a climb into the conduction region from a high negative-bias value. Thus, the grid potential becomes less negative until the bias is low enough to permit conduction. The signal may climb above the conduction region and become slightly positive before it is driven into the cutoff region again, as shown at A. For an *NPN* transistor the signal represents a change from a negative potential (reverse bias) to a positive potential (forward bias) that permits conduction. (For the *PNP*, opposite polarity conditions prevail.)

If a pulse signal is applied to the input circuit slightly ahead of the original signal as shown at B, conduction will occur sooner than the normal free-running signal frequency. Hence, the original signal will be synchronized with the input signal as shown at C because the incoming synchronization pulses now control the *time* when the transistor (or tube) goes into conduction.

For proper synchronization, the sync pulse must be close to t_1 or t_3 at A of Fig. 17-12. If the incoming synchronization pulse appears between the time interval of t_2 and t_3, its amplitude may be insufficient to reach the conduction level (particularly if the sync pulse appears near the bottom of the waveform, near t_2).

One example of such synchronization is the transmission of sync signals by a television station to lock in the sweep circuits of the receiver by remote control as detailed later. In this manner both the vertical sweep and horizontal sweep that traces out the picture on a television screen are synchronized with the sweep circuits at the television station to which the receiver is tuned. Thus, synchronization is maintained between transmitter and receiver, even though separated by a number of miles.

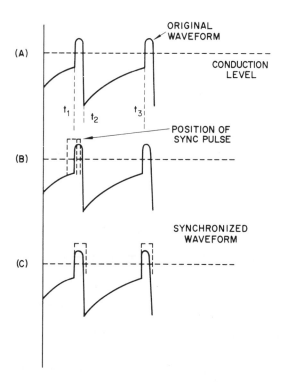

FIGURE 17-12. Synchronization in relaxation oscillators.

Because a sync pulse that occurs halfway between the conduction levels has an insufficient amplitude to overcome the dip in the waveform, a synchronizing signal of twice the frequency than the original signal can also be used to lock in the latter. Thus, a sync frequency of 100 Hz can be used to lock in a relaxation oscillator generating 50 Hz. In such an instance, *every other* sync pulse triggers the oscillator, while alternate pulses are insufficient in amplitude to be effective. The process of using a higher-frequency signal to lock in a circuit producing a lower-frequency signal is known as *frequency division*. Thus, relaxation oscillators are often used as frequency dividers in addition to simple generators of a waveform of a specific frequency.

17-11. Multivibrator

A typical multivibrator-type of relaxation oscillator is shown in Fig. 17-13, using a pair of *NPN* transistors. Again, the intermittent conduction principle

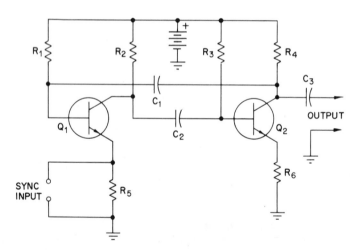

FIGURE 17-13. Multivibrator oscillator.

is used for sustaining oscillations in this *RC*-type signal generator. Instead of a transformer, another transistor (or tube) is used for feedback and phase-inversion purposes. (As with the blocking oscillator, the multivibrator can be synchronized by an external signal for frequency-control purposes.)

For the circuit at 17-13, note that the output signal of each collector is coupled to the base input of the other transistor. Since the common-emitter circuitry provides for a phase reversal of the signal between base and collector, the output from Q_1 is out of phase with respect to the input at the base. This out-of-phase signal is now applied to the base of Q_2 and when the signal appears at the collector of Q_2, it is again out of phase with that at the base, thus bringing it back to the original phase condition existing at the input of Q_1. Thus, by coupling the collector of Q_2 back to the base input of Q_1 (via capacitor C_1), we form an *in-phase* feedback loop necessary for producing oscillations.

The multivibrator of Fig. 17-13 is symmetrical, having identical-value resistors in each base circuit and collector circuit, as well as similar-value capacitors. Transistors are of the same type with equal characteristics. Despite this, however, when power is first applied, current flow rises more rapidly in one transistor than in the other and thus initiates the oscillatory cycle.

Assume, for instance, that Q_1 initially conducts slightly more current than Q_2. When Q_1 conducts, the voltage drop across its collector resistor R_2 rises and capacitor C_1 is charged with a negative polarity at the base of Q_2 and a positive polarity toward the Q_1 collector. As the negative potential at the base of Q_2 increases, conduction through this transistor drops because of the decrease in its forward bias.

As conduction through Q_2 drops, the positive potential at the collector of Q_2 will increase since the voltage drop across its collector resistor R_4 declines, bringing the collector closer to the positive supply voltage. The increase in Q_2 collector voltage is felt at the base of Q_1 because of the coupling capacitor C_1. This increase in the forward bias at the base of Q_1 raises conduction additionally through this transistor. The conduction rise continues until saturation is reached, at which time Q_2 is at cutoff and no longer undergoes a *change* of voltage or current. Since a *change* no longer occurs, capacitor C_2 now discharges and within a short time permits Q_2 to conduct again. When this occurs, Q_2 collector potential decreases as the voltage drop across R_4 increases. This collector change of potential charges C_1 with a negative polarity toward the base of Q_1. This, in effect, reduces the forward bias of Q_1 and hence decreases its conduction. The process continues rapidly until Q_1 is at cutoff and Q_2 conducts at saturation. Now the reverse of the original condition prevails and the process starts over again, repeating the process and thus sustaining oscillations and the generation of a signal.

Since a symmetrical circuit prevails, the synchronization input could be across R_6 instead of R_5 as shown in Fig. 17-13. Similarly, the output could be procured from the collector of Q_1 instead of Q_2, with the only difference relating to the phase of the signal in terms of time. The frequency of the generated signal depends on values of R and C, as well as the internal characteristics of the particular transistors employed, plus the amplitude of supply potentials applied.

17-12. Discharge Circuit

By proper selection of component values and potentials, the output waveforms from relaxation oscillators have sharp rise time and decline time, as well as time intervals when signal voltage remains at a fixed level. Thus, basic square waves are formed that can be refined additionally to obtain the waveshape needed. Square waves are important since they are used in a number of electronic applications as more fully described later. On other occasions, however, a special waveform is required in which the voltage has a *gradual incline* and a *rapid decline*. This is the case with television transmitters and receivers, as well as in such test equipment as oscilloscopes, where the gradual incline of voltage is used to deflect an electron beam either across or down the face of the cathode-ray tube.

When a waveform has a gradual incline and a sudden decline in periodic intervals, it resembles a sawtooth pattern. Hence, this type of signal is known as a *sawtooth waveform*. As shown in Fig. 17-14, a sawtooth voltage can be formed by placing a capacitor from collector to ground of a common-emitter circuit. An input signal obtained from a relaxation oscillator is applied and the amplitude is sufficient to cut off conduction and permit conduction periodically.

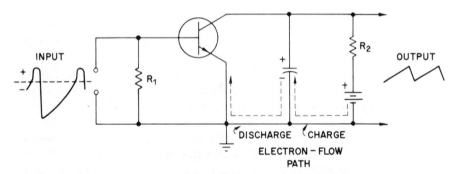

FIGURE 17-14. Discharge method of sawtooth formation.

During the time the transistor is in its nonconducting state, it behaves as an open circuit across the output capacitor. During this time the power supply or battery charges the capacitor, with electron flow as shown by the arrow. Thus the capacitor receives a charge that is negative toward ground and positive toward the collector and output terminal. Thus, there is a gradual rise of voltage across the capacitor as shown by the output-signal waveform of Fig. 17-14. This gradual rise of voltage across the capacitor forms the initial incline of the sawtooth waveform. The current also flows through resistor R_2 in circulating through the conduction loop formed by the capacitor, battery, and resistor. Hence, R_2 limits such current flow and affects the amplitude of the sawtooth that is developed because of the time constant established.

When the input signal swings into the direction that causes the transistor to conduct (positive for the *NPN* type), current from the supply source flows through it. During conduction, the internal resistance of the transistor is very low and it shunts the discharge capacitor. Because of the transistor's low resistance during conduction, a short *RC* constant exists and the capacitor discharges through the transistor, with electron flow to the emitter and collector in its return to the positively charged capacitor side. Thus, the voltage charge across the capacitor drops to zero, forming the sharp sawtooth segment.

Because of the short duration of the positive portion of the input signal, the transistor conducts for only a short period of time. Abruptly, the transistor is once more driven to cutoff and hence offers a high impedance across the capacitor again. The periodical charge and discharge of this capacitor thus forms the sawtooth signal-voltage waveform shown at the output. Such a discharge circuit produces a sawtooth having a linear rise if the time constant of charge does not extend into the curved portion of the capacitor's charge characteristic. The discharge capacitor can be placed directly across the output of a relaxation oscillator, though more critical design is necessary to obtain the required sawtooth amplitude and desired waveshape.

17-13. Square-Wave and Pulse Factors

The square-wave type of signal generated by the relaxation oscillators previously discussed differs radically from the simple sine-wave signal. Because of the inherent differences between the square-wave signal and other special signals, a more detailed analysis is necessary, in order to understand the factors involved in the handling and applications of such signals in electronic circuitry.

The square-wave type of signal is made up of a fundamental frequency plus a number of odd harmonics. This is in contrast to the simple sine-wave that consists of a fundamental frequency only. A square wave not only has a fundamental frequency but a third harmonic having one-third the amplitude of the fundamental, a fifth harmonic having one-fifth the amplitude of the fundamental, a seventh harmonic having one-seventh the amplitude of the fundamental, and successive odd harmonics, up to approximately 15 for a low-frequency fundamental. For higher-frequency square waves, the harmonics may have frequences extending up to the two-hundredth odd harmonic. In some very-high-frequency square waves (1 million square waves per second), harmonics of the fundamental range over 1000. When a waveform has a fundamental frequency plus a number of harmonic frequencies, all the signals combine to make up the square-wave appearance of the composite signal. The manner in which such signal combinations form the square wave may be understood by reference to Fig. 17-15. At A is shown a sine wave that represents the fundamental frequency of the square wave. A waveform of this type, which has a gradual incline and decline, is composed

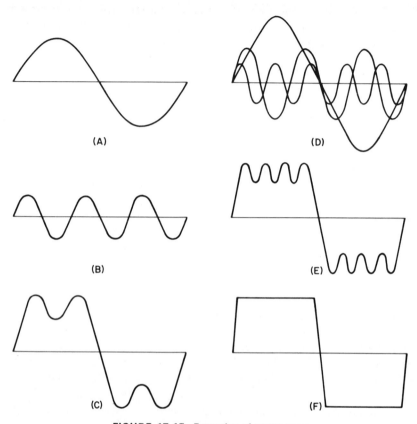

FIGURE 17-15. Formation of square wave.

of one frequency only, provided that the amplitude of each alternation is the same as the others, and also provided that the duration of one alternation is identical to the other alternations. At B, a signal having three times the frequency of the fundamental is shown. Since the frequency of the signal shown at B is three times that at A, the signal at B is known as the third harmonic. If the third-harmonic signal is combined with the fundamental, the signal shown at C will be produced. The production of the signal shown at C is the result of the combination of the in-phase and out-of-phase sections of the signals shown at A and B. At the point of maximum amplitude of the fundamental, an out-of-phase condition exists with respect to the waveform of the third harmonic at that point. Thus, the addition of the two amplitudes at this particular point results in a decrease in the overall amplitude because of the out-of-phase conditions. The harmonic signal, however, will not always be out of phase since it is higher in frequency than the fundamental and, at some points, an in-phase condition exists. Thus, the harmonic signal will result in the addition and subtraction of the amplitude of the fundamental along various points.

At D the fundamental, the third-harmonic, and the fifth-harmonic signals are shown. Note that, at the beginning of the combination waveforms, in-phase conditions exist that will cause the combination to have a steep rise time. Thus, when a number of odd-harmonic components are added to a fundamental frequency, the composite waveform begins to resemble the square wave. If the fundamental, the third, fifth, and seventh harmonics are combined, the waveform shown at E will be formed. Already the composite signal begins to have the sharp rise-time characteristics of the square wave. When all the higher-order odd harmonics are combined with the fundamental frequency, the resultant will be the square wave shown at F.

The square wave has successive positive and negative alternations and, hence, is ac in its characteristics. Pulses, on the other hand, consist of either negative or positive polarity signals and thus have characteristics that differ somewhat from the square wave. The differences between the square wave and the pulse are shown in Fig. 17-16. At A a square wave having a frequency

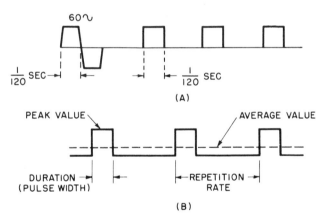

FIGURE 17-16. Pulse characteristics.

of 60 Hz is shown. Hence, each alternation of the square wave occurs in $\frac{1}{120}$ of a second. When pulses are formed from this square wave, each pulse will have a duration of $\frac{1}{120}$ of a second, as shown at A. The pulses, however, need not occur at a rate of 60 per s. In many electronic devices, the pulses may occur at 100, 300, 600, or any other rate per second. The rate at which the pulses occur per second is known as the *repetition rate*. The repetition rate of pulses can be adjusted as desired, regardless of the duration of a pulse. Thus a pulse may have a duration of $\frac{1}{50}$ of a second, or a millionth of a second, but its recurrence may be at a low rate or at a very high rate, as required. This is not the case with an unbroken chain of square waves (or sine waves) because such waveforms have a repetition rate fixed by the duration of the individual alternations of the signals.

From the foregoing, it is obvious that the assignment of a specific frequency to a series of pulses is meaningless because it would not impart the necessary information related to the pulses. If we say the repetition rate is 25 per s, it does not tell us whether the pulse has a duration of $\frac{1}{50}$ of a second or $\frac{1}{1000}$ of a second. Consequently, an evaluation of a pulse waveform must be based on the repetition rate, as well as on the duration of the pulse. When the duration of a pulse is multiplied by the repetition rate, a figure known as the *duty cycle* is obtained. As an example, assume that each pulse in a series has a duration of 4 μs. If the pulses are repeated 600 times per second, the duty cycle would be 0.000004 times 600, or 0.0024. Thus, the duty cycle is a designation that relates to the pulse duration, as well as to the repetition rate. Once the duty cycle of a pulse train is known, the average value of the pulse train can also be ascertained. As mentioned earlier, the effective value for sine waves was obtained by multiplying the peak value by 0.707. A calculation of this type, however, cannot be applied to find the average value in a pulse train because the repetition rate is a determining factor with respect to the power that is present. The average power for a series of pulses is usually considerably below the 0.707 value obtained when solving for the power in sine waves. The average value for a series of pulses is shown at B of Fig. 17-16.

The average power for a train of pulses may be calculated by multiplying the peak power value by the duty cycle. Thus, if the duty cycle of a series of pulses is 0.0024 and the peak power of a pulse is 10 watts, the average power would be

$$0.0024 \times 10 = 0.024 \text{ W}$$

Since the formula for finding the average power is based on the peak power and the duty cycle, the duty cycle can be found by dividing the average power by the peak power. Thus, in the foregoing example, if the average power of 0.024 is divided by the peak power of 10 W, the duty cycle of 0.0024 will be obtained.

As is the case with the square-wave type of signal, the pulse is made up of a number of harmonics, each higher harmonic having a lower amplitude than the previous harmonic. The more narrow the pulse, the greater the harmonic content. In contrast to square waves, however, the pulse waveforms contain both odd and even harmonics. The maximum number of harmonic frequencies that have a bearing on the shape of the pulse varies inversely with respect to the duration of the pulse. Wide pulses may have significant harmonic components only to the fifteenth order, while very narrow pulses may have harmonic frequencies ranging to 1000 or more, as previously mentioned. The rise time of the leading edge of a pulse as well as the decline time (or decay time) of the trailing edge depend for their sharpness on the harmonic content of the pulse. This is an important factor with respect to the amplifiying circuits that must handle pulses (as well as square waves). If the circuits are not well designed and cannot pass some of the higher-frequency harmonics of a square wave or pulse, distortion of the original waveform will result. A pulse that has lost some of its higher-frequency signals will no longer have sharp rise time and decline time with respect to the leading or the trailing edges, and hence the pulse will become distorted because the leading and trailing edges will slope. Thus, amplifier circuits that are used to handle square waves or pulses must be able to pass all the harmonic components contained within the waveform, for an accurate amplification of the original signal.

The least amount of pulse distortion occurs for loss of the *lower*-frequency components. Often, some of the lower-frequency components of a pulse may be removed without seriously altering the pulse shape. Thus, the requirements with respect to low-frequency response in an amplifier are often based on the repetition rate of the pulses to be handled. For retaining the higher order of harmonics, however, the circuit through which the pulses pass must be able to pass signals having frequencies equivalent to

$$\text{Base frequency} = \frac{1}{\text{duration}}$$

The foregoing formula solves for what is known as *base* frequency and is indication of the highest frequency that the circuit should pass for good reproduction of the pulse. Thus, if the pulses in a train of pulses each has a duration of 5 μs, the base frequency is 20,000 Hz (20 kHz). Hence, the circuits handling this pulse train should be capable of passing frequencies from the repetition rate up to the base frequency of 20 kHz. If the repetition rate is 1000 Hz, this would establish the limit of the low-frequency response for the circuit.

The formula for finding the base frequency does not utilize the repetition rate of the pulses because the repetition rate has no effect on the harmonic content of each individual pulse. If the 4-μs pulse mentioned above occurred at a rate of 5000 per s, the frequencies contained within the pulse would occur more frequently than at a low repetition rate, but the number of harmonics contained within the pulse would not change.

17-14. Pulse Distortion

As mentioned earlier, square waves and pulses are composed of a fundamental frequency and a number of harmonics. Consequently, any circuit through which such waveforms pass must be so designed that none of the harmonic components are diminished. If some of the harmonic components are lost as the waveform passes through a circuit, signal distortion will result. Two primary types of distortion with respect to square waves and pulses are

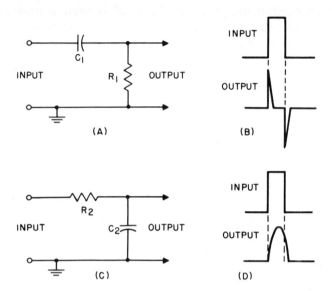

FIGURE 17-17. Differentiation and integration of pulses.

the loss of high-frequency components and the loss of low-frequency components. Figure 17-17 illustrates the types of distortions that result from either high-frequency or low-frequency signal losses. Capacitor C_1 and resistor R_1, at A, represent a typical signal-coupling circuit between two amplifier stages, with R_1 constituting the input resistor and C_1, the conventional coupling capacitor. Usually, C_1 is made sufficiently large in value so that it will have a low reactance and pass all frequencies desired, without undue diminishing of the signal amplitudes. If C_1 has too small a value, its reactance will be high for low-frequency signals and, hence, would diminish the lower-frequency signals to a greater degree than high-frequency signals. If a number of sine-wave signals of different frequencies were applied to the input, the lower-frequency signals appearing at the output would have less amplitude than the higher-frequency signals, even though all had the same amplitude when applied to the input.

When a pulse is applied to the input, the circuit behavior follows the time-constant factors given earlier with respect to the universal time-constant chart. The leading edge of the pulse has a sudden rise in voltage and, according to the exponential curve for a capacitor charge, maximum current flows in the circuit during the initial capacitor charge. Hence, the sudden flow of maximum current through R_1 results in a sharp rise of voltage across R_1. With a small-value capacitor, it only takes a brief moment to become fully charged. Thus, during the flat-top portion of the pulse, a steady-state condition prevails and current flow through the capacitor declines, again according to the exponential curve of the time-constant chart. Hence, voltage across the resistor drops, as shown at B of Fig. 17-17. With the capacitor fully charged, current flow ceases. When the trailing edge of the pulse arrives, the voltage at the input of the circuit drops sharply below the value of the charged capacitor. Consequently, the capacitor suddenly discharges through R_1 and the discharge current flows in the opposite direction from the charge current. Thus, the voltage across R_1 rises in a *negative* direction, as shown at B. When the capacitor is fully discharged, the voltage across R_1 drops to zero again.

From the foregoing, it is evident that the value of the coupling capacitor (and the time constant of the circuit) can be such that a pulse applied to a circuit can be modified considerably at the output. The output pulse shown at B is known as a *differentiated* waveform. The term indicates that the pulse has lost low-frequency components and that the waveform at the output contains mostly high-frequency components. In electronic devices where the pulse must retain its original waveform, it is essential that the capacitor be made adequate in size so that no differentiation occurs. On occasion, however, the time constants of a coupling circuit may be deliberately changed so that differentiation takes place. A differentiation circuit is employed on those occasions when the important factor of a waveform is a sharp leading edge. A differentiated waveform is ideal for synchronization purposes and is used in some computer circuitry, as well as in television receivers.

A differentiating circuit can be considered as a high-pass filter because the high-frequency components of a pulse or square wave pass through the circuit but low-frequency components are diminished. The differentiating circuit has no effect on sine-wave-type signals, other than diminishing their amplitude, depending on the frequency of the signals and the reactance of C_1. The waveshape of sine-waves is not altered when such signals pass through a differentiating circuit.

As mentioned earlier, interelectrode capacities in transistors or tubes, circuit capacities, and the distributed capacities in transformers, all have a shunting effect on signals because of the low-reactance path to ground that prevails due to such capacities. While these capacities will diminish the amplitude of sine-wave signals, their effect on a pulse waveform causes distortion to occur because some of the higher-frequency harmonics of the pulse will be diminished in amplitude. At C of Fig. 17-17, a circuit is shown in which C_2 represents a shunting reactance. If such a capacitance has an appreciable low-value reactance for the higher-frequency components of a pulse, signal distortion as shown at D will occur. The leading and trailing edges of a pulse represent high-frequency harmonic components. At the output of the circuit shown at C, the leading and trailing edges of a pulse will not have the sharp changes contained in the original but will have slopes as shown at D. This type of waveform is kown as an *integrated* waveform, indicating a decrease in high-frequency signal amplitudes but containing the low-frequency signal components. An integration circuit is essentially a low-pass filter, which filters out higher-frequency signals but passes the lower-frequency signals. The various shunt capacities represented by C_2 are usually held at a minimum to prevent pulse distortion. On occasion, however, an integrating circuit is employed deliberately, such as in the vertical sweep systems of television receivers.

17-15.   Clippers and Clampers

Circuits that perform functions of clipping, limiting, and clamping are widely used in all branches of electronics for removing noise or other undesired signals from waveforms, for clipping sine waves to produce square waves, for changing square waves to pulses, for holding waveforms at a constant amplitude, and for establishing a fixed d-c reference level. Such waveform modification is extensively employed in industrial control circuits, automation circuitry, computer systems, communication gear, and other electronic devices.

Often the terms *clipper*, *limiter*, and *clamper* are used interchangeably. Strictly speaking, however, clipper is a circuit wherein the amplitude of the output waveform is proportional to those values of the input waveform *exceeding* a certain critical value. A limiter, on the other hand, is a circuit wherein the amplitude of the output waveform is proportional to those values of the input waveform *up to* a certain critical value. A clamper is a circuit that relates (clamps) the amplitude of the waveform to some fixed reference level.

Clippers are in the form of series or shunt types, depending on whether

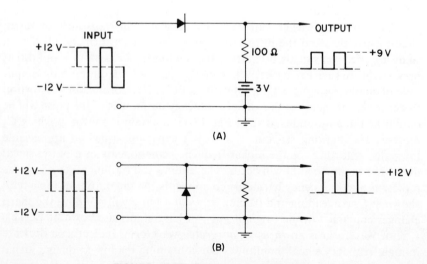

(A)

(B)

FIGURE 17-18. Series and shunt clippers.

the output resistor shunts the clipping diode or is in series with it. Typical series and shunt clippers are shown in Fig. 17-18. For the series clipper at A, the battery supplies a positive bias to the diode which is opposite to that which could permit conduction. Thus, no current flows through the 100-Ω resistor until the positive polarity of the input signal exceeds (and thus overcomes) the 3-V bias. For the negative alternations of the input signal, the diode does not conduct; hence the circuit clips these portions of the input signal. Thus, the output signal has an amplitude that is proportional to that amplitude of the input signal exceeding the bias potential.

For the circuit at B, no bias is used with this shunt clipper. Thus, for positive alternations of the input signal, the diode does not conduct and hence behaves as an open circuit across the output resistor. Since the nonconducting diode has a virtually infinite impedance, it has no effect on the positive-polarity output signal.

For negative alternations of the input signal, however, the diode has proper polarity applied to permit conduction; hence its forward resistance is now so low that it acts as an effective shunt across the output resistor. Thus, no output signal is present when the input signal is of negative polarity. Consequently, the output consists of a series of positive pulses as shown, with the negative portions clipped.

A shunt clipper with positive bias is shown in Fig. 17-19, and again the battery polarity is opposite to that which permits conduction. Thus, since the battery potential is 3 V, diode conduction occurs only when the positive amplitude of the input signal exceeds this value. The effect on several signal

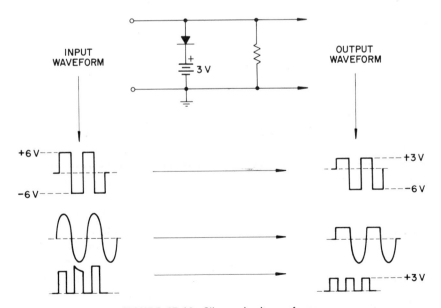

FIGURE 17-19. Clipper-circuit waveforms.

types is illustrated below the schematic. The bias amplitude can, of course, be set at any value necessary to obtain a particular degree of waveform clipping.

As shown, if the input signal is a square wave having a 6-V amplitude (12 V peak-to-peak), an output signal amplitude is obtained that is proportional to the input signal up to a 3-V maximum in the positive direction. Beyond the 3-V amplitude, the diode conducts and hence shunts the signal around the output resistor. For negative alternations of the input square wave, however, the diode does not conduct and the output is thus proportional to the input signal. Similarly, if the diode and the bias potentials are reversed, the negative portions of the output signal would be clipped.

As also shown in Fig. 17-19, the biased shunt clipper can also be used for removing portions of either the negative or positive alternations of sine waves, depending on the polarity of the bias potential. For the circuit shown, the result is a flat-top production of the positive output alternations. Again, reversing the diode and bias polarity will result in clipping the negative-polarity alternations of the input sine wave.

As shown in the lower waveform drawings of Fig. 17-19, the shunt clipper is also useful for limiting pulse amplitudes. Thus, if the input pulse train has a varying amplitude (with some pulses higher than others), the output can be held at a constant amplitude as shown. This limiting function is also useful for removing noise transients or other undesirable distortion forms from the top of a pulse train. Since the positive-polarity bias is 3 V, the output amplitude is held at this level because of the shunt effect of the diode when the input-signal amplitude is sufficient to cause the diode to conduct.

At A of Fig. 17-20, two diodes are used (with opposite-polarity connections) to form a parallel clipper (also referred to as a *slicer*). With dual diodes, both the negative and positive alternations are clipped simultaneously, thus forming a square wave at the output as shown. As with other clippers, the degree of slicing is regulated by setting the bias potentials as required.

At B is shown a conventional common-emitter amplifier used to form square waves from sine waves. Here, the input signal is sufficient in amplitude to drive the transistor into the saturation level as well as into the cutoff region. Consequently, both peaks of the two alternations of the cycle are clipped, as shown. If the amplitude of the input signal is increased, the output square waves will have steeper sides as greater portions of the peaks are clipped.

Clamping circuits are useful for restoring the d-c component of a pulse train after such a d-c reference level is lost because of capacity coupling in amplifiers or other circuits handling pulse signals. The loss of the d-c component when a capacitor-coupled stage is used is shown at A of Fig. 17-21. The amplified pulse train appearing at the collector of transistor Q_1 rises in a positive direction as shown. Assume there is a 2.5-V drop across the Q_1

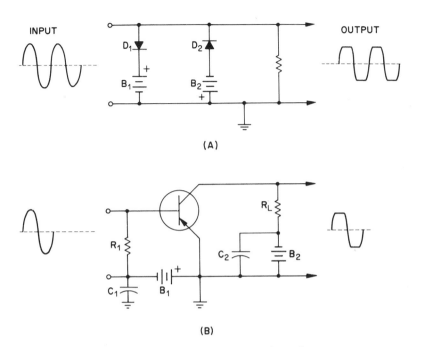

(A)

(B)

FIGURE 17-20. Dual-diode and transistor clippers.

load resistor R_L in the absence of a signal. When a signal appears, the *change* of voltage across R_L constitutes a rise to 7.5 V for the peak of the pulse. Thus, the pulse amplitude is $7.5 - 2.5 = 5$ V. The d-c component exists and is equal to the *average* voltage of the pulse train, as shown.

Because of the coupling capacitor C_1, however, the d-c level is lost because the d-c signal components cannot pass through a capacitance. Hence, the pulse train that now appears at the input of transistor Q_2 has a *zero* reference level (with positive and negative alternations) instead of the d-c average value. Consequently the pulse train has been converted into what is essentially a square-wave signal instead of a series of *single-polarity* pulses.

To obtain the d-c reference level again requires the waveform to be clamped to a d-c or zero level to restore the initial signal characteristics that consist of single-polarity pulses. The clamping process is also called *d-c restoration* and usually consists of the inclusion of a diode in shunt to the input resistor R_1 of Q_2, as shown at B of Fig. 17-21. At time t_1 during which no input signal appears, capacitor C_1 charges through the collector load resistor R_L with a polarity (and electron-flow direction) as shown at A. At time t_2, the input pulse signal appears and voltage rises to 7.5. Because of the sharp rise time of the leading edge of the pulse, the capacitor C_1 is unable to charge any appreciable amount. The high current during the leading edge

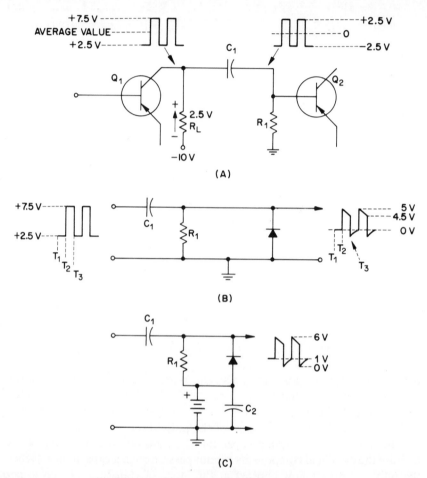

FIGURE 17-21. Clamping (dc restoration).

of the pulse places the full pulse amplitude across the resistor R_1 for this brief interval. Hence, the potential across R_1 is the difference between the 7.5-volt peak of the pulse and the idling voltage change across thhe capacitor C_1. This comes about because the 7.5-V peak pulse amplitude causes the polarity of the input terminals to the clamper to be opposite to the charge on the capacitor, or 5 V. Thus, the output waveform rises to this 5-V peak value as shown at B.

The time constant (RC) of C_1 is long compared to the duration of the input pulses. Hence, C_1 charges only slightly more for the duration of the pulse $(t_2$ to $t_3)$ and current through R_1 is low. If, for instance, the charge on C_1 rises to 3 V at the end of the duration of the pulse, the output pulse amplitude declines by 1.5 V to a value of 4.5 V at t_3 as shown at B.

At time t_3 when the input pulse amplitude declines, the voltage drop across R_1 changes $(-3 + 2.5 = -0.5 \text{ V})$. Hence, the voltage output at t_3 becomes -0.5 V, producing a short negative spike in the output waveform as shown. This negative voltage across the diode causes conduction and the low impedance of the conducting diode shunts R_1, changing the long time constant to a short one for a brief interval of time. Hence, a rapid discharge occurs for the -0.5-V charge of C_1.

Capacitor C_1 discharges the -0.5 value through the diode, leaving the original 2.5-V charge on the capacitor. Since this charge is a static condition for the time interval *between* pulses, there is no current flow through R_1 and hence no voltage drop across it. Thus, the interval between the pulses appearing at the right of C_1 are clamped to zero level as shown at B of Fig. 17-21.

Sometimes it is desirable to clamp the output voltage level at some value above or below zero. To do this, the circuit is modified by adding a fixed potential (battery or other power source) as shown at C. Capacitor C_2 has a bypass effect across the battery, thus placing the bottom of R_1 and the diode at *signal* ground.

To understand circuit function, initially assume a pulse train is to be clamped at a 1-V level above the zero line. To do this, the battery or power supply has its positive terminal connected to the diode anode and the bottom of R_1 as shown at C. At time t_1 the charge across the coupling capacitor C_1 would normally be a fixed 2.5 V but the 1-V bias will oppose the charge voltage and hence C_1 charges only to -1.5 V.

At time t_2 the circuit functions in a manner similar to that described for the circuit at B. The rapid rise of the pulse amplitude is too fast for the long time constant of the RC network to permit any appreciable charge to build up across C_1 during the leading-edge time interval. Thus, maximum pulse current flows through R_1, creating a 5-V drop across it. Electron flow is up through the resistor and the polarity of the voltage drop across R_1 coincides with the bias polarity, producing a rise in voltage at the output to 6 V (5 V + 1 V). During the rapid rise time, C_1 and C_2 have virtually no charge built up across them but permit full current flow through R_1.

The time constant is selected so that an additional 1-V charge develops across C_1 during the flat-top interval of the pulse. This added charge now raises the charge across C_1 to -2.5 V again since it had a -1.5-V charge on it originally. Now the output pulse amplitude (at the base of Q_2) drops to 5 V.

At time t_3 (when the input pulse drops in amplitude) there is an excessive negative charge of 1 V across C_1, causing the diode to conduct. Now the excessive 1-V charge is discharged and the C_1 charge reverts to -1.5 V again (2.5-V source less the 1-V bias). The discharge produces the negative spike to the 0 line as shown at C. If the excessive charge on the capacitor had been only -0.5 V, the spike would not have reached the zero reference line. In either case, however, the waveform is clamped to the 1-V level as shown.

Review Questions

17-1. Briefly explain what is meant by *flywheel effect* and how Lenz's law applies.

17-2. What factors determine the resonant frequency in a sine-wave generator?

17-3. What two factors relating to a tickler coil must be observed with respect to the feedback oscillator?

17-4. Explain why the Hartley oscillator shown in Fig. 17-3 can also be called a *shunt-fed* oscillator.

17-5. What are the essential circuit differences between a Hartley oscillator and a Colpitts oscillator?

17-6. Briefly explain in what manner a crystal stabilizes the frequency of a sine-wave oscillator.

17-7. What is the primary advantage of the electron-coupled oscillator over the basic Hartley oscillator?

17-8. Briefly explain what is meant by a *relaxation* oscillator.

17-9. How does the time constant (RC) relate to the frequency of signals produced by a blocking oscillator?

17-10. What circuit characteristics in relaxation oscillations permit synchronization of frequency by an external source? Explain briefly.

17-11. How are relaxation oscillators converted into frequency dividers?

17-12. Explain the purpose for a discharge circuit, and describe its general function.

17-13. What type harmonics make up a square wave, and how does the combination of such harmonics produce a sharp leading and trailing edge?

17-14. Briefly explain the difference between a square-wave-type signal and a pulse.

17-15. What is meant by the *duty cycle* of a pulse, and how is the duty cycle used for ascertaining the average power of a pulse train?

17-16. Briefly explain the factors that cause distortion of square waves and pulses.

17-17. Draw a differentiated waveform and an integrated waveform, and also draw the circuits that produce such waveforms.

17-18. Are the distorted pulses represented by integrated and differentiated waveforms ever employed deliberately? Explain.

17-19. Explain, in your own words, the differences among limiters, clippers, and clampers.

17-20. Draw the schematic of a shunt limiter, using negative pulses at the input, with each pulse having a different amplitude than the others. Indicate the limiting function by illustrating the output-pulse waveform.

17-21. Draw a circuit showing positive-voltage clamping, with the input pulses having an amplitude from 50 to 150 V, clamped at a +15-V output level.

Practical Problems

17-1. In the design of an oscillator, the parallel-resonant circuit had a total inductance of 0.00005 H and capacity of 0.00025 μF. What is the resonant frequency?

17-2. The engineering specifications for an RF oscillator gave the product of *LC* as 0.0253. If the inductance was in μH and the capacity in μF, what is the resonant frequency in kilohertz?

17-3. An RF oscillator designed to operate above the broadcast band has a total *LC* product of 0.00045, with *L* rated in μH and *C* in μF. What is the resonant frequency in kilohertz?

17-4. In testing a crystal oscillator such as shown in Fig. 17-8, it was found that 150 mA of plate current flowed when operating with a normal load and resonant-circuit conditions. If the voltage drop across the cathode resistor is 6 V, what is the value of the resistor? Would a 2-W resistor be adequate?

17-5. In an experimental electron-coupled oscillator designed to operate at approximately 652 kHz, a capacitor of 85 pF was used with a 704-μH inductance. Interelectrode and stray capacitances, however, contributed 15 pF of capacity to the resonant circuit. What is the actual resonant frequency?

17-6. In the oscillator in Problem 17-5, an RF choke coil of 20 mH was used. What is its reactance?

17-7. A relaxation oscillator used for industrial control produced pulses having a duration of 2 μs each. If the repetition rate is 1000 per s, what is the duty cycle?

17-8. If the oscillator mentioned in Problem 17-7 produced pulses having a peak power of 5 W, what is the average power of the pulse train?

17-9. If the oscillator mentioned in Problem 17-7 were modified to decrease the repetition rate by one-half at the same peak power, what would the average power be?

17-10. The pulses employed in an automation circuit had a repetition rate of 40 kHz and each pulse had a duration of 5 μs. What is the base frequency and the lowest frequency that an amplifier circuit must be capable of passing to handle such pulses?

17-11. What is the duty cycle of the pulses in Problem 17-10?

17-12. In the design of a zero-clamp circuit, the capacitor had a value of 0.0001 μF and the resistor had an ohmic value of 50,000 Ω. What is the time constant of the capacitor-resistor combination?

17-13. In a positive-voltage clamp circuit such as shown at C of Fig. 17-21, capacitors C_1 and C_2 *each* had a value of 0.004 μF. If the resistor value is 50,000 Ω, what is the time constant?

Modulation and Demodulation   18

18-1. Introduction

Low-frequency electric signals such as produced from relaxation oscillators, microphones, and other devices are incapable of being transmitted any great distances without employing such enormous power as to make it impractical. High-frequency (RF) signals, however, can be sent over thousands of miles with power only a fraction of that which would be required to send low-frequency signals over short distances. Thus, the high-frequency signals are utilized to "carry" the low-frequency signals in the manner described in this chapter. When a high-frequency signal is thus utilized, it is known as the *carrier* signal, and the manner in which this is done at the sending transmitter is known as *modulation*. The process is widely used in the various electronic branches. In radar, for instance, the carrier is modulated by a pulse, while in radio the carrier is modulated by audio signals. In television the carrier is modulated by picture-signal information.

When the modulated carrier signal arrives at the receiver, the information is extracted from the carrier by a process known as *detection* or *demodulation*. The various methods for both modulation and demodulation are covered in this chapter.

The basic principles covered in this chapter serve as a foundation for the complete systems described and illustrated in Chapter 19. Chapter 19 includes television, stereo, and multiplex coverage. AVC, since it relates to detectors, is covered in this chapter.

18-2. Amplitude Modulation

There are several methods for modulating a carrier with a low-frequency signal. One of the earliest processes was that known as *amplitude modulation*. This system is still used in radio (AM carrier) and in television (picture carrier). Essentially, AM is a system of modulation where other signal components are created in addition to the carrier to produce a resultant (composite) modulated waveform that has amplitude variations conforming to the characteristics of the audio or other signal information to be transmitted.

Figure 18-1 shows a basic transmitting system. When someone speaks or a musical instrument is played, the varying air pressure on the microphone generates an a-c signal that corresponds in frequency to the original sound. The amplitude of the a-c signal that is generated by a microphone, however, is too low for practical use and must be increased by several audio-amplifier stages, as shown in Fig. 18-1.

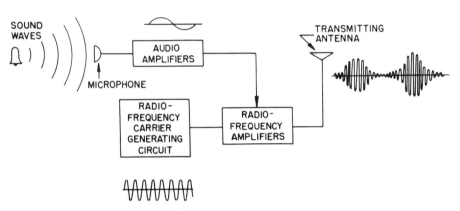

FIGURE 18-1. Simple block diagram of radio transmitter.

The transmitter has an RF carrier-generating circuit that produces a high-frequency signal of constant amplitude. This carrier-generating circuit usually consists of a crystal oscillator, such as described earlier. The oscillator generates the basic *carrier frequency*, as allotted by the Federal Communications Commission for the particular radio station. For AM radio broadcasting, a spectrum range from 550 to 1600 kHz is allocated. Thus, the carrier frequency for a particular radio station may be 700 kHz, or perhaps 1 MHz, or any other single frequency in the radio broadcast band. Much higher carrier frequencies are used in shortwave, television, FM transmission, and radar. Frequency-modulation allocations are between 88 and 108 MHz, while television stations have space in both the VHF and UHF regions (see appendix).

As with the microphone, the oscillator does not furnish sufficient signal output for most transmission purposes, and hence the a-c carrier signal generated by an oscillator is amplified by several RF amplifiers of the Class C variety. Most commercial transmitting stations build up their carrier to many *kilowatts* of power. Such high power is necessary so that the transmitted signal will not only travel over the required distances but will also place a high-level signal on the receiving antenna.

The amplified audio energy from the microphone is combined with the amplified carrier signal. The two combine in such a manner that the amplitude of the resultant RF waveform changes in accordance with the audio waveform. This produces *amplitude modulation*, as represented by the waveform drawing beside the antenna in Fig. 18-1.

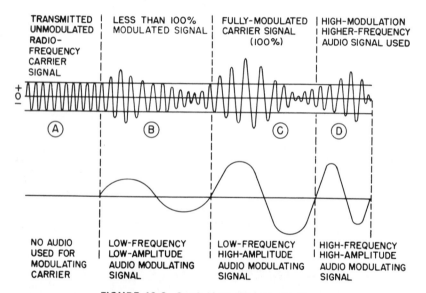

FIGURE 18-2. Graph of amplitude modulation.

The manner in which the carrier is modulated by the audio-frequency signal is more clearly illustrated in Fig. 18-2. Here, the unmodulated RF carrier signal is shown at A. It is an a-c type of signal that has a constant frequency and constant amplitude. When the audio signal is of a low frequency and amplitude, it will modulate the carrier as shown at B. Note that the amplitude of the resultant waveform increases slightly for one alternation of the audio-signal waveform and decreases slightly below its normal level for the second alternation of the audio waveform. Thus, the waveform amplitude has been modulated to a slight degree by the low-frequency, low-amplitude audio signal. If this same frequency audio signal is increased (i.e., if the sound going into the microphone is made louder), it will have a greater amplitude and thus will have more effect on altering the amplitude of the resultant waveform. This is shown at C, where the amplitude now reaches a level twice that of the normal carrier amplitude for the initial alternation of the high-amplitude audio waveform. For the second alternation of the high-amplitude audio waveform, the resultant amplitude drops to virtually a zero level. This is known as a fully modulated carrier, or a carrier modulated 100%.

If the carrier is modulated more than 100%, the amplitude of the resultant waveform will increase beyond twice that of the normal carrier amplitude and, for the second alternation of the audio waveform, the amplitude would reach zero and remain there for an appreciable interval. Since this cuts off the carrier for a short interval, the carrier would not be transmitted during this time, resulting in high distortion. Such distortion tends to decrease the clarity of the signals at the receiver. This undesired condition is prevented by keeping the modulation below the 100% level.

If a high-amplitude as well as a higher-frequency audio signal is employed, it will cause the waveform amplitude to vary more rapidly, as shown at D of Fig. 18-2. Thus, the resultant waveform increases and decreases in amplitude at a more rapid rate (in a shorter time interval) than for a lower-frequency audio-modulating signal.

The modulated carrier cannot be represented accurately in schematic form since the high frequencies employed for carriers prohibit indicating the number on the drawings. If, for instance, the carrier is 1 MHz (1 million Hz) and the audio signal is 500 Hz, the 500-Hz audio tone would cause the carrier to increase and decrease in amplitude 500 times per second. Since, in this instance, there are a million cycles of the carrier signal per second, it means that for every audio cycle there will be 2000 carrier cycles. Thus, one cycle of the 500-Hz audio signal causes an increase and decrease of a 2000-Hz sequence of the carrier signal.

It must be noted that, in amplitude modulation, the audio is not really "carried" by the transmitted signal. Actually, the RF signal waveform has been changed only insofar as its amplitude goes, and the audio component is represented only by such changes in the amplitude of the RF-waveform *envelope*. At the receiver, a special *detector* circuit must be employed to *demodulate* the carrier and derive from it the audio waveform, as described later in this chapter.

When an RF carrier is modulated by an audio signal, the original sine-wave carrier undergoes a form of waveshape distortion. When a pure sine wave is altered with respect to its waveshape (e.g., duration of one alternation different from the other, or amplitude of the first alternation different with respect to the second), additional frequency components are present. During amplitude modulation, the sine wave of the carrier is distorted with respect to the amplitude of the various alternations. If any cycle is analyzed in the modulated waveform, it will be evident that one alternation has a different amplitude from the other, as shown in Fig. 18-2. This distortion process of the sinusoidal waveform during amplitude modulation results in the creation of two additional frequencies, besides the carrier and audio signals. One of these newly developed signals will be lower in frequency than the carrier by an amount equal to the audio frequency. The other newly created signal will be higher than the carrier frequency by an amount also equal to the audio signal. Thus, if the carrier is 1000 kHz and the modulating audio frequency is 500 Hz, a new frequency (999.5 kHz) will be generated below the carrier and another new frequency (1000.5 kHz) will be generated above the carrier. These two additional signals are known as the *side-band* signals. If the modulating audio voltage is 2 kHz, the lower side band will be 998 kHz and the upper side band will be 1002 kHz. Thus, for each audio frequency employed in the modulating process, two side bands are generated. If two audio signals are employed, four side bands will be present. During the transmission of music, for instance, a number of audio signals would be used to modulate the carrier at any particular time, with the result that many side-band signals would be created.

If the side-band signals were filtered from the composite AM waveform, it would be found that the carrier has a constant amplitude. It is only when the side-band signals are combined with the carrier that the resultant RF waveform contains amplitude variations.

From the foregoing, it is evident that a carrier, plus its side bands, will require more spectrum space than an unmodulated carrier since the latter consists of only a single frequency. In the example cited above, the 2000-Hz audio tone would generate the two side bands that, in conjunction with the carrier, would occupy a space of 4 kHz. If an audio frequency as high as 5 kHz is utilized to modulate the carrier, it would result in two side bands, one 5 kHz below the carrier and the other 5 kHz above the carrier. In the latter case, the result is the transmission of a band of frequencies 10 kHz wide, as shown in Fig. 18-3.

Because a *band* of frequencies is generated during modulation, it is

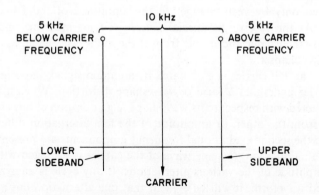

FIGURE 18-3. AM bandpass.

necessary that the receiver provide a band pass in its RF stages that will accommodate the carrier plus the accompanying side bands of the transmission. In radio receivers, the band pass is usually set at 10 kHz since, in normal transmission of AM, the highest-frequency signal components that are transmitted rarely exceed 5 kHz. There are some higher-fidelity AM stations that transmit a band of frequencies extending from 16 Hz to 10 kHz. These, in turn, would require a band pass within the receiver extending to 20 kHz for proper reproduction of the higher-frequency audio components above the 5-kHz range.

Since AM is also used for the transmission of picture signals in television, an extremely wide band-pass range is required because video signals have frequencies from 30 Hz to 4 MHz. For this reason, television receivers must have a band pass of 4 MHz in the RF stages to accommodate the range of frequencies involved. (Most of the lower-side-band components are suppressed, in television transmission, to eliminate the necessity for employing an 8-MHz band pass.)

There are several methods for producing amplitude modulation, and one of these is shown in Fig. 18-4. Here, the plate current for the final carrier

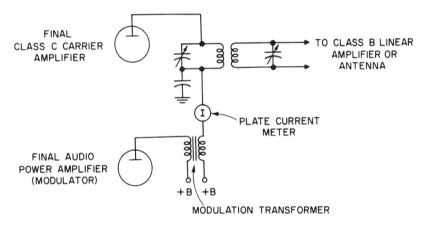

FIGURE 18-4. Plate modulation.

amplifier (Class C) is fed through the secondary of a special transformer known as a *modulation* transformer. The primary of the modulation transformer is part of the anode circuit of the final audio-power amplifier. The final audio amplifier, when transformer-coupled to the anode circuit of the Class C carrier amplifier, is known as a *modulator*. (The final audio amplifier is similar to that used in public-address systems, except that higher power is developed for proper modulation of the carrier. The modulation transformer can be likened to the output transformer feeding the conventional loudspeaker. In modulation, however, the anode circuit of the Class

C amplifier replaces the loudspeaker. The modulation transformer matches the impedances involved in the two circuits.)

The system shown in Fig. 18-4 is known as a *plate modulator* because modulation takes place in the plate (anode) of the final Class C carrier amplifier. In this system, the side-band power is furnished by the modulator since it is the latter that is directly influential in changing the signal-current amplitude and signal voltage in the Class C amplifier. Audio voltages induced across the secondary of the modulation transformer either add to or subtract from the voltage applied to the Class C amplifier, depending on the polarity of the particular alternation developed across the secondary. Thus, the voltages across the secondary have a direct influence on the current amplitude changes and thus amplitude-modulate the carrier.

In plate modulation, the *output power* of the modulator must be one-half of the Class C modulated-amplifier *input* power for 100% modulation. The *input* power to a Class C amplifier refers to the product of the plate voltage multiplied by the plate current. The input power must not be confused with the signal power applied to the input of the Class C amplifier. The signal input power is known as *excitation*. The modulator *output* power refers to the audio-signal power developed by the modulator. For the modulator shown in Fig. 18-4, the final audio-power amplifier usually consists of push-pull tubes, either Class A or Class B. For high-power transmitters, special large-sized audio-power tubes are employed. The anodes of such tubes often dissipate so much heat that forced-air cooling is utilized. In some instances, cooling is accomplished by circulating water through copper tubing around the anode section of the tube. The final Class C carrier amplifier also consists of two tubes in push-pull, in the larger transmitters. For low-power transmitters, a single tube may be employed, as shown in Fig. 18-4.

In transistor circuitry, the equivalent of plate modulation is the collector-modulation process illustrated in Fig. 18-5. Here push-pull circuitry is illustrated, though single-ended amplifiers could also be used. Since the Class C final RF amplifiers (Q_1 and Q_2) are triodes used in resonant circuitry, neutralization is required and each neutralizing capacitor (*NC*) couples the collector of one transistor to the base circuit of the other. Split-stator capacitors are used for tuning the tank (output) resonant circuit, and the ground of the common rotor shaft minimizes the danger of shock during tuning. A radio-frequency choke (RFC) isolates the RF energy of the Class C stages from the modulator and power supply.

Since this modulation is similar in principle to the plate modulation discussed for Fig. 18-4, the same factors apply. Side-band power is supplied by the modulator, and the audio signals induced across the modulator-transformer secondary winding add to or subtract from the amplitude of the voltage potential applied to the Class C push-pull stage. Again, the output power of the modulator must be one-half the Class C modulated-amplifier input power for 100% modulation.

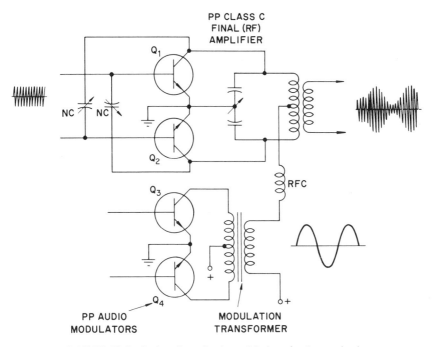

FIGURE 18-5. Push-pull amplitude modulation of collector circuitry.

Besides modulating the output section of the Class C stages, it is also possible to modulate in the input circuitry, by applying the modulating signals to the grid of tubes or the base elements of transistors, as shown in Fig. 18-6. (In tube circuitry it is also possible to modulate at the screen grid or even the suppressor grid. Circuit factors for such modulation are similar to grid modulation.)

The modulator used for such grid-input or base-input modulation need furnish only a fraction of the power required for plate or collector modulation. In grid or base modulation, the modulator signals add to or subtract from the bias voltage applied to the Class C amplifier stages and, in this manner, vary the amplitude of the current changes in the Class C output circuits, in accordance with the audio-modulating signals.

As shown in Fig. 18-6, the bias for the Class C tubes or transistors is initially set by a battery or power supply (bias pack). Signal-voltage variations across the secondary of the modulation transformer add or subtract from this bias, depending on whether the signal-voltage alternations are negative or positive. In such modulation, the signal applied to the input of the Class C amplifier tubes or transistors must have an amplitude as shown in Fig. 18-7. With the Class C amplifier biased beyond cutoff, the input signal (excitation) from the previous stage must be of such an amplitude that it

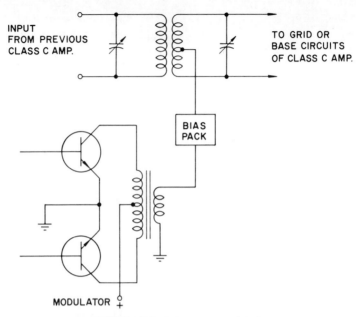

FIGURE 18-6. Grid or base modulation.

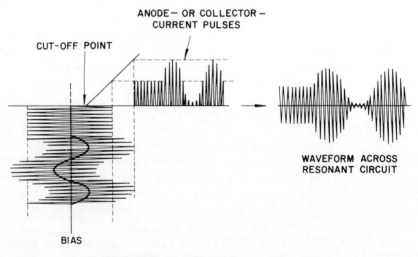

FIGURE 18-7. Graph of modulation by bias variation.

extends from cutoff to approximately half the distance to current saturation. This setting of excitation is essential so that during the modulation process the input signal can swing the carrier signal both *above* and *below* the unmodulated value. Output current (anode or collector), in pulses, produces the modulated carrier in the output resonant circuit because of the flywheel effect inherent to the resonant characteristics. Because the excitation to the Class C amplifier must be reduced for such modulation of the input circuitry, the efficiency and power output of the Class C stage is below what it would be for plate or collector modulation. For input modulation as shown in Fig. 18-7, the Class C amplifier (*not the modulator*) must furnish the side-band power.

The AM carrier, whether produced by output- or input-circuit modu-lation, is represented by the type of drawing shown earlier in Fig. 18-2. In such a representation, the *audio* signal is not present but is represented by the amplitude changes of the carrier, such amplitude changes being proportional to the amplitude of the audio signal. The recurrence rate of the amplitude changes of the carrier are proportional to the frequency of the modulating audio signal. The overall waveform, as shown at A of Fig. 18-8 also con-tains the side-band components in addition to the carrier. If the side-band components were filtered from this waveform, the remaining carrier would be represented as shown at B. The upper side band would be as shown at C, while the lower side band is represented at D. Note that the carrier has a *constant* amplitude, as does each side band. It is only when the side-band

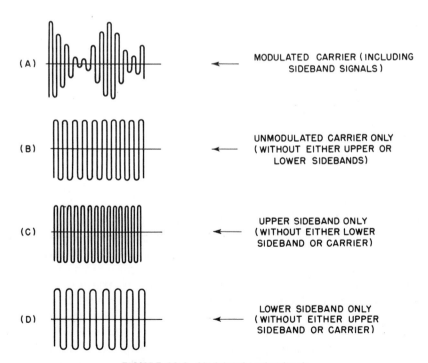

FIGURE 18-8. Modulated carrier signals.

signals are added to the carrier that the resultant amplitude-changing wave-form shown at A is obtained because the point-by-point addition of the wave-forms involves out-of-phase signals at certain times and, at other times, in-phase (or partially in-phase) signal conditions. When either side band is combined with a carrier, the resultant again resembles the composite wave-form shown at A, even though one side band is absent. Since a signal with amplitude changes is necessary for detection of the modulated wave, either one or both side bands, in conjunction with the carrier, must be available for the receiver. Since side-band power contributes to the total signal energy, the reception of a carrier plus two side bands will result in a louder signal than the reception of a carrier plus only one side band. The latter procedure, however, can be employed for spectrum-saving reasons. This is done in the transmission of the television AM carrier. In ordinary radio broadcasts, however, both side bands are transmitted and received.

The two side bands can be sent out without the carrier, in a transmitting system known as suppressed carrier transmission. This method is employed on such occasions where it is not expedient to transmit the carrier. For proper demodulation purposes at the receiver, however, it is necessary to have available a frequency identical to the carrier, in order to detect the transmitted signal information. This system is employed in color television transmission, where the color subcarrier is utilized for the generation of side bands, but the carrier is suppressed and only the side band components are transmitted. At the receiver, however, a separate oscillator must be employed to generate a frequency identical to the carrier frequency employed at the transmitter. This new carrier signal which is generated in the receiver is then mixed with the side-band signals which are received in order to produce the AM signal necessary for detection purposes.

18-3. Frequency Modulation

Another method of modulation other than AM is that known as *frequency modulation* (FM). This particular modulation principle is utilized in FM transmission, between 88 and 108 MHz (standard FM broadcast band). Frequency modulation is also employed with respect to the sound carrier that accompanies the television signal. Besides such standard broadcasting, FM is also utilized in industrial closed-circuit transmission and reception, as well as in special electronic devices employed by the armed forces.

Frequency modulation has several advantages over AM. In FM, the dynamic range is much greater. By dynamic range is meant the difference between loud and soft volume levels of audio. In AM transmission, an excessive amount of modulation can cause severe distortion because the carrier can be overmodulated. In frequency modulation, however, overmodulation is not possible and, hence, the range of soft and loud sounds is much more realistic because the ratio between loud and soft is much greater than can be employed for AM. In AM, the frequency range is also limited since transmission is usually confined to a maximum frequency extending to between 5 and 8 kHz of audio. In FM, however, an audible range of from 30 Hz to 15 kHz can be employed. Actually, the AM process itself does not limit the frequency range. The limitation is imposed by the bandwidth employed in AM transmission. At the same time, most of the intermediate-frequency stages of the smaller radios are set at a 10-kHz bandwidth to minimize interference from adjacent stations. Hence, the maximum audible range is limited to approximately 5 kHz. This is a disadvantage because many of the overtones of music are not transmitted and the reality of the reproduced music suffers when compared to the original. In frequency modulation, the extended audio range permits a much more realistic reproduction of music, which produces what is known as "presence" in high-fidelity terminology. The word *presence* denotes that the reproduced music or sound gives the impression of exact reality since the orchestra or the performer seems to be *present* in the room.

Another advantage with FM is the minimum of interference that results from adjacent stations. Beside interference rejection, one of the primary advantages of FM is the high reduction of static. For normal reception, there is a total absence of static since static is a form of amplitude modulation and a well-adjusted FM receiver will reject AM signals. Hence, FM reception is not marred by crackling noises or squeals and whistles, as is often the case with AM radio reception.

The one disadvantage that FM suffers with respect to AM is the fact that a wide-band type of transmission must be employed. In AM, most stations utilize only a 10-kHz bandwidth, while a few utilize 16 kHz if they can use such transmission without interference to adjacent stations. In FM, however, a 200-kHz bandwidth allocation is employed for each station in the standard FM broadcast band. In the FM utilized for the sound of television, a narrower FM transmission is employed, approximating 50 kHz. In either case, however, a much wider portion of the spectrum is utilized by each FM station than by a standard broadcast AM station. This slight disadvantage, however, is more than outweighed by the many advantages that FM has over AM.

In FM, the carrier is shifted above and below its normal resonant frequency by the modulating audio signal. Thus, if a 400-Hz audio tone is employed for modulation purposes, it will shift the carrier above and below its center frequency 400 times per second. If a 1000-Hz audio tone is used for modulating purposes, the carrier will shift above and below its resonant frequency 1000 times per second. (The resonant frequency of the FM carrier is often referred to as the *resting* frequency.)

Thus, the *frequency* of the audio signal determines the *rate* at which the frequency of the carrier shifts above and below its resting frequency. The *degree* of shift (the extent by which the frequency changes) is related to the *amplitude* of the audio signal. Thus, a *low volume* of audio will cause the carrier to shift *only slightly* above and below its resting frequency, while a *loud* (high-amplitude) *audio signal* will cause the carrier to shift to a *greater extent* on each side of its resting frequency. Consequently, the *extent* of carrier frequency shift depends on the *amplitude* of the modulated audio signal.

A better understanding of how the carrier frequency shifts with respect to an audio signal can be gained by analyzing a simple FM oscillator. One such basic circuit is shown in Fig. 18-9 and consists of a variable-frequency Hartley oscillator similar to the type previously discussed. In place

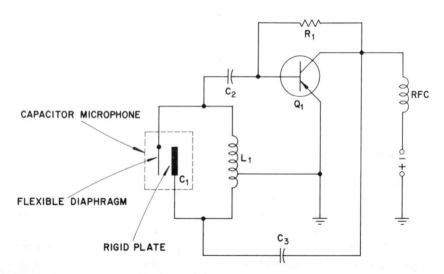

FIGURE 18-9. Simple FM process.

of a variable capacitor, however, a *capacitor microphone* is employed, as shown. This capacitor microphone (C_1) will influence the frequency of the Hartley oscillator in a fashion similar to that of a variable capacitor. With a variable capacitor, an increase in the capacity will result in the production of a lower frequency from the Hartley, while a decrease in the capacity will cause the frequency of the Hartley oscillator to increase. This same ability to change the frequency of the oscillator holds true with capacitor microphone C_1, shown.

The influence of the capacitor microphone on the frequency when sound is impressed on the microphone can be understood by reference to Fig. 18-10. When sound is generated by a musical instrument or by the spoken word, the air pressure in the surrounding area is alternately increased

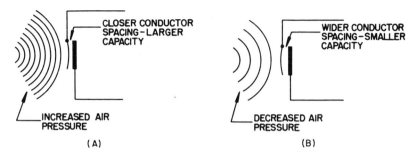

FIGURE 18-10. Effects of varying air pressure on capacitor microphones.

above normal and decreased below normal. (This increase and decrease in air pressure impinge on the eardrum and alternately presses the eardrum inward and pulls it outward. This mechanical motion of the eardrum is translated into the sensation of sound by the auditory nerve in the ear.)

What occurs to the diaphragm of the capacitor microphone when an increase in air pressure is impressed across it is shown at A of Fig. 18-10. The flexible microphone diaphragm is pressed inward and thus is closer to the rigid back plate. This closer spacing of the two conductors (the flexible metal diaphragm and the rigid metal plate) *increases* the capacity of the capacitor microphone. (Bringing the plates of the capacitor close together increases capacity.) Since the capacitor microphone represents the capacity that in conjunction with the inductance (L_1), forms the resonant circuit of the Hartley oscillator, the frequency output of the latter is changed. With a larger capacity, there is a decrease in the frequency output of the Hartley oscillator. Thus, the frequency of the oscillator is lowered by virtue of the increased air pressure. When the air pressure becomes normal, the diaphragm of the capacitor microphone will again assume its regular position and the oscillator is brought back to its normal resonant resting frequency. When the air pressure decreases, the diaphragm of the capacitor microphone and the rigid plate separate, as at B, resulting in a smaller capacity. The decrease in capacity will increase the signal frequency of the Hartley oscillator and the result is that there has been a shift to a higher frequency because of the decreased air pressure. When the air pressure again becomes normal, the diaphragm of the microphone reverts to its regular position and the oscillator is brought back to its normal resonant resting frequency.

The increase and decrease in air pressure represents one cycle of the audio tone produced by a musical instrument or by speech. Thus, the frequency of the oscillator decreased, returned to normal, increased, and returned to normal again. This simple form of FM serves to illustrate the actual formation of an FM wave, though more elaborate methods are employed for the FM process, as more fully explained later. The relationships of frequency change are identical, however, and the simple circuit helps illustrate the process more thoroughly. If the volume of the sound reaching the microphone is increased, it will be obvious that the change in air pressure illustrated at A would also increase, and the diaphragm of the capacitor microphone will be forced inward to a greater extent. This closer capacitor plate spacing results in a much larger capacity than was the case with the smaller degree of air pressure increase. Consequently the frequency of the oscillator shifts to a much lower value. The same degree of shift also holds for the illustration at B. If the air pressure decreases to a greater extent, the capacity is reduced proportionately and a greater increase in oscillator frequency ensues.

The frequency change is represented also in Fig. 18-11, which shows the exact relationships between the audio-modulating signal components and the carrier wave. At A, the initial portion of the carrier wave is shown in its unmodulated state. When a low-volume audio tone is used for the modulating process, the frequency of the carrier shifts slightly lower. The point of greatest frequency shift will be where the audio volume is at its peak level. When the first alternation of the audio tone drops to zero, the carrier returns to its resting frequency. When the second alternation of the audio signal occurs, the oscillator frequency shifts slightly higher, the maximum shift

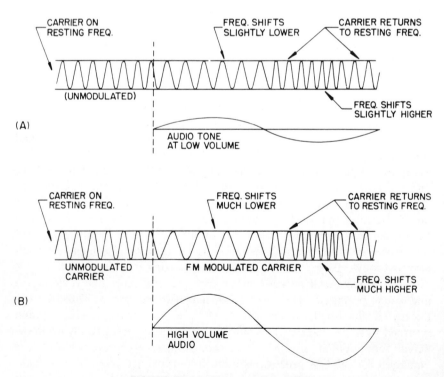

FIGURE 18-11. Characteristics of FM.

being at the point where the peak negative amplitude of the second alternation occurs. When the second alternation of the audio signal drops to zero, the carrier frequency will come back to its resting frequency.

The increase in carrier shift (deviation) for a louder audio tone is shown at B of Fig. 18-11. The higher amplitude audio causes a much greater shift of the carrier frequency to a lower frequency. Again, however, at the zero level of the first alternation, the carrier returns to its resting frequency. When the negative (second) alternation signal occurs, the carrier frequency shifts to a much higher frequency than was the case with the low-level audio signal shown at A. At the zero level of audio at the end of the second alternation, the carrier comes back to its resting frequency.

The *frequency* of the audio signal also has an effect on the carrier shift. If the audio frequency were a 3-kHz tone, the carrier would shift 3000 times per second. If the same audio frequency were used as at B, the carrier would still shift 3000 times per second but would now deviate to a greater extent than was the case at A.

To illustrate the foregoing by actual figures, assume that a station of 90 MHz is on the air. The carrier would then be on its resting frequency for the unmodulated condition shown initially at A. When a low-volume audio signal is employed, the carrier may shift from 90 MHz (90,000 kHz) to 89.975 MHz (89.975 kHz) representing a 25-kHz shift in frequency for one alternation of the audio signal.

When this audio-signal alternation drops to zero, the carrier shifts back to 90 MHz. For the second alternation of the audio signal, the carrier shifts to 90.025 MHz (90,025 kHz) and drops back to 90 MHz when the audio alternation drops to zero. Thus, for one cycle of the audio signal, the carrier is made to deviate a total of 50 kHz, or 25 kHz on each side of resting frequency. At B of Fig. 18-11, where a greater audio-volume level is used, the carrier would shift from 90,000 kHz in its unmodulated state to 89,950 kHz for the first alternation of the audio-modulating signal. At the second alternation, the carrier would shift to 90,050 kHz. Thus, the larger audio volume now causes a total carrier deviation of 100 kHz since it deviates 50 kHz on either side of its resting frequency. In either case, the carrier would shift above and below its resting frequency 3000 times per second if a 3000-Hz audio tone is employed.

From the foregoing, it is evident that the loudest signal will cause the greatest carrier deviation. Since the *extent* of carrier deviation for a given audio-signal amplitude can be established at the transmitter, it is essential that some regulations be imposed on the transmitting industry so that a uniform type of transmission is employed by all. For this reason, the Federal Communications Commission has allocated for standard FM stations a maximum deviation of 75 kHz on each side of resting frequency. Thus, the *total permissible deviation* for each station is a maximum of 150 kHz. Standard FM stations are assigned carrier frequencies between 88 and 108 MHz.

This allocation is just above the lower television stations (Channel 6 has an allocated frequency of 82 to 88 MHz.) The upper FM-band limit of 108 MHz is followed by other broadcasting services extending to 174 MHz. Channel 7 begins at 174 MHz, as shown in the appendix.

As can be seen from an inspection of the FM carrier in Fig. 18-11, the modulation of a carrier by varying its frequency again distorts any cycle of the modulated wave, by varying the time duration of one alternation with respect to another. Any alternation of the time duration of one alternation with respect to another results in the production of frequencies other than the fundamental frequency. For this reason, side bands are generated in FM as is the case with the AM. With FM, however, there is virtually an *infinite* number of side bands generated. As with AM, the first two side bands (one above and one below the carrier) are spaced from the carrier by a frequency equal to the audio frequency producing the modulation. The additional side bands in FM are also spaced from *each other* by a frequency equal to the modulating frequency. Thus, if a 1000-Hz audio signal is employed for modulating purposes, each side band is spaced from the other by 1000 Hz, or 1 kHz. The side bands near the carrier frequency have the greatest amplitude, and subsequent side bands spaced away from the carrier have a decreasing amplitude. Of the many side bands produced, there are only a few that have sufficient amplitude to prove of value during the reception of the signal. The side bands that have a value for detection are known as the *significant side bands.* In standard FM broadcasting, eight significant side bands are present above the carrier and eight below, during maximum permissible modulation. This is based on the ratio of maximum carrier swing versus the maximum audio frequency employed and is known as the *deviation ratio.* Thus, in standard FM broadcasting, the maximum deviation is 75 kHz, and the highest audio frequency employed for modulation purposes is 15,000 Hz. This produces the following:

$$\frac{75,000}{15,000} = 5$$

The *deviation ratio* relates to *maximum values* as opposed to the instantaneous characteristics which occur during actual broadcast (since maximum carrier swing and maximum audio-frequency signals occur rarely). During normal broadcasting, where values drop below the limits set by the deviation ratio, the term *modulation index* is applied to the ratio of carrier shift to modulating signal.

For a modulation index of 0.4 or less, only one significant side band exists above and below the carrier. For a modulation index of 0.5, there are two significant side bands above and below the carrier. For a modulation index of 1, there are three significant side bands above and below the carrier. For any modulation index between 1 and 10, the following significant side bands exist:

Modulation Index	Number of Sidebands above and below Carrier
1	3
2	4
3	6
4	7
5	8
6	9
7	10
8	12
9	13
10	14

In television, the frequency modulation employed for the sound uses a maximum deviation, on each side of the carrier, of 25 kHz. For this reason, the number of significant side bands is much less than in standard FM since the deviation ratio is lower.

The deviation shown above indicates maximum modulating conditions and is not representative of normal transmission. Since no musical instruments produce *fundamental* frequencies in excess of 5 kHz the frequencies above 5 kHz that are generated are only the *overtones* produced when musical instruments are played. Such overtones or harmonics are much lower in amplitude than the fundamental frequency, and for this reason they would not cause as great a deviation of the carrier frequency. At the same time, the side-band components produced by the overtones would be spaced far from the carrier and would have low amplitude. Thus, during normal transmission, the significant side bands do not extend beyond the 75-kHz limit set for deviation on each side of the resting frequency of the carrier. As an added protection, however, 25 kHz are added to each FM station channel to guard against spillover that would interfere with an adjacent station. These 25 kHz-sections illustrated in Fig. 18-12 are known as *guard bands*, and because of them each station has a *total frequency allocation* of 200 kHz.

In frequency modulation, the *carrier* must furnish the *side-band power*. Consequently, the carrier amplitude actually varies somewhat during the modulation process since some of the energy of the carrier is utilized for the generation of the side bands. Thus, during extensive deviations of the carrier due to a loud AM signal, the total power contained in the side bands can exceed the carrier power. The typical FM carrier wave shown earlier in Fig. 18-11 is representative not only of the carrier but also of the side bands. If the side bands were filtered from the modulated carrier, the remaining carrier signal would be a single-frequency signal having a varying ampli-

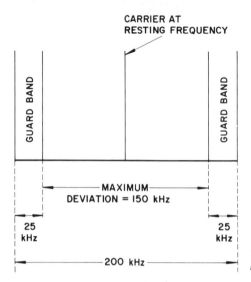

FIGURE 18-12.   FM station frequency allocation.

tude conforming to the power periodically relinquished to create the side bands.

18-4.   Controlled-Reactance Circuitry

The capacitor-microphone method for producing FM is unsatisfactory in terms of linearity (equal deviation above and below the carrier) and, in commercial FM systems, other methods are employed for deviating the carrier frequency in accordance with the modulating signal. The process involves using circuits in which the reactance (and hence either the capacitance or inductance) can be altered by applying d-c or a-c signals to the input. Since such reactance circuits are capable of controlling oscillator frequency, they are also used for stabilizing the carrier oscillator. In FM the necessity for varying the carrier frequency requires use of a variable-frequency oscillator. To keep such a signal generator on its resting frequency, however, its frequency is compared to that of a crystal oscillator. A discriminator circuit (discussed later) can then be used to obtain a correction signal for application to the reactance circuit.

Both the solid-state devices and vacuum tubes can be used to form reactance or frequency-control circuitry. A typical tube type is shown in Fig. 18-13 using a triode. Pentodes could also be used. The functional components making up the reactance are resistor R_2 and capacitor C_3. (Capacitor C_2 is a large-value unit, usually of 0.01 μF to 0.03 μF, for blocking the dc of the reactance tube anode and thus preventing its being shorted to ground via the oscillator coil. At the frequency used, its reactance is too low to contribute anything to the reactive function of the circuitry.)

Resistor R_2 is chosen so that its resistance is approximately 10 times the

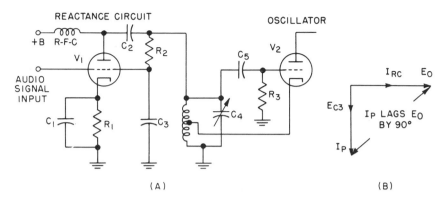

FIGURE 18-13. Reactance-tube circuit.

reactance of C_3. Thus, R_2 may have a resistance of approximately 100,000 Ω, while C_3 has a reactance value around 10,000 Ω. Note that the reactance tube anode-grid circuit is coupled to the oscillator resonant circuit. Such coupling impresses across the R_2-C_3 network the voltage of the oscillator E_0. For a clearer understanding of the phase relationships that are established, reference should be made to the vector diagram shown at B. The oscillator voltage E_0 is designated by the horizontal line. Inasmuch as R_2 is 10 times as high in resistance as the reactance of C_3, the *RC* network is primarily resistive so that the current flow created by E_0 will be virtually in phase with the voltage. Hence, current through the *RC* network (I_{RC}) is also indicated by a horizontal line at B. The grid signal voltage of the reactance tube V_1, however, is derived from across C_3 only (E_{C_3}). Also, in a capacitor, voltage lags current. Thus, the grid signal voltage lags the *RC* network current by 90°, as shown on the vector diagram. In a vacuum tube, plate current is in phase with grid voltage since a negative grid-signal decreases plate-current flow, and a positive grid signal increases current. The plate current (I_p) for the vector diagram is thus drawn along the vertical line, to show the in-phase condition with respect to E_{C_3}. Obviously, then, *the reactance tube plate current lags the oscillator voltage* by 90°. A lagging current (or a leading voltage) is indicative of an *inductive reactance*; hence the reactance circuit shown at A behaves as an inductance. Since this inductance shunts the oscillator resonant circuit, the reactance tube inductance influences the resonance of the oscillator.

Reactance in a vacuum tube depends on current flow, and any change of reactance tube current will affect the reactance value of V_1. Hence, a change of grid potential will alter reactance. The grid potential can be in the form of an audio signal and applied to the grid, as shown at A of Fig. 18-13. (This tube does not have dual grids. The drawing of a grid wire extending out from the tube, both at the left and at the right, is a common expedient used to simplify schematic drawings.)

The value of the inductance originally established by the reactance tube depends on the transconductance (g_m) of V_1, as well as the carrier frequency generated by the oscillator. The inductance can, therefore, be ascertained by use of the formula

$$L = \frac{10}{6.28 f g_m} \tag{18-1}$$

Note that the formula takes into consideration the fact that R is 10 times X_c and also employs the angular velocity figures discussed previously in Chapter 8.

The amount of reactance that the tube is to contribute across the oscillator coil is established initially by use of the formula. The total inductance (reactance tube and oscillator coil) is combined with C_4 to establish the resonant frequency of the oscillator.

In producing FM, an audio signal at the grid of the reactance tube will alternately increase and decrease plate-current flow. The current changes, in turn, alter the zero-signal reactance of the tube by decreasing and increasing the reactive value. This change of inductive reactance at the oscillator resonant circuit shifts the oscillator frequency above and below its carrier frequency, at a rate established by the frequency of the audio-modulating signal. As the amplitude of the AM signal is increased, it creates greater

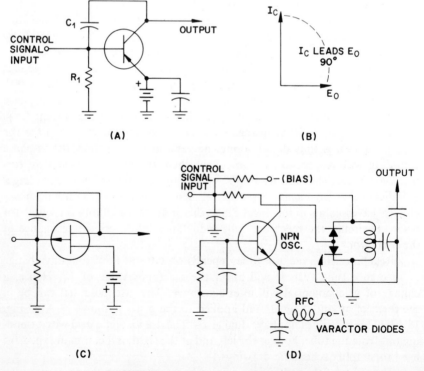

FIGURE 18-14. Solid-state reactive circuitry.

reactive changes in the reactance tube and hence shifts the oscillator frequency to a greater degree (greater deviation of the carrier frequency).

Solid-state reactance circuitry is shown in Fig. 18-14. At A the reactance circuit is very similar to the tube type just discussed, except that the capacitor-resistor combination have been reversed. Now the capacitor is between collector and base and the resistor from input to ground is as shown. The capacitance value of C_1 is selected so the reactance is approximately 10 times the ohmic value of R_1. Consequently, current for the C_1-R_1 network *leads* the signal voltage obtained from the oscillator. Since, however, the signal applied to the base of the transistor is obtained from across R_1 only, the voltage at the base will be in phase with the C_1-R_1 network current since there is no phase shift across a pure resistance. Because the signal current through the transistor is in phase with the voltage at the input (base), the collector current leads the oscillator signal by 90° as shown at B of Fig. 18-14. Since a capacitor produces a current lead, the circuit at A thus simulates a capacitor and thus exhibits capacitive reactance.

At C is shown a field-effect transistor version of a reactance circuit, using a *P*-channel unit to produce the same type reactance as the transistor circuit at A. An *N*-channel could, of course, also be used with appropriate bias changes. Similarly, an *NPN* transistor could be used for the circuit at A, again with proper polarity changes of bias.

A typical circuit using varactor diodes for frequency control of an oscillator signal is shown at D. The varactor diode is a *PN* junction device that takes advantage of the voltage-variable depletion capacitance of the back-biased junction. For variable-capacitor usage, two diodes are usually connected back to back as shown, to minimize distortion. Note the two varactor diodes shunt the oscillator inductor and replace the conventional variable capacitors. As alternations of the input signal cause a reverse-bias voltage change at the input to the junction of the diodes, the capacitance value changes and thus alters the frequency of the signal generated. For signal voltages that reduce the bias, the capacitance changes to swing the signal in reverse of the frequency change that prevailed for the initial signal alternation.

18-5. Demodulation (Detection)

As covered in Chapter 14, the tube or solid-state diode can be used as a rectifier. As mentioned, a rectifier is a device that has the ability to convert either low-frequency or high-frequency ac to dc. Because detection is essentially a rectifying process, the principles of rectification find application in both power supplies and detectors of modulated signals. In the latter application, it follows the RF amplifiers in receivers, as shown in Fig. 18-15, and produces the audio-signal voltages from the modulated RF carrier.

The basic detection process is shown at A of Fig. 18-16. Here, the input signal is applied across the primary winding L_1 and, by mutual inductance,

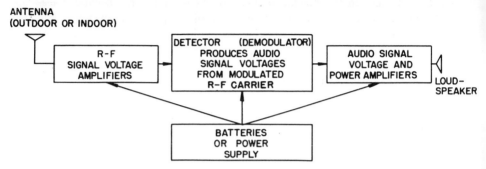

FIGURE 18-15. Basic block diagram of radio receiver.

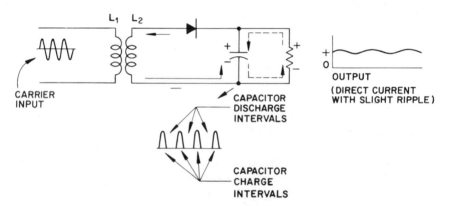

FIGURE 18-16. Basic detection process.

transferred to the secondary winding L_2. For a positive alternation of the input signal, the diode conducts and current flows through the output resistor setting up a polarity as shown. The electron flow as shown by the solid arrows also charges the capacitor. During negative alternations across L_2, the diode does not conduct, thus permitting the capacitor to discharge across the resistor as shown by the broken-line arrows. Thus, the pulsating dc is filtered and appears as shown, representing the results of filtered rectification (see also Fig. 14-4).

The representation in Fig. 18-16 is for an unmodulated carrier signal. When modulation is present, the amplitude of the a-c waveform varies as shown in Fig. 18-17. Hence, the output alternations will also increase and decrease in like fashion as shown. This is the underlying principle of detection, wherein the rectifying action of the diode is used for demodulating purposes in receivers. Both the power-supply principle and the detector principle depend on the rectifying action of the diode and, hence, the circuit arrangements are quite similar. Differences exist only because of the nature of the waveform that is handled. When the ac from the power mains is to

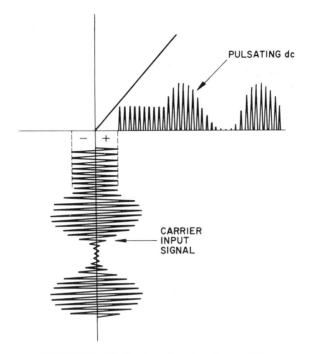

FIGURE 18-17. Graph of detection characteristics.

be changed into dc, a constant-amplitude input signal is applied to the rectifier circuit and a filter arrangement is employed to smooth out the ripple
components. In a detector system, however, a much higher-frequency signal
is handled but the basic circuit still resembles that of a half-wave power
supply.

The pulsating dc produced in Fig. 18-17 is shown again at A of Fig.
18-18 to help make clear how detection derives an audio signal from the RF
carrier. By use of an RF filter capacitor, the pulsating dc shown at A is converted into an average value, as shown at B. This average value varies above
and below a reference level, in the same manner as the carrier amplitude
increases and decreases. When this is applied to a subsequent stage by use of
a coupling capacitor or transformer as shown at C, only the a-c component
is transferred and the dc is kept out of the subsequent stage since neither
a capacitor nor a transformer will pass dc.

The rectifying process is necessary since the a-c RF signal, with its
varying amplitudes, cannot be heard if applied to a speaker *because its
frequency is too high above the audible range.* Even though the amplitude
variations that extend both above and below the reference line of zero
represent the audio information, the average value of pulsating dc, not ac,
is necessary to convert the amplitude changes of the RF carrier into audible
signals.

MODULATED CARRIER CONVERTED TO PULSATING dc
BY THE RECTIFYING FUNCTION OF THE DETECTOR

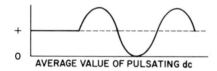

AVERAGE VALUE OF PULSATING dc

FIGURE 18-18. Characteristics of detection process.

18-6.   Diode AM Detectors

Early diode detectors consisted of so-called *galena crystals* wherein a sensitive detecting spot was found by a needle-like probe attached to a spring. Subsequently, tube-type diodes were extensively used as detectors for both AM and FM, later to be replaced almost exclusively by the solid-state junction diodes. Any type diode will detect, as long as it has a good forward-to-back resistance ratio. Detecting diodes are, however, usually smaller types since current-carrying capacity need not be high.

The diode detector can, of course, be utilized in a simple receiving circuit as shown in Fig. 18-19. Circuits of this type were the forerunner of the modern radio and still give good results in areas where signal strength is reasonably high. The loopstick is an RF transformer with a ferrite core (a high-permeability metallic crystal structure) with C_1 made variable for tuning purposes. An outdoor antenna, however, is necessary for good reception of local transmission.

The carrier that appears across the secondary winding is rectified by the diode, with the capacitor C_2 (approximately 0.00025 μF) acting as a filter component. Thus, across the earphones, the audio (represented by

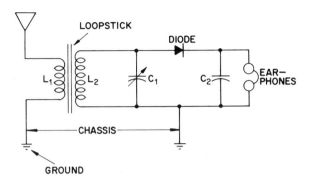

FIGURE 18-19. Basic AM detector-receiver.

variations in the average pulsating d-c voltage) is heard. The bottom of inductance L_1 is usually connected to an outside pipe driven into the ground, or the connection can also be made to a water pipe; hence this terminal is known as the *ground terminal* and the symbol for ground is used as shown. While the "ground" designation originated with this simple type of receiver, as the design of radio receivers improved, the ground connection was not necessary but the symbol for ground is still employed to indicate that a certain wire or terminal is connected to the *chassis* of the receiver. When the chassis is employed for ground purposes, the chassis becomes the interconnecting conductor for the various points shown on the schematic.

Typical of the diode detector circuit used in modern receivers is that shown in Fig. 18-20. The RF modulated-carrier signal is derived from a previous RF amplifier stage and applied across the primary L_1 of the transformer. This signal appears across L_2 and is tuned to resonance by C_1 and in turn is demodulated by the circuit in the manner previously detailed.

Capacitor C_2 filters the ripple component by charging to the peak values and discharging only slightly during the valleys between the peak pulses of the dc, as discussed earlier. The audio-signal voltage is developed across

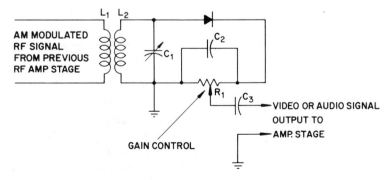

FIGURE 18-20. AM detector and gain control.

the load resistor R_1. Capacitor C_2, because it is a ripple-frequency filter, effectively bypasses the RF carrier signal from across the load resistor R_1. The latter resistor can be in the form of a *potentiometer* so that the degree of signal intensity (volume for audio or gain for video) can be adjusted as desired. Thus, by using a potentiometer for R_1, the latter becomes a *volume control* for regulating the desired output level from the receiver. The signal at R_1, however, is still too weak for operating a loudspeaker, though the audio sounds would be sufficiently audible in earphones. Also, if this is a video (picture signal) detector, the signals would have insufficient amplitude for application to a television-receiver picture tube. In consequence, the signal is coupled via a coupling capacitor C_3 to a subsequent stage or stages for additional amplification.

18-7.   Regenerative and Heterodyne Detection

In tube-type radios of early vintage, a triode was often used for detection purposes, with the grid-cathode portion acting as the rectifying (demodulating) diode portion, and the anode-cathode side adding amplification. While such detection methods are now obsolete, some special types merit consideration since they perform tasks not encountered with basic processes. One of these is the regenerative (positive feedback) type shown in Fig. 18-21.

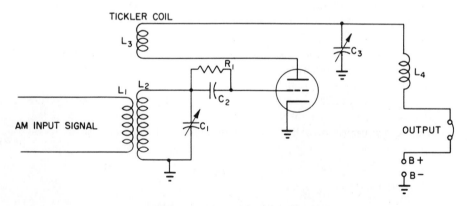

FIGURE 18-21. Regenerative detector.

Here a portion of the signal energy in the anode circuit is fed back, via a coil L_3, to the secondary winding (L_2) of the input transformer. By such coupling, L_3 induces some of the amplified signal energy into the secondary inductance L_2, as with the feedback oscillator previously described. If the polarity of the feedback voltage is the same as that of the signal voltage in the L_2 circuit, the fed-back signal voltage will reinforce the signal voltage existing at the grid so that circuit efficiency will be increased. Since the resistive component of the input circuit is also decreased, selectivity and sensitivity are high. This form of feedback is known as *regeneration*.

The degree of coupling between L_3 (also known as the *tickler coil*) and the secondary coil L_2 can be varied by changing the spacing between the two coils. Another method for controlling the amount of regeneration is to vary the capacitor C_3 in the tickler coil circuit. A third method for controlling the amount of regeneration is to place a variable resistor in series with the B+ lead and thus change the plate current that flows through the tickler coil.

Inductance L_4 is an RF choke that offers a high opposition to the signal energy for the radio frequencies. Hence, this series choke coil keeps the RF frequencies from the output circuit. Capacitor C_3, in addition to being a regeneration control, also bypasses the RF signal components, while filtering the audio components for reproduction at the output.

The regenerative detector can receive its input from a previous RF amplifier stage or from an antenna resonant circuit. The output can be applied to a pair of earphones as shown or to an audio-amplifier stage for additional buildup of the audio-signal components.

When the regeneration and positive feedback are increased, there will be a point where so much signal energy is fed back from the output circuit to the input circuit that the circuit will oscillate. When the circuit oscillates, it has reached an equilibrium with respect to the feedback voltage, and the system becomes self-sustaining; that is, it furnishes its own input signal and produces or generates a given output frequency. Under such a condition, the circuit would not need a signal input. This type of oscillator was discussed earlier.

The regenerative detector, whether tube or solid-state type, is not used in modern receivers, except in experimental types or in shortwave portable types where a high degree of sensitivity and selectivity is necessary. The regenerative detector shown in Fig. 18-21 produces as much audio-signal output as could be obtained from a detector stage with an additional stage of RF amplification preceding it, for gain-increasing purposes.

The regenerative detector has several disadvantages. Adjustments to produce the maximum regeneration without the circuit going into oscillation are critical. When maximum regeneration is once established, it may not remain fixed for long. For this reason, some instability may result when the regeneration controls are set for too high or too critical a regeneration point. When the circuit is permitted to oscillate, squeals will be heard in the output and since an oscillator generates a frequency, it will also radiate such energy and cause interference in nearby receivers. The regenerative detector in its oscillating state finds primary applications in shortwave work where it is necessary to receive code signals that have no modulation characteristics, as discussed next.

There are two types of code signals that are transmitted in shortwave

(A)

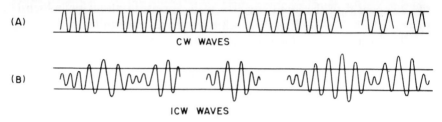

C W   WAVES

(B)

ICW   WAVES

FIGURE 18-22. Continuous waves and interrupted continuous waves.

commercial and government applications. One type is obtained by generating a carrier of a fixed frequency and amplitude and interrupting this carrier to form short and long transmission intervals. This is shown at A of Fig. 18-22, where the short transmission intervals represent dots and the longer transmission intervals represent dashes. In this fashion, the Morse code can be transmitted. [The code is named after Samuel F. B. Morse (1791–1872), the American inventor who originated it for use with respect to the early telegraph.]

Another method for sending code is to use a modulated carrier and interrupt it as shown at B. If the modulated carrier is modulated with a 400-Hz audio tone, for instance, it can be received by any of the detectors previously discussed. The detector will then demodulate the AM waveform, and the detection process will produce a 400-Hz audio tone that will be periodically interrupted to represent the dots and dashes of the Morse code.

When a continuous wave having a constant amplitude and frequency is transmitted, the type of transmission is known as *CW* to indicate *continuous waves*. This designation does not refer to the interruption rate, but rather to the fact that the amplitude is continuous and does not vary. When a modulating tone is utilized and the amplitude varies, the transmission is known as *ICW*, to indicate *interrupted continuous waves*. This term indicates that the continuous amplitude is not maintained but is varied (interrupted), increasing and decreasing at a predetermined rate. The ICW transmission must still be broken up into dots and dashes as with the CW transmission. The CW wave has the advantage that it is a narrow band type of transmission, as opposed to the ICW wave, which occupies more space since side bands are generated whenever a carrier is amplitude modulated.

When the CW type of wave is transmitted, the lack of modulation prevents the use of an ordinary detector to demodulate the carrier. To intercept and detect the CW type of code transmission, it is necessary to have an oscillating detector. The regenerative detector shown in Fig. 18-21 can be employed by adjusting the feedback amplitude to the point where sufficient energy is applied to the grid to sustain and generate oscillations. Such detection is known as *heterodyne* detection.

Heterodyne refers to the process by which two signals of different frequencies are combined in special circuitry, to produce additional signals having frequencies other than the original. Earlier it was pointed out that harmonic distortion is produced when signals are amplified in a circuit having nonlinear characteristics. Such harmonic distortion consists of signals generated within the amplifier and not contained in the original signal. If a circuit is used having severe nonlinear characteristics and *two* signals are injected into the circuit, the output signals will consist of the initial two signals plus signals whose frequencies are the *sum* and *difference* frequencies of the original two signals. The additional signals are generated because, in a nonlinear circuit, the two original signals *beat together*, or, to use the more technical term, *heterodyne*. Thus, if one of the original signals has a frequency of 1 MHz and the other signal has a frequency of 999.5 kHz, the heterodyning process produces a signal having what is known as a *difference* frequency of 500 Hz (0.5 kHz). At the output of the circuit, the original two signals are also present plus the sum frequency of the original two signals (1999.5 kHz).

The heterodyning process is important, not only because it is used in the reception of CW waves, but also because it is used extensively in virtually all modern receivers such as AM radio, FM, television, shortwave, and others that employ the *superheterodyne* principle, as more fully discussed in the next chapter.

For reception of a CW signal, the regenerative detector shown in Fig. 18-21 is adjusted to oscillate and hence it will generate a signal of its own. If the frequency of the oscillating signal is near the incoming CW signal frequency, an audio tone will be generated. Assume that the CW signal is 25 MHz. If the regenerative detector is tuned so that it oscillates at 25,001 kHz, the difference frequency of the signal produced by heterodyning will be an audible signal of 1 kHz (1000 Hz).

When the incoming signal mixes with the signal produced by circuit oscillations, the resultant signal that is produced will have amplitude variations because the progressive point-by-point addition of the two signals results in decreasing and increasing amplitudes, as the phase between the two signals of different frequencies varies. (This signal with varying amplitude is similar to the modulated-type waveform produced by the addition of a constant-amplitude carrier and a constant-amplitude side band, as discussed earlier in this chapter.) These amplitude variations occur at a rate corresponding to the difference frequency between the two signals. Thus, by the combined heterodyning and demodulation process of the detector, the audio-frequency difference component is detected and made audible at the output. Since the regenerative detector also amplifies, the oscillating detector produces a fairly high-volume audio signal at its output.

The oscillating frequency of the detector could also be set at 24,999 kHz and the difference frequency would still be 1000 Hz and, hence, audible. The oscillating frequency of the detector can be altered to give a variety of audio tones, as desired. If the carrier and oscillating frequencies are close together, a low-frequency audio tone will be developed. As the oscillating frequency and the incoming carrier frequency are separated more and more, an increasingly high audio frequency will be generated. A frequency above 15,000 Hz soon becomes inaudible to the average ear. (Very few individuals can hear frequencies above 20,000 Hz, and many can hear frequencies only up to 15,000 Hz.)

The oscillating type of detector is also known as the *autodyne* detector. It has the advantage of excellent sensitivity and selectivity and does not require a separate oscillator for the reception of CW signals. In instances where diodes or other types of detectors are employed that do not have feedback for producing oscillations, a separate oscillator would have to be used, and the signal from the latter injected into the detector. With the autodyne detector, the oscillator and detector are contained in one circuit and employ a single triode or pentode tube.

If such an oscillator is connected directly to an antenna for the input-signal source, the oscillating signals will leak into the antenna system and will be radiated. Hence, such signals can be picked up by receivers tuned to the same frequency, and interference with other stations will result. A regenerative detector should be operated with an isolating RF amplifier stage between the antenna and the oscillating detector. In addition to this, the detector circuit should be well shielded to minimize radiation from the connecting leads and from the inductances and other component parts.

The *oscillating* detector previously mentioned is suitable primarily for reception of CW signals that contain no AM. When in an oscillating state, it is not suitable for the reception of AM or FM signals. For CW reception, however, the oscillating detector has several advantages because, when in the oscillating state, the circuit impedance develops a high order of signal energy since the resistive component is virtually zero and no losses occur through the power consumed by any circuit resistance. The advantages of the oscillating state can be utilized for the reception of a modulated signal, however, by the special super-regenerative circuit discussed next.

The super-regenerative detector was devised by the famed American inventor Major Edwin H. Armstrong (1890–1953), and its basic circuit is shown in Fig. 18-23. The circuit resembles the regenerative or oscillating detector previously described, except that a signal of *supersonic frequency* is injected in series with the $B+$ lead (to the tickler coil) and tube anode. The B voltage is initially set at the point just below where oscillations would occur. The supersonic oscillator generates a frequency above the audible range, and such a frequency can range from approximately 25 to 30 kHz. If the B voltage is set just below the point where oscillation occurs, the super-sonic oscillator signal will alternately add positive and negative potentials to the B voltage impressed on the tube. If, for instance, the B voltage is 45 V, and above this value the circuit would oscillate, a positive alternation of 5 V from the supersonic oscillator would add to the 45 V from the B supply

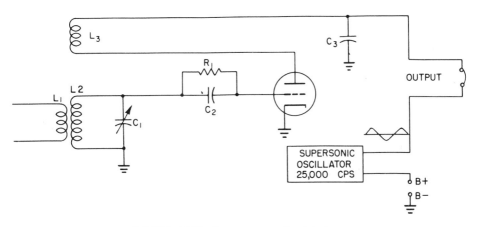

FIGURE 18-23. Super-regenerative detector.

and make the total anode voltage 50 V. With the increased B voltage, the circuit would oscillate. When a negative signal alternation occurs from the supersonic oscillator, it would oppose the B battery voltage. If the negative alternation from the supersonic oscillator is 5 V, it would decrease the 45-V B potential to 40 V and, hence, the circuit would be below the critical regeneration point and would not oscillate. Thus, the super-regenerative receiver is periodically thrown in and out of oscillation at an extremely rapid rate. The rate is above the audible frequency and, thus, the injection of the supersonic signal will not result in the latter being heard at the output. On the other hand, the circuit is in an oscillating state during one-half its operating time, and hence the advantages of high efficiency and sensitivity are realized. Since the oscillating state is not a sustained one, no heterodyning process occurs to produce a beat-frequency signal. Thus, for the reception of voice or ICW, the super-regenerative detector works satisfactorily.

When the receiver is not tuned to a station (no signal input), the electron flow in the tube circuit varies because of the supersonic oscillator's effect on the B voltage. This rapidly changing B voltage produces a hissing sound in the output, which is suppressed, however, during the reception of signals at or above medium strength. The super-regenerative receiver has poor selectivity because of loading effects, though sensitivity and efficiency are high. It finds special applications in the reception of ICW signals or for AM-modulated signals that are broadcast in a frequency spectrum not crowded by other transmission.

18-8.   Discriminator FM Detector

Detectors for FM must be so designed that they will interpret the frequency changes of the modulated carrier and produce the equivalent audio-signal components. The diode detector described earlier can be employed in a dual-

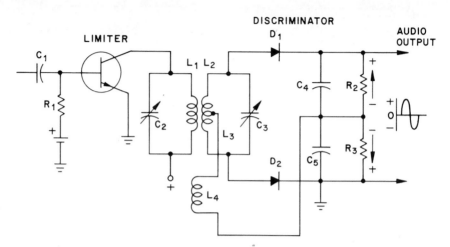

FIGURE 18-24. Discriminator FM detector.

diode arrangement known as a discriminator. Such a detector discriminates with respect to carrier-frequency changes and produces the necessary audio-signal components. The amplitude-detecting characteristics of the diode, however, are undesirable in an FM detector and, hence, some provisions must be made to eliminate all amplitude changes that reach the discriminator type of detector. One method for doing this is to precede the discriminator circuit with a clipper-type circuit, as shown in Fig. 18-24. The clipper, as mentioned earlier, will have a limiting action if the input signal has sufficient amplification to drive the transistor into cutoff and to saturation. With sufficient signal drive, amplitude variations of the signal are limited and, hence, the circuit is known as a limiter, when used in conjunction with a discriminator. The limiter, by eliminating amplitude changes, minimizes static because the latter arrives at the receiver in the form of carrier-amplitude changes.

The discriminator circuit utilizes two diodes, D_1 and D_2, as shown. The device detects because of phase differences that occur in the resonant circuits during the time the FM carrier deviates above and below its normal frequency. Because operation involves such phase changes, it is necessary to employ vector diagrams in explaining circuit function.

For the discriminator shown, three inductors are used between the limiter and the discriminator diodes. Inductance L_1 transfers signal energy to the tapped secondary inductance (L_2 and L_3) because of the transformer arrangement. Inductance L_4, however, also picks up a signal from inductance L_1, such a signal being applied between the center tap of the secondary and the output network of the discriminator. Instead of the inductance L_4, the necessary additional coupling between L_1 and the center of the secondary can be achieved by use of a coupling capacitor.

Note that the anodes of the diodes are connected across the resonant circuit composed of L_2, L_3, and C_3 and that the cathodes are connected to the output circuit. Electron flow through D_1 is from cathode to anode, and hence through R_2 in the direction shown by the arrow. Current flow through D_2 causes a polarity as shown to be established across R_3.

For an understanding of how the discriminator detects FM, an evaluation must first be made of the signal voltage and current relationships that exist between the limiter and discriminator. Such relationships are shown at A of Fig. 18-25. Here E_p refers to the signal voltage of the primary. Because L_1 is an inductance, the primary current (I_p) lags the primary voltage, as shown at A. When the primary current goes through its greatest change (positive peak to negative peak), the lines of force that are produced will induce a voltage across the secondary. As shown at A, the induced voltage (E_{ind}) lags the primary current by 90°. Hence, there is a phase difference of 180° between the *primary voltage* and the *induced voltage*. Inductance L_4 is also coupled to L_1 and picks up an induced voltage. The leads from L_4, however, are transposed so that the voltage obtained from L_4 is in phase with the voltage across L_1. Thus, the voltage across L_4 is also 180° out of phase with the induced voltage.

At B of Fig. 18-25 a vector diagram is shown for the various voltages of the circuit. Note that the voltage of L_4 (E_{L_4}) is shown with a vertical arrow pointing upward to indicate an out-of-phase condition with respect to the

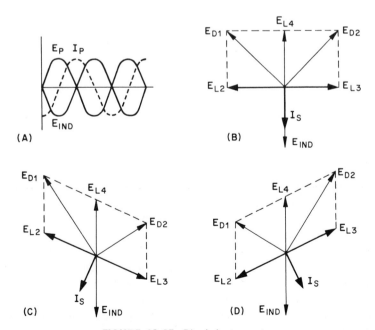

FIGURE 18-25. Discriminator vectors.

induced voltage (E_{ind}), which is shown with the vertical arrow pointing downward.

The voltage induced across the secondary causes current to flow, such current being in phase with the voltage because of the resonant circuit formed by L_2, L_3, and C_3. At resonance, the circuit is purely resistive; hence the secondary current (I_s) is in phase with the induced voltage. Thus, the secondary current is also shown by a vertical arrow pointing downward. The secondary current is represented by a heavy arrow, to distinguish it from the induced voltage.

Because the secondary is center tapped, it actually consists of two coils, L_2 and L_3. Thus, the voltage appearing across the secondary actually consists of two voltages, one applied between D_1 and the center tap of the secondary and the other between D_2 and the center tap. As an individual inductance, the voltage across L_2 will be out of phase with the secondary current by 90°. Also, the voltage across L_3 will be out of phase with the secondary current by 90°. The voltage relationships for L_2 and L_3 are also shown at B of Fig. 18-25, and are represented by the heavy solid horizontal arrows. Signal voltages that appear across the secondary will cause D_1 to conduct at one time and D_2 at another. Even though the diodes conduct alternately, for simplicity both voltages are represented simultaneously in the vector diagram. The voltage for D_1 is obtained from L_2 and the center tap (the voltage obtained from L_4). Because, however, there is a 90° phase difference between the voltage of L_4 and the voltage of L_2, a vector representation of the voltage for D_1 must be drawn with the slanting arrow shown at B. A similar slanting arrow is shown at the right of the same drawing for the voltage across D_2. This composes the vector shown at B and represents the phases of the voltages, when the carrier is at its center frequency. As seen from the drawing at B, the voltage for each diode is identical and, hence, the voltage drops across R_2 and R_3 are also identical. Thus, the output voltage is zero because the voltage across R_2 is equal and opposite to the voltage across R_3.

When the carrier deviates from its center frequency during FM, the carrier no longer finds a resonant condition in the tuned circuits between the limiter and discriminator because such tuned circuits are set at the center carrier frequency. If the carrier shift is to a higher frequency, the resonant circuit composed of the secondary inductance and C_3 will become primarily inductive. This comes about because energy is induced from L_1 into the secondary inductance, by virtue of the magnetic lines of force, and not into the capacitor and inductance combination initially. Hence, the voltage is induced as though into a series resonant circuit. At resonance, the series circuit has a low impedance and the inductive and capacitive reactances are equal, though opposite in phase. When the frequency rises above resonance, inductive reactance increases and capacitive reactance decreases. Hence, the rise in inductive reactance offers the greatest opposition and causes the circuit to become primarily inductive. This causes secondary current to lag, as shown at C of Fig. 18-25. The voltages across the secondary inductances L_2 and L_3 are still 90° out of phase with the secondary current because of the inductive characteristics of L_2 and L_3. Thus, the vector arrows for E_{L_2} and E_{L_3} must also be shown at an angle, to maintain their right-angle relationship with I_s. The parallelograms for the voltage of D_1 and the voltage of D_2 now show a rise of voltage for D_1 and a decline of voltage for D_2. If the voltage across R_3 declines and the voltage across R_2 increases, a positive alternation of the audio signal will be created at the output of the discriminator, as shown in Fig. 18-24. When the carrier shifts back to its normal frequency, the secondary current goes back in phase with the induced voltage, as at B, and the output alternation drops to zero. When the carrier shifts to a lower frequency, secondary current leads because the circuit becomes capacitive, and the vector shown at D prevails. Now, D_2 voltage increases and D_1 voltage decreases. Hence, R_3 voltage rises and R_2 voltage declines, producing a negative alternation at the output.

18-9. Ratio FM Detector

The *ratio detector* is another type of FM demodulator and a typical circuit is shown in Fig. 18-26. The advantage of this system over the discriminator is that no limiter need precede it because the ratio detector will not demodulate AM. As shown, two diodes are again used, though now the secondary transformer (L_2 and L_3) couples to the anode of one diode and the cathode of the other. Consequently, electron flow through the output resistors R_3 and R_4 is in series, as shown, so that the sum of the two voltages appears at the output when no signal is being produced. This d-c voltage, however,

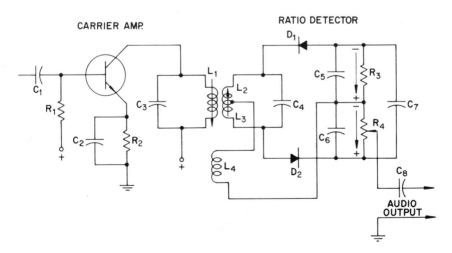

FIGURE 18-26. Ratio FM detector.

can be blocked by the coupling capacitor C_8 and, hence, is not applied to the input of the following amplifier stage.

The manner in which one diode conducts more than the other to produce audio is similar to the discriminator type of detector, and the vector diagrams of Fig. 18-25 also apply to the ratio detector.

During frequency modulation of the incoming carrier, one diode will conduct more than the other and vice versa, as with the discriminator. Assume that, initially, 2 V appear across R_3 and 2 V across R_4, producing a total d-c voltage of 4. If D_1 conducts more than D_2, the voltage across R_3 may rise to 3 V and the voltage across R_4 may drop to 1 V. Now, 4 V dc still appear across the combination of resistors, and no voltage *change* has occurred. When the carrier swings in the other direction and D_2 conducts more than D_1, the voltage across R_4 may rise to 3 V and the voltage across R_3 may drop to 1 V. Again, no change occurs because 4 V still exist across the combination of R_3 and R_4. While the *ratio* of voltages may change, the total voltage does not and, hence, the audio must be derived from one of the two output resistors. Since R_4 is connected to ground, it is more convenient to use the latter as the output resistor. By using a potentiometer for R_4, a volume control is formed. As the carrier deviates, the changing voltage across R_4 will produce the audio signal.

Capacitor C_7 across R_3 and R_4 is a large value, usually several microfarads. This capacitor charges to the value of the d-c voltage across the combination resistors R_3 and R_4. Because a capacitor opposes a change of voltage, C_7 maintains the voltage that appears across R_3 and R_4 at a fairly constant level. In consequence, any sudden changes in the total voltage that might occur because of sharp static bursts are minimized by the action of C_7.

Because C_7 is instrumental in suppressing static and other forms of amplitude modulation, it has an important circuit function. The ratio detector has been extensively used, though on occasion the discriminator type of detector is preferred because of its greater immunity to high noise interference. As with the discriminator, a balanced circuit arrangement gives best performance. A well-balanced circuit means a matched pair of diodes, as well as matched resistors and matched capacitors in the output circuit.

18-10. Gated-Beam FM Detector

The *gated-beam FM detector* is a type that was once widely used but now is found only in older equipment or in special circuitry where its particular advantages outweigh the simplicity and capabilities of the diode detection systems. The gated beam is of particular interest since it converts the modulated carrier to a form of pulse-width modulation before detection. The system was designed by Dr. Robert Adler, of the Zenith Radio Corporation. The circuit uses a special tube and provides a high order of sensitivity and

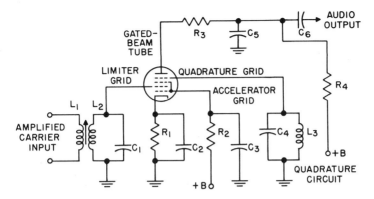

FIGURE 18-27. Gated-beam FM detector.

good amplitude of output audio. Thus, only a single amplifier stage need precede the beam detector and output is sufficient to drive a power-amplifier output circuit.

A typical gated-beam FM detector system is shown in Fig. 18-27. The tube is one wherein the accelerator grid structures of the tube are in reality in the form of plates, which help shape the electrons into a narrow beam. The positive voltage of the accelerator grid structure increases the electron beam velocity and forces the beam through a narrow slot in the accelerator electrode. The electron beam then encounters the limiter grid, which, in conjunction with the quadrature grid, acts to control the electron flow. As with other tubes, the anode is made positive to attract the electrons emitted by the cathode.

The limiter grid has sufficient control over the electron beam to produce cutoff for *any* negative voltage. If the limiter grid has zero voltage or a positive voltage applied to it, however, it will permit current flow within the tube. The quadrature grid, being slightly negative, will also cause plate-current cutoff. Thus, both grids are influential in preventing or permitting current flow within the tube.

With a small value of fixed bias, such as 1 V, an incoming signal has sufficient amplitude to cause the tube to be operated at saturation for the positive peaks of the grid signal or at cutoff for the negative peaks of the incoming signal. Because the grid structure releases current flow rather suddenly, and also stops current flow quickly, a square wave of beam current occurs within the tube, in the region beyond the input grid. Thus, the tube acts as a self-limiting device and will eliminate AM variations in the incoming signal. As shown in Fig. 18-27, a parallel-resonant circuit is connected to the quadrature grid. The quadrature-resonant circuit is tuned to the center carrier frequency of the incoming FM signal. During signal input, the cloud of electrons (space charge) around the cathode varies, and the quadrature grid is also affected by the electron beam because of space-charge coupling.

Hence, the square-wave type of signal generated within the tube is also present at the quadrature grid, and will pulse the quadrature circuit into a resonant flywheel condition. The signal voltage which appears across the quadrature circuit, however, lags the input signal by approximately 90°. The phase lag occurs because of the nature of the space-charge coupling. With a 90° lag between the signal at the quadrature grid and that at the limiter grid, the plate current of the tube is cut off for a greater period of time than would otherwise be the case. This can be seen from an inspection of Fig. 18-28, which shows that the plate current can only flow when neither the limiter grid nor the quadrature grid is negative. Thus, only about one-half of each square-wave alternation reaches the anode during the time the carrier is at center frequency.

When the incoming FM carrier shifts to a higher frequency, the quadrature circuit will be off resonance with respect to the shifted carrier frequency. The quadrature circuit becomes predominantly capacitive because the higher frequency impressed on it increases inductive reactance and decreases capacitive reactance. Since the capacitive reactance is low, the current through the capacitive reactance is higher than that in the inductive reactance. Because a parallel-resonant circuit, with the resonant frequency

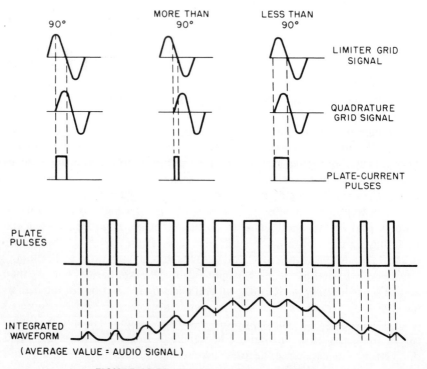

FIGURE 18-28. Gated-beam detector waveforms.

impressed on it, exhibits a high impedance, the reduction of such impedance through decreased capacitive reactance causes the circuit to be predominantly capacitive. The capacitive characteristics of the quadrature circuit will now cause the signal voltage at the quadrature grid to lag the signal at the limiter grid by more than 90°, which is the lag at center carrier frequency. Because of the increased phase difference between the two current-controlling voltages, *less* than one-half of each square-wave alternation arrives at the anode of the tube. Hence, the *average value* of plate current decreases. When the carrier signal at the limiter grid shifts lower in frequency, the quadrature circuit becomes predominantly inductive and the voltage tends to lead. As shown in Fig. 18-28, more than one-half of each square-wave alternation reaches the anode and, thus, the average value of plate current increases.

Capacitor C_5 and resistor R_3 in the anode circuit form an integration circuit of the type described previously. An integration circuit has the ability to produce an average value from a series of pulses having various widths.

Resistor R_4 is the conventional load resistor across which the audio-signal voltages develop. Capacitor C_5 has a low-shunt reactance for the high carrier frequency and thus eliminates the latter from the output circuit.

18-11. Automatic Volume Control (AVC)

Automatic volume control consists in changing the bias on the RF stages preceding the detector, so as to alter the gain characteristics of transistors (or tubes) and, thus, the amplification. The purpose for automatic volume control is to increase the gain of the RF amplifier stages for weak signals and to decrease the gain for strong signals. By automatically altering the gain to suit signal reception conditions, the output from the detector will be maintained at a fairly constant level. Thus, if the radio listener sets the volume control to the level desired, the radio receiver will not blast loudly for a local station when tuning over the broadcast band. Also, AVC will automatically increase the gain for weak stations.

Automatic volume control can be achieved by attaching a lead to the simple diode detector, so as to provide a negative voltage for application to the grid of the previous RF amplifier stages. Figure 18-29 shows the basic method for obtaining an AVC voltage. Since the cathode of the diode conducts electrons through the anode and then through L_2 and to the resistor, the electron flow is in the direction indicated by the arrow beside the resistor. Such flow will develop across R_1 a voltage drop having a polarity as shown (negative toward the L_2 side and positive toward the cathode side). The carrier signal will establish an average value of voltage across this resistor, though this voltage will vary because of the audio components caused by AM. As previously mentioned, this audio-signal component can be taken

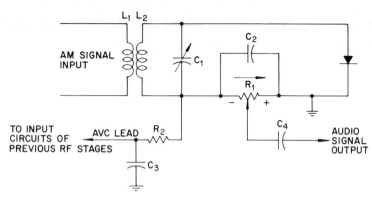

FIGURE 18-29. Typical diode detector and AVC.

from the resistor by a movable potentiometer arm in the form of a volume control. The signal is then transferred, via the series coupling capacitor C_4, to the next stage (the voltage amplifier stage).

When a lead is attached to the *negative* side of the resistor, a voltage is obtained that represents the average value produced by the demodulation process. The *average* value of the carrier represents a specific negative voltage at this point. In order to eliminate the audio-frequency components that cause the average d-c value to fluctuate, a filter resistor, R_2, and a shunt capacitor, C_3, are employed. A voltage drop for the audio signal occurs across resistance R_2 and this audio signal is shunted by the capacitor C_3. Capacitor C_3 has a large value, to provide a low reactance for the audio voltages. The filter capacitor C_3 cannot be connected directly to the negative terminal of R_1 because, in that position, the capacitor would shunt the audio-signal components and, thus, would prevent the transfer of these components to the audio-amplifier stage following the detector. Therefore, R_2 is needed as an isolating resistor.

The voltage beyond R_2 and C_3 is a relatively steady dc, which represents the average value of the received carrier. When a strong station is tuned in, the increase in carrier amplitude will develop a larger voltage drop across R_1, and hence a greater negative potential is applied to the inputs of the RF amplifier stages preceding the detector. When such an increase in negative voltage appears at the base circuits of *NPN* transistors, the forward bias is reduced. (For *PNP* transistors, the diode of Fig. 18-29 is reversed to produce a positive potential for gain reduction.) For discussion purposes, assume the RF stages are *NPN* transistors. Thus, a decrease in forward bias will reduce the gain in such stages and hence reduce the loud audio signals that would be developed. For a weak audio signal, less voltage is developed across R_1 and hence a lower negative-voltage value is applied to the inputs of the RF stages. The *decrease* in negative voltage causes an increase in amplification and hence the signal arriving at the detector will increase. Thus, the AVC system regulates, within limits, variations in carrier signal strength that arrive at the receiver. Automatic volume control does not affect the setting of a volume control, however, and the listener can still regulate the volume to suit his taste. Once he establishes the volume desired, the automatic gain control will maintain this volume at a substantially constant level, regardless of whether a strong or a weak station is tuned in. It is only for an extremely strong station within a few miles of the receiver or an extremely weak station that the AVC characteristics of a circuit become less effective. The automatic volume control can actually be considered as an automatic *gain* control. The term AVC is utilized for radio receivers, however, to differentiate between it and a similar system employed in television receivers; in the latter instance, the system is called an *automatic gain control system* (AGC), because the gain of the picture signal is regulated.

The foregoing AVC system will automatically increase the volume level for weak signals and decrease the volume level for strong signals. There are occasions, however, when it is desirable to omit AVC bias during reception of weak stations to increase the volume to acceptable levels. It is true that, for weak stations, very little AVC voltage is developed but even a small amount will decrease the gain of the RF amplifier stages. For extremely weak signals, this is undesirable since it will reduce the sensitivity of the receiver for such stations. Hence a refinement of the basic AVC circuit is sometimes utilized. This consists of a circuit that *delays* the amount of AVC bias generated until the signal strength from the station has reached a level where it is necessary to reduce its volume by AVC.

A basic delayed AVC circuit is shown in Fig. 18-30. This type of circuit does not develop any AVC bias for the weak stations but will develop such a bias for the stronger stations to be received. Basically, the system contains two detector circuits, one for the demodulation of the AM signal and the other for AVC purposes. For this reason, two diodes are necessary. As shown in Fig. 18-30, the basic detector circuit is still employed. The upper diode, D_2, is attached to the secondary of the input transformer, L_2, and the path of conduction for this diode is shown by the solid arrows. Resistor R_1 is the conventional volume control, as described for the previous diode detector systems.

For purposes of delayed AVC, it is necessary to apply a fixed reference potential on diode D_1, so as to prevent it from conducting until a predetermined level of input signal is reached. For Fig. 18-30 the fixed potential is applied across resistor R_4 as shown (delay bias).

Diode D_1 obtains signal energy from diode D_2, via the coupling capacitor C_2, which bridges both diode anodes. The path for electron flow for the signal is as shown by the dotted arrows and, in consequence, electrons flow through R_3 and R_4 and to diode D_1. Diode D_1 is also across the delay-bias resistor R_4, with the cathode connected to one side and the anode to the other side through R_3. Thus, for the polarity of the voltage drop across R_4, a reverse bias is applied to diode D_1 that prevents conduction.

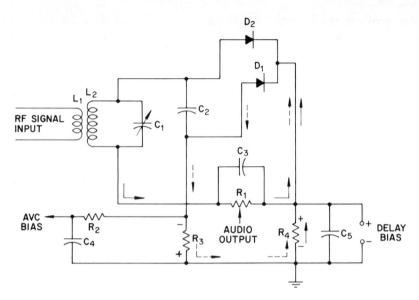

FIGURE 18-30. Delayed AVC.

Assume, for instance, that the voltage drop across R_4 is 2 V. This would make the *cathode* of D_1 *positive* by 2 V with respect to the anode of D_1. Hence, the anode of D_1 will be negative in relation to its cathode by 2 V, and therefore nonconducting. Thus, any weak signal input that does not develop 2 V will be unable to cause diode D_1 to conduct. When, however, a strong signal is tuned in, the positive peaks of the carrier signal will overcome the 2 V of negative potential on the anode of D_1 and will thus permit this diode to conduct. Now electron flow is in the direction shown by the dotted arrows and, consequently, a voltage drop develops across R_3 with a polarity as indicated. This voltage drop consists of the AVC bias voltage. The latter is filtered by resistor R_2 and capacitor C_4 and is then applied to the previous RF amplifier stages.

The capacitor C_5 across resistor R_4 is for the purpose of filtering out audio-frequency components from across this resistor. Such filtering is necessary so that the bias established across R_4 will remain constant and will not vary with signal modulation or audio.

Since the detector diode, D_2, is not connected to the lower section of resistor R_4, no negative potential is applied to the anode of the detector diode. Hence, diode function is not disturbed for the weak signals. If resistor R_1 terminated at the bottom of R_4, instead of at the cathode, it would place a negative bias on the anode of D_2 and, hence, the circuit would not detect weak signals. By placing resistor R_1 as shown, however, the negative voltage at the bottom of R_4 is not applied to the anode of D_2.

The diode characteristics for the delayed AVC tube are shown in Fig.

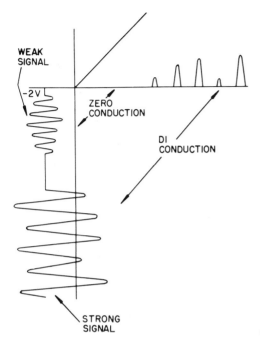

WEAK
SIGNAL

-2 V

ZERO
CONDUCTION

DI
CONDUCTION

STRONG
SIGNAL

FIGURE 18-31. Delayed AVC diode characteristics.

18-31. Here it is assumed that a bias of 2 V has been established across R_4, that makes diode 1 negative by 2 V with respect to the cathode. As seen from this drawing, a weak signal is insufficient to overcome this bias and, hence, the diode cannot conduct. When a strong signal is tuned in, however, it overcomes the 2-V negative potential; hence diode D_1 conducts. The conduction is in the form of variations in amplitude since the input signal is amplitude-modulated. In reality, D_1 will also detect the signal in the same fashion that the demodulator diode D_2 will detect the signal. The audio component that would normally appear across R_3 of Fig. 18-30, however, is filtered by the AVC filter network (R_2 and C_4) and, hence, a relatively ripple-free bias is applied to the RF stages.

Review Questions

18-1. Briefly define the terms *carrier* and *amplitude* modulation.

18-2. What factors determine the percentage of modulation in AM?

18-3. Explain why an AM carrier should not be modulated more than 100%.

18-4. (a) If a 1-MHz carrier is amplitude-modulated by a 3-kHz audio tone, what are the frequencies of the side-band signals produced?
   (b) If the volume of the audio-modulating signals in the foregoing example is increased, what changes occur in the AM carrier?

18-5. When modulating in the output circuit of a Class C amplifier (plate or collector modulation), what must be the relationships between the input power to the Class C amplifier and the output power of the modulator?

18-6. (a) In which type of modulation is the side-band power supplied by the modulator?

(b) In which type of modulation is the side-band power furnished by the Class C amplifier?

18-7. What are the advantages and disadvantages of plate (or collector) modulation versus grid (or base) modulation?

18-8. Briefly explain how frequency modulation of a carrier differs from amplitude modulation.

18-9. List several advantages of FM over AM.

18-10. In FM, an audio signal causes the carrier to deviate 10 kHz on each side of its center frequency. Would a change of audio frequency or a change of audio volume cause the carrier to deviate to a greater extent?

18-11. (a) If a carrier shifts back and forth 1000 times per second, what type of audio signal would cause the signal to shift back and forth at twice the former rate?

(b) What would cause the carrier in part (a) to shift back and forth at half its original rate?

18-12. Briefly explain what is meant by a reactance circuit, and give a typical example of its application to FM.

18-13. Reproduce a simple diode detector, and explain its operation with respect to AM.

18-14. Briefly explain the difference between regenerative detection and heterodyne detection.

18-15. Define the designations CW and ICW.

18-16. Briefly explain why a limiter stage must precede the discriminator type of FM detector.

18-17. What circuit differences are present in the ratio detector and the discriminator detector that aid in identifying the particular type?

18-18. Briefly explain why the ratio detector does not require a limiter preceding it.

18-19. Why is a quadrature circuit necessary in a gated-beam FM detector?

18-20. Briefly explain the difference between conventional AVC and *delayed* AVC.

18-21. How does the inclusion of AVC improve reception in radios?

Practical Problems

18-1. In testing a transmitting system, an AM carrier of 1 MHz was modulated by a 500-Hz signal and a 2600-Hz signal. What side bands were produced?

18-2. What is the required bandpass for the modulated carrier in Problem 18-1?

18-3. In an experimental FM system, the carrier deviated a maximum of 30 kHz and the highest audio signal had a frequency of 15 kHz. What is the deviation ratio?

18-4. How many significant side bands were produced for the system in Problem 18-3?

18-5. A reactance circuit for an FM system operated on 90 MHz and was similar to that shown in Fig. 18-13. If the tube had a g_m of 5000 μmhos, what inductance value was developed?

18-6. In a laboratory experiment, a 1-MHz signal was heterodyned with an 80-kHz signal. What were the frequencies of the signals produced at the output of the circuitry?

18-7. In a radio receiver a 700-kHz signal was heterodyned with a 1155-kHz oscillator signal. All but the lowest frequency signal was filtered out by resonant circuitry. What was the frequency of the remaining signal?

18-8. In a television receiver an oscillator having a signal frequency of 113 MHz formed a mixing process with two signals, one having a frequency of 67.25 MHz and the other of 71.75 MHz. The *difference* signals (each of lower frequency than the original signals) were again heterodyned to produce a still lower frequency signal. What was the frequency of this final signal?

18-9. For CW reception, what signal frequency must be heterodyned with a CW signal of 40 MHz to produce an audio signal of 400 Hz?

18-10. An oscillator producing a signal of 20,230 kHz developed an audio tone of 1 kHz when heterodyned against a CW signal higher in frequency than the oscillator. What is the frequency of the CW signal?

Applications and Components III

Transmitter and Receiver Systems 19

19-1. Introduction

Complete systems making up AM transmitters (radio and television), FM transmitters (monaural or stereophonic), plus associated receivers and other electronic gear are formed by the sequential linkage of specific circuits from among those covered in previous chapters. Many circuits (notably amplifiers) are, of course, used extensively in numerous systems, while others are designed to perform a particular task for virtually exclusive usage in one system only.

Representative examples of transmitter and receiver systems are covered in this chapter to illustrate practical applications of the various circuitry and principles covered in earlier sections of this book. Because of such previous circuit analysis, the systems are shown primarily in block-diagram form. Where special circuitry is involved, or where particular emphasis is required for explanatory purposes, typical schematics are shown.

19-2. AM Transmitter

The basic system for AM transmission was shown earlier in Fig. 18-1. A more comprehensive block diagram is shown in Fig. 19-1. The AM radio transmitter is the most simple transmission system, with FM much more com-

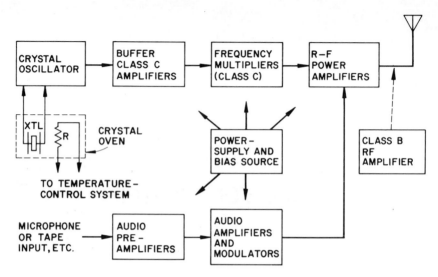

FIGURE 19-1. AM transistor.

plex, and TV utilizing the greatest number of circuits since it embraces both the AM and FM modes.

As discussed earlier, a crystal oscillator is used to maintain frequency stability of the carrier signal, though this generator does not necessarily produce the fundamental-frequency signal. Often the crystal oscillator furnishes a signal which is harmonically related to the frequency which is used for the transmitted carrier. The crystal oscillator's signal is of lower frequency than the final one to permit use of a thicker piezo-quartz crystal so as to achieve greater efficiency in the oscillator circuit and provide some protection against breakage. This usage of a thicker (lower-frequency) crystal applies particularly to high-frequency transmission practices such as found in TV or commercial shortwave services.

With too thin a crystal, operation at reduced ratings would be necessary to minimize the arcing between plates that causes cracking of the crystal or formation of burned spots. As shown in Fig. 19-1, the crystal is housed in a compartment (termed a *crystal oven*) where a heating element (R) maintains a constant temperature for precise frequency control of the generated signal.

Additional buffer Class C amplifiers are used to increase signal power to that required, while also isolating subsequent stages from the initial crystal oscillator. This isolation minimizes effects of load changes at the output stage from altering the stability of the oscillator. Resonant-circuit frequency multipliers raise the frequency to that required for the particular station.

Once the carrier has been modulated, additional amplification by Class C is not possible because of severe distortion. Additional amplification is possible, however, by using resonant-circuit Class B amplification as shown, following the final modulated Class C amplifier.

If the final Class C amplifier is modulated and no additional RF amplification is used for the modulated carrier, the system is known as *high-level* modulation, indicating modulation occurs at the highest RF power level, regardless of whether the modulation occurs in the input of the Class C (grid or base) or at the output (plate or collector). When the modulated Class C amplifier is followed by one or more Class B amplifiers, the system is referred to as *low-level* modulation to indicate that modulation occurs at a lower-power level than the final output power from the last RF stage.

19-3. B/W TV Transmitter

The basic black-and-white television transmitter system is shown in Fig. 19-2. Since the amplitude-modulation process is used for the picture signal, the sequence of circuits from the crystal oscillator to the modulated final resembles that for the AM transmitter. Again frequency multipliers are used to bring the carrier to the desired VHF or UHF allocated frequency.

The TV camera tube converts light and dark areas of a scene to changes in electric-signal amplitude. Since an abrupt change from a dark to a light area produces a sudden change in signal amplitude, the frequency span of video signals is much greater than that of audio and may extend to 4 MHz. Hence, widely spaced side bands are produced and bandwidth requirements are high. Most of the side bands below the carrier are eliminated by special filter circuitry to conserve spectrum space. Despite this, however, TV is still wide band, with each station having an allocated bandwidth of 6 MHz.

As shown in Fig. 19-2, generators are also used to produce what are termed *sweep* signals. These signals (obtained from relaxation-type oscillators) are formed into sawtooth waveforms for sweeping the electron beam of a picture tube across (horizontal rate of 15,750 Hz) and down the face of the tube (vertical rate of 60 Hz). Blanking signals are also produced so that beam retrace (and the picture) is blanked out during the synchronization of sweep circuits.

The composite video signal, a portion of which is illustrated in Fig. 19-3, is used to amplitude-modulate the Class C final as shown. (When this signal arrives at the receiver, the sync pulses lock in similar relaxation oscillators and thus keep the receiver in precise scanning relationship with the transmitter.)

The sound portion of television transmission consists of a frequency-modulated carrier that differs in frequency from the picture carrier and is also independent of it. The FM transmitter portion is, therefore, a separate system and contains circuitry as shown later for standard monaural FM. Both the FM (sound) carrier and the video (AM) carrier are applied to a

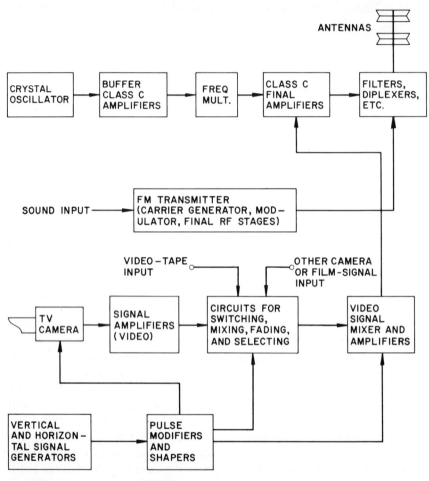

FIGURE 19-2. B/W TV transmission system.

diplexer circuit that isolates the output circuitry of the two and prevents interaction, while at the same time sending both signals to a common antenna. A side-band filter partially suppresses the lower side bands to reduce spectrum span as mentioned earlier.

Specific frequencies for the picture and sound carriers for both VHF and UHF station allocations are given in the appendix. The 6-MHz allocation applies to both the VHF and UHF stations. Allocations of the same channel to two or more stations in a given area are avoided to minimize interference between the two. Often several hundred miles separate stations using the same video and sound carrier frequencies.

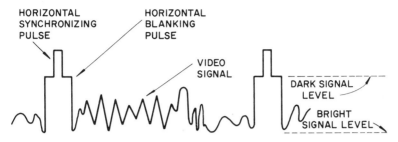

FIGURE 19-3. Portion of composite video signal.

19-4. Color TV Transmitter

The modern color-television system is compatible; that is, color transmission can be received as a black-and-white picture on a black-and-white set (a condition that did not prevail for some of the other color systems proposed originally). Since black-and-white television had been developed and station allocations issued prior to the advent of color, it was necessary to incorporate the color signals in the already-existing 6-MHz station allocations. This necessitated using a separate carrier for color at the transmitter but eliminating that carrier from the transmission after the side bands were produced. Thus, such a missing carrier has to be generated in the receiver and recombined with the transmitted side bands before demodulation can take place.

Also, since the color signals had to be included in the existing spectrum space for a station allocation, the vertical and horizontal sweep rates had to be altered slightly from the 60 and 15,750 Hz used in black and white. Consequently, the color transmission uses 15,734.264 Hz for horizontal sweep and 59.95 Hz for the vertical sweep rate. Both frequencies are sufficiently close to the normal frequencies of the signals generated by the sweep oscillators of the receiver to permit proper synchronization. Thus, no trouble is experienced in receiving color signals on a black-and-white receiver or receiving black-and-white signals on a color set.

For color transmission, however, many of the other specifications for the black-and-white transmission are similar. Thus, the picture is still transmitted with a ratio of width to height as 4 to 3 (termed the *aspect ratio*). The picture carrier is 1.25 MHz above the lower end of the channel spectrum as with black and white. The scanning down the face of the picture tube is interlaced; that is, the beam traces out the picture once (scanning across and down) and then repeats the so-called field by rescanning between the horizontal picture lines traced before. (Two of the fields constitute a *frame*.)

As with black and white, the sound carrier is 4.5 MHz above the picture

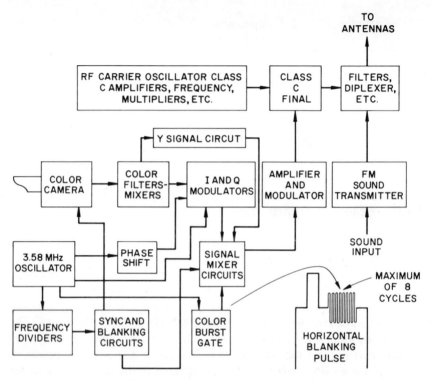

FIGURE 19-4. Color TV transmission system.

carrier frequency. Permissible deviation of the sound FM carrier is 25 kHz each side of center. As with the black-and-white transmitter, an independent sound transmitter is used as shown for the block diagram of the color-TV transmitter in Fig. 19-4.

The color camera, with color filters, reduces the various colors of a televised scene into electric signals representing only three colors: red, blue, and green, with required proportions of each determined by the color mixer circuitry. Thus, one of the signals obtained is the so-called *luminance* signal that corresponds to the black-and-white signal of monochrome transmission and termed the *Y* signal. The three color signals are also combined into two basic signals to conserve space in the channel spectrum.

The two basic color signals produced by the matrix section consist of an *I* (in-phase) signal and a *Q* (quadrature) signal. These are fed to balanced modulators (described later) where they modulate 3.58-MHz subcarriers (one displaced by 90° with respect to the other). The modulators suppress the subcarrier, *I* and *Q* signals, and thus the output consists only of the sideband components.

A burst gate, synchronized by the 3.58-MHz oscillator, produces a minimum of 8 Hz of the 3.58-MHz signal, which is mounted on the horizontal blanking level following the sync pulse, as shown in Fig. 19-4. This burst is necessary at the receiver for synchronization of the 3.58-MHz oscillator that replaces the missing subcarrier, as explained later.

When the complete composite signal is obtained (including video, sync, blanking, color side bands, etc.), it modulates the carrier as shown and, as with black-and-white transmission, the modulated RF from the AM section is combined with the modulated RF of the FM transmitter in the diplexer and fed to a common antenna system.

Various power supplies (not shown) must, of course, also be present to supply voltages and currents to the various stages.

19-5. FM Mono Transmitter

A basic system for FM transmission is shown in Fig. 19-5, where a phase discriminator and reactance circuit are used. As shown, the audio-signal input to the reactance circuit causes an FM deviation as described earlier. In a system of this type the initial deviation is held at a minimum to maintain good linearity and minimize distortion. Thus, if the variable-frequency oscillator has a resonant frequency of 5.833 MHz, a series of doublers and triplers are used to raise the carrier frequency to that required (such as 105 MHz in the FM band) for a particular station. If the deviation is 4 kHz, it would in

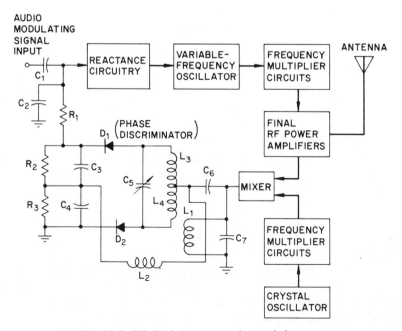

FIGURE 19-5. FM discriminator-control transmission system.

turn be tripled, doubled, and again tripled to reach a value of approximately 72 to 75 kHz for maximum permissible deviation for the 88 to 108 MHz FM band.

Because a variable-frequency oscillator is used to permit frequency modulation, the carrier may drift from its assigned frequency without crystal control. To prevent this, a crystal oscillator is used, as shown, and its frequency is also raised until near that of the final carrier frequency. Thus, the carrier signal and that from the crystal oscillator are combined in a mixer. A heterodyning process produces a difference frequency signal that is applied to the phase discriminator shown. If, for instance, the selected difference frequency is 3 MHz, the resonant circuits (C_5, L_3, L_4) are tuned to this frequency.

A reference voltage is necessary with the discriminator as described earlier; hence another inductor (L_2) is used to establish this potential, tapping the center of the discriminator inductance (L_3 and L_4) and also the junction of resistors R_2, R_3 and capacitors C_3 and C_4. Inductor L_1 is coupled to the resonant-circuit inductance by conventional transformer arrangement. Thus, the output of the mixer develops across the secondary L_3 and L_4, while a reference voltage is established across L_2 also. This sets up voltage distributions similar to those described earlier for the FM discriminator detector.

When the variable-frequency oscillator is generating the correct frequency signal, the output from the mixer will be 3 MHz and since the phase discriminator is tuned to this, it develops no output voltage because the individual voltage drops across R_2 and R_3 would be equal but opposite in polarity, thus canceling out, as described for the discriminator FM detector.

If the variable-frequency oscillator drifts from its proper frequency, the mixing process no longer produces 3 MHz because the difference between the final carrier frequency and the multiplied crystal oscillator frequency have changed. Now one diode of the discriminator conducts more than the other, producing unequal voltage drops across the output and thus producing a correction voltage. Since this is applied to the input of the reactance circuit, it brings the frequency of the variable oscillator back to normal. If the drift were in the opposite direction, the polarity of the correction voltage would also change. Thus, the correction voltage can either raise or lower the frequency of the variable oscillator.

Capacitor C_2 has a low reactance for the RF signals; hence it filters them from the output of the discriminator. This bypass effect prevents RF entry into the reactance circuit, thus preventing the FM components present in the phase discriminator from affecting the performance of the reactance circuit. The d-c correction voltages from the discriminator are not shunted by C_2 because no shunt reactance is created by them. (The variable-frequency oscillator drifts slowly off frequency; hence it does not develop rapid center-frequency changes as is the case with audio modulation.)

19-6.   FM Stereo (Multiplex) System

Stereophonic transmission or recording refers to the process whereby directional characteristics are given to the various sounds that are broadcast or recorded. The basic stereophonic audio system utilizes two separate amplifying systems usually contained in one chassis, each separate channel ending in its own speaker system. For stereophonic record reproduction, for instance, two sound tracks are actually recorded in a single record groove, but at different angles. A special phonograph pickup is employed and the sounds reproduced from each track are then channeled to separate amplifiers and speakers. The latter are placed 6 to 8 ft apart and certain orchestral instruments or singers will appear to be at the left and others at the right, in conformity to the placement of such instruments and singers at the studio. Thus, stereophonic reception not only adds directivity to the recorded music but also widens the sound source. Instruments or sounds emanating from between the two microphones that pick up the stereophonic sound will be reproduced at equal levels and thus will appear to have a central sound source during reception.

In FM stereo broadcasting, two separate channels are also employed but the system is so designed that ordinary FM reception is obtainable with a standard FM receiver. For reception of stereophonic sound, however, a special receiver (or adapter for the ordinary receiver) is necessary. When the receiver separates the two transmitted channels, two separate audio-amplifying systems must again be employed (with separate speakers for each). Because reception is possible with an ordinary FM receiver as well as the stereo type, the system is known to be *compatible*.

As described in more detail later, the FM stereo process is a multiplex system, with two separate channels (left and right) combined into a single transmission. In addition, however, some mono- or stereo-broadcasting stations may also be involved with SCA (*Subsidiary Communications Authorization*). This is another type of multiplex transmission, where the main channel is used for general-public-oriented entertainment and a subchannel is used for SCA to broadcast background music or other private services for paying subscribers.

The basic stereophonic FM multiplex system is shown in Fig. 19-6. The individual left- and right-position audio signals picked up by microphones (or from records, tapes, etc.) have a frequency response of *at least* 50 Hz to 15 kHz, with a 75 μs preemphasis for each channel, as detailed more fully later.

As with color television, stereo FM had to fit into the space already occupied by existing stations while at the same time maintaining compatibility. Thus stereo FM can be received as a mono broadcast on mono receiv-

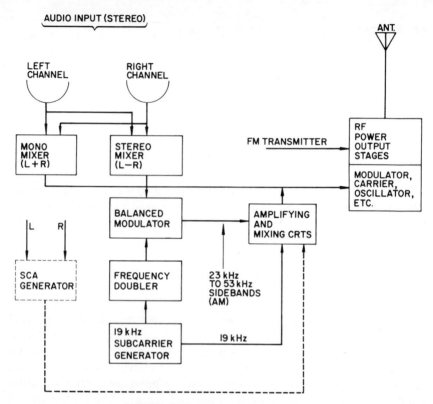

FIGURE 19-6. Stereo multiplex system.

ers and as stereo broadcasts from a stereo receiver, as detailed more fully later in this chapter.

To maintain compatibility, special mixing of left- and right-channel sound signals must be done as shown. A mono signal (L + R) is fed to the modulating section of the main FM transmitter and this circuitry is referred to as the *main channel* transmission. Thus, this main channel contains all sections of a mono FM transmission.

Stereo signals are transmitted by taking advantage of the fact that modulating-signal frequencies can extend well above the specified 15 kHz required for good fidelity. For such higher modulating-signal frequencies, the *rate* of carrier deviation rises but the amount or extent of carrier excursions on each side of resonant frequency depends on the *amplitude* (*not* frequency) of the modulating signal. Thus, in FM multiplex systems, signals having frequencies above 15 kHz are utilized for modulation processes as shown by the spectrum illustration of Fig. 19-7. Note the main channel transmission occupies the initial portion of the complete modulating-signal spectrum required for multiplex transmission.

To obtain the distinct left and right channels needed for stereo broadcasting, an additional signal is used. To multiplex such a signal, a *difference*

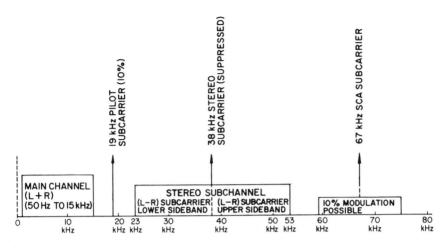

FIGURE 19-7. Modulating-signal spectrum (FM station with stereo and SCA).

signal is obtained by subtracting the R signal from the L signal by feeding both into a mixer circuit, with a 180° out-of-phase displacement for the R signal. This difference signal obtained from L − R modulates an additional carrier (called a subcarrier) by the *amplitude-modulation* process. Thus, side bands are formed, and these in turn share in modulating the transmitted FM carrier as shown in Fig. 19-6.

The subcarrier is suppressed by the *balanced modulators* as described later in this chapter. Thus, the subcarrier is missing from the broadcast and must be reinserted at the receiver before stereo demodulation can take place.

The frequency of the subcarrier signal is 38 kHz and is procured by doubling the frequency of the signal from a 19-kHz oscillator. As shown, this 19-kHz signal is also applied to the mixing circuitry and thus modulates the carrier and is transmitted with it. As discussed later, this signal is used for synchronization of the stereo detection process in the receiver.

The 19-kHz signal is referred to as the *pilot subcarrier* and the frequency is held to ±2 Hz of assigned frequency. It modulates the transmitted FM carrier signal to only 10% but this is enough for the receiver's process of doubling the frequency to 38 kHz to obtain the necessary subcarrier.

As shown in Fig. 19-7, the side bands obtained by modulating the 38-kHz subcarrier with the L − R signals are situated above the mono modulating signals. Note that the side bands occupy a region between 23 and 53 kHz and that the lower (L − R) side bands (as well as the upper) occupy the same bandwidth of 15 kHz as the L + R mono signals, with the same audio-signal frequency range to 15 kHz.

Thus, the entire *multiplex modulating signals* for stereo FM transmission are composed of the mono L + R signal in the audio range (50 Hz to 15 kHz), a supersonic 19-kHz pilot subcarrier signal, plus the L − R supersonic signal (23 to 53 kHz) with the 38-kHz carrier suppressed at the transmitter.

With SCA (discussed later) the span is above 70 kHz as shown in Fig. 19-7.

In the stereo multiplex system, the subcarrier 19-kHz pilot signal is permitted a 10% injection as mentioned earlier, thus providing a carrier-frequency deviation of 7.5 kHz each side of resting frequency. If no SCA transmission is present, a 90% additional modulation is possible for the other channel signals. Thus, this remaining 90% modulation capability must be divided between the mono and the stereo channels. If the mono signal is modulating the main channel up to 90%, the stereo side bands would *not* be modulating the FM carrier. A balance effect is usually present for L and R channel amplitudes since, during stereo broadcasting, volume and hence signal-amplitude differences usually prevail between left and right program material.

The L − R side bands are capable of only modulating up to 45% (and this maximum value only prevails for a single left signal or a single right signal). If the stereo side bands are modulating at the 45% level, the L + R modulating maximum does not exceed this 45% value either.

19-7.   SCA Multiplex System (FM)

To multiplex the SCA transmissions, an SCA generator is used as shown in the broken-line designation in Fig. 19-6. This unit is virtually a complete miniature FM transmitter as compared to the main transmitter, with a center subcarrier frequency of 67 kHz as shown also in Fig. 19-7. The Federal Communications Commission (FCC) does not require the specific 67-kHz subscarrier frequency but this value has been made virtually a standard for stereo-SCA combination stations.

If the station's main channel is mono and the SCA transmission is also used, the 67-kHz signal is applied to the transmitting section along with the conventional modulating signals. Since a 75-kHz deviation each side of carrier is considered equivalent to 100% modulation, the limits thus imposed must not be exceeded when the SCA signal is injected. Hence, the modulation level for the standard FM broadcast is reduced to accommodate the SCA signals. FCC limits the SCA injection to 30% of total modulation. Hence, for a mono station also using SCA, the main channel modulation is set at 70% with 52.5-kHz deviation each side of resting frequency.

Within the SCA generator, the audio signals modulate the 67-kHz subcarrier for a maximum deviation usually held to 7.5 kHz each side of center, with a frequency response range of 30 Hz to 7.5 kHz, and a deviation ratio of 1. Thus, when the SCA system is multiplexed with standard FM, the system is sometimes referred to as FM on FM. Obviously, the system suffers somewhat with respect to dynamic volume range of audio and frequency response in terms of high fidelity because of the channel sharing. The lower deviation ratio also decreases signal to noise ratios, sometimes requiring station muting to silence noises between audio modulation periods.

When the SCA is used with stereo FM, the SCA injection is held to 10%
to prevent exceeding frequency standards. Hence, this injection percentage
(plus the 10% for the pilot subcarrier) reduces the remaining modulation
capability to only 80%. This limit is the maximum for the combined main-
channel modulation *and* the stereo side bands. Thus, without the stereo-
signal modulation, the main-channel modulation is 80% and virtually
monophonic. The stereo side bands, during proper modulation, never
modulate more than 40% for this particular system, with this maximum
also prevailing for the main channel (to add up to 80%).

19-8. Special Circuitry

In both transmitting and receiving circuits, as well as in computer systems and
other electronic areas, the necessity often arises for combining two or more
signals to form a composite signal. On occasion the term *adder* may be used
to indicate the basic function or the circuit may be called a *mixer* or *combining* circuit. Mixers that have nonlinear characteristics produce sum and
difference signals in addition to the original signals as previously described
in Chapter 18. Such heterodyning is a common practice in receivers and is
also used for special purposes in other systems. If circuitry has substantial
linear characteristics, distortion is at a minimum and only the combined
original signals appear at the output, without heterodyning.

A typical nonlinear type mixer is shown at A of Fig. 19-8 and is widely
used in the superheterodyne receiver described later in this chapter. Here
the incoming RF modulated carrier signal is heterodyned with the signal
generated by the oscillator portion of the mixer transistor. Oscillations are
produced because the feedback coil L_3 couples some of the amplified signal
in the collector to the oscillator resonant-circuit inductance L_4. The coupling
is polarized so that positive (regenerative) feedback occurs. The RF signal
in the oscillator's resonant circuit (L_4 and C_5) is coupled back to the emitter
circuit by capacitor C_4.

The mixer at A of Fig. 19-8 is also referred to as a *converter* circuit
because it "converts" the incoming RF and the oscillator signals to a *differ-
ence-frequency* signal referred to as an intermediate-frequency (IF) signal.
Since such a mixer converter also has detection characteristics, it is sometimes
referred to as the *first detector* of the receiver, whereas the diode demodulator
following the IF amplifier stages is termed the *second detector.*

The RF signal appearing between base and emitter heterodynes with
the oscillator signal because nonlinear operational characteristics prevail
(the applied bias operates the transistor near cutoff). After the mixing
process, the sum and difference signals present at the output are rejected by
the resonant circuits of the following stages because those are tuned to
the IF-signal frequency only, as described later.

A linear mixer is shown at B of Fig. 19-8 and two FET are used. Junction

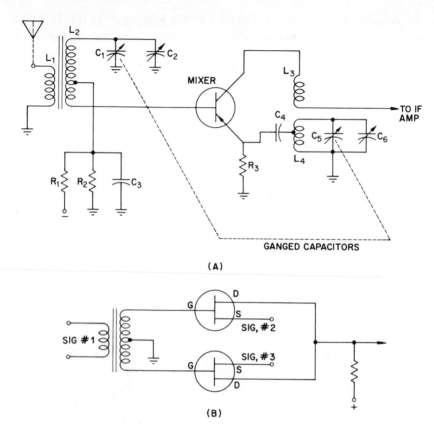

FIGURE 19-8. Mixer circuits.

transistors could, of course, also be used with the secondary of the transformer connected to the base elements and the emitter elements used for inputs of
the second and third signal sources. The tapped secondary of the input
transformer causes the input signal to appear out of phase by $180°$ at the gate
inputs of the two FET units, thus permitting L — R-type mixing, as well as
L + R as required for stereo modulation.

At A of Fig. 19-9 a resistor-transistor (RT) type mixer is shown. Here
two signals may be applied to a simple base-input section of a transistor.
Resistors R_1 and R_2 are for isolation purposes and minimize interaction and
undesired coupling between the two input systems. Resistor R_3 is a balance
control for adjusting the relative amplitudes between the two signals. If
the signal applied to one input is out of phase with that at the other input,
some signal cancellation occurs, depending on the relative amplitudes of
the individual signals applied.

As mentioned earlier, there are occasions where a carrier is to be modu-

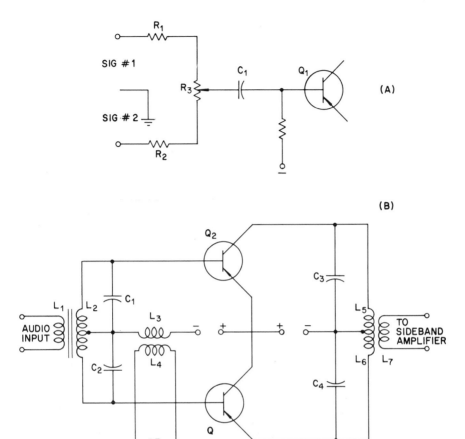

FIGURE 19-9. Mixer and balanced modulator.

lated, but once the side bands are obtained, the carrier must be suppressed. Typical of the circuitry involved is the balanced modulator shown at B of Fig. 19-9. Note that the RF carrier is injected in series with the supply potential and center tap of the input transformer L_2. Hence, this signal is applied in phase to both base elements of the two transistors. Thus, a single RF signal alternation across L_3 causes both base elements to have the same forward-bias *change*. Thus, if the voltage drop across L_3 opposes the normal forward bias (negative for *PNP*), bias for both transistors is reduced and there is a decrease in conduction in both collector-emitter circuits.

As shown at B, the collectors of the balanced modulator are connected in push-pull arrangement and current flow through the L_5 and L_6 windings are in opposite directions. Thus, current changes in this portion of the output transformer are equal and opposite to each other. Thus, cancellation occurs for such a current change representative of the RF signal. (Matched transistors and associated components are, of course, essential in this circuit.)

The audio-modulating signals are applied to the base inputs in typical $180°$ phase difference characteristic of push-pull circuitry. Since these change collector-current flow, carrier-frequency currents within each transistor undergo modulation. Side bands are produced and these find resonant-circuit conditions in the output formed by C_3, C_4, L_5, and L_6. These resonant circuits have a low impedance for the audio signals and hence minimize their appearance at the output. Thus, only side-band-signal information appears from the balanced modulator.

In mono, stereo, or SCA frequency-modulation systems, a noise-reduction process is used involving a system of *preemphasis* at the transmitter and *deemphasis* at the receiver. Interelement transistor and circuit noises are generated at a fixed amplitude level in a given system. Hence, the signal-to-noise ratio is improved by increasing signal level over the constant-level noise. Since noise signals increase for higher-frequency audio, the level of the audio signals is raised to a greater extent as the audio-frequency rises.

The FCC has set the rate of incline for the preemphasis process so that uniformity exists for all FM stations. The rise in the amplitude of the audio signals starts at about 400 Hz and rises gradually. At 1000 Hz the increase is 1 dB; at 1500 Hz the increase is almost 2 dB. At 2 kHz there is a rise of about 3 dB; at 2.5 kHz there is almost a 4-dB increase. From here on the increase in preemphasis is virtually linear, reaching 8 dB at 5 kHz and 17 dB at 15 kHz.

The basic circuit for preemphasis resembles the standard capacitor-resistor input circuit of an amplifier stage with a series coupling capacitor followed by a shunting resistor. Thus, it is similar also to the differentiating circuit shown earlier at A of Fig. 17-17. The time constant (RC) for the preemphasis network is 75 μs. This provides best results without increasing frequency deviation excessively because of the increase of signal amplitude for higher-frequency audio.

At the receiver it is necessary to deemphasise the gradual rise in audio to prevent shrill reproduction. This is accomplished by a series resistor and shunt capacitor as at B of Fig. 17-17, shown earlier for the integrator. RC is again 75 μs.

19-9. Reception Factors

Virtually all types of receivers, when in operation, have a number of signals of various frequencies intercepted by the antenna system and applied to the input circuit. In order to select the desired signal from among all present in the area, it is necessary for the receiver to be tuned to resonance for the modulated carrier signal of the station that is to be received. The carrier signal that is then accepted by the input system of the receiver is amplified, by RF amplifier stages, to the amount required for application to the detector tube. The latter then demodulates the carrier signals and produces the audio signal, or the video (picture) signal in a television receiver. The demodulated signal is then amplified by an additional amount, before application to the loudspeaker or to a picture tube.

From the foregoing, it is evident that the simplest method for achieving this result is to have several RF amplifier stages, each of which includes a variable capacitor for tuning the resonant circuits so that the desired station may be selected. In the early days of radio, such a system was actually employed and the sequence of circuits in this system is shown in the block diagram of Fig. 19-10. Here, the antenna system picked up the desired signals and applied them to the input of the first RF amplifier. This vacuum-tube RF amplifier circuit employed a variable capacitor (designated as C_1 in Fig. 19-10) and an inductance, L_1. Capacitor C_1 was variable and when this variable capacitor was closed (meshed), the maximum capacity thus repre-

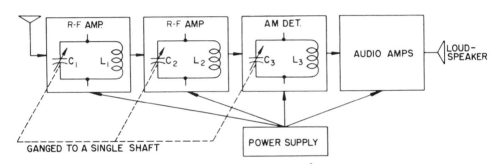

FIGURE 19-10. Tuned RF radio.

sented tuned the receiver to the lowest station in the AM broadcast band. As the variable capacitor was progressively opened, it tuned to higher stations in the AM broadcast band. When the capacitor was fully opened (representing a minimum of capacity), it tuned to the highest-frequency station in the band.

The second RF amplifier stage and the detector stage also had a variable-capacitor arrangement for tuning. The three variable capacitors of the RF and detector stages were ganged together, as indicated by the dashed interconnecting lines shown in Fig. 19-10. A single shaft was employed and when this shaft was turned, it rotated all the capacitors simultaneously.

Such a radio receiver was termed a *tuned* RF receiver (TRF). This receiver, however, presented some serious drawbacks. Since it was necessary to change the value of the variable capacitors for tuning to the desired stations, there was a considerable variation of Q between the closed and open positions of the variable capacitor. Thus, both the sensitivity and the selectivity of the receiver varied as it was tuned over the broadcast band range. On occasion, fixed capacitors were employed and the inductance was varied by employing a movable core slug. Each inductance, L_1, L_2, and L_3, had such a variable core, and all three cores were connected to a common shaft so that they could be moved simultaneously for tuning purposes. Core tuning, however, also suffered the disadvantages of changing circuit Q. For a desired degree of selectivity, the capacitor and inductance of a tuned resonant circuit must be proportioned so that the ratio of coil resistance and reactance produces the desired Q and, hence, the degree of band-pass characteristics desired. In the tuned RF receiver, it was virtually impossible to maintain a constant Q since it is necessary to vary either the capacity or the inductance for tuning purposes.

The unfavorable characteristics of the tuned RF receiver system were overcome by the invention of the superheterodyne receiver by Major Edwin H. Armstrong. In consequence, the superheterodyne receiver principle has superseded the older tuned RF type of receiver and now all modern radio, FM, and television receivers utilize the superheterodyne circuit.

19-10. Superheterodyne Principles

A block diagram of a superheterodyne AM receiver is shown in Fig. 19-11. Here an RF amplifier precedes a mixer circuit. The RF amplifier stage uses a parallel-resonant circuit with a variable capacitor, in a fashion similar to that employed for the TRF receiver. A similar variable capacitor is included in the mixer stage and also in the oscillator stage. These capacitors are ganged together, as shown by the dashed interconnecting lines. Rotation of the common shaft permits tuning over the entire broadcast band. The RF amplifier increases the signal amplitude of the modulated carrier and applies it to a mixer circuit (see A of Fig. 19-8). At the same time, an oscillator (termed a *local oscillator*) generates a frequency that is also injected into the mixer. The local oscillator differs from the carrier frequency by 455 kHz. Thus, if a 1000-kHz station is tuned in, the 1000-kHz carrier will be applied to the mixer. The oscillator is tuned to 1455 kHz and the difference between the carrier and the oscillator signal is 455 kHz. The mixer is operated so that it has pronounced nonlinear characteristics and, hence, it will heterodyne the carrier and oscillator signals. (A discussion of the basic heterodyning principle is given in the preceding chapter.)

The output from the mixer stage, because of the heterodyning process, consists of the original carrier, the oscillator frequency, the sum frequency, and the difference frequency. All these are still RF signals, and the RF amplifier that follows the mixer stage is tuned to resonance for only one of these frequencies, the *difference* frequency. Thus, this RF amplifier stage is tuned to 455 kHz and the selectivity of the tuned resonant circuit rejects

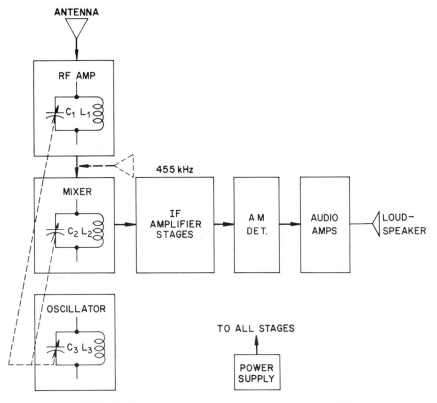

FIGURE 19-11. Ganged capacitors in superheterodyne radio.

the other frequencies. This RF amplifier stage is designated as the *intermediate-frequency* amplifier stage, to distinguish it from the first RF amplifier stage, which handles the carrier signals. The intermediate-frequency (IF) amplifier stage builds up this difference frequency and applies it either to an AM detector or to one or two additional IF amplifier stages. The AM detector then demodulates this IF carrier signal and derives from it the audio signals. The latter are then applied to conventional audio-amplifier stages and impressed on the loudspeaker for sound reproduction.

Initially, it may seem that this receiver has no advantages over the tuned RF receiver since three variable capacitors are still employed, which affect the selectivity. Actually, however, the required degree of selectivity is established in the IF amplifier stages since these IF stages are not retuned when various stations are selected. The IF amplifier stages of the receiver shown in Fig. 19-11 are always tuned to 455 kHz, and the handling of only this specific frequency permits selecting the proper reactance-resistance ratio for the desired band-pass characteristics.

The reason why the mixer will always produce a 455-kHz signal, regardless of the station to which the receiver is tuned, is that the oscillator always maintains a *difference* frequency between it and the RF amplifier stage. One method for achieving this result is to make the variable capacitor C_3 smaller than capacitors C_1 and C_2. By making C_3 smaller, it will tune to a higher frequency. If the capacity of C_3 is designed to tune the oscillator to a frequency higher than the carrier frequency by 455 kHz, it will always maintain this difference frequency because when C_1 and C_2 are varied, C_3 will vary proportionately. Thus, if a station having a frequency of 700 kHz is tuned in by C_1 and C_2, capacitor C_3 will cause the oscillator to generate a frequency of 1155 kHz. This frequency is higher than the 700-kHz station by 455 kHz, and thus the output from the mixer is again 455 kHz.

The oscillator could also be designed to be lower in frequency than the RF amplifier by an amount equal to the difference frequency. Thus, if a 1000-kHz station is tuned in, the oscillator could have a frequency of 545 kHz and the difference would again be 455 kHz. It is preferable, however, for the local oscillator to generate a frequency above the incoming RF carrier. The reason for this preference is that when the local oscillator is below the incoming station signal frequency, the range over which the oscillator must tune is considerably greater than would be the case if the local-oscillator frequency were above the incoming signal frequency. If, for instance, the IF were 455 kHz, the local-oscillator range for the AM broadcast band would be as follows:

550 to 1600 kHz = AM band range

95 to 1145 kHz = local-oscillator range

In such an instance, the local oscillator must have a tuning range from 95 kHz to 1145 kHz. Hence, the local oscillator must be capable of tuning from its lowest value to over 10 times the latter. When, however, the local oscillator is above the incoming frequency, its tuning range need only be approximately 2 times that of its lowest frequency, as shown below:

550 to 1600 kHz = AM band range

1001 to 2055 kHz = local-oscillator range

The local oscillator can also be made to tune above the RF and mixer resonant circuits, by employing a capacitor in series with C_3. This additional capacitor in series with capacitor C_3 will cause the total capacity of the circuit to be less and hence the oscillator will tune to a higher frequency. This method is sometimes employed when the manufacturer prefers each ganged variable capacitor to be of the same physical size.

Many inexpensive table-model radios dispense with the RF amplifier stage and apply the antenna directly to the mixer circuit, as shown by the dotted antenna outline preceding the mixer in Fig. 19-11 and as shown earlier at A of Fig. 19-8. Elimination of the RF amplifier reduces the input selectivity, as well as the sensitivity, so that these receivers do not perform as well as those that have an RF stage preceding the mixer. In locations where the signal strength is high and the stations are not too crowded, satisfactory reception is secured.

The IF does not have to be 455 kHz but can be any frequency between 400 and 600 kHz. Early radio receivers used an IF much lower than modern receivers do, but the low IF caused interference because of the susceptibility of the receiver to *image-frequency* response. The image-frequency interference is a signal that can enter the receiver and heterodyne with the local oscillator to produce the same IF that would be produced by the desired signal. As an example, if the frequency of the desired signal is 1100 kHz and the local oscillator is 1000 kHz, the heterodyning of the two would produce a difference frequency of 100 kHz. A station having a frequency of 900 kHz, however, would also produce a 100-kHz difference frequency. If it heterodynes with the local oscillator of 1000 kHz in such an instance, the undesired 900-kHz station produces an image-frequency response in the receiver. The relationships in the foregoing example are as follows:

$$
\begin{array}{r}
1100 \text{ kHz (desired station)} \\
-\underline{1000 \text{ kHz (local oscillator)}} \\
100 \text{ kHz (difference frequency—IF)}
\end{array}
$$

$$
\begin{array}{r}
1000 \text{ kHz (local oscillator)} \\
-\underline{900 \text{ kHz (undesired station, image frequency)}} \\
100 \text{ kHz (difference frequency—IF)}
\end{array}
$$

The use of a 455-kHz IF in modern receivers has proved satisfactory from the standpoint of image rejection. Such an IF also provides a good signal-amplitude output from the mixer of a receiver that also has provisions for shortwave reception, as well as broadcast band reception.

Most table-model radios employ only one or two stages of IF amplification, which is sufficient since the Q of the resonant circuits is high and, hence, the gain is good. Frequency modulation and television receivers, on the other hand, must tune over a much wider band pass. A wider band pass is secured by reducing the circuit Q, and such a reduction of Q also reduces gain. Hence, receivers that employ wide-band IF amplifiers use several stages of such IF amplification, in order to regain the required amplification that is diminished by virtue of the lower circuit Q.

An FM receiver would resemble the AM receiver shown in Fig. 19-11 in basic block-diagram sequence; the only differences would be in the radio frequencies handled. Standard FM receivers operate between 88 MHz and 108 MHz and, hence, the RF amplifier must be capable of tuning within this range, as contrasted to the 550- to 1600-kHz range usually found in radio receivers. In FM receivers, the IF amplifier stages would be tuned to 10.7 MHz, and one the FM detectors previously discussed would be employed. In television receivers, the local-oscillator signal heterodynes with two incoming carrier signals (the audio and picture carriers), as more fully discussed later.

19-11. Television Receivers

In basic block-diagram form, the modern black-and-white television receiver appears as shown in Fig. 19-12. All present-day television receivers, as with AM and FM receivers, utilize the superheterodyne principle discussed earlier in this chapter. Instead of handling only a single modulated carrier plus its side bands, as in FM or AM receivers, the television receiver handles both the modulated picture carrier (AM) and the modulated sound carrier (FM). In addition to these two carriers, the receiver also utilizes the synchronization signals that are transmitted to keep the scanning beam of the picture tube in perfect synchronization with the scanning beam at the transmitter, also as discussed earlier.

Just like an AM receiver, the television receiver has a tuner that incorporates an RF stage, a mixer, and a local oscillator. The local-oscillator signal heterodynes with the two incoming carrier signals (picture and sound) to produce *two* IF signals. These sound and picture IF signals (plus the synchronization signals) are then amplified in several IF stages, before application to the picture detector circuit.

After they leave the detector, the demodulated picture signals are amplified in the video amplifier (see also Fig. 15-3) and applied to the picture-tube grid or cathode circuit. The sound IF signals are also derived from the video detector and amplified by a special sound IF amplifier, before detection in the FM demodulator. After the latter stage, the audio is amplified and applied to the loudspeaker.

The picture signals, of varying intensity, affect the picture-tube grid and, hence, control the intensity of the electron beam within the cathode-ray picture tube. The beam, in turn, scans the picture-tube face (internally). The internal picture-tube face or screen is coated with phosphor. The phosphor, when struck by the electron beam, will glow (fluoresce) to a degree depending on the beam intensity. The beam sweeps across and down the face of the picture tube, scanning one horizontal line after another to literally "paint" a picture on the screen of the tube.

Synchronization of the beam is accomplished by separating the synchronization signals from the video signal, through use of a sync separator, operated in clipper-circuit fashion similar to the clipper circuits discussed earlier. The signals from the separator circuit are then applied to a vertical oscillator and a horizontal oscillator. The vertical oscillator generates a sawtooth-type signal used to pull the beam downward, while the horizontal oscillator generates a sawtooth-type signal for scanning the beam horizontally across the picture-tube face. In combination, the two oscillators cause the

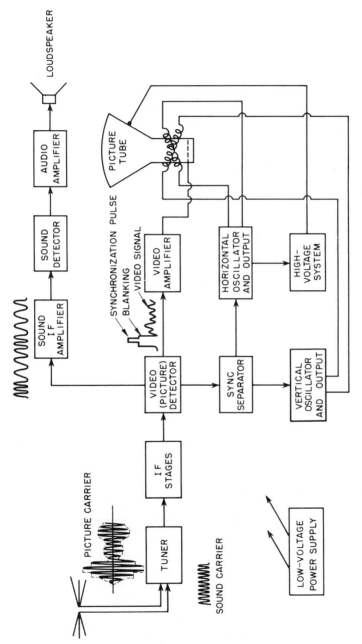

FIGURE 19-12. Block diagram of typical television receiver.

beam to scan the entire tube-face area. Inasmuch as the sync pulses are sent out by the television station, they are instrumental in locking in the vertical and horizontal oscillators of the receiver in perfect timing with the frequencies used at the transmitter. Hence, good picture lock-in is possible with a properly adjusted receiver. (Pulse synchronization of *R-C* oscillators was discussed in Chapter 17.)

A standard low-voltage power supply is used, usually of the full-wave type. A high-voltage power supply is also employed, generating approximately 20,000 V for a 23-in. picture tube. The high voltage is employed to accelerate the scanning beam within the picture tube.

A number of rear and front panel controls must be employed to make proper adjustments of sound, picture, and synchronization. The fine-tuning control is used for precise adjustment of the local-oscillator frequency so that the proper intermediate frequency will be supplied during the heterodyning process between the local-oscillator signal and the incoming signals.

The contrast control regulates the amplitude of the picture (video) signals, just as the volume control regulates the amplitude of the audio signals. The brilliancy control permits manual regulation of the bias on the picture tube and, hence, influences the *background* or *brightness* content of the picture. Hold controls are also present for adjusting the sweep oscillators so their frequency is sufficiently close to the transmitted frequencies of the sync signals for proper lock-in. Other controls include width, height, and linearity, the latter being used for adjustment of vertical and horizontal proportions of the scene to make them conform to normal.

The basic color receiver system is shown in Fig. 19-13. Tuner and video IF stages in the color TV receiver compare to those for black and white except that precautions must be taken to assure a wide band pass in the IF stages for good color detail (about 4.2 MHz). Resonant-circuit traps are necessary to minimize interference; hence the sound carrier is weaker at the video detector than for black-and-white receivers. Consequently a separate sound detector is used as shown.

The luminance (Y) signal, which represents the black-and-white transmission for compatibility, is applied to the cathodes of a three-gun color tube. The inner faceplate of such a tube contains a series of many red, blue, and green phosphor dots that must be struck by the respective electron beams from the cathodes. Thus, for black-and-white reception on a color set, all three phosphors must fluoresce and the blend of the red, blue, and green produces black and white. Hence, if one electron gun becomes inoperative in such a tube, black-and-white reception is no longer possible and color reception will be seriously affected because of the absence of a primary color needed to produce other color shades.

The color (chroma) signals are obtained from one of the video amplifiers and applied to the demodulators where they mix with the missing subcarrier of 3.58 MHz, as shown. The 3.58-MHz oscillator is synchronized by the burst

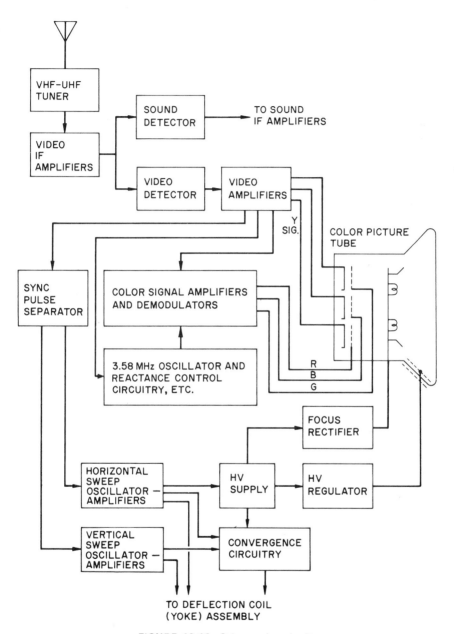

FIGURE 19-13. Color-receiver circuitry.

signal on the rear blanking level as shown earlier in Fig. 19-4. Crystal control is usually used, with reactance circuitry to assure precise stabilization.

The I and Q signals (in relation to Y) are incomplete, and matrix units are used to obtain the true red, blue, and green signals that are applied to the control grids of the picture tube. Conventional sync separation and sweep circuitry are used as in black-and-white receivers. Convergence circuitry is included for exact adjustment of beam alignment within the picture tube. This is essential to cause each beam to hit its respective dot precisely or false colors will appear.

High voltage (ranging up to 25 kV for the 21- to 23-in. tubes) is obtained from the horizontal sweep output system by pulse rectification. Voltage regulation is used to maintain the high voltage at the proper level. This voltage is applied to the inner conductive coating of the picture tube and helps accelerate the electron beams toward the faceplate. (A capacitance is also formed between the inner conductive coating and the outer that is at ground potential. Thus, a filter capacitance prevails for reduction of high-voltage ripple.)

19-12.   Stereo (Multiplex) FM Receiver

The basic circuitry involved with stereo FM reception is shown in Fig. 19-14. The initial stages comprising the tuner, IF amplifiers, and demodulators are identical to the mono FM receivers except deemphasis is provided later. The 19-kHz amplifier uses circuits resonant to this frequency; hence they

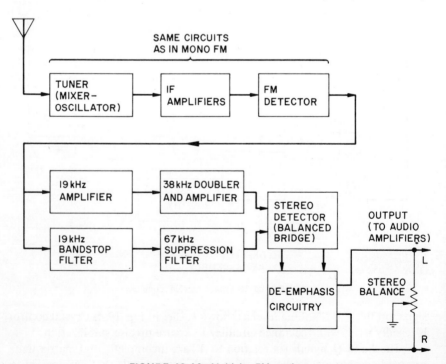

FIGURE 19-14. Multiplex FM receiver.

reject others contained in the composite signal obtained from the FM detector.

The 19-kHz signal is applied to a doubler circuit similar to full-wave diode rectification systems that produce successive alternations at a repetition rate twice that of the pilot carrier. These, in turn, are applied to a 38-kHz resonant circuit and the resultant flywheel effect produces a sine-wave-type 38-kHz carrier for application to the demodulator as shown in Fig. 19-14.

The composite signal is also applied to a 19-kHz band-stop filter as also shown. (Reference should be made to Figs. 10-3 and 10-11.) This filter passes all signals except the 19-kHz pilot carrier; hence its output contains the 50-Hz to 15-kHz demodulated L + R signals, the 23- to 53-kHz L − R sideband signals, and the SCA 67-kHz carrier signals if these are being transmitted by the particular station to which the receiver is tuned. Next, these SCA signals are removed by the suppression filter shown.

The remaining signals are applied to the same balanced-bridge demodulator as the 38-kHz subcarrier. The left- and right-channel signals obtained are individually deemphasized as shown, and a common potentiometer across the two output lines may be used for controlling the relative amplitude of the channels for stereo balance. Standard audio amplification follows.

The balanced-bridge stereo detector shown in block-diagram form in Fig. 19-14 is illustrated schematically in Fig. 19-15. After the 38-kHz signal is amplified, it is applied to the resonant transformer L_1 and L_2 and hence to the diode-bridge rectifier. The composite signal mentioned earlier (without the 19- and 67-kHz signals) is applied to the center tap of L_2. The four diodes thus process the incoming 38-kHz signals in relation to the composite signals,

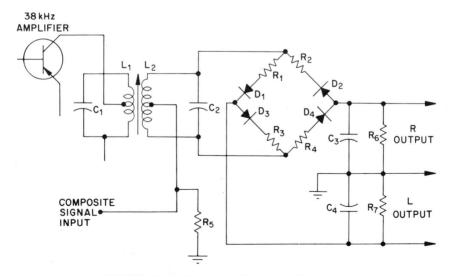

FIGURE 19-15. Balanced-bridge stereo FM detector.

reintroducing the 38-kHz subcarrier into the 23- to 53-kHz side-band structure, and demodulating for the L − R signal components.

The composite signal applied to the center tap of L_2 appears (in phase) at any instant of time at the top and bottom of the bridge rectifier system. The incoming 38-kHz carrier, however, places out-of-phase signals across the top and bottom of the bridge for any instant of time because when the signal potential is positive at the top of L_2, the bottom is negative. Thus, phase relations occur between the signals in similar fashion to the balanced modulators and discriminator circuits discussed earlier. Capacitors C_3 and C_4 convert the rectified pulses to an average-value signal voltage which changes in frequency and amplitude corresponding to the audio-modulating component which had been contained in the original left and right channels.

19-13. High-Fidelity Factors

AM-FM receivers, television sets, phonograph amplifiers, cassette and cartridge recorders, etc., may use either a single-ended power-amplifier stage or push-pull for greater output and better quality sound. Tape *decks* usually contain only audio preamplifiers and are designed to be channeled to the higher-power audio sections of existing receivers or other units. Factors for improving frequency response in amplifiers have been covered in previous chapters. Regardless of the type of system used, however, the final results are often limited by the type of loudspeaker used and the particular housing selected.

The PM (permanent-magnet) speaker is the most widely used type for home entertainment, though horn types may be found in public-address system applications. For the PM type, construction is basically as shown in Fig. 19-16, with a framework having a cross section that often resembles

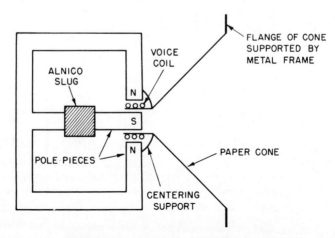

FIGURE 19-16. Basic PM speaker construction.

the letter E. Sometimes a ceramic magnet is used, with a flat circular magnet. Such magnets furnish the necessary magnetic lines of force around the pole pieces at the voice coil. The pole pieces marked N are in the form of a circle so that the N pole pieces surround the center S pole piece and thus create a dense magnetic field in the narrow gap between the S and N pole pieces. Into the narrow gap is inserted an extension of the speaker cone. The voice coil is wound around the outside of this circular cone area. When audio-signal energy from the output amplifier is applied to the voice coil, the latter produces magnetic lines of force. If the magnetic lines of force of the voice coil create a field opposing the field created at the pole pieces, the voice coil is repelled and the cone moves outward. When the voice-coil field changes because of the a-c characteristics of the audio signal, the fields of the coil are attracted by the fields of the pole pieces and the cone moves inward. The cone movement varies the air pressure and sound is produced.

Most speakers have voice-coil diameters of $\frac{1}{2}$ to $1\frac{1}{2}$ in., with some high-fidelity speakers having a 3-in. voice coil to handle greater power. Conventional speakers have cone diameters ranging from a few (in portable receivers) to 15 in. in the larger high-fidelity systems of phonographs, receivers, or tape recorders. Public-address systems may use even larger diameter cones.

The ordinary PM speaker consists of a single paper cone structure coupled to a small-diameter voice coil. The ordinary PM speaker also has a small-sized magnet. Design is such that the speaker can be manufactured economically and still produce a sound level output sufficient for average listening. For high-fidelity applications, however, the speaker is modified to handle a greater amount of audio power, while at the same time providing an extended frequency response range. Besides this, special design factors are incorporated for minimizing undesired resonant peaks and for producing greater speaker compliance for low-frequency sound reproduction.

A single-cone speaker can be designed to have a fairly uniform response from low audio frequencies to over 8000 Hz. For high-frequency reproduction, a stiffer cone material is desirable, while for low-frequency reproduction a large-area soft cone material is preferred. Some speakers corrugate the outer cone areas for better low-frequency reproduction, and the center cone area is often treated with a lacquer compound to stiffen it for better high-frequency signal reproduction.

High-frequency reproduction can also be extended by adding a small thin stiff paper cone section at the center of the large cone section, as shown at A of Fig. 19-17. A dual speaker system can also be employed, where one speaker is utilized for low-frequency signal reproduction, while the other is employed for the high-frequency signals only. When such a combination is employed, the large-cone speaker used for low-frequency notes is known as the *woofer*. The small high-frequency speaker is known as a *tweeter*.

A tweeter can be incorporated with a woofer in a speaker, as shown at B of Fig. 19-17. Such a combination is known as a *coaxial* speaker. The tweeter section can be in the form of a small PM speaker, or a small horn-

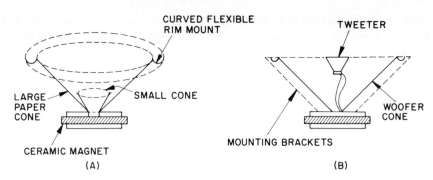

FIGURE 19-17. Speaker types.

type speaker. In either case, the tweeter speaker should be capable of handling high audio-power levels and, for this reason, the ordinary small-cone speaker utilized in table radios is not suitable for tweeter applications. The coaxial speakers have a possible frequency range extending from approximately 50 Hz to 15 kHz, when employed in a suitable enclosure as subsequently described.

A more elaborate type of speaker has also been devised that employs a three-section speaker, known as a *triaxial* speaker. The triaxial speaker has a possible frequency range extending up to 18 kHz or higher, depending on design.

Public-address systems generate a large amount of audio power and, therefore, often use a horn-type speaker to handle the high power. Such a horn type requires a driver mechanism that resembles the PM dynamic speaker, except that a large magnet and small rigid diaphragm are used. The horn section is attached to the driver outlet and helps to reinforce the sounds and bring them to the desired levels. The small output aperture of the horn is referred to as the *throat*, while the flared large opening is known as the *mouth*. The gradual flare of the horn section matches the impedance of the small vibrating diaphragm at the throat to the large-area air medium at the mouth. Most of the larger horns are folded back on themselves to make them more compact.

An unmounted loudspeaker generates acoustical energy from both the front and the back and, for this reason, front and back radiated signals will cancel on those occasions when out-of-phase conditions occur. This is particularly true of low-frequency signals since the wavelength is long and the path short from front to back of the speaker. Because the high-frequency signal components are of a shorter wavelength, the time factor in the travel of signals from the front of the cone to the back and vice versa will result in only partial cancellation of some of these signals. Therefore, an unmounted loudspeaker will give fairly good high-frequency response but lacks low-frequency response to an appreciable degree.

Low-frequency response can be improved by increasing the path length of the signals that emanate from the rear and from the front. This can be done by employing a mounting board that acts as an isolation between the front and back of the speaker. Such a simple device can consist of a panel measuring 4 ft by 4 ft. A hole is cut at the center and the speaker is mounted so that the front of the cone faces through the hole. Under this condition, the path length from front to back is an effective 4 ft. Such a device is known as a *baffle*.

If a larger panel were used, the path length would be effectively increased and, in consequence, low-frequency response would be improved in proportion. A more convenient form of baffle is to utilize a cabinet such as shown at A of Fig. 19-18. The cabinet housing the phonograph or radio device can also be used for speaker baffle. Thus, the baffle is folded back to take up less room, while still providing an effective path length.

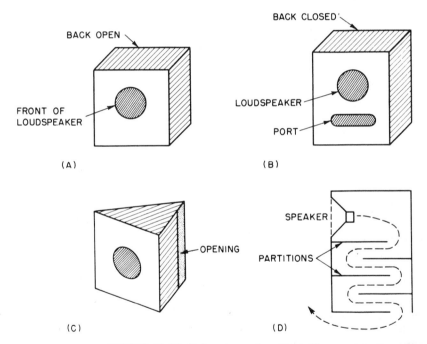

FIGURE 19-18. Various types of speaker baffles.

While speaker cabinets can be designed to function as adequate baffles, the small size encountered in table radios and phonographs provides insufficient path length for satisfactory low-frequency response. At the same time, the larger cabinets which provide additional path length suffer from cabinet resonance and other undesirable effects which prevent the achievement of maximum efficiency and distortion-free-frequency response. Thus, for high-fidelity work, numerous other baffle types have been devised, and these generally outperform the relatively simple baffle arrangement found in conventional radio and television cabinets.

One method for improving the baffle is to employ the bass-reflex principle shown at B. This speaker enclosure is not open at the rear but a hole (port) is provided below the speaker opening. Sound is emitted from both the port and the speaker opening, with good in-phase conditions for bass tones. The bass-reflex cabinet dimensions are based on the size and resonant frequency of the speaker that is utilized. The port size is also related to the resonant frequency of the speaker.

The bass-reflex cabinet gives a fairly uniform response over the audible frequency range and minimizes low-frequency attenuation by utilizing a cabinet that extends the resonant effects of the speaker because of the type of design employed. The cabinet must be padded with a sound-absorbing material.

The bass-reflex cabinet has a relatively high efficiency, as compared to other types that have the rear closed. Cabinets of this type can also be designed for use with dual speakers (separate woofer and tweeter). As with other baffles, the bass-reflex cabinet must be constructed of fairly heavy wood to minimize cabinet resonance, and joints are usually sealed by employing wood glue and using screws to ensure tight connections.

Another speaker enclosure that provides good performance is the folded-horn type illustrated at C. The design of the horn is such that it folds back on itself and thus occupies a minimum of space for the mouth opening secured. If the folded horn is so designed that it can be placed in the corner of a room, the walls of the room act as an extension of the mouth opening so that efficiency and impedance matching characteristics of the enclosure are increased. As with the bass-reflex enclosure, the folded horn can also be employed in conjunction with dual-type speakers. With the folded horns, the efficiency increases as the size is increased and an extended frequency range is possible, besides providing a high degree of efficiency in the low-frequency register.

Theoretically, if a baffle had infinite size, there would be no losses suffered by virtue of the undesired effects that result when rear and front sound radiations intermix. Such a device, while theoretically desirable, is not feasible from the practical standpoint. An approach, however, is the so-called infinite baffle-type enclosure shown at D of Fig. 19-18. Here the path length of the sound leaving the speaker rear has been extended by employing partitions within the speaker cabinet, as shown.

A speaker can be mounted in a cabinet that has no openings for the rear-speaker waves to escape. With a completely closed affair such as this, the back waves are damped and more audio driving power is required. Since cancellation effects are minimized, however, the response characteristics of the system are improved considerably over ordinary receiver cabinets. Such an enclosure has gained favor because it can be made small enough for shelf mounting, thus avoiding the more cumbersome cabinet size of the bass reflex and other similar types.

Other types of speaker enclosures are also found on the market, each having specific characteristics and dimensions designed to extend the frequency response range for a speaker of given size and type. The methods described earlier herein, however, indicate the basic measures generally employed and hence represent fundamental types.

19-14. Quadriphonic Sound

In the early days of radio, phonograph, and tape music reproduction, the mono reception from a single sound source was often compared to a "hole-in-the-wall" type of listening. This was essentially true since a single speaker proffered the sound source as a circular area, and even if a 12-in. speaker were used, it was still comparable to a 12-in. hole in a wall through which music was heard. With woofer and tweeter units separated, but in a single housing, the sound source was spread out a foot or so but was still monaural.

Fidelity of reproduction achieved excellent results, however, because of the advancement of the electronic design principles. Frequency response became more than adequate and *presence* of sound was achieved by remarkable reductions of harmonic distortion and high signal-to-noise ratios. With the advent of stereophonic systems, however, considerable improvement was obtained immediately in several important areas. One factor was the spreading out of the sound source so that instruments could be heard in a panorama fashion from two speakers separated some 8 ft. With properly phased speakers, instrument position along the wide dimension of the sound source appeared pinpointed since the music source appeared to come from the middle when sound output for a particular instrument had equal amplitudes from the speakers. With the advent of high-quality high-fidelity stereo earphones, still another sound-hearing characteristic was obtained. With speakers, for instance, both ears localize a sound source from the right speaker, but with "phones" a direct channel to each ear is obtained. While this is unrealistic in terms of listening to a live performance, it adds a valuable dimension to stereophonic reception.

With a live performance (indoors) the acoustical environment plays a major role in the sum total of the sound characteristics that result. Room dimensions play a part, as do reflecting walls, sound-absorbing sections, reverberation factors, and placement of instruments. All these, particularly the reflected sound from walls, ceilings, and floors, contribute what is known as *ambience*, denoting the atmosphere (environment) of the sound source. When reflected sounds take long to bounce from walls before they reach the ear (because of larger room size), we get a feeling of auditorium-type sound, or spaciousness of environment. With a short bounce, a small-room environment is sensed.

With stereo a considerable degree of ambience was achieved and the effect (as with mono) was often enhanced by introducing reverberation by electronic sound-delay systems (a process often used in electronic organs).

Here some of the sound signal was sent along a spring-type delay line, reconverted into electric signals, amplified, and mixed with the undelayed signals. This gave the illusion of a large concert-hall atmosphere.

After some 12 years of stereophonic tapes, cassettes, tape cartridges and reels, records, and FM, quadriphonic sound was introduced in 1971. This is a four-channel system as compared to the two-channel for stereo. With quadriphonic sound the program material for each channel is somewhat different from the others and speakers are usually placed at the front and rear as shown in Fig. 19-19.

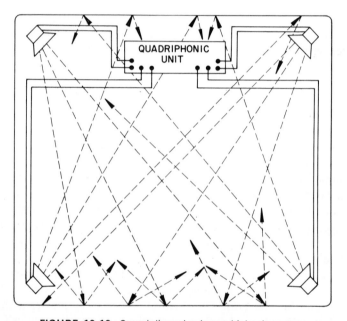

FIGURE 19-19. Sound dispersion in quadriphonic system.

Since four speakers are involved, two additional sound sources are present to add to the ambience and depth achieved by stereo. The reflections are varied and multiple, and only some primary paths are shown in Fig. 19-19. Thus, the realism and presence factors are greatly increased, though of necessity, a total of four separate sound amplifiers and four speakers are now necessary.

Synthesized four-channel sound has been used, where standard two-channel stereo is reprocessed so that two additional channels can be used from sound derivations of the modification process. While this provides for quadriphonic effects, it is not a true system such as obtained when recording four distinct channels and reproducing them by the four-channel amplification system.

Review Questions

19-1. What is the purpose of using a crystal oscillator (in an AM transmitter) with a frequency lower than that finally used for the carrier?

19-2. Briefly explain the difference between *low-level* modulation and the *high-level* type.

19-3. For black-and-white television, what is the allocated bandwidth of a station, and what are the frequencies of the vertical and horizontal sweep rates?

19-4. Define the terms *field*, *frame*, and *luminance signal*.

19-5. In color TV, what is the frequency of the burst gate, and how is this synchronizing signal sent to the color receiver?

19-6. Explain briefly how the variable-frequency carrier oscillator of an FM station is stabilized to prevent carrier drift without interfering with carrier deviation during modulation.

19-7. Define the terms *multiplex* FM and *SCA system*.

19-8. What is the minimum audio-frequency response of mono FM or the main channel of stereo FM?

19-9. How is compatibility achieved in stereo FM systems?

19-10. Briefly explain the purpose for using a pilot subcarrier in stereo FM.

19-11. Briefly explain how the SCA signals are transmitted in multiplex FM.

19-12. Explain the results of mixing two signals in a linear combining circuit versus a nonlinear type.

19-13. What is the purpose for using a balanced modulator in transmitting systems?

19-14. Define the terms *preemphasis* and *deemphasis*, and give the reasons for including these processes in FM.

19-15. Briefly explain the basic principles of a superheterodyne receiver.

19-16. To what color-picture tube elements are the following signals applied: *Y*, red, blue, and green?

19-17. What is the purpose for convergence circuitry in a color receiver?

19-18. How is the 38-kHz carrier obtained in an FM receiver for reinsertion into the side-band structure?

19-19. What factors are essential for obtaining maximum frequency response from a speaker system?

19-20. Define the terms *quadriphonic*, *ambience*, and *presence* in relation to high-fidelity sound reproduction.

Transmission Lines and Antennas 20

20-1. Introduction

Transmission lines are devices that are employed to transfer signals from a receiver to an antenna or vice versa, as well as to transfer signal energy between other devices such as from the output of an audio amplifier to some remote speaker system. The ideal transmission line is one that conveys the necessary signal information without having any adverse effect on the amplitude or characteristics of the signal waveform.

Besides conveying signal information between two points, transmission lines are also used for other purposes. Transmission lines can be employed for introducing a phase shift in sine-wave signals or for producing a delay with respect to pulse signals. Transmission-type delay lines are employed in color-television receivers, computers, radar systems, and other electronic devices. Because transmission lines also exhibit resonant-circuit characteristics, they find application as replacements for the usual capacitor-inductor circuits at the very-high and ultrahigh frequencies (VHF and UHF). From the foregoing, it is obvious that an understanding of transmission-line characteristics is an important phase of electronic knowledge.

Antenna devices, which are used for either transmitting or receiving the signals, also have special characteristics that must be understood in order to

514

obtain the highest degree of efficiency both in transmitting and receiving systems. Basic antenna systems are covered later in this chapter, after the necessary foundation has been laid by an analysis of transmission lines.

20-2.   Basic Transmission Lines

There are a variety of transmission lines employed in electronics. The most familiar types are the conductors used in house wiring, as well as the overhead power distribution cables. The two-wire line from a lamp to a baseplug can be considered as a form of transmission line because it transfers the a-c power from the mains to the incandescent bulb. Telephone wires, television lead-in wires, and the shielded cable from a microphone to a public-address amplifier are other examples of transmission lines.

For higher-frequency work, the most common types of lines are those illustrated in Fig. 20-1. At A is shown the so-called *open-wire line*. This line consists of two wires that are held at a certain spacing by plastic or ceramic insulators. When any two wires are brought into close proximity with each other, capacity exists between them. Thus, it is common to refer to the material that exists in the spacing as the *dielectric*, as with capacitors. For the two-wire line shown at A, the dielectric is air.

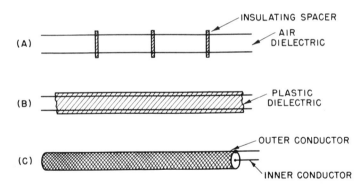

FIGURE 20-1. Types of transmission lines.

Another type of line is shown at B. This is the common *twin lead* used in television reception. Again, two wires are utilized, as with the line at A, but a soft plastic dielectric material is employed in this case. The transmission line shown at C is known as a *coaxial cable*, or *concentric line*. Here an outer conductor in the form of a metal tube or metallic braid encloses an inner conductor composed of either solid or stranded wire. A shielded microphone cable is a form of coaxial line, wherein a flexible plastic is used as the dielectric for spacing the inner conductor from the outer metallic flexible braid. For some high-frequency work, the inner conductor is spaced from the outer conductor by insulating washers composed of ceramic or plastic.

20-3.   Characteristic Impedance

When current flows through the two wires of a transmission line, magnetic lines of force are created around the wire and electrostatic lines of force are set up between the wires, as shown at A of Fig. 20-2. Thus, all transmission lines have both electrostatic fields and magnetic fields, as shown. These fields have an amplitude determined by the amount of current flowing in the lines and by the voltage existing between them. When ac is employed, the fields build up to a maximum for one alternation and then collapse when the zero level of the ac is reached. For the next alternation of opposite polarity, the fields build up again but also have a polarity opposite to what it was originally. Lines of force have an important bearing on transmission-line characteristics, as will be shown later.

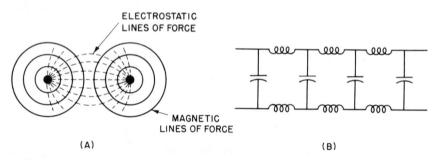

FIGURE 20-2. Transmission-line characteristics.

When two wires are brought into proximity, a certain amount of capacity exists per unit length, as shown at B. Because any length of wire has inductive characteristics, a two-wire line also has a certain amount of inductance per unit length, as shown at B. Each wire of the two-wire line also has some internal resistance, though the latter is usually held at a minimum by making the wire sufficiently large in diameter for the current to be handled. When the dielectric material is other than air, some shunt resistance may also be present across the line, due to the leakage resistance of the dielectric material.

Because a transmission line is composed primarily of the shunt capacitance and series inductance at B of Fig. 20-2, it will have a certain *characteristic impedance* (sometimes called surge impedance). If the resistances present in the wire length and in the dielectric material are not appreciable, the characteristic impedance (Z_o) can be ascertained by use of the formula

$$Z_o = \sqrt{\frac{L}{C}} \tag{20-1}$$

When air insulation is used in a parallel-wire-type line, the following formula can be used for finding Z_o:

$$Z_o = 276 \log \frac{2b}{a} \tag{20-2}$$

where
   b = distance (center-to-center) of the wires
   a = diameter of one of the wires.

The Z_o of a coaxial cable may be found by use of the following formula:

$$Z_o = 138 \log \frac{b}{a} \tag{20-3}$$

where
   b = inside diameter of the outer conductor
   a = outside diameter of the inner conductor.

When such a line is filled with a dielectric material other than air, the result of the formula should be multiplied by

$$\frac{1}{\sqrt{K}} \tag{20-4}$$

where K equals dielectric constant of the material.

For convenience in discussing transmission-line characteristics, reference will be made to a *generator* and a *load*, in a fashion similar to that followed for earlier circuit discussions. Thus, a generator attached to a transmission line may be a transmitter or a receiving antenna since each sends signals along the line. A load can be a transmitting antenna or a receiver since each receives the signal energy sent along the line.

As with other generator-load combinations, maximum signal power is transferred between the generator and the load when impedances are matched. As an illustration, if a television receiver has a 300-ohm input impedance at the tuner, a maximum signal will be transferred to it from the antenna when the latter is also 300 Ω. If a 300-Ω transmission line is also employed, the line will also match the antenna and the receiver. Such a transmission line is referred to as an *untuned* or *nonresonant* line. The designation *flat* line is sometimes employed because power travels along the line from the generator (antenna) until it reaches the load (receiver), where it is completely utilized. In such a line, the impedance is the same at any point since the ratio of voltage to current is identical at any chosen section. This flat voltage-to current ratio is upset when impedances are not matched, as described later.

When an a-c signal flows through a transmission line, current flow in one wire is opposite to that in the other; hence, the fields around the wires oppose each other and tend to cancel. This factor minimizes losses since if the fields did not cancel, some of the energy would leave the transmission line because of the lines of force that would extend beyond the influence of the wires. Some losses take place, however, since complete cancellation of the fields would occur only if each wire of the two-wire line occupied the same space so that the fields would interact perfectly. In the parallel-wire lines, losses are kept at a minimum by close spacing of the wires. In coaxial cables, the complete shielding of the inner conductor by the outer conductor prevents losses by radiation of the fields.

20-4. Standing Waves

Radio, FM, television, and other high-frequency RF signals travel through space with the speed of light (approximately 186,000 mi per s). Thus, if a generator sends out (propagates) only one cycle of an a-c signal, that part of the signal that first left the generator would have spanned a distance of 186,000 mi at the time the end of the cycle leaves the generator. Consequently, a frequency of 1 Hz has a *wavelength* of 186,000 mi, as shown at A of Fig. 20-3. If the frequency is doubled, each cycle occurs in a shorter interval of time so that, at the end of 1 s, two cycles of transmitted energy will span 186,000 mi, as shown at B. Thus, each cycle of the higher frequency has a shorter wavelength. A still higher frequency, as shown at C, shortens the wavelength again for each cycle. With four cycles spanning 186,000 mi, each cycle has a wavelength $\frac{1}{4}$ of 186,000 mi, or 46,500 mi. For television and FM frequencies, the wavelengths are so short that they are specified in feet or in inches rather than in miles. At 200 MHz, for instance, one cycle has a wavelength of less than 5 ft. (The symbol for wavelength is λ, the Greek lowercase letter lambda.) The wavelength of a particular frequency can be found by use of the following formula:

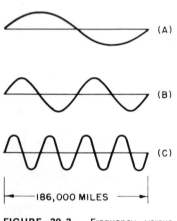

(A)

(B)

(C)

186,000 MILES

FIGURE 20-3.   Frequency versus wavelength.

$$\lambda \text{ (in feet)} = \frac{984}{f} \qquad (20\text{-}5)$$

where f is the frequency in megahertz.

As shown later, however, the half-wavelength dimensions are of primary interest; hence, the formula can be restated as follows:

$$\frac{\lambda}{2} \text{ (in feet)} = \frac{492}{f} \tag{20-6}$$

where f again represents the frequency in megahertz.

For ultrahigh frequencies, where the wavelength is more conveniently expressed in inches, the following formula applies:

$$\frac{\lambda}{2} \text{ (in inches)} = \frac{5904}{f} \tag{20-7}$$

where f still has the same significance as above.

In the study of transmission lines, it is often necessary to refer to a line having a specific wavelength. This is because, in many instances, the wavelength of a line determines whether the line has inductive, capacitive, or resonant characteristics. When a transmission line is employed between a matched generator and load, the transfer of the output power from the generator to the load is a function of the transmission line. If a mismatch occurs, however, between the load and the generator, all the energy will not be transferred between the generator and the load. For instance, the transmission line may be matched to the generator and will accept the full amount of signal energy and transfer it to the load. If the load is not matched to the line or generator, however, all the signal energy will not be accepted and some is reflected back to the generator. The greater the mismatch, the more energy will be reflected back along the wire. Thus, some energy will travel from the generator to the load and, during the same time, energy is reflected back from the load toward the generator. Since the primary signals and the reflected signals are all intermixed along the line, there will be both in-phase and out-of-phase conditions. Hence, at some points along the line, voltages will be in phase and will have a high amplitude, while in other places out-of-phase conditions will occur, resulting in low- or zero-signal-voltage amplitude. Similar conditions occur for signal current. These high and low amplitudes are the fixed or *standing* positions along the line and hence are known as *standing waves* of either voltage or current.

A better understanding of standing waves can be gained by inspection of Fig. 20-4; at A is shown one extreme condition of mismatch since a half-wavelength line is open at the end. The open end represents an infinitely high impedance. Note the current and voltage distribution along such a half-wavelength line. When the wave of current reaches the end of the line, it drops to zero. The current decrease creates collapsing fields and the fields in turn cut the conductor ends and induce a voltage maximum across the ends of the line, as shown. The high voltage and zero current at the end of the line create a high-impedance condition since Z is equal to E/I. Voltage maximum points are referred to as *voltage loops* and current maximums as *current loops*. The voltage or current minimums are referred to as *nodes*. At the

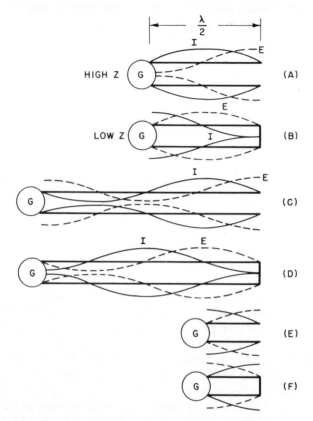

FIGURE 20-4. Characteristics of open and closed lines.

generator, which is removed from the end of the line by one-half wavelength, a high impedance again exists.

The standing waves on the line will have loops and nodes as shown at A, though the standing wave itself is of a-c composition, which means that the amplitude alternately changes from maximum to zero, each maximum alternating from a positive to a negative.

At B, a half-wavelength line is again shown, though now the end of the line is shorted. The closed end of the line again offers a high mismatch because it represents a zero impedance. Current is now high at the generator and at the end of the line, with voltage reaching a loop at the center of the line, as shown. If the short were replaced by a low value of resistance, the ratio of loops and nodes would decrease as the resistance is increased toward the generator resistance. Above the generator resistance, standing waves again are created. The ratio of either voltage or current loops and nodes is known as the *standing-wave ratio*. The ratio can, therefore, be determined by dividing the maximum voltage along the line by the minimum voltage. The standing-wave ratio indicates the degree of mismatch between the generator and the load. When a complete match prevails, the standing-wave ratio is equal to one since loops and nodes no longer exist. The standing-wave ratio can, of course, also be determined by current loops and nodes. For instance, if the standing wave of current reaches an amplitude of 20 mA, and the minimum standing wave of current is 2 mA, the ratio is 10, indicating that the load resistance is either 10 times as large as the generator impedance or one-tenth the value of the generator impedance.

At C and D, full-wavelength lines are shown. Note that the voltage and current relationships at the end of the line correspond to those shown at A and B. At E and F, quarter-wavelength sections of line are shown and, again, voltage is at a maximum at the open end of the line, E, but is zero when the line is closed, as in F. Note, however, that the section of line at the generator has an impedance that is different from that at the end. At F, for instance, the impedance is high at the open end but drops to a low value at the generator because of the voltage decline and the current rise. Because of the impedance difference along such a line, the latter can be used as an impedance-matching transformer. It can be utilized as an impedance step-up device, as at E, or as an impedance step-down device, as at F. The half-wavelength sections shown at A or B can be used as a one-to-one transformer since the impedance at the output is identical to the input impedance. Because this characteristic is repetitive for every half wavelength, a half-wavelength line (or wavelength multiples) behaves as a one-to-one transformer.

The quarter-wavelength section shown at E of Fig. 20-4 behaves as a series-resonant circuit for the generator, due to the low impedance of the line at the point where it is connected to the generator. The generator at F is attached to a high-impedance point; hence the transmission line behaves as a parallel-resonant circuit.

20-5. Resonant Sections

Quarter-wavelength sections of transmission line are actually employed as resonant circuits in VHF and UHF electronic applications to replace the coil and capacitor combinations. At very-high frequencies, the resonant circuits employ smaller values of inductance and capacitance to achieve resonance. As the frequency of the signal is increased, still smaller values of inductance and capacitance are necessary. Consequently, the point is reached where it is more practical to use quarter-wavelength sections of line as the actual resonant circuit. The line sections can be used in either oscillator or amplifier circuits, and typical UHF oscillators using transmission-line sections are shown in Fig. 20-5. At A the oscillator has its collector connected to one wire of the parallel-resonant line, while the second wire of the line is in the base circuit. This oscillator can be likened to the Hartley oscillator described earlier, though it is sometimes called an ultra-audion type. The resonant-line section that is connected to the collector can be considered as the ouput inductance, while the resonant-line section at the base is the input inductor. The voltage-feed point is applied to the movable shorting bar and this point is placed at ground potential for the RF signals, by virtue of the bypass effect of capacitor C_1. Since the emitter is also placed at ground potential, it is at the same potential as the center of the shorting bar. Thus, the emitter taps the inductance in similar fashion to the Hartley oscillator, where the

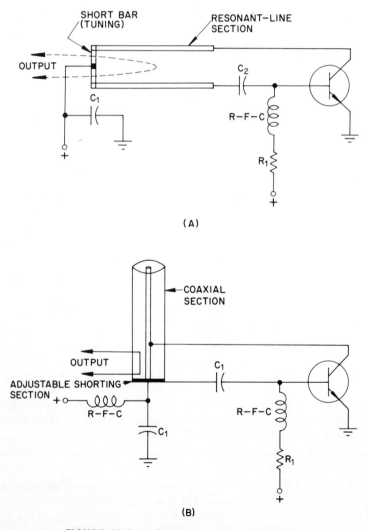

(A)

(B)

FIGURE 20-5. UHF oscillators using parallel lines.

amplified energy in the collector circuit is coupled to the base circuit by the interacting lines of force that are common to both inductances.

Capacitor C_2 blocks the collector dc from being applied to the base circuit since relative amplitudes differ. An RF choke is in series with the base resistor R_1 to minimize the shunting effect this resistor might have on RF energy at the base circuit. The output from the oscillator is obtained by use of a single-loop inductor, shown by the dotted outline at A of Fig. 20-5. This single-loop inductance is often referred to as a *hairpin loop* and is coupled to the parallel section for signal takeoff. The degree of coupling determines the amplitude of signal transfer, as well as the loading effects on the oscillator circuit. Such loading establishes the final Q as well as band-pass characteristics of the circuit. Tuning to proper frequency is done by moving the shorting bar to change the resonant-line length.

When such an oscillator is used in receivers, the resonant section can be composed of small-diameter copper tubing or heavy-gauge solid copper wire (No. 16 to No. 22). The larger-diameter wire is preferred at higher frequencies because of the phenomenon known as *skin effect*, which is present at these frequencies. This factor can be more readily understood by reference to Fig. 20-6, which shows a magnified cross section of a wire carrying electric current. The current flow establishes a series of magnetic lines of force, which are distributed within the wire as well as being present around the outside perimeter. These magnetic lines of force are representative of the inductive factor of the wire length and offer a reactance against the RF current flow. Since inductive reactance increases with higher frequencies, the changing a-c signal, if high in frequency, meets with considerable opposition within the wire core

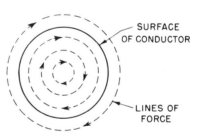

FIGURE 20-6. Skin effect.

and, in consequence, has difficulty flowing through the wire. Since the first electromagnetic line of force external to the wire is spaced a short (but definite) distance from the wire surface, the current flow set up by the pressure of the voltage finds a path of lower resistance on the outside of the wire than it does on the inside. Thus, at high frequencies, the current flow can be considered as being along the surface or "skin" of the wire, hence the term *skin effect*. As higher-frequency signals are employed, the inside-wire opposition is increased and, in order to provide a low-resistance path for the RF energy, it is expedient to increase the wire size so as to decrease resistance for skin effect. Also, a large surface area must be provided for the relatively higher currents encountered in transmitting work. Since the current flows on the outside of the wire, it is not economical to use a solid wire of extremely heavy gauge. For this reason, copper tubing is frequently employed since the center-core area is of no current-carrying importance at very high frequencies.

A coaxial-cable section can also be employed in a high-frequency os-cillator instead of the resonant-line section previously described. A coaxial-cable high-frequency oscillator is shown at B of Fig. 20-5. This oscillator is similar to the one shown at A of the same figure, except for the fact that a coaxial cable instead of a parallel-wire line is used. The inner conductor of the coaxial cable is used for the collector while the outer conductor is at-tached to the base at the shorting section. In this instance, the outer conductor is not at ground potential, except by virtue of the bypass effect of C_1. The output is derived from a hairpin-loop arrangement that is inserted into the coaxial element via two small holes in the outer conductor. Because of skin effect, the RF energy is confined to the inside of the coaxial-cable section and will not penetrate and leak to the outside. Again because of skin effect, the RF energy within the coaxial-cable section flows on the outside of the inner conductor and on the inside of the outer conductor. A conventional RF choke is also employed in the base circuit, and the shorting section for tuning purposes consists of a metal washer that is often mounted on a threaded rod so that exact tuning adjustments can be made.

The parallel-resonant-line sections lend themselves readily to push-pull elements for both oscillators and amplifiers. A typical push-pull tube-type RF amplifier is shown in Fig. 20-7. A balanced arrangement can be employed, as shown, by using a parallel-resonant section for the push-pull plates. The input signals are applied to the grid circuits by use of a hairpin loop, as shown by the dotted-line section at the oscillator input. The output is taken off by a similar hairpin-loop arrangement coupled to the anode resonant-line section.

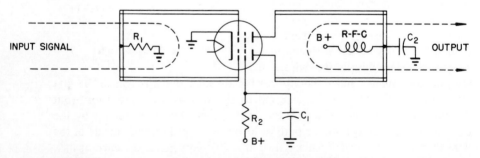

FIGURE 20-7. UHF push-pull amplifier.

Resistor R_1 at the center of the shorting bar at the left is the conventional grid resistor, while the supply voltage is applied to the center of the shorting bar at the anode section. The RF choke, in series with the supply, acts to prevent losses of signal energy to the power supply. The resonant anode circuit is placed at ground potential for RF by capacitor C_2, which returns the RF signal energy to the cathode circuit. This completes the RF circuit, without the necessity for making such a completion through the power supply. Radio-frequency energy that leaks to the power supply will, in turn, be induced to the power mains because of the inductive transfer across the power transformer, with resultant losses of such RF energy.

The tubes shown in Fig. 20-7 are tetrodes, though pentode and beam-power tubes can also be utilized. If special high-frequency tubes of this type are employed, no neutralization will be necessary, provided the circuit is intended to be used as an RF signal amplifier. If triode tubes or transistors are used, the circuit will become a push-pull oscillator, in which instance no input signal would be applied and the output energy would be obtained from the anode circuit, in identical fashion to that shown using the hairpin loop. Basically, the circuit arrangement is similar, whether this stage handles signals in a receiver or in a transmitter except that, in the latter instance, a larger tube and larger-diameter resonant-line sections are employed.

Two-wire lines are particularly suited for push-pull circuits because the parallel lines are *balanced* lines; that is, one line is as much above ground potential as the other. The coaxial cable does not lend itself as readily as the parallel-wire line to balanced-line applications. The twin lead utilized for television receivers is also used in a balanced-line arrangement, as shown in Fig. 20-8. The antenna feeds the signals to the transmission line, which in turn applies them across the primary of the tuner input transformer, as shown. With the primary center-tapped to ground, a balanced arrangement is secured.

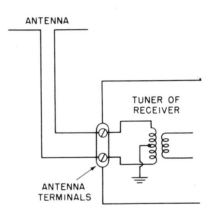

FIGURE 20-8. Balanced input.

20-6.   Sections Forming *L, C,* and *R*

Sections of transmission lines can also exhibit characteristics of either inductance or capacitance instead of circuit resonance. For instance, an open section of line that is less than a quarter wavelength long appears to the generator as a capacity and, hence, has a capacitive reactance. As mentioned earlier, the quarter-wavelength section of line open at the end nearest to the generator appears as a series-resonant circuit (see E of Fig. 20-4). If such a section of line is shortened so as to be less than a quarter wavelength, both the series inductance and the shunt capacitance are decreased. According to the inductive reactance Formula 8-2 given in Chapter 8, when the inductance is decreased, the inductive reactance also decreases. According to the capacitive reactance formula, however, a decrease in capacity causes the capacitive reactance to increase. Because the parallel-line section represents a series-resonant circuit when the end is open, the larger reactance will predominate if the line is less than a quarter wavelength. Since the capacitive reactance now offers greatest series opposition, the line is primarily capacitive. If the line shown at E of Fig. 20-4 is made longer than a quarter wavelength (but less than a half wavelength), inductance and capacity per unit length increase. Thus, inductive reactance increases but capacitive reactance decreases. Hence, the series opposition is primarily that of inductive reactance and the circuit is predominantly inductive. A closed section of transmission line, less than a quarter wavelength long, has opposite characteristics to the open-type line. Hence, a closed quarter-wavelength line shortened to less than a quarter wavelength acts as an inductance. The shorted line behaves as a parallel-resonant circuit and when made less than a quarter wavelength, both inductance and capacity decrease, as with the open line. Inductive reactance also decreases but capacitive reactance increases. The low-shunt characteristics of the inductive reactance cause most of the current to flow through this section and, hence, the current will be inductive. When the closed-end parallel line is made more than a quarter wavelength but less than a half, capacitive reactance predominates. The chart shown in Fig. 20-9 shows the characteristics for both the closed and open lines for various wavelength values.

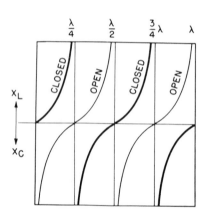

FIGURE 20-9. Reactance versus line length.

20-7.   Filter Sections

When short lengths of transmission lines are used for adding various values of inductance, capacity, or resonance to a circuit, such lines are referred to as *stubs*. Stubs can be used to form circuits that will hold back some frequencies while passing others. Circuits that have such discriminatory characteristics are known as filter circuits and, for audio or RF frequencies, actual capacitors and coils are employed, whereas, for UHF and microwave frequencies, stubs are used. Figure 20-10 illustrates the low-frequency filter systems and their high-frequency counterparts employing stubs. At A a series

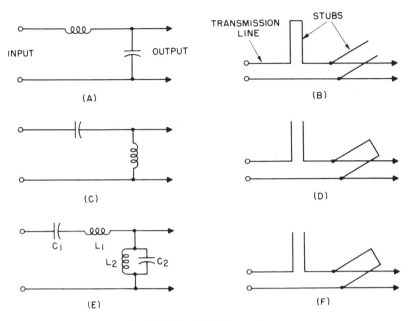

FIGURE 20-10. Filter sections.

coil and a shunt capacitor are shown in a circuit arrangement known as a
low-pass filter. Assume that audio signals of various frequencies are applied
to the input. For the lower-frequency audio signals, there will be a low
inductive reactance established in the coil. For progressively higher audio
frequencies, however, an increasing opposition is established in the series
inductance because of the rising inductive reactance. Hence, the high-
frequency signals do not pass readily through the circuit, and those that do
are shunted by the low reactance they create in the capacitor. The signals of
low frequencies, however, find little opposition in the inductance and hence
pass through. For the low-frequency signals the reactance of the capacitor
is high, and little of the signal energy is shunted. A UHF counterpart of the
low-pass filter is shown at B. Here a closed section of line has been inserted
in series with a transmission line. The closed section of line (the stub) is less
than a quarter wavelength long and, hence, acts as an inductance. An open
section of line less than a quarter wavelength long has been placed across the
main transmission line, and this open stub acts as a capacity, in accordance
with the chart shown in Fig. 20-9. Thus, the circuit shown at B is a high-
frequency counterpart of that shown at A, which handles mostly lower-
frequency signals.

At C of Fig. 20-10, a high-pass filter is shown. If signals of various
frequencies are applied to the input, the higher-frequency signals are coupled
through the capacitor because they find a low reactance. The high-frequency

signals arriving at the output are not shunted by the inductance because of the high inductive reactance. Low-frequency signals, however, have difficulty passing through the capacitor since such signals create a high capacitive reactance. The low-frequency signals that appear across the inductance create a low reactance and, hence, are shunted. Thus, this is a high-pass filter because it passes the high-frequency signals but filters out the low-frequency signals. At D is shown a UHF counterpart of the high-pass filter, using line sections. Again, the open stub acts as a capacitor because it is less than a quarter wavelength long. Hence, the main transmission line is opened up and a stub is inserted, as shown at D. For the inductance, a closed stub less than a quarter wavelength is placed across the main transmission line, as shown.

The circuit at E is known as a band-pass filter because it will pass a narrow band of frequencies centered around the resonance of the circuit but will tend to reject signals of frequencies above and below the resonant frequency. Components C_1 and L_1 form a series-resonant circuit that has a low impedance for the frequency (or narrow band of frequencies) to which it is tuned. Hence, the desired signals pass through this circuit. The resonant circuit composed of L_2 and C_2 is also tuned to the desired signals. This circuit is a parallel-resonant one and, hence, has a high impedance that prevents any of the desired signals from being shunted. For signals having frequencies above and below resonance, the series-resonant circuit will have a high impedance and, hence, will offer opposition. Also, the undesired signals that pass through the series-resonant circuit will be shunted by the parallel-resonant circuit because the latter has a low impedance for signals whose frequencies are below or above resonance. Therefore, this circuit passes only a narrow band of frequencies and filters out all others. A counterpart of the band-pass filter is the circuit shown at F. For the series-resonant circuit, an open stub is employed. The stub is made one-quarter wavelength long and consequently acts as a series-resonant circuit because the impedance is low where the stub is attached to the main transmission line. For the parallel-resonant circuit, a quarter wavelength only closed stub is placed across the main transmission line, as shown at F.

20-8. Basic Antenna Types

If the quarter-wavelength open transmission line shown at E of Fig. 20-4 is opened up so that the two wires are horizontal, an antenna is formed as shown at A of Fig. 20-11. For simplicity, the generator, which would be at the center of the two sections, is not shown. The two line sections, after being opened, now form a single line that is a half wavelength long. Each end still has a voltage loop and current is still high at the center, as was the case for the quarter-wavelength line previously illustrated in Fig. 20-4. The single line represents the shortest length that can be used to form an antenna and, as

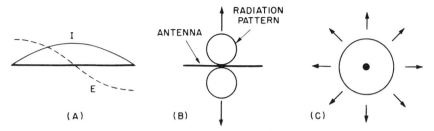

FIGURE 20-11. Antenna patterns.

with the quarter-wavelength open line, it represents a series-resonant circuit. Such a half-wavelength antenna is known as a *dipole* because a popular method of using it is to open it at the center, where the voltage node is located, and to insert the transmission line at that point. When the antenna is opened at the center in such a manner, two quarter-wavelength sections (dipoles) are formed. The center of the antenna is a low-impedance point and when the transmission line is attached, the antenna impedance is approximately 75 ohms. The dipole section shown at A actually has zero impedance at the center because current is high and voltage is zero. When this section is opened, however, capacity is created between the open ends, and the capacitive reactance thus established creates a voltage drop across this section so that voltage is no longer zero at that point. In some applications, the transmission line can also be attached at one end or the other if a high-impedance connection is desired.

The half-wave antenna has magnetic lines of force created by the current and electrostatic lines of force created by the voltage. These are at right angles to each other and, in combination, they form the signal energy that leaves the antenna when the latter is used for transmitting. If the antenna is used for receiving signals, the propagated energy also sets up voltage and current distributions as shown at A, since basic antennas have a reciprocal function with respect to transmitting and receiving. Hence, the characteristics detailed herein apply, whether a transmitting or a receiving antenna is involved.

The dipole antenna is also known as a *Hertz* antenna and it sends or picks up signals at right angles to its length, as shown at B of Fig. 20-11. The circles represent the radiation pattern of the antenna and are also known as *lobes*. The lobes indicate the direction of signal transmission or pickup of the antenna. The radiation pattern shown at B is a cross-sectional illustration. Actually, the same pattern exists whether the antenna is observed from the top, from the bottom, or from the sides. An end view of the antenna is shown at C and here the radiation pattern is indicated as a circle. This indicates that the antenna can receive or send signals in all directions that are at right angles to its length. In transmitting, the high-frequency signals build up lines of force and when the polarity of the signals changes, the lines of force would tend to collapse into the antenna and be reestablished with opposite polarity for the next signal alternation. The rapid change of polarity of the signal at high frequencies, however, does not permit the respective lines of force to collapse back into the antenna so that they leave the influence of the antenna and are propagated at the speed of light into space. The energy that travels through space consists of a composite signal energy made up of electrostatic and electromagnetic lines of force, representing the respective voltage and current factors of the transmitted power. For FM and television transmission and reception, horizontal antennas are used, and the wave that is propagated is said to be *horizontally polarized.* For AM transmission and for some mobile shortwave applications, vertical polarization is employed; that is, the antennas are vertical rather than horizontal.

The figure-eight radiation pattern shown at B of Fig. 20-11 holds only when the antenna is a half wavelength long for the signal to be received. (The half wavelength in this instance refers to the *electrical length* rather than the *physical length*. Because of capacity effects at the ends, known as *end effect*, the antenna is made physically shorter than a half wavelength by approximately 5%.)

The simple dipole has only 75 ohms of impedance, as previously mentioned, and hence creates a mismatch when used with the standard 300-ohm twin lead utilized in FM and television installations. Consequently, most antennas are modifications of the simple dipole so that increased impedance is obtained. One such type is the so-called *biconical* antenna shown at A of Fig. 20-12. Two or three quarter-wavelength rods extend from each side of an insulator, as shown, in order to obtain an impedance close to 300 ohms. Because the antenna is to be used for a number of stations, various signal frequencies will be impressed on the antenna. For frequencies higher than the half-wavelength frequency, multiple lobes appear in the radiation pattern. The antenna is tilted forward slightly so that the multiple lobes combine into a single lobe. The sensitivity of such an antenna is again at right angles to its width and, hence, it will pick up signals in two directions. For increasing signal pickup from one direction, a *reflector* rod is attached, as shown at A. The reflector is a true half wavelength long and is spaced from the antenna by approximately a quarter wavelength, or less. If the spacing is too close, the reflector will decrease the impedance of the antenna appreciably.

Another antenna having 300 ohms of impedance is the folded dipole shown at B. Again, a reflector can be employed for increasing the pickup from one direction. Rods can also be placed in front of the antenna, as shown, such rods being referred to as *directors*. The directors are made shorter than the antenna and are spaced approximately a tenth of a wavelength apart. Commercial types use a number of rods and various spacings in order to achieve broad-band characteristics and maintain the required 300 Ω of impedance. Directors, like reflectors, tend to decrease the antenna impedance

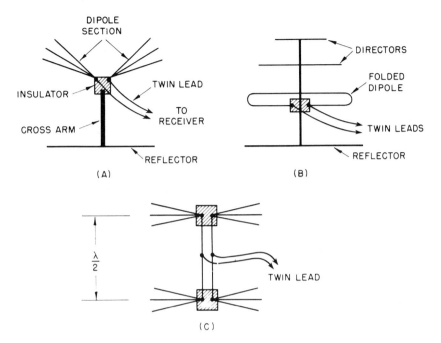

FIGURE 20-12. Basic antenna types.

and hence are spaced farther from the antenna than required for maximum effectiveness. Reflectors and directors, because they are not connected to the transmission line, may be grounded to the supporting crossarm, as shown at A and B of Fig. 20-12. Since a voltage node exists at the center of the reflectors and directors, no signal losses occur because of the grounding of the reflectors and directors to a metal crossarm. Reflectors and directors are also known as *parasitic* elements. The type of antenna shown at B, with one reflector and one or more directors, is known as a Yagi antenna, after the Japanese physicist Dr. H. P. Yagi, its inventor.

When more signal gain is desired than can be obtained from a single antenna, two or more can be stacked, as shown at C. One antenna is spaced above the other by approximately one-half wavelength, and the elements are interconnected by a length of transmission line. The twin-lead line to the receiver is attached to the center, as shown. The two impedances of the antennas are in parallel and would tend to be halved. The transmission-line section interconnecting the two antennas, however, forms two quarter-wavelength stubs and tends to raise the impedance at the point where the twin lead is attached, thus helping to maintain the normal impedance of the antennas.

Standard AM transmission is vertically polarized; hence, for best results, the vertical receiving antenna should be employed. Because of the high degree of sensitivity of modern receivers, however, and the high power used by transmitters, signal reception is usually satisfactory with any antenna type. In most cases, a built-in ferrite-core antenna is sufficient for good reception of all nearby stations.

For transmission, a single quarter-wavelength vertical antenna is used, as shown at A of Fig. 20-13. The bottom of the antenna is grounded and this procedure eliminates the necessity for using a half wavelength. When a

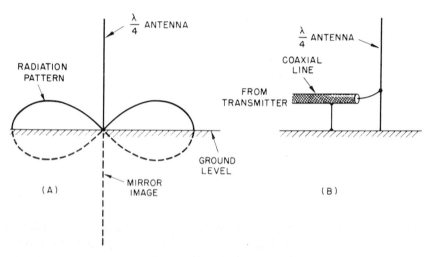

FIGURE 20-13. Marconi antenna.

vertical antenna is grounded, as shown at A, it acts as though it were one-half wavelength long because of the mirror-image effect. A grounded antenna behaves as though it were twice as long, just as a pencil appears twice as long when its end is placed against a mirror. The figure-eight radiation pattern is part of the theoretical mirror image. Hence, the vertical grounded antenna propagates in all directions along a horizontal plane from the antenna proper. This type at antenna is known as the Marconi antenna, after G. M. Marconi (1874–1937) the Italian physicist. The vertical grounded antenna is usually fitted with a coaxial-cable arrangement, as shown at B. Because the antenna is grounded, it is an unbalanced type and the coaxial cable is a suitable transmission line. The outer conductor of the coaxial cable is grounded, as shown, and the center conductor is attached to the antenna a short distance from the ground point. The impedance of the Marconi antenna is zero at ground and rises with antenna height. Hence, if the inner conductor of the coaxial line is moved along the antenna, a place will be found where an impedance match is obtained. With matched impedances, the coaxial line delivers full RF power to the antenna.

Review Questions

20-1. For what other purposes can transmission lines be utilized, other than conveying signal information from an antenna to a receiver?

20-2. (a) How is the characteristic impedance of a transmission line calculated?
(b) Is there a difference in the characteristic impedance of a twin-lead-type transmission line if 200 ft is used instead of 100 ft?

20-3. Briefly explain what is meant by *standing waves* and how they are formed on a transmission line.

20-4. (a) Will standing waves occur if a transmission line matches the receiver but the antenna is mismatched with the line?
(b) Will standing waves occur if an antenna and transmission line are matched but not to the input impedance of a television tuner?

20-5. Briefly explain what information is obtained with respect to impedance matching when the standing-wave ratio is known.

20-6. A quarter-wavelength section of transmission line is attached to a generator and the other end is short-circuited. Briefly explain the type of characteristics such a line presents to the generator.

20-7. Briefly explain what is meant by *skin effect*.

20-8. How can the undesired characteristics of skin effect be minimized in VHF and UHF electronic circuitry? Explain briefly.

20-9. By a schematic drawing, show a typical example of the use of stubs for constructing a low-pass filter.

20-10. (a) What is the shortest electrical length that can be employed for a basic antenna?
(b) What is meant by a dipole type of antenna?

20-11. Briefly explain what is meant by vertical and horizontal polarization of antennas.

20-12. (a) What is the impedance of a dipole type of antenna, with respect to a half-wavelength antenna that has not been opened at the center?
(b) What type of basic antenna has more than 75 ohms of impedance?

20-13. Briefly explain the differences between a Hertz antenna and a Marconi antenna.

20-14. Briefly explain what is meant by a mirror image, with respect to a vertical antenna grounded at one end.

Practical Problems

20-1. In the design of a flexible transmission line, a section was tested and found to have 2.7 μH of inductance and 30 μF of capacitance. What is the impedance of the line?

20-2. In an RF distribution system, an open-wire line was used. Each of the two conductors had a diameter of 0.04 in. and the spacing between them was 2 in. What is the characteristic impedance of this air-dielectric line?

20-3. The output from a UHF transmitter was connected to an antenna by a coaxial cable having air as the dielectric. The inner conductor consisted of a $\frac{1}{4}$-in. diameter copper conductor, and the inside diameter of the outer conductor measured 1 in. What is the characteristic impedance of this coaxial line?

20-4. What would be the impedance of the line in Problem 20-3 if it has a dielectric material with a k of 4?

20-5. What is the wavelength in feet if the frequency is 123 MHz?

20-6. A section of line 6 in. long was to be used as a resonant circuit. Neglecting the reactances contributed by other circuit components, what is the resonant frequency if this section of line is to form a quarter-wavelength circuit?

20-7. What is the length of a quarter-wavelength section of line at 984 MHz?

20-8. In a testing laboratory a dipole was needed as a reference standard for 90 MHz. If 5% of length was reduced for end effects, what was the length of the antenna?

20-9. The testing laboratory required another dipole antenna for the lower UHF band. The antenna was to be designed for use at 500 MHz. What was the total length?

20-10. In a design laboratory a Yagi antenna was needed for television channel No. 3, with the reflector element cut for 60 MHz, the antenna element 5% shorter than the reflector, and the director 4% shorter than the antenna to increase bandwidth. What were the lengths of these elements?

Transducers 21

21-1. Introduction

A number of input and output devices are utilized in conjunction with the circuits found in the different branches of electronics. Such units usually convert one form of energy to another and hence are known as *transducers*. Thus, a microphone is a transducer since it converts sound-wave energy into electric signals. Similarly, a loudspeaker is a transducer because it converts electric signal energy to sound waves. The transducer term, however, does not apply solely to the conversion of acoustical energy to electrical. A playback head on a tape recorder, for instance, is also a transducer because it converts the varying magnetic areas on tape to an electric signal. (An electric light bulb can be considered a transducer since it converts electric energy to visible light waves.)

While speakers, microphones, and recorder playback heads are common items familiar to most, there are a number of other transducers utilized in automation, industrial control, computer systems, and other allied branches of electronics. This chapter covers the variety of transducers (both input and output) that are encountered in these various branches of electronics, and the circuitry connections are illustrated and discussed.

535

21-2.   Solid-State Transducers

The piezo effect with respect to crystals was discussed in Chapter 17. This principle is also utilized in the construction of lower-priced microphones and phonograph pickup cartridges. (The higher-quality and more costly units utilize magnetic principles and these are described later.) One type of crystal that has been widely used is that known as *Rochelle salts* and it also exhibits a piezo-electric effect; that is, it will produce an electric signal when sound waves strike the crystal (or a diaphragm attached to the crystal). When two thin slabs of the Rochelle salt crystal are cemented together, a type of cell is formed that is known as the *bimorph* cell. Such a cell is sensitive to sound waves and will produce an output signal that has a lower sensitivity than if a diaphragm were used. For greater sensitivity, the arrangement shown at A of Fig. 21-1 is used. The larger surface area of the diaphragm picks up

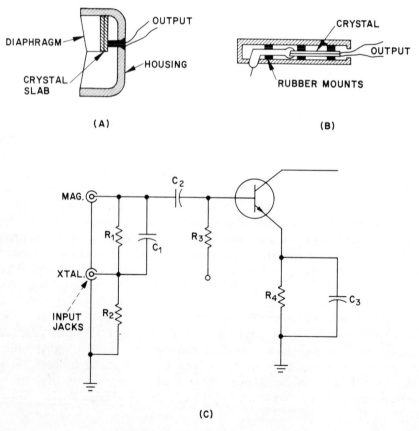

FIGURE 21-1. Input audio transducers.

a greater amplitude of the sound waves that strike the diaphragm and the movement is transferred to the crystal slab with a metal connecting link as shown. The vibrations transferred to the bimorph cell set up mechanical stresses and strains that produce the electric signal output.

When a crystal is also used in phonograph pickup devices, construction is as shown at B. Here the phonograph record groove variations are picked up by the phonograph needle and transferred to the crystal structure. The mechanical distortion that results from the vibrations again produces an output signal as with the microphone.

Some ceramic materials also exhibit piezo-electric characteristics and in recent years have been extensively used in microphones and phonograph pickups. Their operating principles are identical to the Rochelle salts crystal except they are less affected by humidity and temperature. The impedance of both types is high compared to magnetic types; hence resistive or impedance networks are used for lowering the impedance to that required for matching to the lower input impedance of transistors, as shown at C of Fig. 21-1. Here the crystal input is in series with the impedance formed by resistor R_1 and capacitor C_1, while the magnetic-cartridge input is presented directly to the base input of the transistor by coupling capacitor C_2.

In industrial applications, solid-state transducers are often used to sense pressure or to indicate the relative pressure level. One type is manufactured from rare earths that, when processed, undergo a change in resistance when compressed. The rare earths are processed with zirconium tetrachloride and mounted in small metal housings. They are useful for sensing any variable physical tension, strain, vibration, or displacement.

A typical input system using a solid-state pressure transducer is shown in Fig. 21-2. Here, the 120-V a-c line potential is stepped down to that required for the particular pressure transducer utilized. The pressure cell resistance will remain unchanged when no pressure is applied and current flow through resistor R_1 is limited by the value of this resistor, as well as the resistance of the pressure transducer and the transformer secondary.

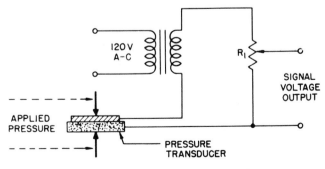

FIGURE 21-2. Solid-state pressure transducer.

When pressure is applied to the cell, its resistance decreases and more current flows through resistor R_1. The increased current through R_1 will raise the voltage drop across this resistor and hence an increase in output voltage is obtained for increased pressure.

Some pressure transducers have ratings from a few grams up to approximately 10 lb of pressure, using a 1.5- or a 3-V supply. Others have ratings to 25 or 50 lb and still others can withstand a pressure of over 5000 lb. In some transducers of this type, a pressure differential of 15 lb may provide a resistance change from zero to 900,000 Ω. Others, have a resistance of 2 or 3 MΩ at 5000 lb pressure. The output may also be sensed in microamperes by placing a milliammeter in series with R_1. One application of such a device is in the sensing of pressure in pipelines. Some types have sufficient sensitivity to indicate oil leakage of 1 lb per 5000 lb of oil pressure.

21-3.   Resistive Transducers

As mentioned in Chapter 2, the amount of resistance within a conductor or a resistor depends on its cross section and length, as well as temperature and the composition of the materials making up the resistor. Thus, if a section of thin resistance wire is stretched, the decrease in its diameter and the increase in its length lowers its resistance value. This principle is employed to construct pressure-sensitive transducers for the measurement of force, weight, or strain. Such devices are called *strain gauges* and are widely used in industrial electronics.

One type of strain gauge is shown at A of Fig. 21-3. Here the resistance wire is cemented to a plastic-treated paper carrier sheet and this is bonded

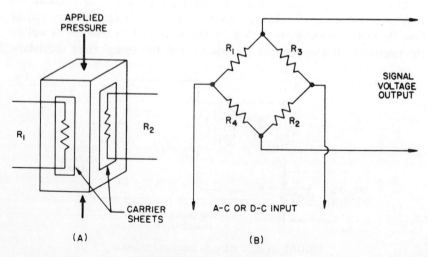

FIGURE 21-3. Strain gauge.

directly to the material that is to be sensed for pressure, strain, or the stress resulting when force is applied. Such a strain gauge is known as a *bonded* type since it is fastened directly to the unit under pressure. When the resistance wire is utilized in this fashion, the wire lengths are usually from a fraction to approximately 6 in., with an average diameter of 0.001 in. The average resistance without stress is approximately 100 Ω.

The resistive change for applied pressure is slight but is detectable in a bridge circuit such as shown at B. The signal output voltage can be amplified additionally as required. When force is applied to the top and bottom of the material that is being tested, the pressure that is created reduces the resistance of the R_1 and R_2 resistor strips. Since these are part of the bridge circuit, as shown at B, the bridge circuit becomes unbalanced with a change for the resistance values of R_1 and R_2 and an output voltage is procured. (For Wheatstone bridge factors, see Chapter 5.)

Another resistive-type transducer is shown at A of Fig. 21-4. This unit is light-sensitive; that is, it changes its resistance value for changes of light intensity striking its surface. Thus, it is basically a photocell transducer. (Other photoelectric devices are discussed in Chapter 23.)

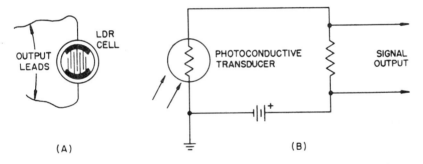

FIGURE 21-4. Photocell transducers.

The photocell transducer shown at A is known as a *light-dependent resistor* (LDR), most types of which are sensitive throughout the entire visible range of light as well as through a portion of the infrared spectrum. Some of these cells have a resistance ratio in excess of 25,000 to 1 for a light-intensity change from total darkness to daylight. They do not follow rapid changes in illumination, though the more intense the incident light, the quicker the response.

Their construction consists of a glass envelope containing a plate of photoconductive material approximately 0.028 in. thick. As shown at A, metal-film electrodes form two interlocking combs that make up one side of the photoconductive coating. These electrodes increase the light sensitivity of the unit and also enlarge the area of sensitivity. The circuit for an LDR is shown at B and since these transducers are resistive in their characteristics, a voltage source is required as shown. Such photocells are known as *photo-conductive* types in contrast to the photovoltaic types that require no additional voltage source but produce an output voltage when light strikes the photoconductive surface.

The thermistor can also be employed as a resistive transducer. The thermistor, because resistance changes with the temperature, will produce an output-signal voltage as shown in Fig. 21-5 that will vary as the thermistor resistance varies. Generally, however, thermistors are utilized in series with some circuit to provide a higher than normal resistance when current is first applied and then have a decreased resistance after other circuit elements have come up to temperature. One such application is in series-filament strings. When voltage is first applied, the resistance of the filaments of the various tubes is low and the filament wires may be overloaded. A series thermistor, however, maintains a fairly high resistance until the filaments have warmed up and increased their resistance. At this time, the thermistor has a decreased resistance. Thus, the thermistor maintains a fairly stable circuit-resistance value during warm-up time.

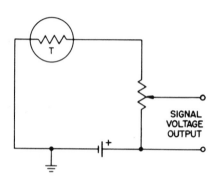

FIGURE 21-5. Thermistor as a transducer.

21-4.   Inductive Input Transducers

Basic construction of the dynamic-type microphone is shown at A of Fig. 21-6. Structurally, this unit resembles the PM speaker discussed in Chapter 19. It has a permanent magnet with E-shaped pole pieces to concentrate a strong magnetic field across the small coil between the air gaps. As shown, the coil is attached to a diaphragm in a fashion similar to a speaker cone. (Actually a PM speaker substitutes for a microphone, though its quality is not comparable to that obtained from a microphone specifically designed as such. Some intercommunicating systems use the speaker for a microphone in the "talk" position.)

The dynamic microphone has a low impedance since there are only a few turns on the coil. Thus, the 5 to 15 Ω of impedance must be stepped up by an input transformer if necessary to match to a higher circuit impedance. Some inductive microphones have a small matching transformer built in, with a switch for selecting low- or high-impedance outputs.

Another type magnetic microphone is that shown at B. This is a *ribbon* microphone (sometimes referred to as a *velocity* microphone). As shown,

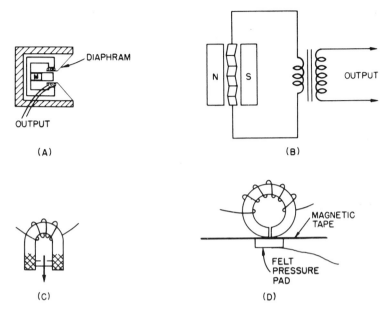

FIGURE 21-6. Magnetic audio input transducers.

a thin corrugated ribbon of aluminum alloy is suspended between two pole pieces. Since there is no diaphragm, a sound wave that impinges on the corrugated ribbon produces a force related to the velocity of the air pressure changes. This comes about because both the front and back of the ribbon are subjected to the sound pressure and the difference between the front- and back-pressure force causes the ribbon to move and cut the lines of force. Thus, a voltage is induced in the circuit composed of the ribbon and the primary of the transformer. As with the dynamic microphone, a step-up transformer is required for matching many input circuits since the impedance of the velocity microphone is also very low.

The magnet-coil principle for phono pickup units is as shown at C of Fig. 21-6. Here, as the needle armature rides in the record grooves and vibrates in accordance with the groove variations, the needle armature cuts across the lines of force produced by the magnet and hence produces an output signal voltage. Here, as with the magnetic microphones, the impedance is low.

The microphone transducers converted variations in air pressure into electric signals, while the magnetic phonograph pickup converted mechanical vibrations into electric signals. In tape recording, variations of magnetic density along a plastic tape (coated with magnetic materials) is sensed by a playback head as shown at D. Here a felt pressure pad presses the magnetic tape against the playback head. The latter has a very small air gap and as the tape passes over this air gap, the alternating magnetic field, represented by varying degrees of magnetism along the tape, will induce a voltage in the playback head pickup coil. The output leads from this coil thus furnishes to the amplifier input circuit the signal information that was originally placed on the tape during the recording process.

The plastic magnetic tape is used not only in home tape recorders but also in industrial electronics and computer systems, among other fields. In computers, it is used to store pulse information representing numerical or alphabetical values and, as such, becomes a storage (memory) section of the computer. In automation, it is used to store a program of sequential steps for milling, fabricating, or other automatic processes. In television, it is used to store picture information for rebroadcasting at a future date.

Reactive transducers are also used in industrial electronics to sense mechanical changes and convert them to electric signals. One such device is the inductive differential transducer shown at A of Fig. 21-7. This operates

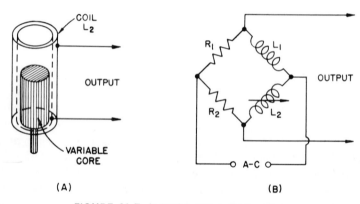

FIGURE 21-7. Inductive differential transducer.

on the principle that when a plunger-type core is moved in or out of a coil, the inductance changes because the permeability is affected by the metallic core. Usually, the plunger-type core is attached to a rod, and this rod in turn is fastened to a sensing rod, lever, or roller as required. A spring is employed to keep the plunger almost out of the coil and hence pressure is required to move the plunger into the coil. If a roller is applied to the bottom of the rod, the transducer can be used for measuring and controlling variations in thickness or curvature of any material over which the roller rides. Similarly, a lever arm could be attached to it to sense lateral variations in assembly-line sheet-metal forming or for other control purposes.

Since the inductive differential transducer has a variation in its inductance value for mechanical movements of the core, the inductance changes can be sensed by utilizing a bridge circuit such as shown at B. When the differential transducer has an inductance equal to L_1, the bridge will be balanced (assuming R_1 and R_2 are also of equal value). Thus, for any mechanical change sensed by the transducer, there will be a change in its inductive value and hence an unbalance of the bridge. The unbalanced bridge will then produce an output voltage in proportion to the inductance change. The latter, of course, is proportional to the mechanical change sensed by the transducer.

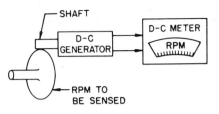

FIGURE 21-8. Tachometer transducer.

Another type of transducer is that known as the *tachometer*. The tachometer principle is illustrated in Fig. 21-8 and consists of a small d-c generator with a shaft extending from its housing and a calibrated d-c meter. When the shaft of the generator is rotated, it will produce a voltage having an amplitude proportional to the shaft rotational speed. Thus, the generator is a transducer since it converts mechanical revolutions to electric voltage. This principle is widely used in industrial electronics for measuring the number of revolutions per minute of a variety of rotating devices such as motors, machine shafts, and spindles. The output from the d-c generator is applied to a d-c meter but the dial of the latter is directly calibrated to read rpm.

21-5. Capacitive Transducers

Shaft rotations can also be sensed by a variable capacitor as shown at A of Fig. 21-9. Here the shaft which is to be tested for rotation is connected to the rotor of a variable capacitor which can turn completely without stop. The output is then applied to a capacitive bridge and when the shaft rotates from its normal position, the degree of rotation is sensed by the proportionate bridge unbalance. The capacitor can also be utilized to sense pressure changes as shown at B. Here two metal plates are separated by a flexible dielectric material as shown. With no pressure applied, the capacitor repre-

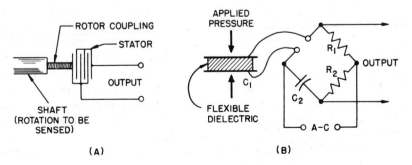

FIGURE 21-9. Capacitive differential transducers.

sented as the differential transducer (C_1) has the same value as capacitor C_2 in the bridge circuit shown. When pressure is applied, it compresses the dielectric material and brings the plates of the capacitor closer together. The result is an increase in capacity and an unbalancing of the bridge. The output voltage produced when the bridge is unbalanced will be in proportion to the pressure applied.

The fact that the amount of dielectric in the capacitor affects the capacitance value is also utilized in industrial electronics to sense fluid levels, as shown at A of Fig. 21-10. Here capacitor plates are suspended in a container

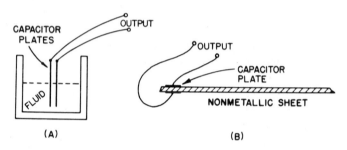

FIGURE 21-10. Liquid level and thickness sensing.

and when the fluid level is near the top, the fluid between the capacitor plates furnishes a maximum amount of dielectric material and hence the maximum capacity is the result. As the fluid level drops, the dielectric constant also changes and hence the capacitance decreases. Thus, the output capacitance reading can be calibrated in proportion to the fluid level in the container. Again, a bridge network can be utilized to produce a voltage in proportion to the changes in fluid level.

Bridge networks need not necessarily be employed to sense reactive changes. The reactance circuit discussed earlier in Chapter 19 can also be used for specific sensing of reactance changes. Meters which read reactance can also be utilized to give a visual indication by having the meter dials calibrated in terms of the mechanical changes which are to be sensed.

The dielectric factor of capacitors can also be utilized for sensing changes in thickness of nonmetallic material as shown at B of Fig. 21-10. In this case, capacitor plates are placed above and below the material and the thickness that the material should have is the reference point with respect to the capacity value obtained. Thus, as the nonmetallic sheet moves through the capacitor plates, an increase in thickness will result in a decrease in capacitance and a decrease in thickness will produce a higher capacitance. Here again various circuits or measuring devices can be employed and calibrated to indicate actual material thicknesses along the sheet.

In industrial electronics some special-type capacitors are also utilized having what is known as *ferroelectric* or *nonlinear* characteristics. Such capacitors have dielectric materials that include titanium dioxide, barium titanate, strontium titanate, etc. When any one of these materials is used as the dielectric of a capacitor, the capacitance value is subject to change over 50% when a voltage change across the capacitor occurs from approximately zero to 200 V. Such capacitors not only undergo a capacitance change for voltage but also for temperature.

This is shown in Fig. 21-11, where the dielectric constant of the ferroelectric capacitor is plotted against temperature on the left and voltage on the right. At normal operating temperatures, the dielectric constant is at its highest point, and this peak level is known as the *Curie* point. For either a decrease or increase of temperature around the Curie point, there is a decrease in the dielectric constant and hence a decrease in the capacitance of the unit. The temperature at which the Curie point is established depends on the composition of the dielectric material. For the graph at the right of Fig. 21-11, the dielectric constant reaches its highest value when no voltage is applied across the capacitor. For the

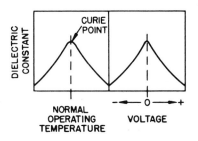

FIGURE 21-11. Characteristics of ferroelectric capacitors.

application of either a negative or a positive potential, however, there is a decrease in the dielectric constant and again a decrease in the capacitance. Thus, such a nonlinear capacitor can be utilized to sense temperature changes as well as voltage changes and hence these devices have transducer characteristics.

21-6. Inductive Output Transducers

There are a variety of output load units utilized in electronics. Speaker systems have already been covered in Chapter 19, and these represent load systems that have transducer characteristics since they convert electric signals into sound energy. In addition to speakers, some of the other devices include those shown in Fig. 21-12. At A is shown a magnetic tape-recorder head. This is similar in construction to the magnetic playback head discussed earlier, except that this device is used to magnetize sections of the tape to conform to the signal current circulating through the recording-head inductance. The output signals from the power amplifier are coupled to the recording head with a step-down transformer, to match the impedance of the output amplifier to the low impedance of the recording head. The signal

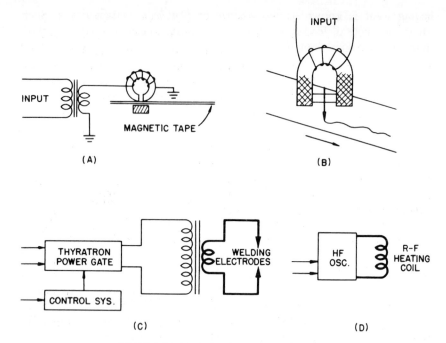

FIGURE 21-12. Industrial output transducers.

energy which then circulates in the coil sets up a magnetic field which varies in intensity to conform to the signal energy. Thus, across the air gap, the changing magnetic lines of force magnetize sections of the tape.

Another type of output transducer is shown at B and is extensively used in industrial electronics. Here a stylus mechanism is used that glides over a moving sheet of paper to graph visually the signal energy applied to the recording head. The stylus or recording pen moves in accordance with the signal energy and hence traces out the signal as shown on the paper. If a mechanism is also employed with the recording head to move it laterally, a plot of both the x and y axes can be obtained. Crystal-type output transducers could also be used, although the magnetic type is capable of handling more power.

Another type of output transducer is that found in industrial welding processes. Arc and spot welding, however, are primarily resistive with respect to the load imposed on the amplifier. When an arc is drawn, the current flow through the arc represents power dissipated in a very low resistance with high currents. Similarly, in spot welding, when two metal plates are to be welded, they provide sufficient surface-contact resistance to generate heat when passing current from the welding electrodes. The heat causes a fusion of the sheets at the spot to be welded. A typical output circuit for welding is shown at C of Fig. 21-12. The line voltage is applied through gated-thyratron circuits and stepped down by the transformer so that a low voltage and high current are available. The voltage may range from approximately 2 to 10 V but the current that flows may range from a few thousand to well over 50,000 A for the larger welders. The thyratron control circuits assure proper power control and minimize burn-through or insufficient welding characteristics.

In arc welding, a high-intensity arc is established between the two electrodes and thus applies heat to the metal to be welded. The electrodes are composed of special welding rods that are melted by the arc so that the melted rod material flows onto the metal being welded.

An inductive-type load is established in induction heating, illustrated at D. Here a high-frequency oscillator produces an RF output signal that is applied to the material to be heated by use of the large, high-power-handling coil, as shown. The RF fields generate heat by induction to the material. In some instances the high-power induction-heating energy is obtained directly from the power mains by using motor-driven generators that have signal frequencies of several thousand hertz. These may range from a few to over 1000 kW. Industrial applications include the annealing of metals, heating of metals for specific purposes, hardening, and brazing processes.

21-7. Tube-Type Output Transducers

When a vacuum or gas-filled tube is designed to display some sort of visible pattern in proportion to signal voltages applied to it, a transducer is again formed. One such transducer is the tuning-indicator tube used in receivers, some test equipment, and other applications where precise adjustments are necessary. Such tuning-indicator tubes have a florescent material within them that glows when in operation. The area of glow is varied in accordance with the voltage amplitude applied to the grid of the tube.

The basic circuit for one type of tuning-indicator tube is shown in Fig. 21-13. The control voltage (such as AVC) is applied to the control grid of the tube through a filter network made up of a series resistor and a shunt capacitor. The purpose for the filter network is to eliminate any a-c type of signals. The tube itself is composed of a triode section shown at the left and

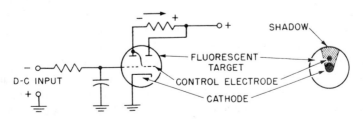

FIGURE 21-13. Tuning indicator circuit.

a fluorescent-target section shown at the right. A control electrode is between the common cathode and the fluorescent target. The latter is in the form of a circular area as shown at the right of Fig. 21-13. The cathode emits electrons that are attracted by the positive potential of the fluorescent target. The latter can be considered as an anode that attracts the electrons from the cathode. The fluorescent coating, however, creates a glow when the electrons strike it. A control electrode is placed between the cathode and the target so that it can repel the electrons that strike one area of the target. The shaded area that is created when the tube is functioning resembles the human eye and hence the tube has also been called a Magic Eye tube.

When the negative potential that is applied to the control grid is of a low value, electron flow from the cathode to the anode of the left section of the tube will be of high value. The high current flow through the plate resistor establishes a polarity as shown. The voltage drop across the plate resistor establishes a potential difference between the anode of the triode section and the fluorescent target. The control electrode, however, has the same potential as the triode plate. Because the control electrode is less positive than the fluorescent target, it tends to repel the electrons from the cathode and a shadow area is created because few electrons go past the control electrode to strike the fluorescent target. With a negative potential applied to the d-c input, however, current flow from the cathode to the plate of the triode section decreases and the voltage drop across the plate resistor declines. With a low value of voltage drop across the plate resistor, only a small difference in potential exists between the control electrode and the target. Consequently, the less negative potential of the control electrode does not repel as many electrons and the shadow area becomes more narrow. Thus, the variation in shadow area is an indication of the change of potential applied to the input of the tuning-indicator circuit, and the tube acts as a visual indication of circuit potentials that are to be used as a reference.

There are a variety of other tubes also used as output transducers that convert electric signal energy to some form of visible light. For instance, a television picture tube is a form of transducer as was discussed in Chapter 19. In Chapter 22, the oscilloscope tube is discussed and this is another form of transducer converting electric energy to visible light information.

Another type of tube transducer is that which displays numerically the numbers from a computer or from some measuring instrument. One such tube is the *Nixie* manufactured by Burroughs Corporation. This tube comes in several sizes, from a 1-in. diameter tube to some almost 2 in. in diameter. The basic tube has 10 cathodes and 1 common anode and is a cold-cathode gas tube that displays a number by ionization within the tube and a glow around a particular cathode. Each cathode within the tube is an element shaped like a number. When a voltage is applied at one of the pins in relation to the anode, the number corresponding to that pin will become visible by gas ionization glow associated with the pin. If, for instance, the

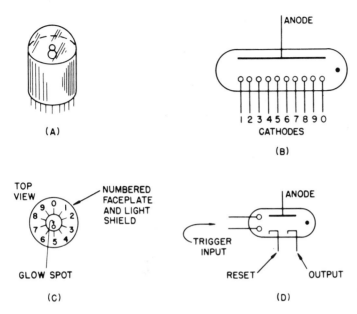

FIGURE 21-14. Nixie and Dekatron tubes.

cathode representing the numeral 8 had a voltage applied to it, this number would appear as shown at A of Fig. 21-14. A number of such indicating tubes can be used side by side to display a number of any desired length. The schematic for the Nixie is shown at B.

Another indicating tube is the Dekatron shown at C. The design of this tube is such that successive pulses applied to the input terminals produce a glowing spot that moves progressively around the top of the tube. The glow moves one segment for each pulse applied and thus can indicate any digit from 0 through 9. A faceplate (bezel) is placed over the top of the tube as illustrated at C.

The Dekatron is also a basic transfer device because an output pulse is produced from an output terminal, shown at D, for complete progression from 0 to 9 (decade). Thus, if several such tubes are coupled together, they will form a counter for a pulse train. If pulses are applied to the trigger input, the glow spot will move around the first tube, and when it reaches the count of 10 at the 0 indication, an output pulse from the first tube triggers the second tube so that its glow spot moves to the 1 position. Thus, an indication of the number to be read out is visible up to the limit of the tubes employed.

21-8. Solid-State Output Transducers

As mentioned earlier, crystal output transducers are found occasionally in audio devices. A typical example are the crystal-type earphones that are available for either monaural or stereophonic reception. Output crystal recording devices have also been utilized on occasion, though as mentioned earlier, the magnetic types produce greater output. Another type of output transducer having crystal characteristics is the ferrite-core assembly utilized in the storage systems of digital computers. These ferrite cores are composed of tiny ferrite rings smaller than a pin head. Ferrite has a crystal spinel composition that is ferromagnetic and brittle; the cores have a very good magnetization factor and the hysteresis loop is virtually a rectangle. Such a characteristic lends itself to rapid switching from one magnetized state to another and hence is valuable in the storing of signal pulse data.

Because of the rectangular-loop characteristic, the ferrite core is essentially a bistable device. For computer storage purposes, a magnetic field is applied to cause the ferrite core to become magnetized in one direction. Such a direction is assigned either the representation "zero" or "one" as desired, with the opposite field representing the opposite designation. For instance, if a north and south pole is chosen to represent "one," a south and north pole (magnetized in the opposite direction) represents "zero."

The cores are wired as shown in Fig. 21-15. While only four cores appear

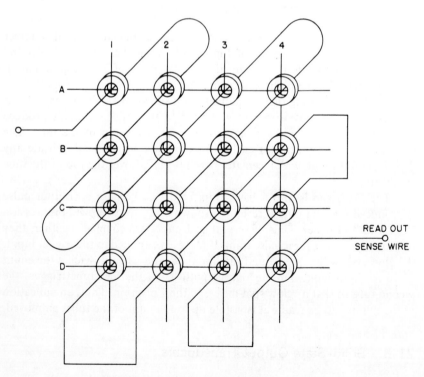

FIGURE 21-15. Ferrite-core computer storage.

in the vertical and horizontal plane, practical ferrite-core storage planes may have as many as 50, 100, or more in a single line.

As shown, individual insulated wires are threaded through each core, with the vertical wires intercepting the horizontal wires at the core centers. In switching the cores, the amplitude of the signal voltage applied to any single wire is held at a value sufficiently low so that it represents approximately one-half of the amplitude necessary to change the magnetic field. Thus, if such a half-signal voltage were applied to the first horizontal wire, designated as A, all the cores in the vertical A plane would be subjected to a slight magnetizing field but not enough to reverse the polarity of the magnetism. The magnetic field that is produced, however, is sufficient to move the flux density along the upper portion of the hysteresis loop to the point marked P in Fig. 21-16. Since this is not sufficient to reach the lower-left saturation point of the hysteresis loop, the direction of magnetism is unchanged. If, now, vertical line No. 2 is also energized by such a half-amplitude signal, the combinations of the magnetic fields from the A wire and the No. 2 wire become sufficient to switch the magnetic state of the ferrite core intercepted by the A and No. 2 wire into the opposite direction. This comes about because the No. 2 wire half-energized the second vertical row of cores and the A wire half-energized the first horizontal row of cores. Only the top second core was energized sufficiently (by virtue of the combined fields) to switch its magnetized state into the opposite direction. When the signal energy leaves, the magnetized state will be as shown

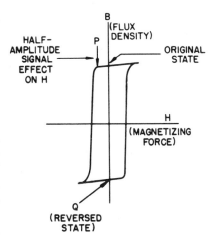

FIGURE 21-16. Ferrite-core reversal.

at Q of Fig. 21-16. Thus, by a proper selection of any two wires, a particular core can be pinpointed for a reversal of its magnetic state from its representative "zero" condition to a representative "one" condition. Thus, any core on the unit can be reversed in magnetic polarity to represent the digit "one." By applying a reverse-polarity signal voltage to the appropriate horizontal and vertical lines, the core can be energized in the opposite direction to represent zero. When a core is switched from the "one" state to the "zero" state, it is known as *clearing*.

The stored information for any particular core is read out by the diagonal wire shown in Fig. 21-15. This wire is sometimes referred to as a *sense* wire. Upon the application of the clearing voltage, the shift of the magnetic field of the core which is being cleared sets up a changing magnetic field which induces a voltage into the sense wire. This is utilized at the terminals of the sense wire for readout purposes and applied to appropriate amplifiers and sensing devices.

The ferrite cores are strung on frames and a number of frames may be stacked one above the other to increase the number of pulse signals that can be retained.

Review Questions

21-1. Why are transformers unnecessary when coupling crystal input transducers to the grid circuit of a vacuum-tube amplifier?

21-2. Reproduce the circuit for a solid-state pressure transducer, and explain its function.

21-3. Explain the principles involved in a bonded strain gauge.

21-4. Why is an LDR cell a photoconductive device?

21-5. Why is a ribbon microphone also called a *velocity* microphone?

21-6. Reproduce the drawing of the magnetic playback head, and explain how it picks up signals. Why is this a transducer?

21-7. What is an inductive differential transducer used for in industrial electronics?

21-8. Explain the basic principles of a tachometer.

21-9. Why must ac be used for the bridge circuit of the capacitive differential transducers shown in Fig. 21-9?

21-10. Explain how liquid level and thickness sensing is accomplished electronically in industry.

21-11. Discuss the characteristics of ferroelectric capacitors.

21-12. Reproduce drawings at two industrial output transducers (other than tube types), and explain their function.

21-13. Reproduce the drawing of a tube-type output transducer, and explain its function.

21-14. If the lower-right ferrite core in Fig. 21-15 is to be energized, what amplitude signal voltages must be applied, and to what terminals?

21-15. Explain the purpose of a *sense wire* in a ferrite-core computer storage.

Test Instruments   22

22-1.   Introduction

In all activities involving design, maintenance, or the modification and improvement of electronic devices, the technician must utilize various units of test instruments as functional tools. By use of test equipment, one can evaluate circuit function in terms of the voltage, current, and resistance units that are present. Test equipment also expedites troubleshooting since it permits readings that can localize circuit faults. This is done by ascertaining which component values or unit measurements do not coincide with the pre-established values indicated by the manufacturer. Test equipment also permits the proper adjustment and alignment of various electronic circuits encountered in transmitting, receiving, and industrial-electronic applications.

Test instruments are available in a variety of models, from the basic combination meters that read voltage, resistance, and current to the more complex oscilloscopes, signal generators, and other devices. With, however, even the simplest equipment, a competent technician familiar with the gear can gather much information regarding the performance of a specific circuit. In such readings, however, when the necessity arises for delving deeper into complex circuitry or when more precise measurements are needed, more elaborate test equipment must be utilized. Thus, even though the technician may be exposed only to a minimum of test equipment in the initial aspects of his professional activity, he should be familiar with the various units of test equipment that are generally available for the completion of various tests and measurements. A knowledge of the manner in which such equipment is used, plus an understanding of the limitations and applications of each particular piece of test equipment, will not only enable the technician to utilize such instruments but will also enable him to evaluate the merits of the test equipment, and thus be in a better position to judge whether or not it will perform the precise functions that the occasion demands.

22-2. Multimeters

When it becomes necessary to read voltages and currents and to take resistive measurements, the basic individual meters for such purposes, as described in Chapter 5, can be employed. The acquisition of an individual meter for each type of test, however, becomes a costly affair and a single meter with proper circuitry can be utilized for making all such tests. The most inexpensive type of test equipment that performs the measurement functions involved in voltage, current, and resistance is the *multimeter*. Since this is a combination meter reading *volts-ohms-milliamperes*, it is often referred to as a "VOM." In its most basic form, the multimeter utilizes a 0-1-mA meter. With proper combinations of shunt resistances and switches, the basic milliammeter becomes a multirange instrument capable of reading current values from a few milliamperes to several amperes. (Virtually all basic measuring devices use the so-called *D'Arsonval* meter movement that basically consists of a small inductance pivoted so it can turn freely between the poles of a permanent magnet. The indicating needle is fastened to the coil.)

Besides being capable of current measurements, this instrument may also be converted to a voltmeter by switching various resistors in series with the meter. The meter also functions as an a-c voltmeter, by employing a germanium- or silicon-diode rectifier to convert the *ac* to *dc* for application to the meter.

Figure 22-1 shows a typical volt-milliampere switching circuit. The range selection switch is the two-pole rotary type as shown. For voltage readings, the switch selects various series resistors for the range required. When in the current-reading position, appropriate resistors shunt the meter as described in Chapter 5. In the position shown in Fig. 22-1, the switch shunts resistor R_2 across the meter and converts the latter to a 0- to 10-mA scale meter. With the switch at the 50-μA range, no resistance shunts the meter because this is the basic meter full-scale current range.

The basic a-c voltmeter switch circuit for a VOM is shown in Fig. 22-2. Even though two diodes are used, the rectification is of the half-wave type, with diode D_1 being the meter rectifier. Diode D_2 shunts the opposite

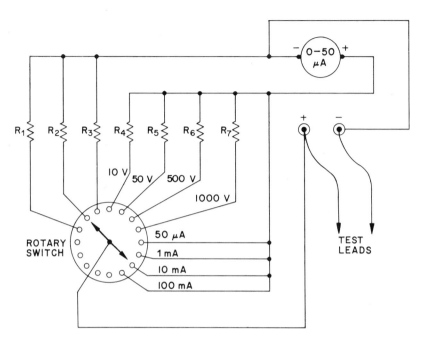

FIGURE 22-1. Volt-milliampere switching circuit.

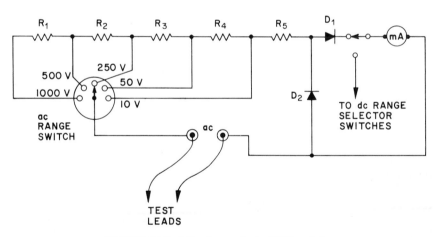

FIGURE 22-2. AC voltage selector VOM switch.

polarity alternation of the ac being measured and thus removes the reverse-polarity voltage from across the D_1 rectifier and the meter. This inverse peak voltage could alter meter accuracy and performance. In the switch position shown, 250 V is the maximum a-c scale range, using resistors R_3, R_4, and R_5

in series to provide the necessary voltage drops. In the 10-V range, only R_s is in series, while on the 1000-V range, all resistors are in series.

For a-c meters the scales are calibrated so that rms values of current and voltage are indicated. The average current (or voltage) for an alternation is $0.636 \times$ peak. For half-wave rectification the average value occurs only for one-half of each cycle; hence the average value over a complete cycle (and successive cycles) is half of 0.636. Thus, for a 10-mA peak value (rms = 7 mA), the average value is $6.36/2 = 3.18$ mA and meter calibration must take this into consideration. With multimeters using full-wave rectification, both alternations of the ac are rectified and the average value is the full 0.636 of peak value. When full-wave rectification is utilized, the common method is to employ the bridge-type rectifier described earlier in Chapter 14.

If a multimeter has a 1000-Ω-per-V unit (see Chapter 5), the input resistance is rather low for voltage measurements so that the device loads down the circuit to which it is attached. Such loading occurs because the amount of current that flows through the multimeter represents a current *shunt* across the resistor or other device where voltage is to be measured. If, for instance, the 50-V scale is utilized, the ohmmeter presents a resistance of 50×1000, or 50 kΩ. If the meter is used to measure the voltage across a 50-kΩ resistor, its application to the resistor would be the same as though the resistor were shunted by another having the same 50-kΩ value. Thus, the 50 kΩ represented by the resistor in the circuit would be reduced to 25 kΩ and, in consequence, double the amount of current would be drawn than would otherwise be the case. For this reason, such a multimeter gives rise to some inaccuracy when making voltage measurements. The resistance of the meter, however, rises with higher scales and, for the 2500-V scale, the shunting resistance as represented by the meter would be 2.5 MΩ. A better multimeter is formed if it uses a 20,000-Ω-per-V meter, which has considerably less loading effect on circuits. Even less loading effect is produced by the costlier multimeters employing 50,000-Ω-per-V meter movements.

The ohmic ranges of the multimeters are obtained by the methods described earlier in Chapter 5. In most multimeters several flashlight cells are used for the ohmmeter circuitry. Again, appropriate switches are used for selection of ranges or for changing from ohmic readings to voltage or current.

From the foregoing, it is evident that multimeters can be employed for the measurement of direct currents flowing in the cathode, plate, screen, and other vacuum-tube circuits, as well as in base, emitter, and collector transistor circuits. The tester can also be employed for measuring the various voltage drops across resistors and at the bleeder sections of power supplies. Besides such measurements, the meter can also be employed to read the voltage across power-supply transformers, as well as the a-c filament voltages of the various tubes. In addition, the ohmic values of resistors, transformer windings, and other units can be measured. The ohmmeter scale is also useful for checking continuity in circuits. While this process is essentially that of taking resistance readings, it will establish whether or not an open or short circuit exists and, hence, proves useful in troubleshooting work. The limitations of the multitester are imposed by the ranges of the various scales, as well as the loading effect.

Variations among commercial VOM models that are encountered consist of scale ranges, sensitivity, and a rearrangement of the terminals. The instructions which accompany the multitester should be consulted with respect to the loading which is to be expected. The instruction sheets also indicate the ranges of the scales and explain procedures recommended for proper measurements of voltages, currents, and resistances.

22-3. Typical Applications

When the multimeter is employed as a continuity-checking device, the ohmmeter scale is utilized, and the equipment to be checked *is shut off or disconnected from the power mains.* If, for instance, the continuity of a power supply is to be checked, the ohmmeter can be placed across the a-c plug, as shown in Fig. 22-3. With the switch open, no reading should be obtained and when the switch is closed, a low ohmic value should be present. This is a continuity check of both the primary of the transformer and the switch. It indicates whether or not the switch is operating properly and also shows whether or not the primary winding is open. This is the extent of a continuity check and if there is reason to suspect that some of the transformer windings are shorted, the ohmic value as read by the meter would have to be compared with the value shown in the service notes or indicated by the manufacturer.

With the power shut off, a similar continuity check can be made of each secondary winding. An additional continuity check can then be made from the cathodes of the rectifiers to the output terminal, as shown in Fig. 22-3. Such a check will establish whether the filter choke or filter resistor is open. By placing the ohmmeter across the output of the power supply, a continuity check will indicate whether the filter capacitors are establishing an abnormal load (due to leakage) and are thus creating a partial short. An open bleeder section can also be ascertained by this procedure.

Similar continuity checks can be made on other circuits in identical fashion. With the receiver shut off, an amplifier stage, such as shown in Fig. 22-4, can be checked for continuity by placing the ohmmeter from the collector of the transistor to the chassis. A meter-needle deflection will indicate that no open-circuit condition exists in the decoupling resistor or load resis-

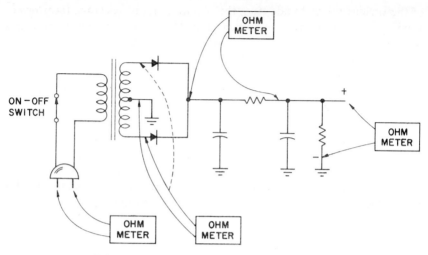

FIGURE 22-3. Continuity checks on power supply.

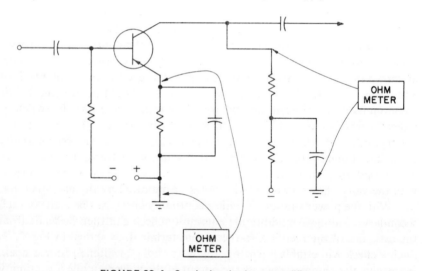

FIGURE 22-4. Continuity checks on amplifier.

tor, while a short may indicate a defective decoupler capacitor or internal short within the transistor. A check from emitter to ground will indicate if the emitter resistor is open or the bypass capacitor shorted.

Continuity checks are made to establish whether an open or short circuit exists, and the actual ohmic values obtained are of little consequence since in most instances the continuity for several series resistors or inductances may be involved. When a continuity check has localized the defect to a particular circuit, individual ohmic checks can be made of the various resistors, inductances, and other units involved.

22-4.   Vacuum-Tube or FET Voltmeter

By taking advantage of the rectifying-amplifying characteristics of tubes or transistors, effective isolation is maintained between the test instrument and the circuit under investigation, while sensitivity and versatility are increased. Early versions were tube types, hence the term *vacuum-tube voltmeter,* or *VTVM.* As with other electronic systems, however, solid-state units such as junction transistors, silicon diodes, and MOSFET units are widely used. For convenience in this discussion, the abbreviations VTVM will be used.

The VTVM is similar to the VOM since it consists of a single instrument that performs a number of measuring functions. The VTVM, however, has much greater ranges in the voltage and resistance readings, plus a high input impedance that minimizes loading effects.

In some units, only the voltage ranges that are dc use the tube or solid-state principles, while a-c readings are obtained in similar fashion to that for the VOM. The more costly units, however, have a number of refinements including the VTVM principle for the a-c scale, as well as provisions for reading peak-to-peak voltages.

Since the vacuum-tube or FET voltmeter has such a high input impedance, with little loading on the circuits under test, the more accurate readings that are obtained have made this instrument the most popular with technicians for troubleshooting hi-fi equipment, industrial-electronic circuitry, and radio and television receivers, as well as for laboratory work involving research and design. Because of the high sensitivity of the VTVM instrument, the ohmmeter range is capable of measuring resistances up to 1000 MΩ.

Special probes are also available for use with this type of voltmeter, for extending the useful range of the various scales to permit a greater scope of tests. A conventional VTVM does not incorporate a milliampere or ampere scale. Current values can, however, be ascertained by use of Ohm's law. Thus, a voltage reading across a resistor, if the resistance value is known, will indicate current when the voltage is divided by the resistance. Since the VTVM also incorporates an accurate ohmmeter, the unknown resistance can also be read when it is necessary to establish the amount of current flowing therein.

Other probes are also available as accessories to the vacuum-tube voltmeter, one of which is the RF probe. This probe contains a germanium- or silicon-diode crystal that converts RF energy to dc so that the calibrated values can be read on the VTVM scale. The RF probe permits direct measurement of audio, supersonic, and RF voltages. Most such probes are of the peak-indicating type; that is, they will indicate a value almost equal to the positive peak voltage of the signal that is measured. Some probes of this type will read amplitudes of signals having frequencies ranging between 20 Hz and 300 MHz, which means that readings can be obtained for voltages from the low audio range to the upper limits of the 13-channel television frequencies. Other probes extend the voltage-measurement range of the VTVM to as high as 50 kV dc, which is useful for reading the high picture tube anode voltages encountered in television receivers.

As with the multimeter, the instruction sheets accompanying the VTVM should be referred to so that the full capabilities of the instrument can be utilized. The VTVM can also be employed for continuity checking, in similar fashion to the methods detailed for the multimeter.

22-5.   Tube and Transistor Checkers

Transistor and diode checkers are much more simple than tube checkers since lower voltages, smaller socket mounts, and the absence of filament checks are involved. Versatility depends on what circuit refinements are included to obtain a better check of transistor performance, how well parameters meet the specifications for the transistor under test, and the overall accuracy of the readings obtained. In selecting such testers, the manufacturer's descriptions should be compared to similar types made by others to obtain one providing the most satisfactory service and to meet particular testing requirements.

Tube checkers are available in many models, with the simplest and most inexpensive using the emission-test process. Such a tube tester connects all the grids (control grid, screen grid, suppressor grid, etc.) together with the anode, and the cathode emission is then compared to the value considered as a standard for the particular tube under test. A more costly tube tester is the transconductance type wherein standard voltage values are applied to each tube element. The resultant anode-current value then indicates the transconductance (g_m) of the tube under test for static conditions. A *dynamic* transconductance test consists of setting up the tube elements as with the static test but also applying an a-c signal to the tube so that the anode current compares to that experienced in actual electronic circuitry. Thus, the anode current varies in relation to the signal applied to the grid and represents a dynamic transconductance condition.

Most tube checkers have provisions for modernizing the checker, as newer tubes are produced by the various manufacturers. Where a roll chart is used for ascertaining the various settings for a particular tube, a new roll chart may be issued periodically to keep the instrument up to date.

Since the physical makeup of the various tube checkers differs to a considerable extent, it is essential that the instruction manual that accompanies the tube checker be referred to, in order to utilize the instrument to its fullest capabilities. Unfamiliarity with the manner in which the tube checker should be employed may result in damage to the tubes. An abnormal application of filament potentials to the heaters of the tube to be checked can mean a tube burnout, while excessive plate potentials could impair the emission characteristics of the cathode structure. For this reason, a careful study should be made of the operational principles before a tube is inserted.

22-6. Capacitor Checker

A capacitor checker is a useful device for servicing and laboratory work since it gives a reading of the value of a capacitor as well as indicating its power factor in terms of leakage resistance.

Capacitor checkers usually employ a balanced-bridge circuit that compares the value of the unknown capacitor to that of precision-value capacitors in the instrument. A Magic Eye type of indicating tube or meter is used to show balance.

The probes of the checker are applied across the capacitor to be tested and the approximate range is selected. The dial is then rotated until the shadow area of the Magic Eye tube indicates a balanced bridge. The value of the capacitor, in microfarads or picofarads, is then read directly on the scale. Most capacitor checkers also permit the application of voltage to the capacitor, for breakdown and leakage testing. The leakage is read by the power-factor method. Paper, ceramic, and mica capacitors should have a very low power factor and, in most instances, no reading will be obtained for these when they have no internal leakage. Electrolytic capacitors, such as employed for power supplies and for large-capacitance values in transistor circuitry, have a much higher power factor (ranging from 3 to 15). An excessive power-factor reading indicates an abnormal leakage and, in such an instance, the capacitor should be replaced by one having a power-factor reading within the range specified by the manufacturer.

Some capacitor checkers also utilize the bridge circuit for measuring inductance values. The usual method, in such an instance, is to employ an inductance of known value and utilize it for balancing the bridge, to ascertain the value of the unknown inductance.

22-7. The Oscilloscope

The oscilloscope is an electronic instrument that is capable of giving a visual indication of a signal waveform. For this reason, it is a valuable instrument for research, design, test, and maintenance. With an oscilloscope, the waveshape of a signal can be studied with respect to amplitude, distortion, and deviations from the normal. In addition, the oscilloscope can also be employed as a voltmeter and a signal tracing device.

Shortly after the oscilloscope was first manufactured, its usefulness was confined to laboratory applications but since the advent of industrial electronics and control, television, radar, and other specialized branches of electronics, the oscilloscope has become a valuable instrument for wide-band IF alignment, troubleshooting, signal waveform observation and inspection, and other similar applications.

The heart of the oscilloscope is the cathode-ray tube. The internal construction of a typical tube of this type is shown in Fig. 22-5. As with other vacuum tubes, the filament heats the cathode to the degree where the latter emits electrons. The control grid influences the amount of current flow, as in standard vacuum tubes. Two anodes are employed, each having a positive d-c potential applied to it. These anodes accelerate the electrons and form them into a beam. The intensity of the beam is regulated by the potential applied to the control grid.

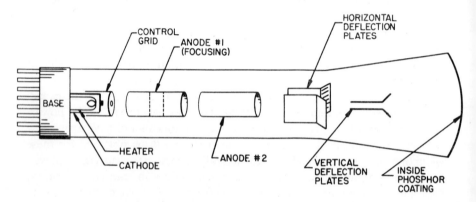

FIGURE 22-5. Electrostatic cathode-ray tube.

The cathode consists of a nickel cylinder, at the end of which an emitting element is fused. This element is made of either barium or strontium oxide and permits a sufficient release of electrons for the formation of an electron stream.

The grid structure, while controlling electron flow as in conventional tubes, differs from the wire mesh of receiving tubes and consists of a cylinder with a tiny circular opening to keep the electron stream small in size. The beam is focused into a sharp pinpoint by controlling the voltage on the first anode. The two anodes of the cathode-ray tube can be compared to a glass lens system, such as is employed in movie projectors, because the anodes focus the beam to a pinpoint at the face of the tube. A high voltage is applied to the second anode so that the electron stream will attain high velocity for increased intensity and visibility when it strikes the tube face. The beam-forming section of the tube is known as an *electron gun*.

As shown in Fig. 22-5, two sets of plates are present within the tube, beyond the second anode. These plates are for deflecting the electron beam both horizontally and vertically. If, for instance, a voltage is applied across the horizontal deflection plates, it will influence the beam because the negative potential on one of the horizontal plates will repel the electron stream, while the positive potential on the other horizontal deflection plate will attract the beam. If such a voltage is a sawtooth type, the gradually rising potential of the sawtooth will pull the beam gradually toward the positive horizontal deflection plate. Hence, the electron beam is made to *scan* across the face of the tube. Also, any potential applied to the vertical deflection plates will cause the beam to move vertically.

As shown in Fig. 22-5, the inside of the tube face is coated with phosphor so that when electrons strike this coating, it will fluoresce and emit light. After the exciting electron stream has left the area, the fluorescing characteristics that emit light rapidly decay and the light level declines. The chemical composition of the coating, however, can be such that the emitted light persists for an appreciable interval so that visual observation can be made of the light. Since the beam is swept across the tube at a fairly rapid rate, the light must persist for a sufficient time interval so that it leaves a complete trace of the waveform drawn on the picture-tube face. At the same time, the persistence of the phosphorescent coating should be sufficiently short so that if the beam stops, the pattern that is traced on the tube will disappear very rapidly. The regular standard numbering system is utilized within the tube designation code, for ready identification of the phosphor characteristics. If 3AP1 is a cathode-ray tube's numerical designation, for instance, it indicates that the tube has a 3-in. face because the first number identifies the face diameter. The P1 designation indicates a medium persistence phosphor, which has a green glow when excited. The A in the 3AP1 numerical designation refers to the internal construction of the tube and indicates that this particular tube has some structural changes with respect to a 3P1 cathode-ray tube.

A typical front panel of an oscilloscope is shown in Fig. 22-6. The various knobs are for controlling the size, brilliance, and number of images that appear on the screen. One knob regulates the bias on the control grid and, as mentioned earlier, affects the electron stream intensity. Hence, such a control is known as the *intensity* control and permits adjustment of the image, with respect to making it brighter or dimmer. A focus control is also provided that permits focusing the beam into a sharp pinpoint of light. The focus control regulates the potential applied to the first anode of the cathode-ray tube. Another knob, the horizontal position control, regulates the amplitude of the d-c potential that is applied to the horizontal deflection plates, in addition to the usual sawtooth voltage. The d-c potential permits positioning the image that appears on the screen so that the image can be moved to the right or left, as required. A vertical position control regulates the amplitude of a d-c voltage applied to the vertical plates, in addition to any signal voltage that may be impressed thereon. The d-c voltage on the vertical plates permits movement of the image in a vertical plane.

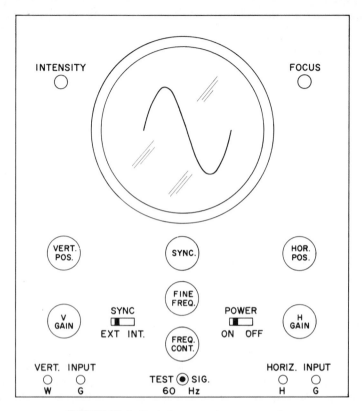

FIGURE 22-6. Basic front-panel scope controls.

The signals to be observed are applied to the vertical input terminals and are amplified within the oscilloscope, before such signals arrive at the vertical deflection plates. The gain of the internal vertical signal amplifier can also be regulated from the front panel of the oscilloscope. An internal oscillator is present that generates a sawtooth of voltage for application to the horizontal plates of the cathode-ray tube. The amplitude and the frequency of this horizontal sweep oscillator can also be regulated by the front panel controls of the oscilloscope. The relative position of the various controls on the front of an oscilloscope varies from one manufacturer to another. All models, however, have the basic controls just mentioned.

Figure 22-7 shows how an oscilloscope traces out a sine wave on the screen when such a signal is applied to the vertical input of the scope. At A the sine wave applied at the vertical input of the oscilloscope is shown. If the internal horizontal sweep generator is turned off, the rising positive potential of the first alternation of the input signal would cause the electron beam to move upward, as shown at B. When the negative alternation of the input signal arrives at the vertical deflection plates, the electron beam is pulled

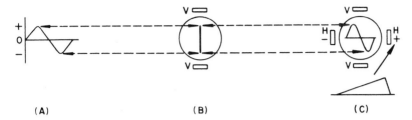

FIGURE 22-7. Formation of sinewave on oscilloscope tube.

downward, as also shown at B. The rapid rise and fall of the signal alternations cause the electron beam to move up and down the face of the cathode-ray tube very rapidly, leaving a vertical line trace, as shown.

If the horizontal oscillator within the oscilloscope is now turned on, a sawtooth of voltage, as shown at C, will be applied to the horizontal deflection plates. As shown at C, the rising potential of the sawtooth voltage causes the right horizontal plate to become positive and the left horizontal plate to become negative. The negative left plate and the horizontal right plate cause the beam to move from left to right because the beam is repelled by the negative plate and attracted by the positive plate. If one sawtooth occurs for each cycle of the sine-wave signal, the beam will be pulled across the face of the tube once for each cycle. Thus, as the first alternation of the input cycle rises in amplitude, the electron beam would normally rise in a vertical plane, as shown at B. The sawtooth on the horizontal plates, however, gradually pulls the beam from left to right and, hence, traces out in visual form the waveshape of the input signal. Similarly, waveshapes of square waves, pulses, or any other types of signals can be observed on the face of the oscilloscope screen. If the input signal has a frequency twice that of the sawtooth applied to the horizontal plates, two cycles will appear on the screen because the beam is pulled across the screen only once for each two cycles of the input signal. Hence, by regulating the ratio of the input-signal frequency to the sawtooth sweep frequency, portions of the input signal, or a number of cycles of the input signal, can be made visible on the screen. By calibrating the frequency of the horizontal sweep waveform so that its exact frequency is known, the frequency of the input signals to the oscilloscope can be ascertained. For instance, if four cycles of a sinewave appear on the screen and the sawtooth generator is set for 100 Hz, the frequency of the signal applied at the input of the oscilloscope is 400 Hz.

The oscilloscope can be utilized for reading the peak-to-peak voltages of a-c signals, just as some of the more expensive VTVM units can be utilized for this purpose. For reading peak-to-peak voltages, a transparent plastic screen is attached to the face of the oscilloscope. The screen is marked off with vertical and horizontal lines in the form a graph. To calibrate the oscilloscope, a voltmeter or VTVM of known accuracy is employed initially.

A low-voltage a-c signal must be available for calibration purposes. Some oscilloscopes have a terminal, on the front panel, that supplies such an a-c reference voltage. If the reference voltage is 5 V, for instance, this voltage is applied to the vertical input terminals of the oscilloscope. The internal horizontal sweep generator is shut off so that a vertical trace, as shown at B of Fig. 22-7, is visible on the screen. This vertical line represents the peak-to-peak voltage of the input signal. The vertical height control is now adjusted so that the line is two or three squares high and the control is left in this position after calibration. Knowing the rms value of the applied a-c calibrating voltage, the peak-to-peak voltage can be ascertained by multiplying the rms value by 1.41 to obtain the peak value and then doubling the peak value to obtain the peak-to-peak value. Once the oscilloscope has been calibrated, it can be utilized to read the peak-to-peak value of other voltages applied to the two vertical input terminals of the scope. Ordinary a-c voltmeters are accurate only for 60-Hz signals but a calibrated oscilloscope can read peak-to-peak voltages of signals having other frequencies. If the vertical amplifier of the oscilloscope has a fairly flat response up to 50 kHz, it will maintain accuracy with respect to peak-to-peak readings of signals whose frequencies range up to 50 kHz.

22-8.   Signal Generator

A signal generator is a calibrated variable-frequency oscillator producing a simulated carrier signal, which can be employed in place of the type of signal obtained from a broadcast station. Thus, the signal generator is a useful device when it is necessary to align a receiver or to track the tuner.

Alignment refers to the tuning of the resonant circuits of the IF amplifiers of superheterodyne receivers. Such tuning is made to "align" the resonant frequencies of the tuned circuits to the necessary IF specified for the receiver. *Tracking* refers to tuning the RF, mixer, and oscillator circuits of the receiver to the proper frequencies so that the right IF signal frequencies are produced, regardless of whether a low-frequency or a high-frequency station is tuned in.

The signal generator can also be employed for signal-tracing purposes since a signal can be injected into the input of a stage and its presence (or absence) checked at the output. A signal generator also incorporates an amplifier, in addition to the oscillator, to bring the signal levels up to those desired. Such signal-level output is adjusted by a knob on the front panel. Figure 22-8 shows a typical signal generator panel layout for AM and FM standard broadcast receiver applications.

Signal generators also have a means for modulating the carrier signal that is produced so that a simulated AM waveform can be obtained. The audio signal employed for modulation usually consists of a 400-Hz tone that modulates the carrier produced by the oscillator. As a rule, the modula-

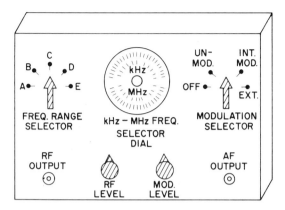

FIGURE 22-8. Typical signal generator panel layout.

tion is no more than approximately 30% and a switch is provided for turning the modulation on or off. This is necessary because, in some applications, a single signal is desired that utilizes only a narrow portion of the spectrum. On other occasions, the modulated signal is useful for producing an audible tone in the receiver during test procedures. In some units, the degree of modulation can be regulated.

The more elaborate signal generators also have an extended frequency range and, in addition, provide a means for frequency-modulating the carrier signal generated by the oscillator so that the instrument can also be employed in conjunction with FM receivers.

The most general use of the signal generator is for alignment purposes. When employed for alignment, the generator should be calibrated accurately if any doubt exists regarding its accuracy. Calibration necessitates checking the frequency of the output signal against some standard signal from a special calibrator device or from a broadcast station.

Alignment procedures for various AM and FM radio receivers as well as television sets differ cousiderably because of the characteristics of the receiver and the manner in which it was designed. For this reason, the service sheets for the receiver to be aligned should be referred to when this procedure (as well as tuner tracking) is necessary.

22-9. Sweep and Marker Generators

A sweep generator is a signal generator that has provisions for frequency-modulating the signal generated within the unit. Thus, the frequency of the signal produced by the generator is alternately swept above and below the frequency established by the oscillator within the unit, in similar fashion to the FM signal that is generated by FM alignment generators.

Sweep generators are useful for obtaining a visual indication of the band-pass characteristics of the RF or IF stages of FM and television receivers. They can also be used for observation of the band-pass characteristics of other electronic amplifying devices.

The sweep generator is applied to the IF stages, for instance, and will furnish an output signal that will vary in frequency sufficiently to span slightly below and above the IF band pass of the receiver. An oscilloscope is then placed across the detector circuit of the receiver under alignment. The horizontal sawtooth sweep of the oscilloscope is synchronized with the sweep generator so that the oscilloscope beam is swept across the face of the cathode-ray tube as the sweep generator sweeps from its lowest to its highest frequency. Since the gain of the IF stages is such that the various frequencies are amplified to a different degree, the beam of the oscilloscope will rise and fall to correspond to the amplification of the sweep signals being injected into the IF stages. Thus, the oscilloscope traces out the band-pass characteristics of the IF stages. Figure 22-9 shows a typical sweep waveform for a television receiver.

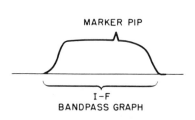

FIGURE 22-9. Sweep waveform.

To identify frequency points along this waveform, an additional generator known as a *marker* generator is employed. The marker generator is a single-signal AM type of generator such as is incorporated in the unit shown in Fig. 22-8. By injecting this single signal, a small section of a waveform on the oscilloscope is broken into a slight ripple known as a *marker pip.* This can be seen in the sweep waveform shown in Fig. 22-9. By varying the output frequency of the calibrated marker, the relative position of the pips that are produced will indicate the respective frequency of such points on the waveform. Hence, the bandwidth, as well as the points of increased or decreased amplification along the response curve, are visually indicated. Alignment can then be undertaken to correct the waveform so that it corresponds to the pattern illustrated in the service notes for the receiver under alignment. The pips supplied by the marker produced by a marker generator are equivalent guideposts of frequency along the band-pass waveform traced on the face of the oscilloscope tube.

22-10.   Dot and Bar Generators

Other generators are also employed for testing alignment and for design purposes. For television alignment and adjustment procedures, a cross-bar or dot-bar generator is also useful. Horizontal and vertical bars, or a combination of both, are available for adjusting the linearity of the television scanning process producing the picture. Adjustments to the receiver consist of manipulation of the vertical and horizontal linearity controls, in conjunction with the height and width controls, until both the horizonal and vertical bars produced on the television screen by the generator are equally spaced. For convergence adjustments in a color-television receiver, white dots are produced on the screen by the dot-bar generator.

Another type of generator useful for color-television servicing, maintenance, and adjusting is the color-bar generator that produces vertical bars having various colors. The proper color sequence of such bars is used as a reference for adjusting the color rendition of the receiver and for ascertaining the circuits that are involved when improper hues or colors are observed.

22-11. Industrial Instruments

The instruments used in the fields of industrial control, computers, automation, and other allied branches of electronics have the same basic purposes as the instruments previously described. In research and design laboratories the test equipment is usually of the more costly type since a greater accuracy is required than encountered in general tests, measurements, and circuit analysis. Oscilloscopes, for instance, may have a much wider frequency response range and sensitivity for accurate measurements of pulse signal durations, rise time, and repetition rates. Special circuits are often used to control more precisely the frequency of internal sweep.

Some oscilloscopes also have *triggered* sweep control, whereby the sweep signal is triggered to start at an exact time in relation to the signal waveform that will be observed. With automatic triggering, the internal sweep circuits of the oscilloscope will synchronize automatically with the frequency of the input signal with no manual adjustments of the sweep frequency control necessary. Thus, even though the input-signal frequency varies, the pattern on the screen will remain locked in.

Besides the basic type VOM and VTVM units discussed earlier, digital-type voltmeters are also encountered in industrial electronics. The digital voltmeter automatically displays the value of the voltage under measurement directly in large numbers. Some units have internal selection of ranges so that no range selection by switches is necessary. The test probes are placed across the voltage to be measured and the device senses the polarity and range and switches internally to the proper circuitry for visual display of the voltage values.

In some industrial-electronic measurement practices, counter-type instruments are also used, with visual display of the count performed by the

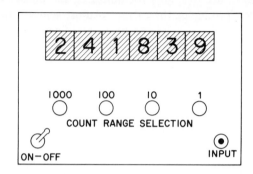

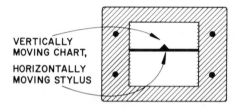

FIGURE 22-10. Counter and plotter.

beam-switching and display tubes described earlier in Chapter 21. (Additional counting devices are described in Chapter 24.)

A typical front-panel layout of a counter is shown at A of Fig. 22-10. Such a device will total the number of times certain industrial processes are repeated or will count pulse rates and other periodic changes that need to be evaluated. Selector knobs are provided for establishing a present quantity to be counted.

In industrial electronics there are various plotters used to provide indication of variations (both electrical and mechanical) along x and y axes, thus performing graphing functions. Most have an appearance as shown at B, where a horizontal bar holds a recording pen. One signal causes a vertical movement of a strip-chart paper roll, while the other input signal initiates horizontal movement of the recording pen. Both the chart movement and pen positioning respond to either a positive or a negative signal, thus being capable of shifting direction of movement. Hence, the chart can be driven up or down in response to changes in the variable, while the recording pen moves to left or right for changes in the second signal variable. Both chart and pen can be energized by any d-c source.

Plotters of this type are useful for providing visual displays of hysteresis loops, speed-versus-torque variations, stress-versus-strain relationships, temperature-versus-pressure values, force-versus-motion ratios, and any other X- and Y-axis graph plottings.

Review Questions

22-1. What type of measurements can be made with a multimeter?

22-2. Briefly explain the advantages of a 20,000-Ω-per-V meter as compared to a 1000-Ω-per-V meter.

22-3. How can a multimeter be used for continuity checking?

22-4. Briefly explain the differences between a VTVM and a VOM.

22-5. Does the standard VTVM employ a means for measuring current?

22-6. Briefly explain what measurements can be made with a capacitor checker.

22-7. What type of tube is employed in an oscilloscope?

22-8. What is meant by the "gun" of a cathode-ray tube?

22-9. Describe at least three oscilloscope controls, and explain their purpose and function.

22-10. To what oscilloscope terminals is a signal applied for observation on the screen?

22-11. Briefly define "persistence" with respect to the phosphorescent coating of a cathode-ray tube.

22-12. How can an oscilloscope be used as a voltmeter? Explain briefly.

22-13. What type of waveform is generated within an oscilloscope?

22-14. What is the purpose of a single-signal generator?

22-15. For what purpose are sweep generators used?

22-16. Explain the basic difference between an AM signal generator unit and a sweep generator.

22-17. What is the purpose of a marker generator?

22-18. Briefly explain the purpose of a cross-bar generator.

22-19. What are some of the refinements often included in industrial oscilloscopes?

22-20. Briefly define what is meant by *triggered sweep*.

22-21. Explain briefly the features of a digital voltmeter.

22-22. How does a counter visually display a total count?

22-23. List some uses for a counter unit.

22-24. What are some of the variables for which an *x-y* plotter is used?

Switching and Gating   23

23-1. Introduction

In addition to the amplification and modification of signals, there are numerous instances where signals or voltages must be switched or gated to other circuits or devices. This is particularly true in such branches of electronics as automation, computer systems, nuclear reactor control, electronic organs, closed-circuit commercial television, radar, and numerous other applications.

The purpose for the switching and gating systems is not to originate signals but rather to hold them in abeyance until such a time that they are to be applied to other circuits or devices. Included among the switching and gating circuits is the so-called *logic* type that is extensively used in computer systems but also finds applications in other branches of electronics where it is necessary to gate signals in or out of subsequent stages as required. This chapter covers the mechanical switching devices, as well as the electrical and electronic ones.

23-2. Electromechanical Switching

The relay at A of Fig. 23-1 is a basic electromechanical switch. It has a coil, with a metal core (forming what is known as a *solenoid*), and is part of the switching assembly as shown. When dc is applied to the solenoid, it energizes the latter and an electromagnet is formed. The electromagnet now pulls down the spring metal contact section that is part of the T_1. The other terminal T_2 attaches to the other contact point represented by the arrow. Thus, when the relay is energized, it closes the circuit for T_1 and T_2 and in consequence a closed switch condition is initiated when the current circulates in the solenoid.

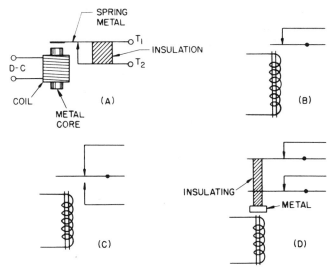

FIGURE 23-1. Electromechanical relays.

The relay shown at A is known as an *open-circuit* type since terminals T_1 and T_2 are in an open-circuit condition until the relay is energized. At B is shown a *closed-circuit* relay representation. Both the open circuit and the closed circuit shown are also known as a *single-pole single-throw switch* since only one spring metal section is thrown (the pole). At C of Fig. 23-1 is shown a single-pole double-throw switch. Here the movable section, the pole, can be thrown to the lower or the upper position; hence it is a double-throw type. At D is shown a double-pole single-throw-type relay. Here, two movable pole sections are thrown only to a single contact point, establishing the single-throw factor.

A typical commercial relay is shown in Fig. 23-2. This is a single-

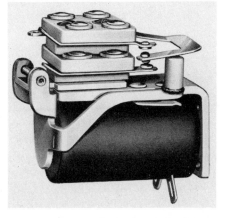

FIGURE 23-2. SPDT relay. (Courtesy Allied Control Co., Inc.)

pole double-throw relay. Terminal lugs at the bottom are for the application of the energizing voltage to the solenoid, while terminals at the rear and side of the top portion are for making contact with the switch section. Abbreviations are usually used for the switch identification. Thus, a single-pole single-throw switch is abbreviated SPST, a triple-pole, double-throw switch is indicated as TPDT, etc.

There are occasions in industrial electronics, computer systems, and other services when a-c or d-c signals are to be switched to a load circuit only when specific conditions exist. One of these conditions, for instance, could be that no power or signal waveforms of any type may be applied to the load circuit until two other voltages occur or are applied simultaneously. Another condition may be that the load gets continuous power (or signal waveforms) unless two other voltages or signals are present. One such circuit is shown at A of Fig. 23-3. Here, an a-c signal or a d-c voltage is impressed across the input terminals at the left as shown. In series with the load resistor are two open-circuit SPST switches controlled by the two relays.

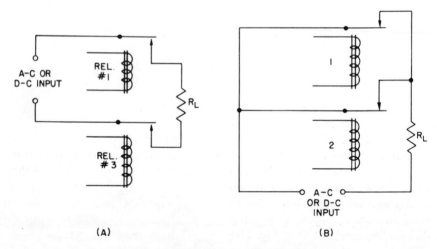

FIGURE 23-3. Coincidence relay switching.

If relay No. 1 is energized, the upper switch will close, but since the lower relay switch is open, on signal energy is applied to the load circuit. Similarly, if the lower switch is closed but the upper open, no energy is applied to the load. When, however, both relays are energized, a closed-circuit loop is formed and the input signal is applied to the load circuit. Thus, we need a *coincidence* of energy applied to both relays in order to close the circuit. Hence, this type of circuit is often referred to as a *coincidence switching system*. In computers, an electronic counterpart (nonmechanical) is utilized as described later. When such a coincidence circuit is used in computers, it is known as an AND circuit. The coincidence or AND must have both the upper *and* the lower switch closed for application of power to the load circuit.

Another coincidence relay switching circuit is shown at B. Here signal energy is applied to the load circuit continuously unless relay No. 1 *and* relay No. 2 are both energized simultaneously. As shown, normally closed relays are utilized here in contrast to the normally open relays shown at A.

23-3. Electronic Switching

The zener diode described in Chapter 14 can also be used as a switching device, as shown at A of Fig. 23-4. Here a zener diode is in series with the relay solenoid as shown. Upon the application of a negative signal having an amplitude sufficient to reach the zener region, the diode will conduct heavily and thus energize the relay. A normally closed relay is shown, though a normally open type could, of course, be used.

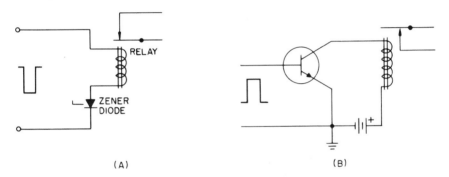

FIGURE 23-4. Solid-state switching.

A tube or transistor can also be used as a switching device. If the bias is at the cutoff region, a positive pulse can be used to trigger the relay switch as shown at B. Here there is no forward bias applied to the transistor; hence the transistor is functioning with a class B characteristic. Since class B operation is at the cutoff point, the solenoid of the relay is not energized and the switch is in the normally opened position as shown. When a positive pulse is applied, it makes the base positive with respect to the emitter, and thus the signal energy supplies the required forward bias for transistor conduction. Since the collector side already has the necessary reverse bias, the collector-emitter circuit conducts and the relay is energized.

The thyratron or the solid-state silicon-controlled rectifier can also be used in switching circuitry. A basic thyratron-tube switching system is shown in Fig. 23-5. Here, the negative potential applied to the grid circuit prevents the thyratron from conducting. Thus, no energy is applied to the load circuit. For rapid triggering of the energy to the load circuit, a differentiating input

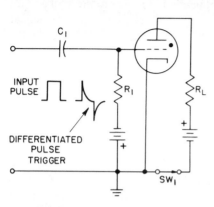

FIGURE 23-5. Thyratron switch.

is employed, consisting of capacitor C_1 and resistor R_1 as shown. These have the long time constant as previously explained, and the resultant sharp positive spike appearing at the grid causes thyratron conduction and the application of the d-c power to the load circuit. As detailed in Chapter 14, once the thyratron conducts, a negative potential at the grid is ineffectual in stopping conduction. Hence, the negative spike of the differentiated pulse will not remove the power applied to the load circuit. To cause the thyratron to stop conducting, the switch SW_1 must be opened to remove the anode-cathode potential.

An extensively used circuit capable of switching, gating, and counting is the Eccles-Jordan flip-flop. One of its primary applications is in the field of digital computers, where it performs arithmetic operations, stores numbers, and provides open- or closed-circuit conditions. Superficially, it resembles the multivibrator relaxation oscillator discussed earlier. Actually, however, it is not a generator or oscillator producing a continuous output but instead develops no output except when input pulses are applied.

The flip-flop circuit is also referred to as a *bistable* device because it has two stable states, known as the "off" and the "on" states. In the "off" state the circuit represents 0 and a so-called "set" pulse is applied to place the circuit in the "on" or 1 state. A reset pulse brings the circuit back to the 0 state.

A basic flip-flop circuit is shown in Fig. 23-6. For the 0 state, a positive pulse had been applied to transistor Q_2 (base), thus overcoming any forward bias that existed between emitter and base and driving this transistor into the nonconduction state. Note that the power-supply potentials are across resistors R_1, R_2, and R_3, and also across the series string of resistors R_4, R_5, and R_6. Forward bias for Q_1 is obtained from across R_6, and the voltage drops across the series resistors makes the top of R_6 minus with respect to ground. Thus, forward bias is applied to the base of Q_1.

When the reset pulse drove Q_2 into nonconduction, the high currents that flowed through R_4 and between the collector and emitter of Q_2 dropped, and only the small currents flowing through the rest of the series string (R_5 and R_6) remained. Thus, the negative potential applied to R_4 does not drop appreciably across it (or R_5) and appears at the base of Q_1, sustaining conduction for Q_1.

With Q_1 conducting, there is a high current flow through R_1 and a large

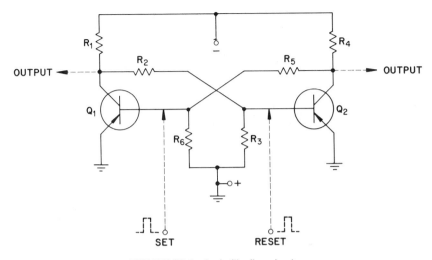

FIGURE 23-6. Basic flip-flop circuit.

voltage drop across it. Also, the conduction of Q_1 creates a low impedance between its collector and emitter, thus placing a low-resistance shunt across the series resistors R_2 and R_3. Consequently, the negative potential at the base of Q_2 is too low to permit conduction and this transistor remains at cutoff. The circuit will now remain in this off state until it is triggered into its other state.

If a *set* pulse of positive polarity is now applied to the base of Q_1, it overcomes the negative forward bias, thus driving Q_1 into cutoff. The voltage drop across R_1 declines, and Q_1 no longer acts as a low-impedance shunt across R_2 and R_3 because in the nonconducting state it behaves as an open circuit. Now Q_2 obtains the necessary forward bias and starts to conduct. During this time it acts as a low-impedance shunt across resistors R_5 and R_6, which, in conjunction with the increased voltage drop across R_4, reduces bias to Q_1 and keeps it in the nonconducting state. Thus, the other state (1) has been reached and the circuit will again maintain this state until a reset pulse is applied.

When the flip-flop changes its state, the current through one transistor drops and the current in the other rises. Thus, the *change* of current through the collector resistor affects the collector voltage, thus producing an output pulse of one polarity from Q_1 and another pulse of opposite polarity from Q_2. Depending on the particular usage, the output is obtained from one of the collectors and applied to successive flip-flop stages, as discussed more fully in Chapter 24.

An indicator light can be included in one of the collector circuits so it will light during the "on" state to indicate a representation of the number 1. The flip-flop will store this number representation for as long as power is applied to the circuit. A series of flip-flop stages will thus store a series of counts and accumulate them as additional pulses trigger the various stages as described in Chapter 24.

Just as an indicator light is turned on and off, so can a relay be tripped by wiring the latter into either collector circuit of the flip-flop. A transistorized flip-flop using dual relays is shown in Fig. 23-7. Normally open relays are shown, though normally closed relays could also be used. Also, one relay could be a normally closed type and the other a normally open type for meeting specific switching needs.

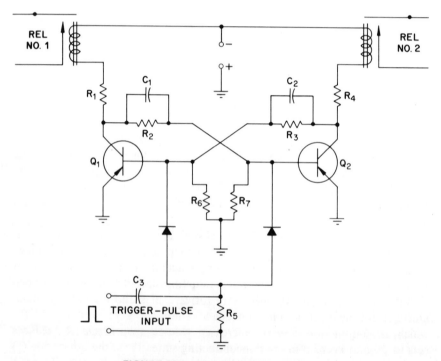

FIGURE 23-7. Flip-flop relay switching.

For purposes of analysis, assume the flip-flop shown in Fig. 23-7 is in its "on" state. With the normally open relays shown, relay No. 1 remains open since the transistor circuit Q_1 is nonconducting. For the second relay in the collector circuit of Q_2, however, the current circulating in the solenoid closes the relay. Now when the flip-flop stage is triggered to the "off" or reset state, the current flow in Q_1 closes relay No. 1 and at the same time Q_2 reaches cutoff, and relay No. 2 assumes its normally open condition.

If the relays are to perform alternate switching functions, that is, with one providing an open circuit during the time the other provides a closed circuit, then relay No. 1 could be a normally closed type and relay No. 2, a normally open type or vice versa, depending on switching requirements.

In industrial electronic switching, plate-sensitive relays (relays very sensitive to small changes of current values) are employed and often these in turn open or close larger relays where higher power is to be switched ultimately. The output terminals of the relays shown in Fig. 23-7 can also be applied to silicon-control rectifiers for triggering these devices and thus switching high power to other circuits or devices as required.

The operation of the flip-flop of Fig. 23-7 is similar to that of Fig. 23-6, except that series diodes have been included in the base circuits so that a common input performs functions of both set and reset. The diodes only permit the entry of positive pulses to the base elements and also isolate the bases from each other. When a positive pulse is applied (now termed a *trigger* pulse), it will appear at both base elements. Since, however, one base is already positive, it will have no effect. For the base that is negative, though, the forward bias is eliminated and that transistor is driven into cutoff. Thus, successive pulses alternately turn the flip-flop to its 0 state and to its 1 state.

Capacitors C_1 and C_2 permit faster switching rates since they are in series (to ground) with interelement capacitances and thus effectively decrease total capacitance factors and charging rates.

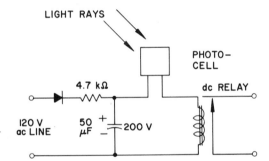

FIGURE 23-8. Photocell relay circuit.

Electronic switching can also be done by using light to change the resistance of a photocell, as shown in Fig. 23-8. The light-dependent resistor, a photoelectric transducer, has already been described in Chapter 21. In this unit, however, the resistive change for variations in light intensity do not have the rapid switching capabilities of transistor and other solid-state photoelectric devices. Thus, in industrial-electronic applications where fast switching rates are essential, the modern rapid-switching photocells are extensively employed.

For the circuit of Fig. 23-8 a half-wave supply is sufficient to provide the necessary dc to operate the photocell and the relay. The latter must be a sensitive type, with a fairly high impedance (10 kΩ or so, depending on the type of cell used).

Photosensitive transistors can also be employed in the same applications as photodiodes. The phototransistor functions on the principle of the conductivity of germanium being altered by incident light energy. This permits

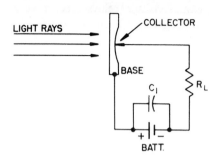

FIGURE 23-9. Photoelectric transistor.

the fabrication of an extremely small photocell that operates on the point-contact transistor principle. Instead of the emitter acting as an input circuit for an electric signal, however, the emitter input is a beam of light that causes current flow. The basic circuit is illustrated in Fig. 23-9.

The concave-shaped transistor is mounted in a metal shell that forms the base connection. The electron flow from the negative side of the battery goes through the load resistor and to the collector and then to the base. The phototransistor has only an extremely small area of light-sensitive surface. The light area centers around the section opposite the collector contact point, and the light-sensitive area is usually 0.01 in. in diameter.

23-4. Gating Circuits

Gating circuits are usually employed to regulate the interval of time during which signal information is either applied to another circuit or device or prevented from reaching another circuit or device. As such, they can also be considered as switching units as with the previous systems discussed. Switching circuits, however, are often used to turn devices on and off in addition to handling or transferring signals. The gating circuits are primarily concerned with the handling of pulse or square-wave signals and can be used to gate such signals in or to gate them out as required. A typical diode coincidence gate is shown in Fig. 23-10. Here two solid-state diodes are used and in the absence of a signal at the input terminals T_1 and T_2 both diodes conduct fully because they are in parallel with the load resistor and source voltage. The electron flow for both diodes is from the ground terminal upward, and the electrons for both flow through the load resistor in the direction shown by the arrow. Because of the current flow through the load, there is a large voltage drop across the latter, establishing a steady-state d-c output in the absence of a signal input.

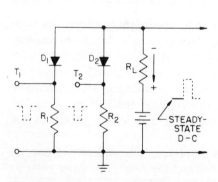

FIGURE 23-10. Diode coincidence gate.

When a pulse of sufficient amplitude is applied to terminal T_1, its negative-polarity voltage drop across R_1 overcomes the battery potential across the series diode network composed of D_1 and R_1. Since this potential is opposite in polarity to the battery potential, it stops conduction of D_1. Diode D_2, however, conducts and hence only a slight change in voltage drop occurs across the load resistance. The same small change would occur if a negative pulse were applied to terminal T_2 only.

When input pulses appear simultaneously at terminals T_1 and T_2, both diodes stop conducting and hence current flow through the load resistor drops to zero. Now the supply potential from the battery predominates, and a rising positive polarity representative of the signal-amplitude change occurs at the output. The output signal will have a flat-top waveshape equaling the duration of the pulse input signal. Instead of a pulse waveform input, steady-state d-c potentials can also be applied simultaneously to the input terminals to cause a change in output potential.

The coincidence, or so-called AND gate, can be made up of two or more transistors as well as several input terminals as desired. A typical three-input AND gate is shown in Fig. 23-11. Here reverse bias for the collectors of the transistors is provided by the battery in series with the load resistance as shown. Forward bias, however, is omitted for each transistor. Consequently, each transistor acts as a class B system since the absence of forward bias causes a cutoff state to prevail. Only when forward bias is applied will the high-potential barrier in the collector side permit current flow through the transistor. Hence, the gating signal must have a negative polarity so the base will be negative with respect to the emitter.

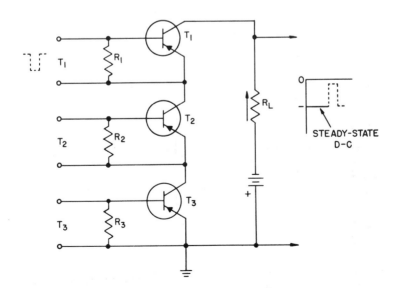

FIGURE 23-11. Three-input AND gate.

When a negative voltage or pulse signal is applied to the input terminal T_1 as shown by the dotted outline, the upper transistor has the necessary forward bias to conduct but is unable to do so because its conduction path is also through the collector and emitter circuits of the lower transistors. Hence, if no signals are applied simultaneously to the lower terminals, no conduction can occur through any of the transistors. Similarly, the application of signals to T_2 only, or T_3 only, will not cause conduction. When negative-polarity signals are applied simultaneously to all three terminals, the necessary forward- and reverse-bias conditions prevail and all transistors conduct. The current flow that now occurs through the load resistance causes a voltage drop to occur across the latter and hence the normal steady-state d-c output declines from its high negative value to a lower value. This represents a positive change and thus the output signal has a polarity opposite to that of the input signal.

In computers, the phase reversal of a signal across a circuit is sometimes referred to as NOT characteristic. This means that the phase of the output signal is *not* the same as that of the input signal. Thus, the circuit shown in Fig. 23-11 is sometimes known as a NOT-AND circuit and hence commonly called a NAND circuit.

Another type of gate circuit is that known as the OR gate illustrated in Fig. 23-12. This circuit will have an output developed when a pulse input or other signal waveform is applied to one, *or* two, *or* all inputs at the same time. For the circuit shown, the necessary reverse bias for the collectors of the three transistors is supplied by the battery in series with the load resistor R_L. In contrast to the three-input AND gate shown earlier in Fig. 23-11, the

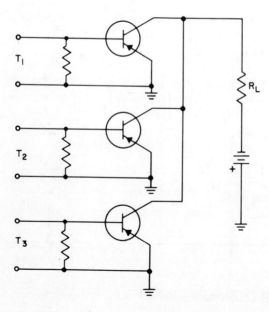

FIGURE 23-12. Three-input OR gate.

collectors for the OR gate are in parallel instead of in series as for the AND gate. Again, no forward bias is applied to any transistor. Thus, if a signal of negative polarity is applied to terminal T_1, it will make the base of the top transistor negative with respect to the emitter. Thus, the forward-bias requirements are supplied by the input waveform, and this transistor then conducts, causing a voltage to appear across the load resistor. Similarly, if a signal of negative polarity is applied to terminal T_2, the center transistor conducts and an output signal is developed across the load. If signals are applied to all inputs at one time, an output signal is again developed. Thus, an output is developed for a signal applied to T_1, *or* to T_2, *or* to T_3, *or* to all three inputs.

When transistors are used to form OR circuits and a phase inversion occurs (as with the common emitter circuit), a NOT function is again obtained as with the NAND circuits. Thus, the NOT-OR combination is usually referred to as a NOR circuit. Some of the logic-gate functions are illustrated in Fig. 23-13, where the standard symbols are shown.

At A is shown a NOT circuit wherein the polarity of the input signal is reversed at the output. The same logic would apply if a negative pulse were entered. In such an instance, a positive-polarity pulse would appear at the output. For the AND circuit at B, coincidence must prevail for input pulses to obtain an output. Thus, since no pulse is present at the lower input while the second pulse appears at the upper input, the output waveform does not contain the second pulse.

At C a typical OR input is shown. The output consists of any input at the top terminal or the lower terminal, or both. The NAND circuit is shown at D, with a reversal of polarity at the output. Since no coincidence prevails at time t_1 but does occur at t_2, an output pulse appears only at t_2. For E the standard NOR-gate function is illustrated, with an inverted output for any input pulse at either or both input terminals.

Negative-polarity input pulses could be used for any of the gates shown, though the NOT function for some would then produce a positive-polarity output pulse. Input signals could be of long duration, or dc to perform certain tasks. At F, for instance, the lower input terminal of the AND circuit has a pulse with a duration equal to three of those at the top input terminal. Thus, the lower pulse, by providing coincidence of voltages at the inputs, will permit three pulses to appear at the output. Thus, it is in effect a *gating-in* circuit. Conversely, it could also function as a *gating-out* circuit since any zero or negative potential at one input will inhibit any output since coincidence does not prevail. (The fourth pulse at the top input at F, for instance, can be considered as being gated out and hence does not appear at the output.)

Logic circuits of this type are widely used in computer systems and other electronic circuitry where switching and gating practices must follow logical sequences. In digital computers various combinations of such gates are used

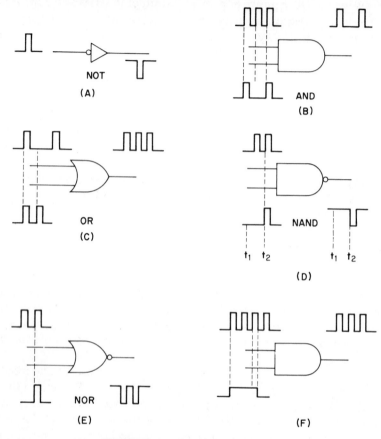

FIGURE 23-13. Logic gate functions.

for proper routing of numerical values, for initiating readin and readout procedures, and for performing other arithmetical and data-processing functions.

Review Questions

23-1. Draw a schematic of a TPDT relay.

23-2. Explain how relays can be used to perform coincidence switching. Reproduce the schematics of the two types discussed in the text.

23-3. Explain how a zener diode may be used in a switching circuit.

23-4. Reproduce the schematic of a flip-flop circuit, and indicate which transistor is conducting when the device is in the "1" state.

23-5. For the circuit shown in Fig. 23-7, which relay is closed when the flip-flop is in the "1" state?

23-6. Briefly explain the functions of flip-flop terminals designated as *set*, *reset*, and *trigger*.

23-7. If a *set* pulse has been entered, what occurs if another *set* pulse is applied to the same terminal? What would occur if a trigger pulse were now applied?

23-8. Reproduce the schematic shown in Fig. 23-10, and modify it to accept positive pulse inputs. Briefly explain how this circuit functions.

23-9. Why is a coincidence gate also called an AND circuit?

23-10. Draw a schematic of a two-input transistor AND gate, using *NPN* transistors. Explain how the circuit functions, and show the polarity of the input pulses as well as the polarity of the output pulses.

23-11. Draw a two-input transistor OR gate using *NPN* transistors. Explain how the circuit functions, and illustrate the type of waveform applied to the input and that produced at the output.

23-12. Briefly explain what is meant by a NOT circuit and a NOR circuit.

23-13. Designate which of the following circuits forms a NOT circuit: common base, common emitter, common collector.

23-14. Using numbers to represent pulses, what output would be obtained from an AND circuit if the input to one terminal is 10111 and the input to the other terminal is 00111?

23-15. What would be the output from an OR circuit if the same input pulse signals were applied to it as in Problem 23-14?

23-16. Explain how a long-duration pulse can be used with an AND circuit for gating in several pulses of a pulse train applied to one of the terminals.

Miscellaneous Circuits 24

24-1. Introduction

In the preceding chapters the fundamental aspects of a number of circuits and components have been covered with respect to specific categories. In amplifiers, for instance, both the voltage and power amplifiers were analyzed and illustrated, and the characteristics relating to the broad field of electronics were covered. There are, however, other types of amplifiers used in automation and industrial control that do not have the linear characteristics of those previously discussed and hence are not suitable for the reproduction of sound or the amplification of carrier signals. Instead, such special amplifiers are primarily used to control the amount of power applied to industrial hardware, such as motors, control systems, and other units such as power supplies. Also, there are numerous occasions in industrial electronics where a specific frequency of a signal, or the pulse-repetition rate, must be altered by either increasing or decreasing the frequency to meet specific requirements. All of these circuits are utilized in some phases of automation, computer systems, nuclear reactor control, closed-circuit commercial television, radar, and numerous other applications. In this chapter the circuits and operating principles of such amplifiers and frequency modifiers are covered to round out the circuit and component discussions in the preceding chapters.

As material related to industrial control is also contained in Chapter 21, Chapter 21 should be reviewed for a more comprehensive coverage of the subject.

24-2.   Control Amplification

In industrial-electronic devices, it is often necessary to vary a large amont of
electric power in accordance with variations that occur at low-power levels.
This is essentially an amplifying process similar to that performed by vacuum
tubes and transistors. Instead of using the latter, however, a magnetic amplifier is employed because of its large power-handling capability and its simplicity. A basic magnetic amplifier circuit is shown in Fig. 24-1. The heart of the
magnetic amplifier is a device that resembles a transformer but actually
functions as a variable inductance. A coil is wound on each of the three legs
of the core as shown. Inductance L_2 acts as a control winding, and sufficient

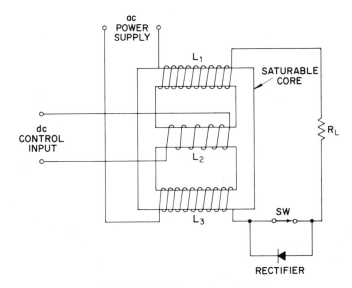

FIGURE 24-1. Magnetic amplifier.

current is circulated in this coil for operation around the saturation level of
the core material in a manner similar to the swinging choke described in
Chapter 14. The a-c power that is to be controlled is applied to the terminals
shown, one of which is connected to coil L_1 in series with coil L_3. Each of
these coils is wound in a direction opposite to the other so that the fields
established by the individual coils are effectively canceled. This lessens the
influence of these coils on the core and also minimizes interaction between
these coils and the control winding coil L_2. The a-c power is fed through
these coils and is applied to the output load, shown as a resistor R_L in
Fig. 24-1. The actual load can consist of a large motor or other device in
which the amount of a-c power that is applied must be regulated and controlled.

   When a control voltage is applied to coil L_2, a certain degree of core saturation is produced, depending on the amount of current flow created by the control voltage. With the core at a certain saturation level, the inductances of L_1 and L_3 are lower in value than would be the case with no core saturation. In consequence, the inductive reactance is also lower than would normally be the case. Because inductances L_1 and L_3 are in series with the load circuit and the power source, the power factor is altered by the amount of inductance present. Hence, the power applied to the load will be determined by the amount of inductance that is in series. If the voltage applied to the control winding is increased, more current flows through the latter and the core flux density increases toward the saturation level. Permeability (and inductance) decreases and hence more power is applied to the load circuit. For a reduction of the control voltage, less current circulates through the control winding and flux density drops to a level farther below the saturation level. Consequently, inductance values of L_1 and L_3 increase and less power is applied to the load circuit. From the foregoing, it is evident that variations in the low d-c input voltage produce high-value variations in the output and hence this device is an amplifier.

The magnetic amplifier is sometimes known as a control device since a large amount of power can be controlled by a relatively low-power value. If d-c output is required, the switch shown at the bottom of the load resistance can be opened so that the rectifier is in series with the load circuit. Because of the rectifying action, pulsating dc will be applied to the load resistance instead of ac. The pulsating dc can be filtered to obtain relatively pure dc if such is required.

As with tube and transistor amplifiers, feedback can also be employed with the magnetic amplifier. The feedback can be regenerative or degenerative as required. Thus, a portion of the amplified signal is fed back to the magnetic amplifier in order to alter its characteristics. With positive feedback, regeneration occurs and the sensitivity as well as the gain of the magnetic amplifier is increased. When negative feedback is used, degeneration occurs and the amplified signal waveform is more faithfully reproduced, although, as with tube and transistor amplifiers, the overall gain is decreased in proportion to the amplitude of the signal that is fed back.

A typical magnetic amplifier with feedback is shown in Fig. 24-2. Here a four-legged type of core is employed in contrast to the three-legged type shown earlier in Fig. 24-1. The additional core leg permits the inclusion of a feedback winding marked L_4 in the illustration. The bridge rectifier provides an output consisting of pulsating dc that can be filtered, as with conventional power supplies, to provide relatively ripple-free dc.

The operation of the amplifier can be more easily understood by assuming instantaneous values of ac applied to the input terminals T_1 and T_2. Assume that the instantaneous value of the ac is negative at terminal T_1 and positive at terminal T_2. With such a polarity at the input terminals, electron flow is from terminal T_1 and through the inductance L_1 and L_2, to the junction

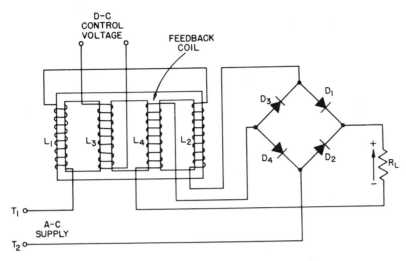

FIGURE 24-2. Magnetic amplifier with feedback.

of the solid-state rectifiers D_1 and D_3. The electron flow must complete its circuit to the positive terminal T_2; hence it flows through diode D_3, and through the feedback coil L_4 to the bottom of the load resistance R_L. The electron flow through the load resistor establishes a voltage with a polarity as shown, and the electron flow than completes the circuit to terminal T_2 by flowing through diode D_2.

When the a-c signal at the input terminal changes polarity, a negative potential appears at T_2 and a positive potential at T_1. Now, electron flow is from terminal T_2 to the junction of diodes D_2 and D_4. Since terminal T_1 is positive, the direction of the electron-flow path must be toward this terminal. Hence electrons flow through diode D_4, through the control winding L_4, and again to the bottom of the load resistor R_L. Again the direction of electron flow through the load resistor is the same as before, producing the second pulsating d-c alternation. Now the electron flow is through diode D_1 and through inductors L_2 and L_1 to terminal T_1. (Note that when electrons arrived at the junction of diodes D_1 and D_2, it would appear that either diode could conduct. The electrons will not flow through D_2, however, because the return path must be to the positive terminal T_1. Hence, conduction for this instantaneous ac is through D_1.)

If the feedback is degenerative, the circuit can be changed to a regenerative one simply by transposing the leads from the feedback coil L_4. If the circuit has been set up to be regenerative, the feedback current fields will cause the current flow through the control winding to increase and hence a greater control change occurs. Thus, an increase in the current through the control winding will raise the flux density to the near-saturation point, and the saturation level increase over and above that which would occur without feedback provides a greater change in the amplified power output. If the current through the control winding decreases, the signal energy at the load circuit decreases and hence the current through the feedback coil L_4 also decreases, causing a reduction in flux density to a greater degree than would occur for a change of current in the control winding L_3 without feedback.

Similarly, if the feedback-coil polarity is such that the circuit is degenerative, the current flow through L_4 would establish magnetic lines of force that oppose the control winding L_3. Consequently, the core saturation level would be decreased below the level that prevails for current through the control winding L_3 without feedback. For instance, a decrease in the control current flow through L_3 would mean that the flux density of the core also decreases to a given level. The feedback current through L_4, however, also causes an additional decrease, with the end result that the flux density would not decline to as low a level as would be the case without degeneration.

Another type of power-control amplifier used in industrial electronics is the so-called *Hall effect* amplifier shown in Fig. 24-3. The heart of the amplifier is a semiconductor crystal slab, as shown at A, composed of intermetallic components such as indium arsenide and indium antimonide. These

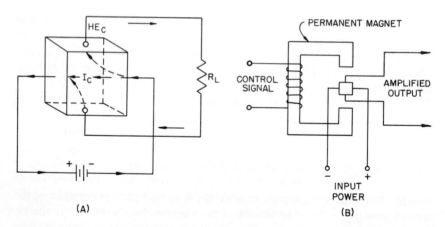

FIGURE 24-3. Hall-effect amplifier.

components have *magnetoresistive* characteristics and have what is known as a high *mobility* of the conduction material. With ordinary conductors the Hall effect is very slight and the voltages set up are insufficient in amplitude to be of any practical use.

As shown at A, the conductive slab has attached to it two independent circuits. The circuit containing the battery or other supply causes electron flow as shown by the arrows and identified as I_C, *control current*. When a magnetic field is applied to the conductive crystal slab so that the magnetic lines of force are at right angles to the slab, the electrons are forced aside as shown by the dotted outline at A of Fig. 24-3. This results in an excessive amount of electrons on that side (negative charge), while on the other side there is a deficiency of electrons creating a positive charge. Thus, the magnetic field creates a potential difference across the edges of the crystal opposite to the potential difference applied by the battery. Thus, electrons flow up through the crystal slab and this Hall-effect electron flow (HE_c) goes down through the load resistor in the direction shown by the arrows, with its return path to the plus terminal of the supply source. Hence this characteristic can be employed in industrial measurements of magnetic field strength, instruments, and power-control amplification.

The basic amplifier circuit is shown at B of Fig. 24-3. With the permanent magnet arrangement as shown, the high magnetic-flux density is impressed across the crystal slab as shown. The control signal is applied to an inductor wound around the permanent magnet. Thus, the magnetic flux between the pole pieces of the permanent magnet can be increased or decreased by the control-signal voltage applied to the coil. If the applied signal voltage produces a magnetic field that aids that of the magnet, the increase in magnetic flux between the pole pieces will cause a greater deflection of the electron stream within the slab, with an increase in the current flow through the load circuit. If, however, the voltage applied to the coil is reversed in polarity, the magnetic field produced by the coil will oppose that of the field of the magnet and hence the flux density between the pole pieces decreases. Thus, there is a decrease in the electron-stream deflection and the current through the load circuit will decrease. As with other amplifiers, a signal of low power applied to the coil will produce a signal of considerably greater power at the output. The input signal can be either dc or ac to produce an amplified output signal of either type as required.

24-3. Signal-Frequency Multiplication

In AM, FM, and TV broadcasting, as well as in radar and other electronic applications, it is often necessary to increase the frequency of a particular RF signal. A common frequency multiplier used is the Class C amplifier previously discussed. Figure 24-4 shows the Class C triode amplifier in a frequency-doubler application. Here an oscillator generating a frequency of 2 MHz is applied to the Class C doubler circuit so that the oscillator frequency may be increased to 4 MHz, as shown.

Oscillators, Class C amplifiers, and other similar devices operate on the nonlinear portion of the transistor's (or tube's) characteristics and, hence, generate a high order of harmonics in addition to the fundamental frequency. In consequence, a Class C amplifier can be employed that is tuned to one of the harmonics of the fundamental so that is will develop across its resonant circuit a primary signal of the harmonic rather than of the fundamental impressed at the input. Such is the case with the doubler circuit shown in

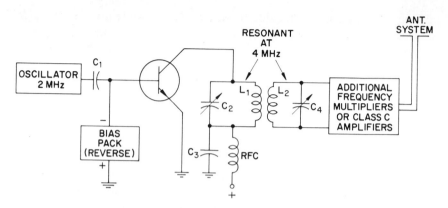

FIGURE 24-4. Multiplication of signal's frequency.

Fig. 24-4. The oscillator is generating a signal having a frequency of 2 MHz but the resonant circuit of the doubler, composed of C_2 and L_1, is tuned to 4 MHz. Because the harmonics of the oscillator are lower in amplitude than the fundamental-frequency signal, the harmonics provide less driving power to the base input of the transistor doubler circuitry. Consequently, the doubler circuit lacks the high efficiency of the single-frequency Class C amplifier.

The input resonant circuit to a subsequent stage (L_2 and C_4) is tuned to the multiplied-frequency signal of 4 MHz. Additional doubler stages can be employed to increase the signal frequency to a higher value if desired. The Class C amplifier will also function as a tripler, multiplying the input frequency three times. As a tripler, however, the efficiency is increasingly lower than in the normal Class C amplifier or the doubler circuit.

Frequency multipliers, because of the difference in the frequency of the signals applied to the input and those generated at the output, are not subject to undesired oscillations and hence rarely require neutralization as is the case with triode Class C amplifiers.

24-4.   Frequency Dividers

Often, in industrial electronics, the pulse frequency (repetition rate) must be divided for purposes of industrial control, synchronization of oscillator circuits, counting, timing, pulse-rate comparisons, and the operation of circuits operating at lower frequencies than the pulse repetition rate. One method for accomplishing frequency division is by use of the relaxation oscillators described in Chapter 17. If, for instance, a blocking oscillator is employed, its free-running frequency can be set at one-half of the repetition rate of the pulses applied to its sync input. This is shown in Fig. 24-5. Here, assume the sync input pulses have a repetition rate of 6000 pulses per second (pps). The blocking oscillator (whether a tube type or transistor), however,

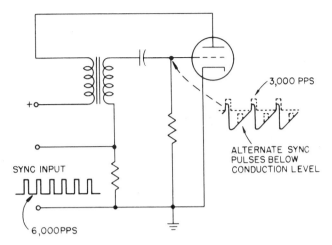

FIGURE 24-5. Blocking oscillator frequency division.

has its characteristics designed so that it has a free-running frequency of 3000 pps. Hence, alternate sync input pulses will keep the oscillator in synchronization as shown by the grid-signal waveform in Fig. 24-5. Every other sync pulse will drop below the tube conduction level and hence will be ineffectual in causing tube conduction. Only the pulses near the free-running frequency are effective in causing tube conduction and thus permitting synchronization. Thus, this system acts as a pulse-repetition-rate divider since the output pulses are synchronized by pulses having a repetition rate twice that of the output.

The flip-flop circuits discussed in Chapter 23 are also capable of pulse-repetition-rate division and are also utilized in computers and industrial-electronic applications. As shown earlier in Fig. 23-7, positive pulses can be used to trigger the circuit into alternate states of zero and one. For contrast, however, we shall discuss another design, wherein the input diodes are reversed, thus operating so that negative pulses are the gating signals as shown in Fig. 24-6. Thus, when a flip-flop is in its zero state and triggered to its one state, a positive pulse develops at the output of the flip-flop, but such a pulse polarity has no effect on a second-stage flip-flop because only negative pulses are instrumental in triggering. When a second pulse is applied to a flip-flop stage, it will trigger the latter into the zero state again, while at the same time a negative output pulse is developed. The latter, when applied to the next flip-flop stage, will trigger it to its one state.

Assume, for instance, that 5 flip-flop circuits are connected as shown in Fig. 24-7. If 2 pulses are entered into the first flip-flop at the left, the output negative pulse will trigger the second stage to its one state. The latter will then produce a positive output pulse, but again such a pulse polarity has no effect on the third stage because of the base diodes. Upon the application of

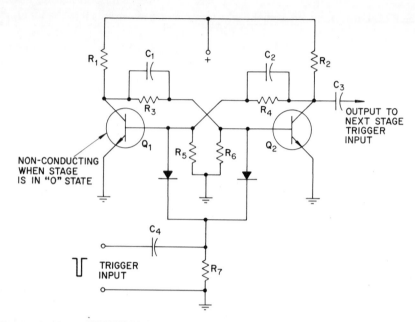

FIGURE 24-6. Flip-flop with negative-signal trigger.

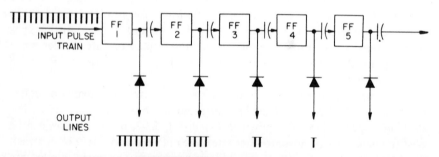

FIGURE 24-7. Flip-flop scaling.

a third pulse to the first flip-flop stage, it will again be triggered into its one state, but again the positive output pulse has no effect on the second flip-flop. Upon the application of a fourth pulse to the first stage, however, the negative output pulse that is applied to the second-flip-flop will trigger the latter to its zero state, and hence this second stage will send a negative pulse to the third stage that is then triggered to the one state.

Thus, it is obvious that successive pulses applied to the first flip-flop will produce one negative output pulse for every two entered. So, if 4 are entered into the first stage, 2 pulses will be produced at the output. If 8 are entered, 4 output pulses will be produced, and if 16 are entered, 8 will be available at the output. Hence, a series of flip-flop stages has the ability to *count down*, or *scale*. As shown in Fig. 24-7, when 16 pulses are entered into the first flip-flop, the 8 that are produced at its output terminal are also applied to the input of the second flip-flop. (The diodes in the output leads also prevent any positive pulses from appearing.) Since 8 pulses are applied to the second flip-flop, 4 pulses are produced at its output and applied to the third flip-flop. The third produces 2 output pulses that are applied to the fourth flip-flop, and the latter produces one output pulse as shown.

This pulse-repetition-rate dividing characteristic of a series of such flip-flop stages is not only a useful characteristic in digital computers but also in industrial control applications. As discussed in Chapter 23, relays can be placed in the output of circuits in the various stages to perform countdown functions and close or open circuits at predetermined intervals. For industrial applications, the input pulses to the first flip-flop stage can be obtained from photocells or other transducers described in Chapter 21. The photocell, with a suitable light beam impinging on it, will produce a series of pulses when the beam of light is interrupted periodically. Such pulse signals, when applied to the series flip-flop stages, can be used in control applications for purposes of counting, fabrication, sorting, and various other processes.

The flip-flop stages shown in Fig. 24-7 operate on what is known as the *binary* system, sometimes referred to as the "8-4-2-1 system." Here the scaling is in terms of 2 raised to some power. For instance, 2 to the zero power equals 1; 2 to the first power equals 2; 2 to the second power equals 4; 2 to the third power equals 8; 2 to the fourth power equals 16, etc. On occasion, however, it is preferable to have the division in terms of our decimal system having a base 10 instead of a base 2. This is done by appropriate circuit changes and modifications that usually consist of using feedback loops within four flip-flop stages so that such a four-stage group will trigger to zero at the count of 10. When this is done, an equivalent decimal system is established as shown in Fig. 24-8.

The flip-flop stages with circuit modifications, and in the arrangement shown in Fig. 24-8, are known as *decade counters*. If 10 pulses are applied to the input terminal of the first flip-flop, an output pulse would be procured

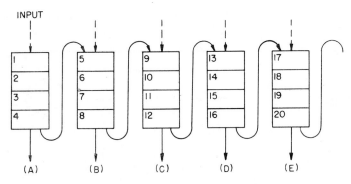

FIGURE 24-8. Flip-flop decade counters.

from terminal A, and such an output pulse will also be applied to the input of the fifth flip-flop as shown. At the same, the first four flip-flop stages will clear to their zero stage. Thus, this series of flip-flop stages will produce an output from any decade group of four, for 10 pulses that enter at the top of the flip-flop decade. For an output to be procured from terminal B, 100 pulses would have to be applied to the input terminal of the first decade counter. With this number applied to the input, there would be 10 pulses procured from A and since these are also applied to the second decade counter, the input to the fifth flip-flop would receive 10 pulses, and one output pulse would be obtained from B. At the same time, the output pulse from the eighth flip-flop would also be applied to the ninth. Thus, the system operates as a decade device and divides down by 10 for each successive flip-flop decade group.

The output terminals from the individual decade counters can be indicated visually by utilization of the indicators described in Chapter 21. As with the flip-flops previously discussed, any numbers that are entered are stored for as long a time as power is supplied to the circuits. The various stages can be cleared of the stored number, and in computer applications the numbers can be channeled to readout devices such as electric typewriters or printers.

The beam-switching tubes can also be utilized as decade counters or dividers by placing them in series as shown in Fig. 24-9. As discussed in Chapter 21, when pulses are applied to such tubes, the internal construction is such that there is successive switching from zero to nine. At the application of the tenth pulse, the tube switches to zero and an ouput pulse is produced. Thus, the tube is basically a 10-to-1 divider. As shown, if 400 pulses are applied to the first beam-switching tube at the left, 40 pulses would be produced at the output and applied to the second beam-switching tube. At the output from the second tube, 4 pulses would be produced and also applied to the input tube. There would be no output from the latter until the count of 10 was reached.

The beam-switching tubes are compact and versatile and will operate at frequencies ranging up to several megahertz. As many tubes as necessary can, of course, be connected together in sequence as shown in Fig. 24-9, or in reverse order for counting purposes.

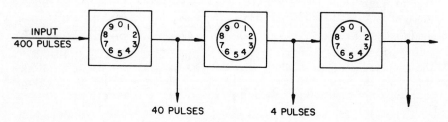

FIGURE 24-9. Beam-switching tube divider.

Another dividing system using a blocking oscillator is shown in Fig. 24-10. Here an integrator circuit is employed, the output of which is applied to the input of a blocking oscillator. As detailed in Chapter 16, an integrator circuit has a long time constant in relation to the pulse duration. Thus, successive pulses are required to charge the integrator capacitors to an amplitude equal to the peak pulse amplitude. For purposes of analysis, assume that the firing level of the blocking oscillator requires a signal amplitude at the input of approximately 8 volts. The incoming pulses have a 10-volt peak level as shown. Upon arrival of the first pulse identified by (x), the integrator capacitors will start to charge for the duration of the incoming pulse. Because of the long time constant, however, the charge that builds up is only a small

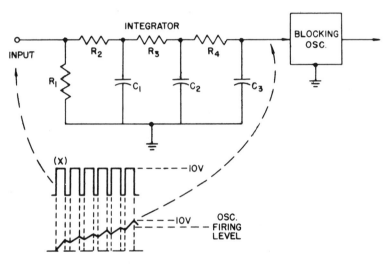

FIGURE 24-10. Integrator-oscillator divider.

percentage of the peak voltage of the incoming pulse. During the interval between pulses, the capacitors will discharge through the resistive network composed of R_1 and R_2. Since the duration between the pulses is shorter than the pulse width, however, the capacitors only discharge a small percentage of the charge that they obtained when the pulse was present. Thus, the second pulse will cause the integrator capacitors to charge an additional amount and during the interval between the second and third pulses there again is a slight discharge, as shown in Fig. 24-10. Eventually, the incoming pulses will build up an amplitude sufficiently high to reach the firing level of the blocking oscillator and hence will cause blocking-oscillator conduction and thus synchronize the free-running frequency of the blocking oscillator.

Thus, if this device has a free-running frequency of one-sixth the repetition rate of the input pulses, a six-to-one division occurs. When the blocking-oscillator tube or transistor conducts, the integrator capacitors discharge through the low impedance of the conducting tube or transistor.

In this circuit the amount of pulse-repetition rate division depends on the time constant of the integrator circuit, the pulse duration, repetition rate, and similar factors. An adjustment of the bias or the free-running frequency of the blocking oscillator will also influence the division rate. This system is utilized in the vertical sweep systems of television receivers and some industrial control applications.

For the system shown in Fig. 24-10, the input must consist of a pulse train without long time intervals between groups of pulses. Any appreciable time interval will result in the discharge of the integrator capacitors through the shunting R_1 and R_2 resistor. If the decline in the charge between pulse intervals is to be avoided, the circuit shown in Fig. 24-11 is utilized. This is

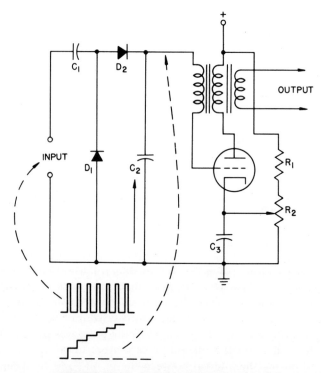

FIGURE 24-11. Staircase divider-counter.

a *staircase*-type divider or pulse counter because the signal waveform appearing at the grid of the tube (or the base of the transistor) appears as an ascending, sharply defined waveform resembling a staircase.

Note that the input pulses are of short duration; the time interval is longer than it was in the previously discussed circuit. Here the input circuit consists of a coupling capacitor C_1, a shunting diode D_1, a series diode D_2, and the staircase charging capacitor C_2. The tube can be a part of a blocking-oscillator circuit or it can consist of a thyratron or silicon-controlled rectifier that will be fired when the signal amplitude applied to the control grid reaches a predetermined amplitude.

Upon the arrival of a pulse to the input of the circuit shown in Fig. 24-11, the positive rise of the pulse waveform will cause diode D_2 to conduct and electron flow will be toward capacitor C_2 in the direction shown by the arrow. Thus, capacitor C_2, as well as C_1, starts to charge. Since, however, the pulse duration is short compared to the long time constant of the circuit, only a small charge builds up across C_2 and C_1. During the interval between the first and second pulse, the charge remains on C_2; hence the signal waveform developed between grid and ground levels off as shown in the staircase waveform drawing. Capacitor C_1, having been charged with a negative polarity toward the top of the diode and a positive polarity toward the bottom of D_1 (through the input coupling network), will now cause diode D_1 to conduct and thus discharge C_1. Capacitor C_2, however, is unable to discharge because the top plate of C_2 is positive with respect to the bottom plate; hence it has no conduction path through either D_1 or D_2. If the input waveform is of the square-wave type instead of the pulse type, diode D_1 would conduct for the negative halfcycle of the square waves, again discharging C_1 but not affecting the charge on C_2.

During the second positive input pulse, the charge on capacitor C_2 is increased by an additional amount in step fashion as shown by the staircase waveform. Even though the intervals between pulses were substantially longer in duration, there would be no discharge of C_2 during such intervals (provided the reverse resistance of D_2 approached that of the vacuum-tube open-circuit condition). Since the capacitor C_2 charges exponentially, the charge-voltage amplitude decreases for successive steps of the staircase waveform as shown.

Because capacitor C_2 is coupled to the grid of the tube through the transformer primary, it will cause tube conduction when the staircase signal amplitude reaches a sufficiently high level to overcome the negative-bias potential of the tube. When the tube fires, the conduction establishes a low-impedance path for discharge of the staircase capacitor. The negative bias for the tube is set by resistor R_2, which applies a positive potential to the cathode and thus establishes the point at which the tube is cut off. Hence, an adjustment of R_2 will alter the operational characteristics of the circuit since it will establish the bias level and also the amplitude of the staircase necessary for overcoming bias. Since R_2 adjusts the firing level, it regulates the amount of division that is procured. Hence, one output pulse is delivered for a predetermined number of input pulses. As discussed in Chapter 17, the output waveform may be modified by clipping or other waveshaping procedures to procure the type of signal required. This may be necessary since the output waveform from a blocking oscillator does not resemble the input pulses shown in Fig. 24-11. If a thyratron or silicon-controlled rectifier is used instead of a blocking oscillator, the staircase amplitude will fire the thyratron when it reaches an amplitude sufficient to release the holding characteristics of the control grid. Once capacitor C_2 has discharged, however, the thyratron will still conduct, as explained earlier, and means must be provided to interrupt the plate-current flow to cause the thyratron to stop conducting again.

Review Questions

24-1. Reproduce a schematic of a magnetic amplifier, and summarize its operating characteristics.

24-2. In what manner may a magnetic amplifier be modified to permit either regenerative or degenerative feedback?

24-3. Explain what is meant by the *Hall effect*.

24-4. Explain in what manner a doubler circuit produces an output that is higher in frequency than the input signal.

24-5. When triode tubes or transistors are used for doubler circuits, why is neutralization unnecessary?

24-6. Why is the efficiency of a Class C doubler RF stage lower than a Class C single-frequency amplifier? Explain briefly.

24-7. Briefly explain in what manner a blocking oscillator can be used as a divider of a pulse-repetition rate.

24-8. How is a series of flip-flop circuits used to perform scaling functions?

24-9. If negative triggering pulses are used, how are positive-polarity pulses prevented from giving incorrect counts in flip-flop circuitry?

24-10. Explain the basic differences between a binary scaler and a decade scaler.

24-11. Draw a block diagram of four beam-switching tubes wired to perform dividing functions. Indicate the number of input pulses that must be applied to the first tube in order to procure one pulse from the output of the last tube.

24-12. Explain how an integrator can be utilized with a relaxation oscillator for pulse-rate division.

24-13. Draw a schematic of a staircase-type divider that is utilized in conjunction with a gas-tube thyratron.

24-14. Draw another schematic of a staircase-type divider, except use a solid-state silicon-controlled rectifier in place of the rectifying tube.

Appendix

Table of International Atomic Weights

Element	Symbol	Atomic Number	Atomic Weight
Actinium	Ac	89	277
Aluminum	Al	13	26.98
Americium	Am	95	*243
Antimony	Sb	51	121.76
Argon	A	18	39.944
Arsenic	As	33	74.91
Astatine	At	85	*210
Barium	Ba	56	137.36
Berkelium	Bk	97	*249
Beryllium	Be	4	9.013
Bismuth	Bi	83	209.00
Boron	B	5	10.82
Bromine	Br	35	79.916
Cadmium	Cd	48	112.41
Calcium	Ca	20	40.08
Californium	Cf	98	*249
Carbon	C	6	12.011
Cerium	Ce	58	140.13
Cesium	Cs	55	132.91
Chlorine	Cl	17	35.457
Chromium	Cr	24	52.01
Cobalt	Co	27	58.94
Copper	Cu	29	63.54
Curium	Cm	96	*245
Dysprosium	Dy	66	162.51
Einsteinium	Es	99	*255
Erbium	Er	68	167.27
Europium	Eu	63	152.0
Fermium	Fm	100	*255
Fluorine	F	9	19.00
Francium	Fr	87	*223
Gadolinium	Gd	64	157.26
Gallium	Ga	31	69.72
Germanium	Ge	32	72.60
Gold	Au	79	197.0
Hafnium	Hf	72	178.50
Helium	He	2	4.003
Holmium	Ho	67	164.94
Hydrogen	H	1	1.0080
Indium	In	49	114.82
Iodine	I	53	126.91
Iridium	Ir	77	192.2
Iron	Fe	26	55.85
Krypton	Kr	36	83.80
Lanthanum	La	57	138.92
Lawrencium	Lw	103	*257
Lead	Pb	82	207.21
Lithium	Li	3	6.940
Lutecium	Lu	71	174.99
Magnesium	Mg	12	24.32
Manganese	Mn	25	54.94
Mendelevium	Md	101	*256
Mercury	Hg	80	200.61

Table of International Atomic Weights (*Continued*)

Element	Symbol	Atomic Number	Atomic Weight
Molybdenum	Mo	42	95.95
Neodymium	Nd	60	144.27
Neon	Ne	10	20.183
Neptunium	Np	93	*237
Nickel	Ni	28	58.71
Niobium	Nb	41	92.91
Nitrogen	N	7	14.008
Nobelium	No	102	*253
Osmium	Os	76	190.2
Oxygen	O	8	16.000
Palladium	Pd	46	106.4
Phosphorus	P	15	30.975
Platinum	Pt	78	195.09
Plutonium	Pu	94	*242
Polonium	Po	84	210
Potassium	K	19	39.100
Praseodymium	Pr	59	140.92
Promethium	Pm	61	*145
Protactinium	Pa	91	231
Radium	Ra	88	226.05
Radon	Rn	86	222
Rhenium	Re	75	186.22
Rhodium	Rh	45	102.91
Rubidium	Rb	37	85.48
Ruthenium	Ru	44	101.1
Samarium	Sm	62	150.35
Scandium	Sc	21	44.96
Selenium	Se	34	78.96
Silicon	Si	14	28.09
Silver	Ag	47	107.880
Sodium	Na	11	22.991
Strontium	Sr	38	87.63
Sulfur	S	16	32.066
Tantalum	Ta	73	180.95
Technetium	Tc	43	*99
Tellurium	Te	52	127.61
Terbium	Tb	65	158.93
Thallium	Tl	81	204.39
Thorium	Th	90	232.05
Thulium	Tm	69	168.94
Tin	Sn	50	118.70
Titanium	Ti	22	47.90
Tungsten	W	74	183.86
Uranium	U	92	238.07
Vanadium	V	23	50.95
Xenon	Xe	54	131.30
Ytterbium	Yb	70	173.04
Yttrium	Y	39	88.92
Zinc	Zn	30	65.38
Zirconium	Zr	40	91.22

* Denotes isotope weight value for most stable type.

Resistor, Capacitor, and Diode Color Codes

Resistors and capacitors have bands or dots of color imprinted on them that represent numbers and letters. By relating these colors to the assigned numerical or alphabetical value, identification and component values can be found. Other information provided includes the tolerance of the value specified and temperature coefficients. Standards have been set up for color coding of electronic components by engineering and manufacturing groups, and they are generally adhered to by the manufacturing industry.

Some surplus components originally manufactured for specific equipment are often available at wholesale houses, however, and these do not conform to the standard color coding. Similarly, some imported items may also lack proper identification. Hence, calibrated test equipment will have to be utilized for ascertaining component values in such instances.

The color coding that follows represents modern standardization practices and, hence, serves as a guide for conventional resistor and capacitor identification and evaluation. For resistor and capacitor coding illustrations specific examples are given to indicate application of the color codings that apply.

Modern resistors of the molded composition type have colored encircling bands grouped at one end, as shown in Fig. A-1. The color coding for the resistors shown is from left to right. Usually four bands of color are present for the carbon composition types, and five bands are used for the film types.

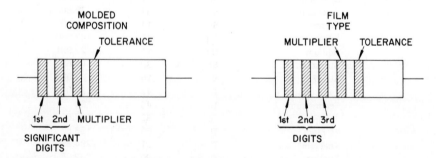

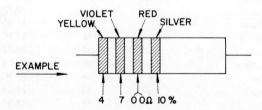

FIGURE A-1. Resistor color coding.

In either case, the last color denotes the tolerance that must be applied to the value obtained. Thus, if a rated 100-Ω resistor has a tolerance of 10%, it could have an actual value between 90 and 110 Ω.

The listing that follows shows the color-code values and tolerances that apply to the carbon and film resistors illustrated in Fig. A-1. The letter group GMV stands for *guaranteed minimum value*. Where a value is marked *alternate*, it indicates a coding that may have been used on occasion in the past but is generally coded in modern components by the *preferred* coding designation.

Resistance in Ohms

Color	Digit	Multiplier	Carbon ± Tolerance	Film-Type Tolerance
Black	0	1	20%	0
Brown	1	10	1%	1%
Red	2	100	2%	2%
Orange	3	1000	3%	
Yellow	4	10,000	GMV	
Green	5	100,000	5% (alt.)	0.5%
Blue	6	1,000,000	6%	0.25%
Violet	7	10,000,000	12.5%	0.1
Gray	8	0.01 (alternate)	30%	0.05
White	9	0.1 (alternate)	10% (alt.)	
Silver		0.01 (preferred)	10% (pref.)	10%
Gold		0.1 (preferred)	5% (pref.)	5%
No color			20%	

In solid-state component designations, the number-letter combination 1N indicates a diode. In some low-current smaller solid-state diodes, standard color coding supplies the missing numbers and letters that identify the complete diode type. For the earlier types that had two center digits (such as the 1N58A), three color bands are employed, as shown in Fig. A-2 (upper left). The color bands are grouped toward the cathode end and, for the diode shown, are read from left to right. The first two color bands indicate the significant figures, and the third color band identifies the letter. For diodes having three central digits (such as the 1N249C), four color bands are employed as shown at the upper right of the drawing. The first three bands from left to right identify the first three significant digits, and the last band provides the letter identification. When the diode has four central digits (such as the 1N1184A), there are five identifying color bands as shown at the lower left. The listing that follows indicates the letter and number values for the color coding.

Ceramic capacitors are generally of two basic shapes: the tubular type

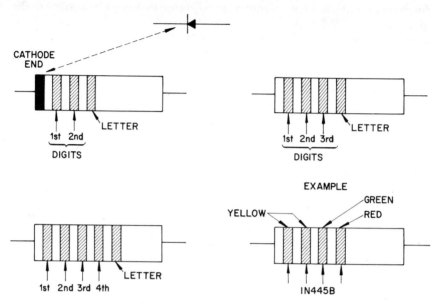

FIGURE A-2. Solid-state diode color coding.

Color	Digit	Suffix Letter
Black	0	—
Brown	1	A
Red	2	B
Orange	3	C
Yellow	4	D
Green	5	E
Blue	6	F
Violet	7	G
Gray	8	H
White	9	J

and the flat disc type. The tubular type may have axial leads (i.e., leads emanating from the ends) or the radial lead connection may be employed where the leads are connected at right angles to the length of the capacitor. The values obtained from the digit color bands are in picofarads (pF). In addition to the tolerance color band and the digit bands, a temperature-coefficient band is also present. The temperature coefficient of the ceramic capacitors is given in parts per million per degree centigrade (ppm/°C). A preceding letter N denotes negative-temperature coefficient (capacity decrease with an increase in operating temperature). The P designation indicates a positive-temperature coefficient, and NPO indicates a negative-positive-zero type.

An N220 designation indicates a capacity decrease with a temperature rise of 220 ppm/°C, and it shows by how much the value changes during the warm-up time of the device in which the capacitor is utilized. The NPO types are stable units with negligible temperature effect on capacitance.

For the general-type ceramic capacitor, five identification bands or markings are used, as shown in Fig. A-3. For the axial-lead type, the identification starts at the end toward which the color bands are grouped, and the first band is the temperature-coefficient marking. The following two bands are the significant digits; the fourth band is the multiplier; and the fifth band is the tolerance, as shown.

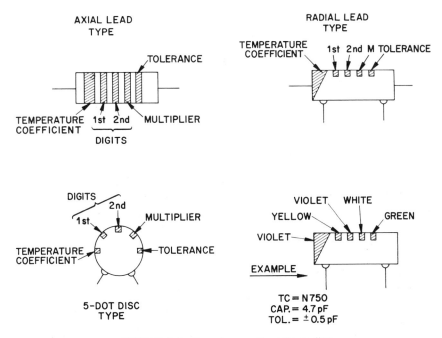

FIGURE A-3. Ceramic capacitor color coding.

Often, the axial-type ceramic capacitors have a wider first band for identification purposes. The radial types have the initial band of greater color area, as shown in Fig. A-3. For the five-dot disc type shown, the lower left color dot gives the temperature coefficient, and the other dots in clockwise sequence have the same coding as for the axial- or radial-lead types shown.

The six-dot ceramic capacitor shown at the upper left of Fig. A-4 is of the extended-range temperature coefficient type. Here, the first color sector indicates the temperature coefficient as with the five-dot types, but the second color segment represents the temperature-coefficient multiplier. The third and fourth colors are the digit values, and the fifth is the multiplier for the significant digits obtained. The sixth color indicates the tolerance range for

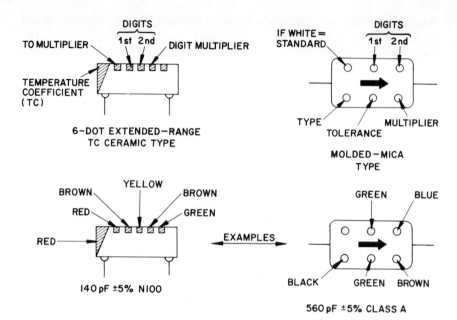

FIGURE A-4. Extended TC and MICA capacitor codings.

the capacitance values obtained. The following listing applies to the five-dot and six-dot ceramic types (including the disc type). For those disc types that have only three color dots, the temperature coefficient and tolerance values are not given. In these, the first two dots (clockwise) are the digits, and the last dot is the multiplier.

Capacitance in Picofarads (pF)

Color	Digit	Multiplier	Tolerance (±) 10 pF or Over — less	Tolerance (±) 10 pF or Over — 10 pF	5-Dot Temp. Coeff.	6-Dot Extended Range TC — Sig. Fig.	6-Dot Extended Range TC — Mult.
Black	0	1	2.0	20%	NPO	0.0	−1
Brown	1	10	0.1	1%	N033		−10
Red	2	100		2%	N075	1.0	−100
Orange	3	1000		3%	N150	1.5	−1000
Yellow	4	10,000			N220	2.2	−10,000
Green	5		0.5	5%	N330	3.3	+1
Blue	6				N470	4.7	+10
Violet	7				N750	7.5	+100
Gray	8	0.01 (alt.)	0.25		*	†	+1000
White	9	0.1 (alt.)	1.0	10%	**		+10,000
Silver		0.01 (pref.)					
Gold		0.1 (pref.)					

*General-purpose types with a TC ranging from P150 to N1500,
**Coupling, decoupling, and general bypass types with a TC ranging from P100 to N750.
†If the first band (TC) is black, the range is N1000 to N5000.

The flat, rectangular-shaped molded mica type capacitor shown at the right of Fig. A-4 has a color-coding sequence as shown. An arrow, or arrowhead, is imprinted on the capacitor face to indicate the direction of color sequence. The lower left-hand color dot indicates the type or classification of the particular capacitor according to the manufacturer's specifications, and it includes temperature coefficient, Q factors, and related characteristics. The following listing applies to such molded mica types.

Capacitance in Picofarads (pF)

Color	Digit	Multiplier	Tolerance (%)	Type Classification
Black	0	1	20 ($\pm$)	A
Brown	1	10	1	B
Red	2	100	2	C
Orange	3	1000	3	D
Yellow	4	10,000		E
Green	5		5	
Blue	6			
Violet	7			
Gray	8			
White	9		10	
Silver		0.01		
Gold		0.1		

Trigonometric Relationships

Trigonometry is the study of various angles, and that branch of trigonometry relating to *right angles* is reviewed here. As shown at A of Fig. A-5, the sides of a triangle are assigned specific names, *using the angle as the reference.* Thus, if we are calculating for the angle at the lower right of A, then the horizontal line extending from this angle is designated as the *side adjacent* to the angle. The vertical line opposite the angle is called the *side opposite*, and the diagonal line from the side opposite to the angle is known as the *hypotenuse*. The angle is the reference point and, as shown at B, the designations follow a similar pattern as at A of Fig. A-5.

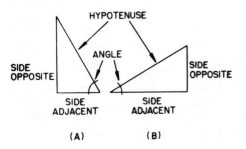

FIGURE A-5.

Table of Trigonometric Ratios

(Sines, Cosines, and Tangents)

Degrees	(opp)(hyp)Sine	(adj)(hyp)Cosine	(opp)(adj)Tangent	Degrees	(opp)(hyp)Sine	(adj)(hyp)Cosine	(opp)(adj)Tangent
0	0.0000	1.0000	0.0000	45	0.7071	0.7071	1.0000
1	.0175	.9998	.0175	46	.7193	.6947	1.0355
2	.0349	.9994	.0349	47	.7314	.6820	1.0724
3	.0523	.9986	.0524	48	.7431	.6691	1.1106
4	.0698	.9976	.0699	49	.7547	.6561	1.1504
5	.0872	.9962	.0875	50	.7660	.6429	1.1918
6	.1045	.9945	.1051	51	.7771	.6293	1.2349
7	.1219	.9925	.1228	52	.7880	.6157	1.2799
8	.1392	.9903	.1405	53	.7986	.6018	1.3270
9	.1564	.9877	.1584	54	.8090	.5878	1.3764
10	.1736	.9848	.1763	55	.8192	.5736	1.4281
11	.1908	.9816	.1944	56	.8290	.5592	1.4826
12	.2079	.9781	.2126	57	.8387	.5446	1.5399
13	.2250	.9744	.2309	58	.8480	.5299	1.6003
14	.2419	.9703	.2493	59	.8572	.5150	1.6643
15	.2588	.9659	.2679	60	.8660	.5000	1.7321
16	.2756	.9613	.2867	61	.8746	.4848	1.8040
17	.2924	.9563	.3057	62	.8829	.4695	1.8807
18	.3090	.9511	.3249	63	.8910	.4540	1.9626
19	.3256	.9455	.3443	64	.8988	.4384	2.0503
20	.3420	.9397	.3640	65	.9063	.4226	2.1445
21	.3584	.9336	.3839	66	.9135	.4067	2.2460
22	.3746	.9272	.4040	67	.9205	.3907	2.3559
23	.3907	.9205	.4245	68	.9272	.3746	2.4751
24	.4067	.9135	.4452	69	.9336	.3584	2.6051
25	.4226	.9063	.4663	70	.9397	.3420	2.7475
26	.4384	.8988	.4877	71	.9455	.3256	2.9042
27	.4540	.8910	.5095	72	.9511	.3090	3.0777
28	.4695	.8829	.5317	73	.9563	.2924	3.2709
29	.4848	.8746	.5543	74	.9613	.2756	3.4874
30	.5000	.8660	.5774	75	.9659	.2588	3.7321
31	.5150	.8572	.6009	76	.9703	.2419	4.0108
32	.5299	.8480	.6249	77	.9744	.2250	4.3315
33	.5446	.8387	.6494	78	.9781	.2079	4.7046
34	.5592	.8290	.6745	79	.9816	.1908	5.1446
35	.5736	.8192	.7002	80	.9848	.1736	5.6713
36	.5878	.8090	.7265	81	.9877	.1564	6.3138
37	.6018	.7986	.7536	82	.9903	.1392	7.1154
38	.6157	.7880	.7813	83	.9925	.1219	8.1443
39	.6293	.7771	.8098	84	.9945	.1045	9.5144
40	.6428	.7660	.8391	85	.9962	.0872	11.4301
41	.6561	.7547	.8693	86	.9976	.0698	14.3007
42	.6691	.7431	.9004	87	.9986	.0523	19.0811
43	.6820	.7314	.9325	88	.9994	.0349	28.6363
44	.6947	.7193	.9657	89	.9998	.0175	57.2900
45	.7071	.7071	1.0000	90	1.0000	.0000	—

In electronic work, the angles involved in ac are often of importance in circuit analysis and design since the opposition to ac offered by capacitors and inductances must be calculated on the basis of simple trigonometry.

A useful form of trigonometry consists in solving for unknown lengths of a triangle, by employing the known angle in conjunction with the length of one of the sides. The relationships of the sides can also be employed to ascertain the angles. For instance, the ratio of the length of the side opposite to the length of the hypotenuse determines the angle. Relationships also hold for the ratio of the opposite side to the adjacent side, etc. When such ratios are set down in table form (see the trigonometric table of ratios that follows), the various angles of triangles can be ascertained by calculating the ratio of lengths of the sides. The following are some of the most frequently used ratios:

$$\frac{\text{Opposite side}}{\text{Hypotenuse}} = \sin \theta \text{ (sine of the angle)}$$

$$\frac{\text{Opposite}}{\text{Adjacent}} = \tan \theta \text{ (tangent of the angle)}$$

$$\frac{\text{Adjacent}}{\text{Hypotenuse}} = \cos \theta \text{ (cosine of an angle)}$$

Another characteristic of the right angle is the Pythagorean theorem, which states that the sum of the squares of the side adjacent and the side opposite of a right triangle is equal to the square of the hypotenuse. This theory was formulated by the Greek scholar *Pythagoras* (about 500 B.C.) and the formula related to this theorem may be stated as follows:

$$a^2 + b^2 = c^2$$

In the foregoing formula, a represents the side opposite; b, the side adjacent; and c, the hypotenuse. Thus, by mathematical rearrangement of the symbols and powers shown above, the hypotenuse can be ascertained from the following formula:

$$c = \sqrt{a^2 + b^2}$$

This equation, except for the substitution of electronic symbols instead of the a, b, and c symbols, is used extensively.

Logarithms

In Chapter 2, under the discussion on abbreviated forms, it was shown that 10 could be expressed as 10^1, 100 as 10^2, 1000 as 10^3, etc. Obviously, intermediate power values for numbers between 10 and 100 or between 100 and 1000 also exist. Similarly, powers for numbers below 10 are also in existence and can be represented by 10 to some fractional power. Inasmuch as 10^0 is equal to 1, and 10^1 is equal to 10, the exponent of a number between 1 and 10 would be a fractional exponent since it must be less than 1. A *logarithm table* shows the powers for these intermediate numbers. A logarithm (abbreviated "log") is the exponent of the intermediate number, without the base written in. The left-hand vertical column of numbers in the log table represents the intermediate numbers for which the logarithm is to be found, and the succeeding vertical columns show the logarithm or exponent without the base. For instance, the logarithm of the number 3 is 0.4771. This means that, instead of expressing 3 as 10 to the 0.4771 power, the 0.4771 is written without the base 10 and hence is known as the logarithm or "log of 3." Since the base is 10, the log of 30 or 300 or 3000 would contain the same digits as for log 3, except for the decimal point. When 10 is expressed as 10^1, and 100 is expressed as 10^2, it is obvious that the exponent of numbers between 10 and 100 must fall between 1 and 2. For the same reason, then, the logarithm of a number between 10 and 100 must be between 1 and 2. Hence, to express the log of 30, the original 0.4771 number is employed, but it must be added to the original exponent in 10^1 and, therefore, would equal 1.4771. Since the power of 10 for the number 100 is 10^2, the power of 10 for 300 would be $10^{2.4771}$. Hence, the logarithm of 300 is 2.4771. Therefore, the logarithm at the right of a decimal point does not change when additional zeros are added to the original number. The number at the left of the decimal point, however, increases by a numerical value of 1 each time the original number is multiplied by 10. The fractional portion of the logarithm (at the right of the decimal point) is known as the *mantissa*. The whole number to the left of the decimal point is known as the *characteristic*.

Numbers may not be single digits followed by zeros. For instance, to find the logarithm of 2720, the number 27 is located in the left-hand column of the table. Along the horizontal row of columns, the logarithm under 2 is now found, to complete the logarithm of 272. This is found to be 4346. To place the decimal point, the decimal point in the original number is moved to the left until the remaining number lies between 1 and 10. Thus, the decimal point is placed in the original number so that the latter is 2.720. Since the decimal place was moved three places, this number forms the characteristic of 3. Hence, the logarithm of 2720 is 3.4346. As another example, to find the logarithm of 60,000, locate the number 60 in the left-hand column. Since the original number has no additional digits, read the number 7782 in the zero column next to the original number 60. Point off the original number to the left until only one digit (between 1 and 10) remains. This requires the pointing off of four places. Hence, the characteristic is 4, and the log of 60,000 is 4.7782.

The logarithm of fractional powers can also be ascertained. As an illustration, assume that the logarithm of the number 0.0015 is to be found. Number 15 is located in the left-hand column and since there are no additional digits following this number, the initial logarithm obtained is 1761. This is expressed as 0.1761; the decimal place in the original number is moved to the right until a single digit remains to the left of the decimal number. Thus, the decimal number is moved to the right three places in the original number, indicating that the characteristic is 3. Hence, the log of 0.0015 is −3 + 0.1761. The latter number can also be expressed as 7.1761 − 10.

Knowing the logarithm for a given number, the original number can be ascertained from the log table. This process is known as finding the *antilogarithm* and, in such a procedure, the original number that produced the logarithm is known as the antilog. For instance, to find the antilog of 2.6599, the .6599 portion of the number is located in the logarithm table. This provides the number 457. The latter number must be assumed to be a number between 1 and 10 and, hence, it is considered to be 4.57. Since the characteristic of the number for which the antilog is to be found is 2, the decimal place is moved over two points and gives 457 for the antilog. The foregoing represents a summary of the basic logarithm principles for use in this book. For more advanced work with respect to interpolation, and to the use of logarithms for multiplication and division, reference should be made to a standard mathematical text.

Common Logarithm Table

N	0	1	2	3	4	5	6	7	8	9
10	0000	0043	0086	0128	0170	0212	0253	0294	0334	0374
11	0414	0453	0492	0531	0569	0607	0645	0682	0719	0755
12	0792	0828	0864	0899	0934	0969	1004	1038	1072	1106
13	1139	1173	1206	1239	1271	1303	1335	1367	1399	1430
14	1461	1492	1523	1553	1584	1614	1644	1673	1703	1732
15	1761	1790	1818	1847	1875	1903	1931	1959	1987	2014
16	2041	2068	2095	2122	2148	2175	2201	2227	2253	2279
17	2304	2330	2355	2380	2405	2430	2455	2480	2504	2529
18	2553	2577	2601	2625	2648	2672	2695	2718	2742	2765
19	2788	2810	2833	2856	2878	2900	2923	2945	2967	2989
20	3010	3032	3054	3075	3096	3118	3139	3160	3181	3201
21	3222	3243	3263	3284	3304	3324	3345	3365	3385	3404
22	3424	3444	3464	3483	3502	3522	3541	3560	3579	3598
23	3617	3636	3655	3674	3692	3711	3729	3747	3766	3784
24	3802	3820	3838	3856	3874	3892	3909	3927	3945	3962
25	3979	3997	4014	4031	4048	4065	4082	4099	4116	4133
26	4150	4166	4183	4200	4216	4232	4249	4265	4281	4298
27	4314	4330	4346	4362	4378	4393	4409	4425	4440	4456
28	4472	4487	4502	4518	4533	4548	4564	4579	4594	4609
29	4624	4639	4654	4669	4683	4698	4713	4728	4742	4757
30	4771	4786	4800	4814	4829	4843	4857	4871	4886	4900
31	4914	4928	4942	4955	4969	4983	4997	5011	5024	5038
32	5051	5065	5079	5092	5105	5119	5132	5145	5159	5172
33	5185	5198	5211	5224	5237	5250	5263	5276	5289	5302
34	5315	5328	5340	5353	5366	5378	5391	5403	5416	5428
35	5441	5453	5465	5478	5490	5502	5514	5527	5539	5551
36	5563	5575	5587	5599	5611	5623	5635	5647	5658	5670
37	5682	5694	5705	5717	5729	5740	5752	5763	5775	5786
38	5798	5809	5821	5832	5843	5855	5866	5877	5888	5899
39	5911	5922	5933	5944	5955	5966	5977	5988	5999	6010
40	6021	6031	6042	6053	6064	6075	6085	6096	6107	6117
41	6128	6138	6149	6160	6170	6180	6191	6201	6212	6222
42	6232	6243	6253	6263	6274	6284	6294	6304	6314	6325
43	6335	6345	6355	6365	6375	6385	6395	6405	6415	6425
44	6435	6444	6454	6464	6474	6484	6493	6503	6513	6522
45	6532	6542	6551	6561	6571	6580	6590	6599	6609	6618
46	6628	6637	6646	6656	6665	6675	6684	6693	6702	6712
47	6721	6730	6739	6749	6758	6767	6776	6785	6794	6803
48	6812	6821	6830	6839	6848	6857	6866	6875	6884	6893
49	6902	6911	6920	6928	6937	6946	6955	6964	6972	6981
50	6990	6998	7007	7016	7024	7033	7042	7050	7059	7067
51	7076	7084	7093	7101	7110	7118	7126	7135	7143	7152
52	7160	7168	7177	7185	7193	7202	7210	7218	7226	7235
53	7243	7251	7259	7267	7275	7284	7292	7300	7308	7316
54	7324	7332	7340	7348	7356	7364	7372	7380	7388	7396
N	0	1	2	3	4	5	6	7	8	9

Common Logarithm Table (*Continued*)

N	0	1	2	3	4	5	6	7	8	9
55	7404	7412	7419	7427	7435	7443	7451	7459	7466	7474
56	7482	7490	7497	7505	7513	7520	7528	7536	7543	7551
57	7559	7566	7574	7582	7589	7597	7604	7612	7619	7627
58	7634	7642	7649	7657	7664	7672	7679	7686	7694	7701
59	7709	7716	7723	7731	7738	7745	7752	7760	7767	7774
60	7782	7789	7796	7803	7810	7818	7825	7832	7839	7846
61	7853	7860	7868	7875	7882	7889	7896	7903	7910	7917
62	7924	7931	7938	7945	7952	7959	7966	7973	7980	7987
63	7993	8000	8007	8014	8021	8028	8035	8041	8048	8055
64	8062	8069	8075	8082	8089	8096	8102	8109	8116	8122
65	8129	8136	8142	8149	8156	8162	8169	8176	8182	8189
66	8195	8202	8209	8215	8222	8228	8235	8241	8248	8254
67	8261	8267	8274	8280	8287	8293	8299	8306	8312	8319
68	8325	8331	8338	8344	8351	8357	8363	8370	8376	8382
69	8388	8395	8401	8407	8414	8420	8426	8432	8439	8445
70	8451	8457	8463	8470	8476	8482	8488	8494	8500	8506
71	8513	8519	8525	8531	8537	8543	8549	8555	8561	8567
72	8573	8579	8585	8591	8597	8603	8609	8615	8621	8627
73	8633	8639	8645	8651	8657	8663	8669	8675	8681	8686
74	8692	8698	8704	8710	8716	8722	8727	8733	8739	8745
75	8751	8756	8762	8768	8774	8779	8785	8791	8797	8802
76	8808	8814	8820	8825	8831	8837	8842	8848	8854	8859
77	8865	8871	8876	8882	8887	8893	8899	8904	8910	8915
78	8921	8927	8932	8938	8943	8949	8954	8960	8965	8971
79	8976	8982	8987	8993	8998	9004	9009	9015	9020	9025
80	9031	9036	9042	9047	9053	9058	9063	9069	9074	9079
81	9085	9090	9096	9101	9106	9112	9117	9122	9128	9133
82	9138	9143	9149	9154	9159	9165	9170	9175	9180	9186
83	9191	9196	9201	9206	9212	9217	9222	9227	9232	9238
84	9243	9248	9253	9258	9263	9269	9274	9279	9284	9289
85	9294	9299	9304	9309	9315	9320	9325	9330	9335	9340
86	9345	9350	9355	9360	9365	9370	9375	9380	9385	9390
87	9395	9400	9405	9410	9415	9420	9425	9430	9435	9440
88	9445	9450	9455	9460	9465	9469	9474	9479	9484	9489
89	9494	9499	9504	9509	9513	9518	9523	9528	9533	9538
90	9542	9547	9552	9557	9562	9566	9571	9576	9581	9586
91	9590	9595	9600	9605	9609	9614	9619	9624	9628	9633
92	9638	9643	9647	9652	9657	9661	9666	9671	9675	9680
93	9685	9689	9694	9699	9703	9708	9713	9717	9722	9727
94	9731	9736	9741	9745	9750	9754	9759	9763	9768	9773
95	9777	9782	9786	9791	9795	9800	9805	9809	9814	9818
96	9823	9827	9832	9836	9841	9845	9850	9854	9859	9863
97	9868	9872	9877	9881	9886	9890	9894	9899	9903	9908
98	9912	9917	9921	9926	9930	9934	9939	9943	9948	9952
99	9956	9961	9965	9969	9974	9978	9983	9987	9991	9996
N	**0**	**1**	**2**	**3**	**4**	**5**	**6**	**7**	**8**	**9**

Standard Resistor Values

Ohms	Ohms	Ohms	Ohms	Ohms	Ohms	Megs	Megs	Megs
—	1.0	10	100	1,000	10,000	0.1	1.0	10
—	1.1	11	110	1,100	11,000	0.11	1.1	11
—	1.2	12	120	1,200	12,000	0.12	1.2	12
—	1.3	13	130	1,300	13,000	0.13	1.3	13
—	1.5	15	150	1,500	15,000	0.15	1.5	15
—	1.6	16	160	1,600	16,000	0.16	1.6	16
—	1.8	18	180	1,800	18,000	0.18	1.8	18
—	2.0	20	200	2,000	20,000	0.20	2.0	20
—	2.2	22	220	2,200	22,000	0.22	2.2	22
0.24	2.4	24	240	2,400	24,000	0.24	2.4	—
0.27	2.7	27	270	2,700	27,000	0.27	2.7	—
0.30	3.0	30	300	3,000	30,000	0.30	3.0	—
0.33	3.3	33	330	3,300	33,000	0.33	3.3	—
0.36	3.6	36	360	3,600	36,000	0.36	3.6	—
0.39	3.9	39	390	3,900	39,000	0.39	3.9	—
0.43	4.3	43	430	4,300	43,000	0.43	4.3	—
0.47	4.7	47	470	4,700	47,000	0.47	4.7	—
0.51	5.1	51	510	5,100	51,000	0.51	5.1	—
0.56	5.6	56	560	5,600	56,000	0.56	5.6	—
0.62	6.2	62	620	6,200	62,000	0.62	6.2	—
0.68	6.8	68	680	6,800	68,000	0.68	6.8	—
0.75	7.5	75	750	7,500	75,000	0.75	7.5	—
0.82	8.2	82	820	8,200	82,000	0.82	8.2	—
0.91	9.1	91	910	9,100	91,000	0.91	9.1	—

Properties of Copper Wire Conductors*
(American Wire Gauge)

Size (gauge no.)	Diam in mils at 20°C, 68°F	Area in Circular Mils	Ohms per 1000 ft 25°C, 77°F
16	50.82	2583.0	4.094
17	45.26	2048.0	5.163
18	40.30	1624.0	6.510
19	35.89	1288.0	8.210
20	31.96	1022.0	10.35
21	28.46	810.1	13.05
22	25.35	642.4	16.46
23	22.57	509.5	20.76
24	20.10	404.0	26.17
25	17.90	320.4	33.00
26	15.94	254.1	41.62
27	14.20	201.5	52.48
28	12.64	159.8	66.17
29	11.26	126.7	83.44
30	10.03	100.5	105.2
31	8.93	79.70	132.7
32	7.95	63.21	167.3
33	7.08	50.13	211.0
34	6.31	39.75	266.0
35	5.62	31.52	335.5
36	5.00	25.00	423.0
37	4.45	19.83	533.4
38	3.96	15.72	672.6
39	3.53	12.47	848.1
40	3.14	9.89	1,069.0

* Only wire sizes found in conventional electronic circuitry are listed. For sizes larger or smaller than shown, refer to Circular 31, National Bureau of Standards, or to the National Electrical Code pamphlet of the National Board of Fire Underwriters.

Standard Broadcast Station Allocations

AM Broadcast Band
550 kHz to 1600 kHz
(Nominally 10 kHz per station)

FM Broadcast Band
88 MHz to 108 MHz
(200 kHz per station)

VHF Television Station Frequencies

Channel Number	Frequency in Megahertz	Video Carrier	Sound Carrier
1		(Not allocated)	
2	54–60	55.25	59.75
3	60–66	61.25	65.75
4	66–72	67.25	71.75
5	76–82	77.25	81.75
6	82–88	83.25	87.75
7	174–180	175.25	179.75
8	180–186	181.25	185.75
9	186–192	187.25	191.75
10	192–198	193.25	197.75
11	198–204	199.25	203.75
12	204–210	205.25	209.75
13	210–216	211.25	215.75

UHF Television Station Band

Channels 14 to 83 inclusive.
Total frequency span: 470 MHz to 890 MHz.
Each station allocated a 6-MHz spectrum space, as with VHF stations.

Frequency Spectrum Designations

VLF	(very low frequencies)	3 to	30 kHz
LF	(low frequencies)	30 to	300 kHz
MF	(medium frequencies)	300 to	3,000 kHz
HF	(high frequencies)	3,000 to	30,000 kHz
VHF	(very high frequencies)	30 to	300 MHz
UHF	(ultrahigh frequencies)	300 to	3,000 MHz
SHF	(superhigh frequencies)	3,000 to	30,000 MHz
EHF	(extra-high frequencies)	30,000 to	300,000 MHz

Military Frequency Spectrum Designations

P-band	225 to	390 MHz
L-band	390 to	1,550 MHz
S-band	1,550 to	5,200 MHz
X-band	5,200 to	10,900 MHz
K-band	10,900 to	36,000 MHz
Q-band	36,000 to	46,000 MHz
V-band	46,000 to	56,000 MHz

Answers to Practical
Problems

Chapter 1

1-1. 7,200 dynes.   **1-2.** 1,800 dynes.   **1-3.** Quadruple.   **1-4.** 3 cm

Chapter 2

2-1. 0.65 A.   **2-2.** 4.5 V.   **2-3.** 800 Ω.   **2-4.** 125 W, 250 V.
2-5. 0.2 W, 0.05 A.   **2-6.** 40 Ω, 200 V.   **2-7.** 20 mV.   **2-8.** 5 kV.
2-9. 30 mA.   **2-10.** 0.4 mA.   **2-11.** 50,000 W.   **2-12.** 0.2 W.
2-13. 0.00025 V.   **2-14.** (a) 7 zeros. (b) 9180 nanos.   **2-15.** 15 pico units.
2-16. 90,000,000.   **2-17.** 1,200 kHz, 1,200,000 Hz.   **2-18.** 30 Hz to 20 kHz.

Chapter 3

3-1. 0.5 A, 240 Ω.   **3-2.** 10 kΩ, 2,500 Ω total = 12,500 Ω.
3-3. $R_1 = 4$ W, $R_2 = 1$ W.   **3-4.** 10 mA.
3-5. $I = 0.03$ A, $R_1 = 50$ Ω, $R_2 = 100$ Ω.   **3-6.** 20 kΩ, 0.05 A.
3-7. $R_1 = 0.075$ A, $R_2 = 0.3$ A, $R_3 = 0.15$ A, $I_T = 0.525$ A, $R_T = 285.7$ Ω.
3-8. 285.7 Ω.   **3-9.** 50 W.   **3-10.** 78.75 W.   **3-11.** 315 W.
3-12. $R_1 = 5$ Ω, $R_2 = 10$ Ω, $R_T = 3.33$ Ω.   **3-13.** 15 cells in series.
3-14. 3.75 V.   **3-15.** 2.5875 Ω.   **3-16.** 4.115 Ω.

Chapter 4

4-1. $R_T = 400$, $E_{R_L} = 200$ V, $E_{R_2} = 600$ V.
4-2. $I = 0.5$ A, $R_T = 1$ kΩ, $E_T = 500$ V.   **4-3.** 125 Ω.
4-4. 156, 78, and 52 Ω (one solution of many).

4-5. 40 Ω total; E across 18 Ω resistor = 36 V.

4-6. I_1 = 0.147 A, I_2 = 0.3236 A, I_3 = 0.4706 A.

4-7. 20 Ω = 4.4 W, 8 Ω = 0.72 W.   **4-8.** 9.4 V.

4-9. R_e = 1,078 Ω, I_L = 0.01 A.   **4-10.** 0.01 A.

Chapter 5

5-1. 0.04 V.   **5-2.** 2 kΩ.   **5-3.** 20 Ω.   **5-4.** 49,900 Ω.   **5-5.** 8 kΩ.

5-6. 6 dB.   **5-7.** −10 dB.   **5-8.** 18 dB.   **5-9.** 4.77 dB.   **5-10.** −9.54 dB.

5-11. 18 dB.   **5-12.** by one-eighth.

Chapter 6

6-1. 0.003 H.   **6-2.** 3.14 gilberts.   **6-3.** 12.56 gilberts.   **6-4.** 50.24.

6-5. 7.536.   **6-6.** 18 μF.   **6-7.** 0.05 μF.   **6-8.** 0.006 μF.

6-9. 20 pF, 1.8 pF, 135 pF.   **6-10.** 0.05 second.   **6-11.** 300 μs.

6-12. 150 V.

Chapter 7

7-1. 2 MHz.   **7-2.** 20,000 and 2,000,000.   **7-3.** I = 0.7 A, E = 169 V.

7-4. 32 V.   **7-5.** 212 V.   **7-6.** 176.25 V.   **7-7.** 240.38 V.   **7-8.** 211.5 V.

7-9. 44.54 V.   **7-10.** 338.4 V.

Chapter 8

8-1. 7,536 Ω.   **8-2.** 17 mA.   **8-3.** 7 W.   **8-4.** 12.5 Ω.   **8-5.** 7 kΩ.

8-6. 30 V.   **8-7.** 40 Ω.   **8-8.** 3.5 kΩ.   **8-9.** 8 mH.   **8-10.** 60 to 1.

8-11. 6.5 mH.   **8-12.** 21.5 mH.   **8-13.** 25 μsec.   **8-14.** 50 μF.

8-15. 24 μF, 450 working volts.   **8-16.** Yes, 1 Ω.   **8-17.** No, 600 Ω.

8-18. Two capacitors of 0.02 μF in series, shunted by 0.05 μF.

8-19. 63 V, 1.65 W approx.   **8-20.** 31.5 V, 20 mA, 1,575 Ω.

8-21. 680 mA, 200 Ω.

8-22. 45 pF total capacity, voltage rating of group = 200 V total.

8-23. 12 H, L_x = 16 (1500/2000).   **8-24.** 21 μF (approx).

Chapter 9

9-1. 8 A, 36°.   **9-2.** 120 V.   **9-3.** (a) X_L = 18,840 Ω, X_C = 200 kΩ,
(b) X_L = 37,680 Ω, X_C = 100 kΩ, (c) X_L = 75,360 Ω, X_C = 50 kΩ,
(d) X_L = 150,720 Ω, X_C = 25 kΩ, (e) X_L = 301,440 Ω, X_C = 12,500 Ω.

9-4. 155 Ω.   **9-5.** 78 W apparent power, 46.8 W true power.

9-6. X_L total = 1,256 kΩ, I = 10 mA, Z = 500 Ω.   **9-7.** 0.04 W.

9-8. Z = 192 Ω, I = 50 mA.   **9-9.** 0.288 W.   **9-10.** Z = 360 Ω.

Chapter 10

10-1. 2 mA, 0.2 W, X_L = 23 kΩ; at 180 Hz, X_L = 69 kΩ.

10-2. 200, 20, the 10-Ω resistor.

10-3. 15.92 Hz.   **10-4.** 4 MHz.   **10-5.** 10 MHz.

10-6. Series resonant circuit of 45.75 MHz followed by a shunting series-resonant circuit tuned to 47.25 MHz.   **10-7.** At 5 MHz.   **10-8.** 1 MHz.

10-9. 600-kHz resonance and frequency of signal at load.   **10-10.** 4.4, 335 kHz.

Chapter 11

11-1. mu = 7.3   **11-2.** r_p = 1,333 Ω (approx).   **11-3.** g_m = 0.0025 mho.
11-4. mu = 20.   **11-5.** r_p = 6,250 Ω.   **11-6.** 0.001 mho.   **11-7.** 800 Ω.
11-8. 0.002 mho or 2,000 μmhos.   **11-9.** mu = 25.
11-10. No. 1 at mu of 100. No. 2 has a mu of only 50.

Chapter 12

12-1. 450 V.   **12-2.** 150 kΩ.   **12-3.** 398 kΩ.   **12-4.** 141 V.   **12-5.** 60°.
12-6. 0.6 microsecond.   **12-7.** 115 Ω.   **12-8.** 65 to 1.   **12-9.** 0.01 second.
12-10. 67.5 Ω.   **12-11.** 2,860 Ω, 30 mA.   **12-12.** 20.   **12-13.** 3 MHz.
12-14. 50.   **12-15.** 10 kΩ.

Chapter 13

13-1. current gain = 32.   **13-2.** *beta* = 25.   **13-3.** (a) 30 gain, (b) 125 gain.
13-4. alpha = 0.83.   **13-5.** A_i = 66.6.   **13-6.** R_i = 2,100 Ω.
13-7. A_e = 230.   **13-8.** A_p = 15,318.   **13-9.** A_i = 80.   **13-10.** A_i = 100.

Chapter 14

14-1. Two 16 μF, 450 working-voltage capacitors in series. (This provides 8 μF at 900 V. If the other 16 μF were used in series, the working voltage would only be 300 and this would not allow for any voltage increase variations.)
14-2. Choke No. 3. (No. 1 and No. 3 are the only ones with adequate current ratings and the choice would be No. 3 because of its higher henry rating.)
14-3. R_1 = 4 kΩ, R_2 = 20 kΩ, R_{L_1} = 5 kΩ.   **14-4.** Two 50 μF, 30 working-volt capacitors in series, paralleled by a duplicate series section.
14-5. 211.5 V (75 × 1.41 × 2) charged input capacitor contributes an additional peak voltage.   **14-6.** 15,072 Ω. (With full-wave rectification the ripple frequency is 120 Hz, hence X_L = 6.28 × 120 × 20).   **14-7.** At 60 Hz, X_L = one-half the value in Problem 14-6, or 7,536 Ω.   **14-8.** 25%.
14-9. 25%, calculated on the basis of (2000 − 1600)/1600 = 0.25 × 100 = 25%. Regulation was improved, because at 33%, the 2,000 V dropped to approximately 1,500 V with same load current.   **14-10.** 50 kΩ.   **14-11.** (a) 48.4 W, (b) 22 V.
14-12. (a) 10 W, (b) 250 Ω, (c) yes.

Chapter 15

15-1. Gain = 60.   **15-2.** 66.6   **15-3.** 80.   **15-4.** 980.   **15-5.** 7,000.
15-6. −2.7.   **15-7.** 260.8 V.   **15-8.** 8.7 W.   **15-9.** 2 kΩ.   **15-10.** 1.4 W
15-11. By 10 mW.

Chapter 16

16-1. 10 to 1.   **16-2.** 40 Ω.   **16-3.** 8.3%.   **16-4.** 59 = 60%.
16-5. 20 Ω.   **16-6.** 32 turns.   **16-7.** 6.4 W.   **16-8.** 53%.   **16-9.** 1.3%.
16-10. 200 Ω.

Chapter 17

17-1. 1,428.5 kHz.   **17-2.** 1 MHz.   **17-3.** 7500 kHz.
17-4. 40 Ω. Power = 0.9 W, hence 2-watt resistor is adequate.   **17-5.** 600 kHz.
17-6. 75,360 Ω.   **17-7.** 0.002.   **17-8.** 0.01 W.   **17-9.** 0.005 W.
17-10. 200 kHz base frequency (1/dura. = 1/0.000005 = 200,000) 40 kHz is lowest
frequency.   **17-11.** 0.2.   **17-12.** 5 microseconds.   **17-13.** 100 microseconds.

Chapter 18

18-1. 997.4 kHz, 999.5 kHz, 1,000.5 kHz, 1,002.6 kHz.   **18-2.** 5.2 kHz.
18-3. 2.   **18-4.** SSB = 4.   **18-5.** 3.5 μH.
18-6. 200 kHz, 800 kHz, 1 MHz, 1800 kHz.   **18-7.** 455 kHz.
18-8. 4.5 MHz.   **18-9.** 40,000.4 kHz or 39,999.6 kHz.   **18-10.** 20,231 kHz.

Chapter 20

20-1. 300 Ω.   **20-2.** 552 Ω.   **20-3.** 82.8 Ω.   **20-4.** 41.5 Ω.
20-5. 8 feet.   **20-6.** 492 MHz.   **20-7.** 3 inches.   **20-8.** 5.2 feet.
20-9. 11.2 inches.   **20-10.** Reflector = 8.2 feet, antenna = 7.8 feet, and
director = 7.5 feet.

Index

Donor atom, 227
Dot generator, 568
Dynamic characteristics, 201
Dyne, 14

E

Eccles-Jordan, 576
Eddy currents, 129
Edison, 190
Effective value, 110
Electrolyte, 25
Electrolytic capacitors, 143
Electromagnetism:
  fields, 11, 13, 89, 91
  induction, 95
Electromotive force, 24
  back emf, 117
  counter, 117
Electron coupled oscillator, 407
Electrons, 3
  movement, 9
  spin, 16
Electrostatic fields, 11, 13
Elements, 4
  inert, 6
Emitter follower, 253, 323
Energy levels, 8, 223
Ewing, 15
Excitation, 384
  field voltage, 112
Exponent, 37

F

Farad, 100
Faraday, 12, 95, 100
Feedback factor, 376
Feedback oscillator, 398
Ferrite, 93
Ferromagnetic, 36
FETVM, 559
Field, 3, 11, 483
  coils, 111
Field-effect transistor, 248
  lead connections, 265
Filters, 286, 293, 526
  parallel, 180
  RF, 172
First detector, 491
Fleming, 190
Flip flop, 576
  decade, 595
  frequency divider, 593
Flux, 35
Flywheel effect, 177, 396
FM detector, 461

FM transmitter, 485, 487
Folded horn, 510
Force, 13
Forward bias, 231
Four-channel sound, 512
Frame, 483
Frequency, 109
  resonant, 168
Frequency dividers, 592
Frequency modulation, 440
Frequency multiplication, 591
Frequency spectrum
  designations, 618
Full-wave supply, 278

G

Gain, 245, 254, 324, 337
Galvanometer, 85
Gas tubes, 216
Gated-beam detector, 466
Gating, 572
  circuits, 580
Gauss, 35
Generator, 107
  practical, 111
Gilbert, 92
Grid, 196
  bias, 197
Ground connection, 193, 274, 456
Grounded base, 252, 343, 346
Grounded collector, 252, 323
Grounded drain, 252
Grounded emitter, 312, 343
Grounded gate, 252
Grounded grid, 346
Guard bands, 447

H

Hairpin loop, 523
Half-power points, 170
Half-wave antenna, 529
Half-wave supply, 274
Hall effect, 590
Harmonics, 328
Hartley oscillator, 399
Heat sink, 233
Henry, 117
Hertz, 30, 38, 109
  antenna, 529
Heterodyne detector, 456
High fidelity, 506
High-pass filter, 527

Hole flow, 224, 243, 266
Hum reduction, 317
Hybrid parameters, 261
Hydrometer, 27
Hysteresis, 93

I

ICW signals, 458
Ignitron, 302
Impedance, 123, 139, 154, 357
  characteristic, 516
  matching, 357
Impurities, 226
Indirectly heated, 191
Induced current, 95
Inductors:
  practical factors, 128
  reactance, 116
  stored energy, 97
Infinite baffle, 510
Instantaneous values, 119
Insulator, 10, 45
Integrated circuits, 267
Integration, 421, 597
Intermediate frequency, 497
Inverse feedback, 373
Ionization, 10
  bonds, 18
  deionization, 299
  in tubes, 216, 299

J

Joule, 33

K

Kelvin, 40, 192
Kirchhoff, 69

L

Laminations, 129
L-C-R line sections, 525
Lead dress, 146
Leading edge, 418
Left-hand rule, 90
Length, 40
Lenz, 96
Level controls, 309
Light-dependent resistor, 539
Light-emitting diode, 233
Limiter, 422
Lines of force, 12, 29